AF293608

ASTRONOMY AND ASTROPHYSICS LIBRARY

Series Editors:
I. Appenzeller, Heidelberg, Germany
G. Börner, Garching, Germany
A. Burkert, München, Germany
M.A. Dopita, Canberra, Australia
T. Encrenaz, Meudon, France
M. Harwit, Washington, DC, USA
R. Kippenhahn, Göttingen, Germany
J. Lequeux, Paris, France
A. Maeder, Sauverny, Switzerland
V. Trimble, College Park, MD, and Irvine, CA, USA

Springer

Berlin
Heidelberg
New York
Hong Kong
London
Milan
Paris
Tokyo

D. Boccaletti   G. Pucacco

Theory of Orbits

Volume 1:
Integrable Systems and Non-perturbative Methods

With 71 Figures

 Springer

Dino Boccaletti
Università degli Studi di Roma "La Sapienza"
Dipartimento di Matematica "Guido Castelnuovo"
Piazzale Aldo Moro, 2, 00185 Roma, Italy
e-mail: boccaletti@uniroma1.it

Giuseppe Pucacco
Università degli Studi di Roma "Tor Vergata"
Dipartimento di Fisica
Via della Ricerca Scientifica, 1, 00133 Roma, Italy
e-mail: pucacco@roma2.infn.it

Cover picture: A wide-field view of the Carina region in the Southern Sky, kindly supplied by ESO (European Southern Observatory), with an insert from a miniature of the XII century *God architect of the cosmos*, miniature from "Bible moralisée", Cod. 2554 f. 1v (Österreichische Nationalbibliothek, Vienna)

Cataloging-in-Publication Data applied for.
A catalog record for this book is available from the Library of Congress.

Bibliographic information published by Die Deutsche Bibliothek
Die Deutsche Bibliothek lists this publication in the Deutsche Nationalbibliografie; detailed bibliographic data is available in the Internet at http://dnb.ddb.de

Corrected Third Printing 2004
ISSN 0941-7834
ISBN 978-3-642-08210-8

This work is subject to copyright. All rights are reserved, whether the whole or part of the material is concerned, specifically the rights of translation, reprinting, reuse of illustrations, recitation, broadcasting, reproduction on microfilm or in any other way, and storage in data banks. Duplication of this publication or parts thereof is permitted only under the provisions of the German Copyright Law of September 9, 1965, in its current version, and permission for use must always be obtained from Springer-Verlag. Violations are liable for prosecution under the German Copyright Law.

Springer-Verlag Berlin Heidelberg New York
a member of BertelsmannSpringer Science+Business Media GmbH

http://www.springer.de

© Springer-Verlag Berlin Heidelberg 2010
Printed in Germany

The use of general descriptive names, registered names, trademarks, etc. in this publication does not imply, even in the absence of a specific statement, that such names are exempt from the relevant protective laws and regulations and therefore free for general use.

Cover design: *design & production* GmbH, Heidelberg
Printed on acid-free paper

Preface to the Corrected Third Printing

As a third edition of this book has been called for, we have taken the opportunity of making a few corrections and additions throughout the text. Bibliographical Notes have been left unchanged.

Our thanks are due to a number of our collegues and students for pointing out errors and misprints, and in particular we are grateful to Prof. D. Schlüter of the Christian–Albrechts–Universität Kiel for the trouble which he has taken in supplying us with a somewhat thorough list.

Rome, July 2003

Dino Boccaletti
Giuseppe Pucacco

Preface to the First Edition

Half a century ago, S. Chandrasekhar wrote these words in the preface to his celebrated and successful book:[1]

In this monograph an attempt has been made to present the theory of stellar dynamics as a branch of classical dynamics – a discipline in the same general category as celestial mechanics. [...] Indeed, several of the problems of modern stellar dynamical theory are so severely classical that it is difficult to believe that they are not already discussed, for example, in Jacobi's Vorlesungen.

Since then, stellar dynamics has developed in several directions and at various levels, basically three viewpoints remaining from which to look at the problems encountered in the interpretation of the phenomenology. Roughly speaking, we can say that a stellar system (cluster, galaxy, etc.) can be considered from the point of view of celestial mechanics (the N-body problem with $N \gg 1$), fluid mechanics (the system is represented by a material continuum), or statistical mechanics (one defines a distribution function for the positions and the states of motion of the components of the system).

The three different approaches do not of course exclude one another, and very often they coexist in the treatment of certain problems. It may sound obvious if we state that the various problems are reduced and schematized in such a way that they can be looked at from one of the above viewpoints and with the tools that can be provided by the relevant discipline. However paradoxical it might appear (given the enormous amount of work produced by mathematicians on the N-body problem), it is our opinion that it is the first kind of approach which has received the least attention from the researchers on stellar dynamics or, at least, has received much less attention than the progress in the field could allow. If, from the publication of Chandrasekhar's book up to the present this has indeed happened it is due (in our opinion) mainly to two things.

The first is to do with the belief that the results of celestial mechanics always refer to only a few bodies and therefore cannot be applied to stellar dynamics (many bodies); in more concrete terms the situation has been such that the dialogue between those who have dealt with the problems of celestial

---

[1] S. Chandrasekhar: *Principles of Stellar Dynamics* (University of Chicago Press, 1942; reprinted by Dover, New York, 1960), p. VII.

mechanics (the mathematicians) and those who have dealt with the problems of stellar dynamics (the astrophysicists) has been minimal. We shall come back to this last point, later on.

The second and much more recent, concerns the ever-growing and somehow overwhelming use of computers. In the last decade, in particular, numerical simulations have a more and more important position with respect to analytical studies. Doubtless computers are an exceedingly powerful tool for investigating certain problems. However, in our opinion they should be used to single out those points on which to concentrate analytical study rather than as a short cut to avoid it. That is, computers should promote analytical study rather than replace it. This was the purpose of the well-known paper by Hénon and Heiles[2], which has opened up new horizons to a branch of mathematical physics. In addition, as proof of how essential the above-mentioned dialogue is, the paper deals with a stellar dynamical subject.

Furthermore, we are convinced that knowledge of the problems of the celestial mechanics (at a non-elementary level) is indispensable for those dealing with stellar dynamics: moreover, we think that there should not be a sharp boundary between the two disciplines. This led us to make the whole area the subject of a single book, albeit in two volumes. Our purpose is therefore to provide researchers in astronomy and astrophysics with an as thorough and clear an exposition as possible of the problems which constitute the foundations of celestial mechanics and stellar dynamics. It is therefore intended that for the latter the chosen approach is the first of those listed above.

There is now a general agreement that "mathematical" and "physical" cultures are quite distinct and that they also have difficulties, sometimes, in understanding each other. In our opinion, this situation, owing to the ever-increasing specialization of scientific learning, causes damage that is particularly severe in the field of astronomy and astrophysics.

Whereas in the past (we are speaking of a "golden age" that ended in the 1920s) the astronomer and the astrophysicist could take advantage of current work in mathematics and physics, today this is not only impossible but even unthinkable. The university education of astronomers and astrophysicists is overwhelmingly of the "physical" type: the mathematical tools acquired are inadequate for tackling the reading of any mathematical paper whatsoever (of mathematical physics, the theory of differential equations, etc.), which turns out to be necessary in some research. This is also because there is an irresistible tendency for everybody (and therefore mathematicians also) to retire into their own special language. Astronomers and astrophysicists, owing to the nature of the things they are dealing with, continually need to resort to results obtained by physicists and mathematicians. In the latter case, for the reason given above, that turns out to be exceedingly difficult and sometimes impossible. It is obvious, and we are convinced of this, that

---

[2] M. Hénon, C. Heiles: The applicability of the third integral of motion: some numerical experiments, *Astron. J.* **69**, 73–79 (1964).

the times and cultural environments in which personalities like Poincaré, Jeans or Eddington were present cannot come back again. However, we are also convinced that one can and must do something to better the present situation.

Our aim in planning, and in writing, this book has been to contribute to ferrying from the "mathematical" side to the "astronomical-astrophysical" side some of the results achieved in the last few decades, which we consider essential to anyone dealing with solar system, stellar systems, galactic dynamics, etc. It is clear that in an operation of this kind it may happen that some of the things to be ferried fall overboard, whether the boat was overloaded or the boatman not expert enough: we hope, however, to have kept the losses within acceptable limits. To continue with the methaphor, to the people living on the side at which the boat lands, we assume preparation to the intermediate graduate level (calculus, differential equations, vector calculus, ...).

We have done our best to provide a self consistent treatment, at least at a first level of understanding, to spare the reader continuous jumps from one textbook to another; at the same time we have also endeavoured to facilitate the deepening of individual arguments supplying indispensable information, including bibliographic details.

The point of view we have assumed is that of discussing the *problems* and not of going into the details of different applications: we have tried to single out the fundamental problems (i. e. the mathematical models) and to present them in as clear and readable a way as possible for a reader having the mathematical background assumed above. We also consider the reader to be fully acquainted with celestial mechanics at undergraduate level, to the extent that can be obtained, for instance, from an excellent book such as Danby's[3].

By tradition, the old textbooks on celestial mechanics used to include a chapter devoted to Hamiltonian mechanics, an indispensable tool for perturbation theory. We have not escaped from the tradition and the first volume includes a chapter devoted to selected topics of dynamics and dynamical systems. The second chapter, devoted to the two-body problem, is not meant to replace traditional expositions (which are assumed known to the reader) but simply to emphasize features of the problem which can prompt further developments. The third and the fourth chapter (the N-body problem and the three-body problem) follow on in the same spirit, giving much space to results so far to be found only in the original papers. The fifth chapter, to our mind, is intermediate between celestial mechanics and stellar dynamics as usually agreed upon. In all four chapters (from the second to the fifth), besides some novelties (we believe) in the planning of the material and the exposition of recent results, classical arguments sanctioned by tradition remain. For the

---

[3] J. M. A. Danby: *Fundamentals of Celestial Mechanics*, 2nd Revised & Enlarged Edition (Willmann–Bell, Richmond, 1988).

latter, we have sometimes drawn our "inspiration" from the expositions of authors whose works can now be considered "classics" and whom the reader will found mentioned in the notes to each chapter.

In the second volume the first three chapters are devoted to the theory of perturbations, starting from classical problems and arriving at the KAM theory and to the introduction of the use of the Lie transform. A whole chapter treats the theory of adiabatic invariants and its applications in celestial mechanics and stellar dynamics. Also the theory of resonances is illustrated and applications in both fields are shown. Classical and modern problems connected to periodic solutions are reviewed. The description of modern developements of the theory of Chaos in conservative systems is the subject of a chapter in which is given an introduction to what happens in both near-integrable and non-integrable systems. The invaluable help provided by computers in the exploration of the long time behaviour of dynamical systems is acknowledged in a final chapter where some numerical algorithms and their applications both to systems with few degrees of freedom and to large N-body systems, are illustrated.

Acknowledgments

We wish to acknowledge here all those people who have helped and advised us in writing, editing and publishing this book. Warm thanks goes to Carlo Bernardini for encouraging us to pursue the work and to Giancarlo Setti for recommending that it be published. We are also very happy to thank Catello Cesarano, Giovanni De Franceschi and Maria Rita Fabbro for the help they provided during the revision and editing of the final version, although the responsibility for any mistakes remains with us.

Finally we thank I.C.R.A. (International Centre for Relativistic Astrophysics), Rome for having supported us with computer and printing facilities.

Rome, August 1995

Dino Boccaletti
Giuseppe Pucacco

Contents

Contents of Volume 2

Introduction – The Theory of Orbits from Epicycles to "Chaos"

The subject of this book is the study of the orbits followed by a body (mass point) subjected to the gravitational attraction of a given number of other bodies (mass points). We shall start from the minimum number (the two-body problem) and proceed to the case where it is convenient to represent the action exerted by a great number of bodies by means of a mean potential. Although the concept of the orbit as a continuous line drawn in three-dimensional space by the subsequent positions of a moving mass point may appear elementary and hence be considered a primitive notion which should only require intuition, nevertheless it has, over the centuries, undergone some kind of evolution. In view of future developments, we shall choose as a starting point for the evolution of the concept of the orbit the formulation given by the Greeks in the third century B.C.

From the Theory of Epicycles to Newton

Although historians of ancient astronomy have evidence of a notable development in astronomy among both the Egyptians and the Babylonians, it was only among the Greeks, and in a sophisticated mathematical environment, that the theories were born which, with some adjustment and improvement, would dominate for about two thousand years, i.e. up to the Copernican revolution.

The Greeks formalized their ideas about the motion of celestial bodies in the *theory of eccentrics* and in the *theory of epicycles*. The fundamental fact to keep in mind is that according to the Greeks the celestial bodies could not move on any kind of curve unless it was a circle (the perfect curve and therefore the only one worthy of a celestial body). This implied that any motion could only be either a uniform circular motion or a combination of uniform circular motions. We should also add that, in so doing, they renounced the possibility of considering seriously the problem of investigating the true nature of the physical system of the world, namely of finding the causes of the motion of celestial bodies. The various mathematical elaborations were basically oriented towards the *description* of motions. Through a crude schematization, we can say that from Hipparchus (second century

B.C.), through Ptolemy (second century A.D.), to Newton (the *Principia* was published in 1687) attention has been given mainly to the *kinematics* and not to the *dynamics* of celestial bodies.

The theory which almost certainly appeared first is that of the *eccentrics*. At the basis of this theory the Earth is situated at the centre of the universe and the Moon rotates around it on a circular orbit with a period of 27 days; the Sun also rotates around the Earth with a period of one year; the centre of these orbits does *not* coincide with the position of the Earth. The inner planets, Mercury and Venus, move on circles whose centres are always on the line joining the Sun and the Earth; the Earth remains outside these circles. The outer planets (Mars, Jupiter, Saturn) also move on circles whose centres are on the line joining the Earth and the Sun, but these circles have so great a radius as to always encircle both the Earth and the Sun. This theory was employed by Hipparchus to represent the motion of the Sun.

The *eccentric* (see Fig. 1) is a circle whose centre (C) does not coincide with the position of the observer[1] on the Earth (E). If the Sun (S) moves with uniform motion on this circle, and hence the angle ACS increases uniformly, it will not be the same for the angle AES: this will increase more slowly when S is close to A (the apogee, or farthest point from the Earth) and more quickly when S is close to B (the perigee, or closest point to the Earth).

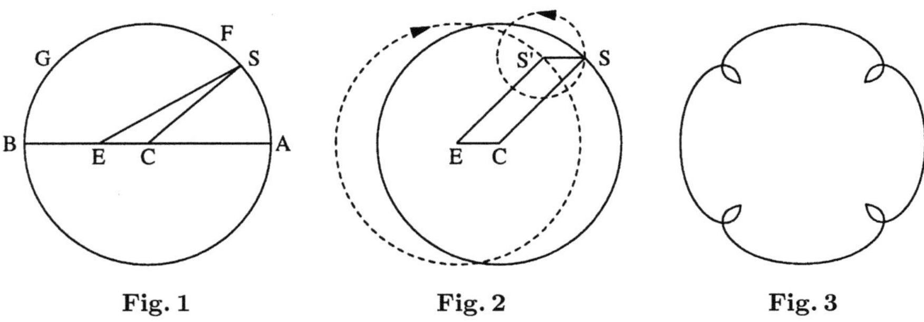

Fig. 1   Fig. 2   Fig. 3

The motion of the Sun could equally be obtained by means of the epicycle model, which dates back to Apollonius (end of the third century B.C.). In such a model, the body whose motion must be represented is considered to be moving uniformly on a circle (epicycle). The centre of the epicycle moves uniformly on a second circle called the *deferent*. To demonstrate the said equivalence, we shall refer to Fig. 2. If one takes as the deferent a circle equal to the eccentric but with its centre at E (the dashed circle), and if one takes the point S' on it such that ES' is parallel to CS, then $S'S$ is equal and also parallel to EC. Then the Sun S, which moves uniformly on the

---

[1] As the observer is in a geocentric system, motions should be reproduced as seen from the Earth.

eccentric, can be considered in the same way to be in uniform motion on a circle of radius $S\,S'$ whose centre S' moves uniformly on the deferent. In this case, therefore, the two methods lead to the same result. The arrows in Fig. 2 show that the motion on the epicycle must occur in a direction opposite to that of the motion on the deferent.

Similar considerations apply for the motion of the Moon, although in this case the theory turns out to be very rudimentary and fails to give an account of the greater complexity of the lunar motion compared to the solar one. Hipparchus found a way to overcome these difficulties by taking an eccentric whose centre describes a circle around the Earth in a period of nearly 9 years (corresponding to the motion of the apses). The epicycle theory was later improved by using as a deferent an eccentric circle and by introducing a new point called the *equant*, placed symmetrically to the Earth with respect to the centre of the deferent. This point replaced the centre of the deferent as a point from which to see the centre of the epicycle moving with constant angular velocity. Somehow, this violated the rule that the considered circular motions should be uniform. Later still, the theory was improved by adding secondary epicycles, epicycles which had an epicycle as a deferent.

We shall not deal with this matter further, since our purpose has been only to capture the spirit of the epicycle theory. It would be the basis of the Ptolemaic system which would reach, substantially unchanged in its fundamental structures, the age of the Renaissance and beyond. It is well known, in fact, that, during the thirteen centuries following the composition of the *Almagest*, the models worked out by Ptolemy were modified and improved several times, particularly by the Arabian astronomers, and therefore instead of the Ptolemaic system one should speak of Ptolemaic systems; however, the basic concepts have never been a matter of controversy. The Earth was immobile at the centre of the world and all the other bodies moved around it with motions that, however complicated, would have to be explained as suitable compositions of circular motions.

The idea of circular motion continued to obsess even Copernicus, who with his work *"De Revolutionibus orbium coelestium"* (published in Nuremberg in 1543, the year of his death) founded the new cosmology and established a new theory of planetary motion. The title of Chap. 4 of the first book of *De Revolutionibus* in fact reads:[2] "The reason why the motion of celestial bodies is uniform circular and perpetual or composed of circular motions", and later on, with the need to employ rectilinear motion, he deemed it necessary to demonstrate that a rectilinear motion could be generated by the composition of two circular motions. By moving the centre of the system from the Earth to the Sun, Copernicus succeeded in obtaining for all the planets, as for the Earth, quite simple orbits (circular in a first approximation) and no longer curves like that of Fig. 3, which represents the orbits (seen from the Earth!) obtained from the epicycle theory. Copernicus granted to the Earth,

---

[2] Nikolaus Kopernicus: *Gesamtausgabe* (Munich, 1949).

nevertheless, a prime role, by assuming in his planetary theories that the centre of the terrestrial orbit, rather than the barycentre of the Sun-planet system, was the centre of every motion. Of course, this led to mistakes which still required the use of epicycles as correctives in order to be in agreement with the results of the observations (albeit still rather inaccurate). This, however, did not worry the astronomers, who were used to a veritable crowd of epicycles; in fact, Copernicus, in that little work known as *Commentariolus*, which he had circulated handwritten and which seems to precede the editing of the *De Revolutionibus*, concluded almost in triumph: "And in this way, Mercury moves altogether by seven circles, Venus by five, the Earth by three and, around it, the Moon by four; at last Mars, Jupiter and Saturn, each by five. As a consequence, 34 circles are sufficient to explain the entire structure of the universe, as well as the dance of the planets". Notwithstanding the error of having assumed the centre of the terrestrial orbit to be the centre of every motion (an error which, as later emphasized by Kepler, is considerable in the case of Mars's motion), Copernicus marked the beginning of a new era for astronomy and for the determination of the orbits of celestial bodies. The world had to wait a little less than 70 years from the publication of *De Revolutionibus* to reach, with Kepler, the very high point of what we have called "the kinematics of celestial bodies" and at the same time the sunset of the epicycles.

However, the disappearance of the epicycles was only temporary: some time later, like the Phoenix, they rose again from their ashes. In fact, as remarked (perhaps for the first time) by G. V. Schiaparelli in the last century, Fourier series expansions brought the epicycles[3] back again, in modern dress, in celestial mechanics.

Between 1609 and 1619 Kepler announced those results which have been known ever since as his "three laws". Although he also addressed his attention to the causes of the motion, the results in this field had never been very positive (for instance, he ascribed the motion of the planets around the Sun to a magnetic force different from gravity). From our point of view, therefore, it is convenient to take into consideration Kepler's work solely as the end point of the kinematics of celestial bodies: it is from Kepler's three laws that Newton could eventually deduce the law of universal gravitation, and thereafter it became truly possible to speak of the dynamics of celestial bodies.

Although according to some historians the first intuition of the law of universal gravitation dates back to as early as 1666, when Newton was only 24, it is twenty years afterwards that Newton presented to the Royal Society of London a copy of the paper that would be printed one year later (1687) in its ultimate form and with the title "Naturalis Philosophiae Principia Math-

---

[3] See S. Sternberg: *Celestial Mechanics*, Part I (W. A. Benjamin, 1969), Chap. 1; and D. G. Saari: A visit to the Newtonian N-body problem via elementary complex variables. *Amer. Math. Monthly* **97**, 105–119 (1990).

ematica". Newton's law[4] is deduced from Kepler's three laws and Kepler's three laws are deduced from Newton's law. With Newton, not only was a kinematics of the orbits made available, but at last also a quite general physical law, applicable according to the particular problem of interest. Newton's law refers to the gravitational interaction of two bodies (to be considered either mass points or with a spherical mass distribution), and thence if we apply it to the planetary motion we must refer to the ideal case, which implies only the Sun and the planet under examination: *the two-body problem.*

Any deviation from the calculated Keplerian orbit has to be attributed to the action of the other bodies of the solar system; action which in principle can be evaluated, since we are able to express through Newton's law the force acting between any two bodies: the *N-body problem.*

The substantial change which occurred with respect to what was done before Newton lies just in this: up to that time, every orbit, as seen from the Earth, was reproduced with some approximation by combining a certain number of circular motions whose presence was justified only by the results they enabled to be obtained and there was no aknowledgement whatsoever of the fact that the entire solar system was "held up" by the mutual interactions of the various components. With Newton, on the other hand, the orbit followed by each celestial body could be exactly calculated, in principle, if one were to take into account all the bodies involved and their mutual interactions. One was then in the position where the mathematical model could completely reproduce the physical phenomenon. Possible inadequacies did not derive from failures of the model, but on calculational difficulties.

In the *Principia*, Newton applied only geometrical methods and because of that the evaluation of the perturbation of the Keplerian orbits became particularly cumbersome. The most important case, of course, was the motion of the Moon, to which he devoted almost all of the third book as well as great energy, since it appears[5] he told Halley that the problem of the motion of the Moon *made his head ache and kept him awake so often that he would think of it no more.*

The influence of the Sun on the motion of the Moon around the Earth is so important that the problem of the lunar motion must be considered a three-body problem. Just to give some idea of how the perturbing force of the third body varies in a complicated manner, by using only geometrical tools à la Newton, let us consider the Earth–Moon–Sun system,[6] as shown in Fig. 4 (where the radius of the lunar orbit – approximated by a circle – is exaggerated with respect to the distance from the Sun). In the figure, m_1 and m_2 represent two positions of the Moon in its circular orbit around the Earth (E) and S represents the position of the Sun. The vectors $m_1 K_1$, $m_2 K_2$ and

---

[4] See, for instance Sir Isaac Newton: *Principia*, translated by Motte and revised by Cajori. (University of California Press, 1934).

[5] The quotation is extracted from F. R. Moulton: *An Introduction to Celestial Mechanics*, 2nd edition (Macmillan, 1914), p. 363.

[6] The example and the figure are taken from Moulton: op cit. p. 337.

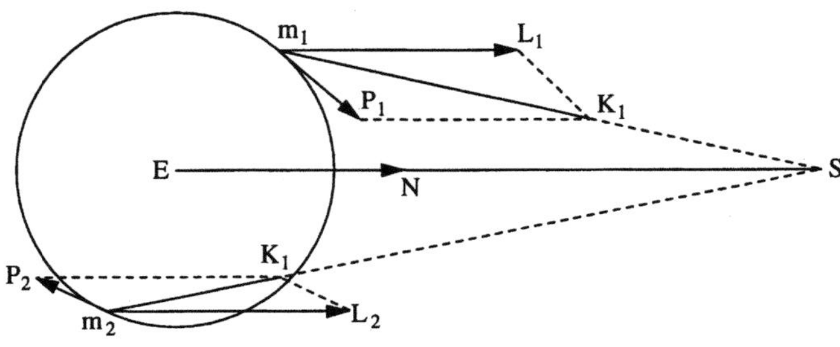

Fig. 4

EN give on suitable scales the acceleration on the Moon and on the Earth due to the action of the Sun. By decomposing m_1K_1, m_2K_2 in such a way that one of the components is equal and parallel to *EN*, one has the result that the perturbing acceleration due to the Sun is given in the two positions by m_1P_1, and m_2P_2 respectively. The perturbing acceleration in P varies then from point to point and is directed toward the line *ES*, or its extension beyond *E* when the distance *mS* is greater than *ES*.

The evaluations of the various effects that could be obtained through a study of the Newtonian geometrical constructions correspond roughly to the first order[7] of the perturbative theories which would be developed in the following century. After Newton, almost only analytical methods would be applied, the geometrical ones being reserved for elementary explanations of the causes of the various types of perturbation.

The Theory of Orbits from Newton to Poincaré

From the third and last edition of the *Principia* (1726) to the publication of the *Traité de Mécanique Céleste* of Laplace (the first volume was published in 1799), the theory of orbits progressed significantly through the work of essentially five persons: Euler (1707–1783), Clairaut (1713–1765), D'Alembert (1717–1783), Lagrange (1736–1813) and Laplace himself (1749–1827).

In a little more than half a century, one goes from the geometrical proofs of Newton to a generalized use of analytical tools which, very often, were created just to solve the problems of celestial mechanics (it is enough to mention the case of the method of variation of arbitrary constants in the solution of differential equations). We have chosen the work of Laplace as a means of marking the process of formal elaboration of Newtonian mechanics, even if this may appear objectionable from the point of view of the historian of

---

[7] For the motion of the lunar perigee, in fact, Newton had obtained a value which was only half that given by observations. In 1872, in his unpublished manuscripts, the exact evaluation was found in which Newton considered perturbations of the second order.

science, since in the last century a kind of general model for the formalization of a physical theory was identified in the *Mécanique Céleste*.

In 1803, the third volume of the *Mécanique Céleste*, dedicated to Napoléon Bonaparte, was published. Laplace wrote in the preface as follows (in translation):

"We have given in the first section of this work the general principles of the equilibrium and the motion of the matter. Their application to the celestial motions led us, *without hypotheses and through a series of mathematical arguments,*[8] to the law of universal gravitation, of which gravity and the motion of projectiles on the Earth are but particular cases. Then, taking into account a system of bodies submitted to this great law of the nature, we have reached, by means of a particular analysis, the general expressions of their motions, of their figures and of the oscillations of the fluids which cover them; expressions from which one has seen all the observed phenomena of the tides, of the variations of the degrees and the weight at the surface of the Earth, of the precession of equinoxes, of the libration of the Moon, of the figure and rotation of the Saturn rings, and their permanence on its equatorial plane derived. [...] In our century, the successors of this great man [Newton] have built up the construction of which he had laid down the foundations. They have improved the infinitesimal Analysis, they have invented the differential Calculus and the finite difference Calculus, and have turned into formulae the whole Mechanics. By applying these findings to the law of gravity, they have explained by this law all the celestial phenomena and given an unexpected accuracy to the theories and astronomical tables [...] It is mainly in the applications of the Analysis to the System of the world where the power of this wonderful tool, without which it had been impossible to understand a mechanism as complicated in its effects as simple in its causes, is shown".

What might be called the "the Laplacian belief" is encapsulated in the above quotation. All the motions of celestial bodies are explained "without hypotheses and through a series of mathematical arguments", starting from the general principles of mechanics. The "wonderful tool" of analysis is the weapon which allows one to win even the most difficult challenges of astronomy. The future of any system of gravitating bodies can be exactly foreseen for eternity as long as we know its present state: it is merely a matter of calculation, that is of application of the "wonderful tool". In fact, to quote Laplace,[9] once more, this time from "Essai philosophique sur les Probabilités" (in translation):

"We shall then imagine the present condition of the universe as the effect of its previous condition and as the cause of the forthcoming one. An intelligence which, at a given instant of time, would know all the forces which act on the nature and the natural conditions of the elements which compose it, if in addition he were sufficiently powerful to modelize these data with the Analysis, would include in the same formula the motions of the greatest bodies of the universe as well as those of

---

[8] The italics are ours.

[9] For the quoted excerpts, see P. S. Laplace: *Oeuvres Complètes* (Gauthiers-Villars, Paris, 1878).

the lightest atoms; nothing would be uncertain to him, and the future, as well as the past, would be present at his sight".

This "absolute" determinism of Laplace would later on find its expression and formal endorsement in the theorem of existence and uniqueness for ordinary differential equations. Before reaching this point, let us pause for a while in order to evaluate the situation. In the days when Laplace was writing, the gravitational constant was known with a certain accuracy (the Cavendish experiment in 1798 had given a result of 6.75; the error with respect to today's measurements was about 1%); by knowing the masses involved in each problem, one could write the related system of differential equations. The immediately following step was to consider the problem as the sum of two others: two-body problem and a perturbation due to the action of every other body (in some cases only one).

The solution of the first problem being known (*the unperturbed problem*) mathematical techniques had to be developed to handle the perturbation. Generally, an expansion in a power series of a small quantity (the eccentricity, the ratio of one of the masses and that of the Sun, etc.) was used and the problems of the convergence of the series were neglected. As a consequence, it was thought possible to approximate the "true", physical orbit more closely the more series terms were taken into account. The fact that two new planets, Uranus (1781) and Neptune (1846), were discovered through the use of this kind of calculation only confirmed what we have called the "Laplacian Belief".

Let us go back to the theory of ordinary differential equations, which we have already mentioned, to locate something very strange. In a number of books,[10] mostly concerning dynamical systems, the existence and uniqueness theorem for ordinary differential equations is not only considered the formal expression of determinism in mechanics, but also appears to be so intimately connected with it that determinism could not even exist without that theorem. As a matter of fact, the two questions from the very beginning have grown independently from each other[11] and only towards the end of the nineteenth century, when through the work of Peano the question of the uniqueness of the solution was clarified, could one begin to overlap the theorem and the deterministic principle so that one became the supporting arm of the other. In any case, the theory of orbits, by the theorem of existence and uniqueness for the solution of ordinary differential equations and the further theorem of continuous dependence of this on the initial conditions, achieved at a certain point the enviable condition of "perfect science" where the conformity of the mathematical model to the physical reality and the rigour of the mathematical tool gave to the predictions of celestial mechanics the features of infallibility and precision.

---

[10]See, for instance, V. I. Arnold: *Mathematical Methods of Classical Mechanics* (Springer, 1978).

[11]See G. Israel: *Il determinismo e la teoria delle equazioni differenziali ordinarie*, *PHYSIS* **28**, 305–358, (1991).

In fact, once a given problem was set up in the correct manner (that is, the correct system of differential equations was written down), the theory guaranteed (apart from the existence or otherwise of singularities) that the solution existed because of the properties of the regularity of the functions representing the components of the given forces. Furthermore, the solution corresponding to given initial conditions, was also unique. Since the initial data were the result of measurement operations and therefore subject to errors, the corresponding solution consequently provided predictions subject to errors; the above theorem of the continuous dependence of the solution on the initial conditions also guaranteed that the error implied in the solution was small, provided the initial data measurement error was small.

The state of affairs appeared to be such as to make Moulton write,[12] about one century after Laplace:

"At the present time Celestial Mechanics is entitled to be regarded as the most perfect science and one of the most splendid achievements of the human mind. No other science is based on so many observations extending over so long a time. In no other science is it possible to test so critically its conclusions, and in no other are theory and experience in so perfect accord".

But even at the time Moulton wrote this, the situation was not such as to justify that tone of triumph. From 1890,[13] when the *mémoire couronné* of Poincaré was published, the seeds had been spread of many plants destined to grow in our century, and some of them with plenty of thorns.

From Poincaré to the Deterministic Chaos: The Sensitive Dependence on the Initial Conditions

Poca favilla gran fiamma seconda
Dante: *Paradiso* I, 34

We have already mentioned that, for the problems of celestial mechanics, a "strategy", generically called perturbation theory, had been developed. Although this name covered and still covers very different theories, one feature is common to all these theories: each problem was solved by "perturbing" a problem whose solution was already known. The perturbation was represented by a series expansion. These series, heirs, as we have already said, of the ancient epicycles, contained however, a lot of contradictions. First of all, the problem of convergence. In fact, although at the beginning the series were used freely without difficulty, at a certain point it began to be asked whether one was really dealing with convergent series. The answer was provided by Poincaré, as already mentioned: in general, these series did not converge;

---

[12]See F. R. Moulton: op. cit. p. 430.
[13]H. Poincaré: Sur le problème des trois corps et les equations de la dynamique, *Acta Mathematica* **13**, 1–271 (1890).

as was later clarified, they converged only for some frequencies. A second problem was given by the use of approximated expressions obtained by truncating the expansions: they contained "secular" or "mixed" terms, that is, terms proportional to t (the variable indicating the time) or to t multiplied by a periodic term. In both cases, it was clear that these expressions would lose any meaning if they were used over a fairly long interval of time.

A. Lindsted (1882) found the method to overcome this inconvenience. Starting from the observation that the non-linearity of the equation involved certainly modified the frequency ω_0 of the unperturbed problem, he expressed the variable t through a new independent variable multiplied by a power series (of a "small" quantity ε), where the coefficients were given by different frequencies ($\omega_1, \omega_2, \ldots, \omega_n$, etc.). By fixing afterwards, at each perturbative order, the value of the corresponding ω_i in a suitable manner, he eliminated the terms having t as a factor.

Besides the presence of secular terms, the perturbative series were affected by another pathology: *small divisors*. For systems of two or more degrees of freedom, even at the first perturbative order, linear combinations of the frequencies, which could take a very small value, close to zero, appeared in the denominators of the terms of the expansion. When these divisors were as small as zero, the frequencies were "in resonance".

It is clear from what has been said so far that the perturbation theory, notwithstanding its successes, was close to entering into a deep crisis. This had also an impact on the old problem of the stability of the solar system, which had been formalized in the answer to a prize question which the King of Sweden had proposed in 1885 (the prize was awarded to H. Poincaré) as follows: *For an arbitrary system of mass points which attract each other according to Newton's laws, assuming that no two points ever collide, give the coordinates of the individual points for all time as the sum of a uniformly convergent series whose terms are made up of known functions.*

The situation was eased by A. N. Kolmogorov,[14] who proposed in 1954 a theorem, later on demonstrated by V. I. Arnold and J. Moser and ever since universally known as the "KAM theorem", where the conditions which had to be verified in order that a motion undergoing a "small perturbation" could preserve the features of the unperturbed motion were laid down. As we shall see in Chap. 1, if one considers the phase space of a system with n degrees of freedom, for an integrable system one has that the trajectories of the representative point are constrained to remain on the surface of an n-dimensional torus defined by n constants of motion of the system itself.

The "KAM theorem" states that, under suitable conditions the perturbations do not "destroy" the tori but only deform them, and then the trajectories are still constrained to wind around n-dimensional surfaces which are

---

[14]A. N. Kolmogorov: *Doklady Akad. Nauk.* **98**, 527 (1954). English translation in *Stochastic Behavior in Classical and Quantum Hamiltonian Systems* ed. by G. Casati, J. Ford (Springer, 1979), pp. 51–56.

close to the initial tori. The conditions required by the theorem are however very stringent; besides allowing only very small perturbations and excluding the cases of resonance, the theorem also established how far from possible resonances we must remain. The KAM theorem is not applicable, therefore, to the solar system, where various cases of resonance occur; nevertheless its appearence brought new life to studies of the perturbation theories and once again caught the attention of mathematical physicists to celestial mechanics. Almost at the same time as the demonstration of the KAM theorem, a further topic appeared, which has raised an enormous interest in the last few years under the appealing title of "deterministic chaos". The development and the improvement of computing systems, in the 1960s, made available calculations which up to that time were not feasible. It was then possible to develop a method for studying the non-integrable systems already suggested by Poincaré and Birkhoff: the "surfaces of section". Without going into the details, which will be described in Volume 2, let us imagine drawing in phase space a surface which intersects the trajectories of the representative points of the system. If the system is integrable, and consequently the trajectories are wrapped around the n-dimensional tori of which we have already made mention, the intersection with the above surface shall give as a result points (more and more numerous with increasing time) which will lie along well-defined curves (intersections of the tori with the surface). Obviously, in the case of closed orbits (simply periodic motions) one will have as intersections isolated points that cross at each period.

In 1964, M. Hénon and C. Heiles[15] investigated by this method the motion of a star in a galaxy, representing the gravitational attraction of the galaxy by a potential having cylindrical symmetry. The chosen potential was given by the potential of a planar oscillator to which two terms were added (of third degree in the coordinates). Those terms made the problem non-linear and non-integrable. Hénon and Heiles found that, for very low values of the total energy (the minimum value chosen by them was $E = 1/12$), the system was very close to being integrable, since the vast majority of intersection points were placed along well-defined curves corresponding to the intersections with the tori. By raising the energy, the regular curves progressively disappeared, through a phase where they were reduced to tiny isolated zones. For $E = 1/6$, in practice the whole available area (with the exception of extremely small "islands") was filled thickly by points generated by a single trajectory (see Figs. 5, 6 and 7).

The result of Hénon and Heiles emphasized two features. The first was that the system was "more integrable" than one would expect from the use of purely analytical methods: in fact, at low energies, the system was "practically" integrable. Numerical methods could then provide much help in guessing if a system was very close to the condition of integrability or not (it was this belief, among others, that gave reason for such research). The second fea-

---

[15]M. Hénon, C. Heiles: See Footnote 2 in the Preface.

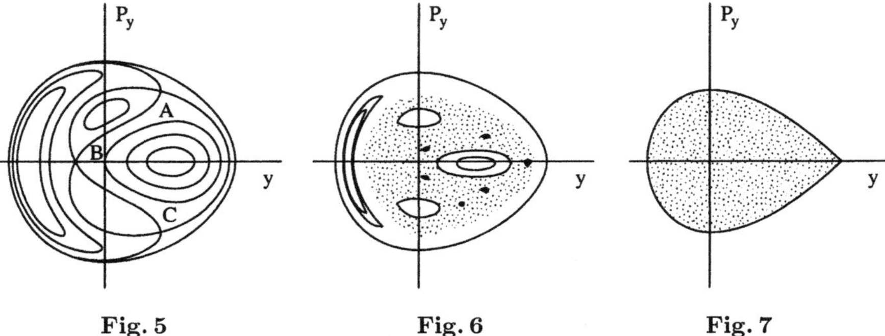

Fig. 5   **Fig. 6**   **Fig. 7**

ture was that the more the energy increased and the system departed from integrability, the more the motion became "chaotic".

We shall now try to explain the meaning of this adjective. We have already recalled the theorem of the continuous dependence of a solution (in our case a trajectory) on the initial data. Quantitatively, if $x_i(t)$ and $y_i(t)$ are two solutions of the system $\dot{x} = X(x,t)$ defined in the closed interval $[t_0, t_1]$ and K is the constant introduced in the Lipschitz condition (see Footnote 1, Sect. 1.1), then

$$|x_i(t) - y_i(t)| \leq |x_i(t_0) - y_i(t_0)| e^{K(t-t_0)}.$$

This inequality does not prevent two curves which were close at the time $t = t_0$ exponentially diverging afterwards within a very short time; for some systems this is what really occurs, and is what is called a sensitive dependence on initial conditions. Until recently, physicists thought that this could only occur in very particular cases (unstable equilibrium positions, etc.). What is new is the realization that in some systems one has sensitive dependence on initial conditions, whatever the initial conditions, that is, the dependence is not the exception but the rule. In this case, one speaks of chaotic motions and deterministic chaos.

In the Hénon–Heiles problem, the system, as the energy grows, becomes a (partially) chaotic system: the intersections of the trajectories with the surface of section thickly fill the area determined by energy conservation, but points corresponding to trajectories originating in "close" points are here very "far". The exponential divergence of the trajectories characterizes the chaotic motion; in spite of this, the chaos is "deterministic", in the sense that the divergence of the trajectories is "determined" by the initial conditions according to the equations of the system. This "determinism", however, does not help at all in predicting the system's behaviour, because in practical cases one has to deal with experimental errors. With regard to this, we cite what J. Hadamard wrote in 1898; together with Poincaré, he can be rightfully considered the forerunner of those who characterize the present trend:[16]

---

[16] See J. Hadamard: Les surfaces à courbures opposées et leurs lignes géodesiques, *J. Math. et Appl.* **4**, 27–73 (1898).

"Certainly, when a system moves under the action of given forces and the initial conditions of the motion have *given values, in the mathematical meaning of the word*, the subsequent motion and, as a consequence, its behaviour, when t increases indefinitely are for these very reasons known. However, in the astronomical problems, it will not be like that: the constants which define the motion are *physically given*, that is with some errors whose magnitude is reduced as long as the power of our observation means increases, but which is impossible to cancel. If one follows the trajectories only for a given time, on the other hand arbitrary, one could think that the errors on the initial conditions have been reduced to such an extent that they cannot appreciably modify the shape of these trajectories during the time interval under consideration. [...] it is not by all means correct to draw a similar conclusion regarding the *final* behaviour of the same curves. This may very well depend [...] on *arithmetical* discontinuity properties of integration constants" (in translation).

If we consider that even the simplest systems (and with only two degrees of freedom) are in general *mixed* systems, that is, they can have both regular solutions and chaotic behaviour, then we realize that the Laplace universe, this gigantic and perfect watch, has disappeared for ever in the haze of the past. But if, as Poincaré states,[17] "... the real aim of Celestial Mechanics is not to calculate the ephemerides, because one could be satisfied of a short term forecast, but to recognize if the Newton law is sufficient to explain all the phenomena", the task of determining the orbits of celestial bodies still remains to challenge the researcher.

Unlike Hipparchus and Ptolemy, we are now able to calculate ephemerides with great accuracy, to foresee when and at which point of a planet a space probe will land, but we do not yet know if the solar system (or, better, its Newtonian model) is stable, nor are we able to foresee if one of the thousands of asteroids which fill it will some day end up hitting the Earth. There are still many questions without answers.

---

[17]H. Poincaré: *Les Méthodes nouvelles de la Mécanique Céleste* (Gauthiers-Villars, Paris, 1892) – Vol. I, p 5.

Chapter 1

Dynamics and Dynamical Systems – Quod Satis

In some cases it is easier to describe what something is not, rather than what it is or what it wants to be. We shall therefore adhere to this principle when introducing this chapter.

First of all, this chapter is not a summary of the subjects mentioned in the title. Only a few topics from mechanics appear, and furthermore they are reviewed while we constantly keep in mind the problem of integrability and the search for first integrals. If only a selection of topics from mechanics is given, and moreover aimed at a specific goal, still less is said about dynamical systems and stability, in fact the minimum sufficient to make a few indispensable concepts familiar. Obviously, one could argue that the reader could find elsewhere the subjects explained here, by consulting various textbooks on analytical dynamics, dynamical systems and some review articles. It was important to us, however, that some subjects were explained in a particular manner, with certain features emphasized and consequently treated in such a fashion as to be applicable according to a specific conceptual scheme. In essence, we have committed ourselves to providing the reader with those tools, and only those, indispensable to the understanding of subsequent chapters. The sentence *Quod Satis* also stems from here, as in ancient recipes: the ingredients are mixed, we hope, in sufficient quantities for the aim to be satisfied. We have also tried to emphasize clarity without sacrificing rigour too much. Furthermore we have sought to maintain the language at a "moderate" level, so as not to discourage the reader who is not too keen on appreciating the charms of abstraction.

A. *Dynamical Systems and Newtonian Dynamics*

1.1 Dynamical Systems: Generalities

In this book we shall deal mainly with Hamiltonian dynamical systems, which represent a very peculiar class of continuous dynamical systems. Therefore, instead of starting from the definition of abstract dynamical systems, we shall limit ourselves to the case that will be the object of our applications. Consequently, we shall define "dynamical system", a set of m first-order differential equations in the variables $x_1, x_2, \ldots, x_m$:

$$\frac{dx_i}{dt} = X_i(x_1, x_2, \ldots, x_m, t), \qquad i = 1, 2, \ldots, m, \tag{1.A.1a}$$

or, in vector notation:

$$\dot{\boldsymbol{x}} = \boldsymbol{X}(\boldsymbol{x}, t). \tag{1.A.1b}$$

In (1.A.1a), the independent variable t represents the time and one assumes that the m functions X_i are defined in a suitable $m + 1$-dimensional domain and there satisfy the conditions under which the theorem of existence and uniqueness of the solution holds.[1] The m-dimensional space $(x_1, x_2, \ldots, x_m)$ is called the *phase space* and, for the time being, coincides with the Euclidean space $\mathbb{R}^m$. If the functions X_i, which define the vector field $\boldsymbol{X}$, do not depend on time, the system is said to be *autonomous*. In general, a solution of the system (1.A.1a) will be a vector $\boldsymbol{x} = \boldsymbol{x}(\boldsymbol{x}_0, t_0, t)$ verifying the initial condition $\boldsymbol{x}(\boldsymbol{x}_0, t_0) = \boldsymbol{x}_0$. For an autonomous system, the *translation* property holds: if $\boldsymbol{x}(t)$ is a solution, then $\boldsymbol{x}(t - \alpha)$, with α a constant, is also a solution. The proof is straightforward, because in this case the right-hand side of (1.A.1b) does not contain the time. In the case of an autonomous system, we can use one of the x_i, for instance x_1 (if $X_1 \neq 0$), as the new independent variable, in place of t, and obtain the $m - 1$ equations

$$\frac{dx_2}{dx_1} = \frac{X_2(\boldsymbol{x})}{X_1(\boldsymbol{x})}, \quad \frac{dx_3}{dx_1} = \frac{X_3(\boldsymbol{x})}{X_1(\boldsymbol{x})}, \quad \ldots, \quad \frac{dx_m}{dx_1} = \frac{X_m(\boldsymbol{x})}{X_1(\boldsymbol{x})}. \tag{1.A.2}$$

The solutions of the set (1.A.2) in phase space are called *phase curves*. If the existence and uniqueness theorem holds for the set (1.A.1), it will hold for (1.A.2) also; therefore the curves in phase space will not intersect. One must keep in mind that different solutions, $\boldsymbol{x}(t)$ and $\boldsymbol{x}(t - \alpha)$, correspond to the same phase curve.

---

[1] The m functions $X_1, \ldots X_m$ must be continuous in all the $m + 1$ variables in the assumed domain together with their partial derivatives with respect to the x_i. The last condition can be replaced by the weaker condition

$$\left| \frac{\partial \boldsymbol{X}(\boldsymbol{x}, t)}{\partial x_j} \right| \leq K, \qquad \text{with } K > 0, \quad j = 1, 2, \ldots, m$$

or even $|\boldsymbol{X}(\boldsymbol{x}, t) - \boldsymbol{X}(\boldsymbol{y}, t)| \leq K|\boldsymbol{x} - \boldsymbol{y}|$ (the *Lipschitz condition*).

If a point $\boldsymbol{x} = \boldsymbol{a}$ exists for which $X_i(\boldsymbol{a}) \equiv 0$, then the point $\boldsymbol{a} \equiv (a_1, \ldots, a_m)$ is called a *critical point* or *singular point*. Obviously at such a point there is no possibility of reducing the system to a set of $m - 1$ equations by eliminating the time, as we did in (1.A.2). A critical point can also be considered as a phase curve degenerated into a point. Moreover, being there $\dot{x}_i = 0$ $(\forall t)$, a critical point corresponds to an *equilibrium solution*; $x_i(t) = a_i$ $(\forall t)$ is a solution of the system and, for the theorem of existence and uniqueness, is the unique solution through $\boldsymbol{a}$; thence, if $\boldsymbol{y}(t)$ is a distinct solution , it will be $\boldsymbol{y}(t) \neq \boldsymbol{a}$, $\forall t$. However, it may happen that, for $t \longrightarrow \infty$, some solution tends to an equilibrium solution; in this case the equilibrium solution is called an *attractor*. More precisely, a critical point $\boldsymbol{x} = \boldsymbol{a} \in \mathbb{R}^m$ is called a *positive attractor* if a neighbourhood $\Omega_a \subset \mathbb{R}^m$ of the point $\boldsymbol{a}$ exists so that $\boldsymbol{x}(t_0) \in \Omega_a$ entails that in the limit $t \to +\infty$, $\boldsymbol{x} \to \boldsymbol{a}$. If this happens for $t \to -\infty$, $\boldsymbol{x} = \boldsymbol{a}$ is said to be a *negative attractor*.

Let us assume now that a solution $\boldsymbol{x} = \boldsymbol{x}(t)$ of our system, defined in some interval $I \subset \mathbb{R}$, with values $\boldsymbol{x}(t) \in D \subset \mathbb{R}^m$, is such that $\boldsymbol{x}(t) = \boldsymbol{x}(t + T)$, with $T \geq 0$. In this case $\boldsymbol{x}(t)$ is said to be a *periodic solution* with period T. It is easy to see that $\boldsymbol{x}(t) = \boldsymbol{x}(t + T)$ implies also that $\boldsymbol{x}(t) = \boldsymbol{x}(t \pm nT)$, with n any integer. If we consider the phase space of the system $\dot{\boldsymbol{x}} = \boldsymbol{X}(\boldsymbol{x})$, since for a periodic solution we have that $x_1, x_2, \ldots, x_m$, at the time $t_0 + T$, again assume the values they had at the time t_0, it follows that a periodic solution gives rise to a *closed phase curve* (or *cycle*). Making use of the *translation* property of autonomous systems and of the fact that several *translated* solutions correspond to the same phase curve, one can demonstrate also the inverse: to a closed phase curve there corresponds a periodic solution. As an example, let us consider the system

$$\dot{x}_1 = x_2, \qquad \dot{x}_2 = -x_1. \tag{1.A.3}$$

The solutions are linear combinations of $\sin t$ and $\cos t$ (one solution is obtained from the other by means of a translation of $\pi/2$) and we have for $\boldsymbol{x} \equiv (x_1, x_2)$ a periodic solution of period $T = 2\pi$. The phase curves are obtained by transforming the set (1.A.3) into

$$\frac{dx_2}{dx_1} = -\frac{x_1}{x_2}, \tag{1.A.4}$$

as in (1.A.2). Integrating (1.A.4), we have:

$$x_1^2 + x_2^2 = \text{const.}, \tag{1.A.5a}$$

i.e. the phase curves are concentric circles (Fig. 1.1). From $x_1 = a\cos t + b\sin t$, $x_2 = -a\sin t + b\cos t$, $a, b \in \mathbb{R}$, it is evident that the critical point $x_1 = x_2 = 0$ is not an attractor.

If instead we consider the system

$$\dot{x}_1 = x_1, \qquad \dot{x}_2 = x_2, \tag{1.A.6}$$

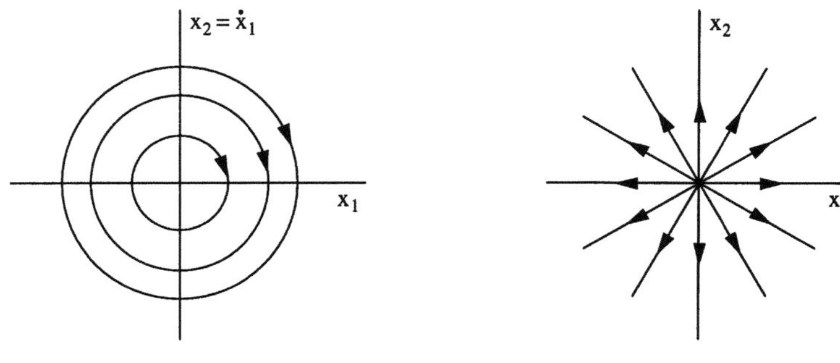

Fig. 1.1   **Fig. 1.2**

the phase curves will be

$$\frac{dx_2}{dx_1} = \frac{x_2}{x_1},\qquad(1.\text{A}.7)$$

from which $x_2 = \alpha x_1$, $\alpha \in \mathbb{R}$. The phase curves are straight lines through the origin (Fig. 1.2) and $x_1 = c_1\exp(t)$, $x_2 = c_2\exp(t)$ so that the equilibrium solution $(0,0)$ is a negative attractor. The arrows in the figures indicate the direction of the motion of the phase point along the curves with increasing t. This motion is called the *phase flow* and the plot of a set of phase curves (corresponding to different initial conditions) the *phase portrait.*

In the two examples proposed, it has been possible to integrate the system (1.A.2) and to obtain a relation among the components of $\boldsymbol{x}$; this relation defines an $(m-1)$-dimensional manifold, a subset of the m-dimensional phase space: it represents the level set of a function of $x_1, x_2, \ldots, x_m$. Such a level set, which we can in general indicate with

$$F\left(x_1, x_2, \ldots, x_m\right) = \text{const.},\qquad(1.\text{A}.8)$$

is called an *integral manifold* and the function $F(\boldsymbol{x})$ the *first integral* of the system $\dot{\boldsymbol{x}} = X(\boldsymbol{x})$. Generally speaking, we say that a real differentiable function $F(\boldsymbol{x})$ is a first integral for the system $\dot{\boldsymbol{x}} = X(\boldsymbol{x})$ if it stays constant along a solution of the system itself, that is,

$$\frac{dF}{dt} = \frac{\partial F}{\partial x_1}\dot{x}_1 + \frac{\partial F}{\partial x_2}\dot{x}_2 + \ldots + \frac{\partial F}{\partial x_m}\dot{x}_m = 0.\qquad(1.\text{A}.9)$$

The existence of a first integral has the result of reducing by one the dimensions of the phase space available for the motion of the system. In turn, the phase curves will be subsets of this $(m-1)$-dimensional manifold. In the two cases we have seen, being $m = 2$, the phase curves coincide with the family of integral manifolds.

Since the examples we have considered up to now were extremely simple (linear systems with constant coefficients like (1.A.3) and (1.A.6)), the method of integration based on lowering the order of the system by means of the first integral may appear an unnecessarily cumbersome procedure.

However, it is essential in the more complex cases, so it is worth explaining its application. Consider the system (1.A.3), the *harmonic oscillator*. As we know, the solution can also be written as $x_1 = x = A\sin(t + \phi)$, $x_2 = \dot{x} = A\cos(t + \phi)$, with $A = \sqrt{a^2 + b^2}$, $\phi = \arctan(a/b)$. From the first integral (1.A.5a), which we rewrite as

$$\frac{1}{2}(x_1^2 + x_2^2) = I_1, \tag{1 A.5b}$$

and the first equation of (1.A.3), we get

$$\frac{dx_1}{dt} = \pm\sqrt{2I_1 - x_1^2}. \tag{1.A.10}$$

And, integrating,

$$\int dt = \pm \int \frac{dx_1}{\sqrt{2I_1 - x_1^2}}, \tag{1.A.11a}$$

that is,

$$t + I_2 = \arcsin\left(\frac{x_1}{\sqrt{2I_1}}\right), \tag{1.A.11b}$$

where I_2 is an integration constant. By inverting (1.A.11b), one finally[2] has

$$x_1 = \sqrt{2I_1}\sin(t + I_2), \tag{1.A.12}$$

which coincides with the above-mentioned solution if we set $A = \sqrt{2I_1}$, $\phi = I_2$. The integration constants are evaluated by means of initial conditions; if we put $x_1(0) = x_1^0$, $\dot{x}_1(0) = \dot{x}_1^0$, we have:

$$I_1 = \frac{1}{2}(x_1^0)^2 + \frac{1}{2}(\dot{x}_1^0)^2 = \frac{1}{2}(a^2 + b^2),$$

$$I_2 = \phi = \arctan\frac{x_1^0}{\dot{x}_1^0} = \arctan\frac{a}{b}. \tag{1.A.13}$$

Summarizing, the procedure we followed consists of four steps:

a) to find a first integral;
b) to use the found integral to reduce the order of the system;
c) to integrate the reduced system (if the original system were of order two only a quadrature remains to be performed);
d) to invert and so to obtain the solution.

From what we have said, it is obvious that to integrate a system of order m explicitly (i.e. to reduce it to quadratures), one needs $m - 1$ first integrals.

---

[2] The ambiguity in sign of (1.A.10) is inherent to the problem: during the oscillatory motion between the two values of x_1 (points of inversion of the motion) given by $\pm\sqrt{2I_1}$, $\dot{x}_1$ changes sign going through zero at the inversion points. The integral of (1.A.11a) will be evaluated by taking the $+$ sign for positive dx_1 and the $-$ sign for negative dx_1. Then the value of the integral increases with increasing t. We shall meet the same problem again later on.

Up to now we have considered first integrals which were functions only of the variables $x_1, x_2, \ldots, x_m$ and not of the time t, according to the definition (1.A.8). However, sometimes integrals exist which depend also on time. One of the best known cases is given by the system describing a damped oscillatory motion:

$$\dot{x}_1 = x_2, \qquad \dot{x}_2 = -x_1 - \lambda x_2, \tag{1.A.14}$$

with λ a positive constant. The first integral is given by:

$$e^{\lambda t}\left(x_1^2 + x_2^2 + \lambda x_1 x_2\right) = I_1 = \text{const.} \tag{1.A.15}$$

We are not interested in continuing our discussion of the system (1.A.14), but we shall employ the systems (1.A.3) and (1.A.14) to anticipate a classification which is of paramount importance: the distinction between "conservative" and "non-conservative" systems. If we put $x_1 = x$ and $x_2 = \dot{x}_1 = \dot{x}$, the two systems are equivalent respectively to $\ddot{x} = -x$ and $\ddot{x} = -x - \lambda \dot{x}$. We call a system conservative for which

$$\ddot{x} = f(x) \tag{1.A.16}$$

and non-conservative for which

$$\ddot{x} = g(x, \dot{x}). \tag{1.A.17}$$

The first case, in mechanics, represents a force deriving from a potential (and depends only on the position) and the second, a force depending also on the velocity. In the first case we have conservation of the total energy.

1.2 Classification of Critical Points – Stability

Let us assume that the system (1.A.1) has a critical point at $\boldsymbol{x} = \boldsymbol{a}$; in this case, the Taylor expansion in the neighbourhood of $\boldsymbol{x} = \boldsymbol{a}$ is

$$\dot{\boldsymbol{x}} = \boldsymbol{X}(\boldsymbol{x}, t) = \left.\frac{\partial \boldsymbol{X}}{\partial \boldsymbol{x}}(\boldsymbol{x}, t)\right|_{\boldsymbol{x}=\boldsymbol{a}} (\boldsymbol{x} - \boldsymbol{a}) + \mathrm{O}(|\boldsymbol{x} - \boldsymbol{a}|^2). \tag{1.A.18}$$

If we limit ourselves to considerering only the first term of the expansion (1.A.18), this means that we replace the system (1.A.1) by its linear approximation: i.e. we have *linearized* the system. As we shall see later on, the linear approximation is much more meaningful and rich in information about the solutions of the complete system than one could expect. In the case in which the complete system is autonomous, the square matrix $m \times m$, which we have indicated symbolically by $(\partial \boldsymbol{X}/\partial \boldsymbol{x})|_{\boldsymbol{x}=\boldsymbol{a}}$, will have constant elements. In this case it is convenient to perform a translation so as to shift the critical point to the origin.

Therefore, without loss of generality, we shall limit ourselves to studying an autonomous system with a critical point at the origin. Then the linearized system will be

$$\dot{\boldsymbol{x}} = \boldsymbol{A}\boldsymbol{x}, \tag{1.A.19}$$

$\boldsymbol{A}$ being a constant matrix with elements given by

$$A_{ij} = \left.\frac{\partial X_i}{\partial x_j}\right|_{\boldsymbol{x}=\boldsymbol{0}}.$$

We shall assume that $\det \boldsymbol{A} \neq 0$ ($\boldsymbol{A}$ is a non-singular matrix) and look for solutions of the system (1.A.19) of the form

$$x_1 = \alpha_1 e^{\lambda t}, \quad x_2 = \alpha_2 e^{\lambda t}, \quad \ldots, \quad x_m = \alpha_m e^{\lambda t}, \tag{1.A.20}$$

with $\alpha_1, \alpha_2, \ldots, \alpha_m$ and λ to be determined. By substituting in (1.A.19), one gets the set:

$$\begin{aligned}
(A_{11} - \lambda)\alpha_1 + A_{12}\alpha_2 + \ldots + A_{1m}\alpha_m &= 0, \\
A_{21}\alpha_1 + (A_{22} - \lambda)\alpha_2 + \ldots + A_{2m}\alpha_m &= 0, \\
\vdots \qquad\qquad\qquad \vdots \\
A_{m1}\alpha_1 + A_{m2}\alpha_2 + \ldots + (A_{mm} - \lambda)\alpha_m &= 0,
\end{aligned} \tag{1.A.21}$$

which will have solutions different from $\alpha_1 = \alpha_2 = \ldots = \alpha_m = 0$ only in the case in which the determinant of the coefficients will be zero:

$$\det\left(\boldsymbol{A} - \lambda\mathbf{1}\right) = 0, \tag{1.A.22}$$

$\mathbf{1}$ being the identity matrix. Equation (1.A.22) is called the *characteristic equation* of the system (1.A.19) and $\det(\boldsymbol{A} - \lambda\mathbf{1})$ the characteristic polynomial. Equation (1.A.22) is an algebraic equation of degree m and will have m roots, real or complex, and not all necessarily distinct. The values of λ, roots of (1.A.22), are the *eigenvalues* of $\boldsymbol{A}$. When the m roots $\lambda_1, \lambda_2, \ldots, \lambda_m$ are distinct, by substituting them into the set (1.A.21), one obtains corresponding to each of them an m-tuple

$$\left(\alpha_1^{(j)}, \alpha_2^{(j)}, \ldots, \alpha_m^{(j)}\right) = \boldsymbol{\alpha}^{(j)}, \tag{1.A.23}$$

where the m-tuple corresponding to the eigenvalue λ_j is labelled by j; $\boldsymbol{\alpha}^{(j)}$ is said to be the eigenvector corresponding to the eigenvalue λ_j. Since we are dealing with a homogeneous system, one of the components of the vector $\boldsymbol{\alpha}^{(j)}$ is arbitrary, i.e. it is defined up to a scalar factor. In the case of distinct eigenvalues, one can demonstrate that the m eigenvectors $\boldsymbol{\alpha}^{(j)}$ are linearly independent.

With the vectors $\boldsymbol{\alpha}^{(j)}$, one then constructs the m solutions of the system (1.A.19), obtaining

$$\boldsymbol{x}^1(t) = \left\{ \alpha_1^{(1)} e^{\lambda_1 t}, \alpha_2^{(1)} e^{\lambda_1 t}, \ldots, \alpha_m^{(1)} e^{\lambda_1 t} \right\},$$

$$\vdots \qquad\qquad \vdots \qquad\qquad\qquad (1.A.24)$$

$$\boldsymbol{x}^m(t) = \left\{ \alpha_1^{(m)} e^{\lambda_m t}, \alpha_2^{(m)} e^{\lambda_m t}, \ldots, \alpha_m^{(m)} e^{\lambda_m t} \right\}.$$

Therefore, the general solution of the system (1.A.19) will have the form

$$\boldsymbol{x} = \sum_{j=1}^{m} C_j \boldsymbol{x}^j(t), \qquad\qquad (1.A.25)$$

where C_j are m arbitrary costants. However, in practice, instead of determining the vectors $\boldsymbol{\alpha}^{(j)}$ from the set (1.A.21), it is preferable to look for the general solution of the system (1.A.19) in the form

$$x_1(t) = \sum_{j=1}^{m} a_1^{(j)} e^{\lambda_j t}, \quad \ldots, \quad x_m(t) = \sum_{j=1}^{m} a_m^{(j)} e^{\lambda_j t}, \qquad (1.A.26)$$

where the λ_j are the (distinct) roots of the characteristic equation and the $\boldsymbol{a}^{(j)}$ are determined by substituting the $x_j(t)$ in the system (1.A.19) and equating the coefficients of the same exponentials $e^{\lambda_j t}$. The $\boldsymbol{a}^{(j)}$ will depend on m arbitrary constants. We shall not deepen the general case (any m) further, limiting ourselves to the case $m = 2$, which allows the use of a graphical representation and thus a more immediate understanding.

In this case the system (1.A.19) will become

$$\begin{aligned}
\dot{x}_1 &= A_{11} x_1 + A_{12} x_2, \\
\dot{x}_2 &= A_{21} x_1 + A_{22} x_2,
\end{aligned} \qquad\qquad (1.A.27)$$

and the equation for the phase curves,

$$\frac{dx_2}{dx_1} = \frac{A_{21} x_1 + A_{22} x_2}{A_{11} x_1 + A_{12} x_2}. \qquad\qquad (1.A.28)$$

In general (1.A.28), which in any case can always be integrated, is too complicated to allow a discussion of the geometric nature of the solution in a simple way. It is more convenient to rely on a non-singular linear transformation, given by a suitable matrix $\boldsymbol{S}$, such that

$$\boldsymbol{x} = \boldsymbol{S}\,\boldsymbol{\xi}, \qquad\qquad (1.A.29)$$

$\boldsymbol{\xi}$ being a new vector. A transformation of this type may consist of a rotation and a dilation (or contraction): curves get deformed but the behaviour in the neighbourhood of $(0, 0)$ does not change.

From (1.A.29), $\boldsymbol{\xi} = \boldsymbol{S}^{-1}\boldsymbol{x}$ and $\dot{\boldsymbol{\xi}} = \boldsymbol{S}^{-1}\dot{\boldsymbol{x}} = \boldsymbol{S}^{-1}\boldsymbol{A}\boldsymbol{x}$. Then

$$\dot{\boldsymbol{\xi}} = \boldsymbol{S}^{-1}\boldsymbol{A}\boldsymbol{S}\,\boldsymbol{\xi} = \boldsymbol{B}\,\boldsymbol{\xi}. \qquad\qquad (1.A.30)$$

The matrices A and B are said to be *similar* and have the same characteristic polynomial.[3] Now we must determine S so as to obtain the simplest matrix B as possible. One can show[4] that there is a real non-singular matrix S such that B is equal to one of the following six matrices:

$$\text{a)} \quad \begin{pmatrix} \lambda_1 & 0 \\ 0 & \lambda_2 \end{pmatrix}, \qquad \text{b)} \quad \begin{pmatrix} \lambda_1 & 0 \\ 0 & \lambda_2 \end{pmatrix},$$

where in (a) $\lambda_2 < \lambda_1 < 0$ or $0 < \lambda_2 < \lambda_1$, in (b) $\lambda_2 < 0 < \lambda_1$,

$$\text{c)} \quad \begin{pmatrix} \lambda & 0 \\ 0 & \lambda \end{pmatrix}, \qquad \text{d)} \quad \begin{pmatrix} \lambda & 1 \\ 0 & \lambda \end{pmatrix}, \qquad (1.A.31)$$

where $\lambda > 0$ or $\lambda < 0$.

$$\text{e)} \quad \begin{pmatrix} u & v \\ -v & u \end{pmatrix}, \qquad \text{f)} \quad \begin{pmatrix} 0 & v \\ -v & 0 \end{pmatrix},$$

where in (e) $u, v \neq 0$ and $v > 0$ or $v < 0$; in (f) $v \neq 0$; λ_1 and λ_2 are the roots of the characteristic polynomial ($\lambda_1 = u + iv$, $\lambda_2 = u - iv$ when they are complex conjugate).

Different phase portraits correspond to the cases listed in (1.A.31). Case (a) is pictured in Fig. 1.3, in which the arrows indicate the direction of time and correspond to negative roots λ_1, λ_2: the origin is called an *improper stable node*.[5] If λ_1, $\lambda_2 > 0$, the picture is obtained by inverting the arrows and exchanging ξ_1 with ξ_2: we have an *improper unstable node*.

Case (b) corresponds to Fig. 1.4 ($\lambda_2 < 0 < \lambda_1$) and the origin is called a *saddle point*. For case (c), the picture is given by Fig. 1.5 and we have a *stable proper node* corresponding to $\lambda < 0$; an example of the case $\lambda > 0$, an *unstable proper node*, is given by the system (1.A.6) of Sect. 1. Case (d) gives a further case of *improper node* (Fig. 1.6). The node is *stable* (if $\lambda < 0$, the case pictured in our figure) or *unstable* (if $\lambda > 0$). For complex conjugate roots, case (e), we have a *spiral point*: stable if $u < 0$, *unstable* if $u > 0$. In Fig. 1.7 we have pictured an unstable spiral point. Case (f), $u = 0$, corresponds to imaginary roots and the origin is called the *centre*; in Fig. 1.8 we have considered $v > 0$. The case of the linear oscillator with unit frequency considered in (1.A.4) is the same but with inverted arrows (it corresponds to $v < 0$).

Let us continue, now, with the autonomous system

$$\dot{x}_i = X_i(x_1, x_2, \dots, x_m), \qquad i = 1, 2, \dots, m, \qquad (1.A.32)$$

---

[3] See, for instance, F. R. Gantmacher: *The Theory of Matrices* (Chelsea, New York, 1959), Chap. III.

[4] For this, and other assertions given without proof, see F. Bauer, J. A. Nohel: *The Qualitative Theory of Ordinary Differential Equations. An Introduction* (Dover, New York, 1989), Chap. 2 and Appendices.

[5] The meaning of the term "stable" will be made clear later on in this section.

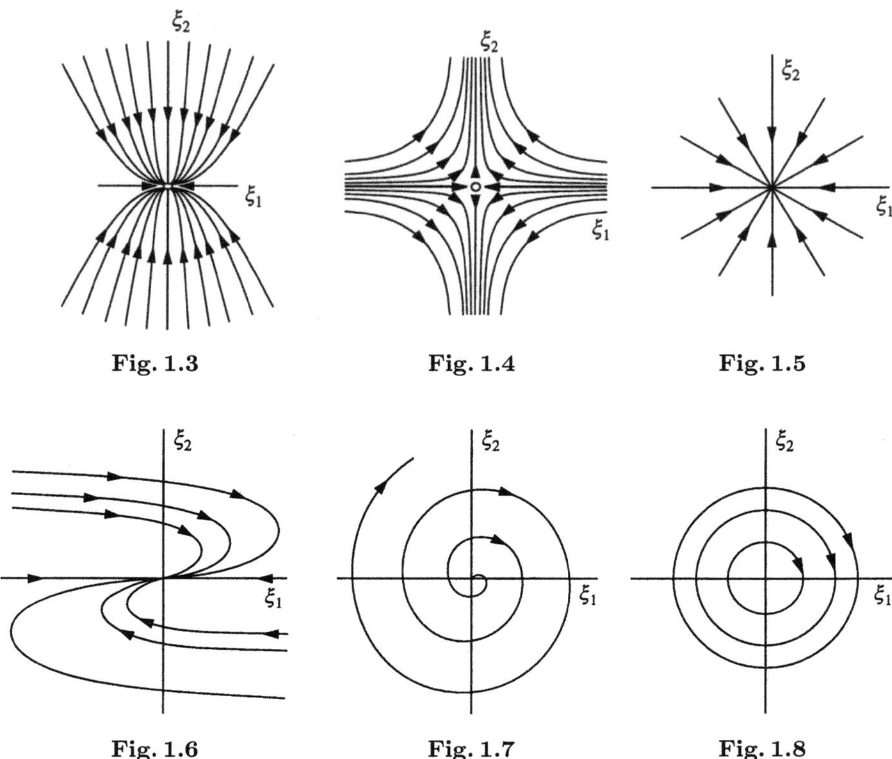

Fig. 1.3   Fig. 1.4   Fig. 1.5

Fig. 1.6   Fig. 1.7   Fig. 1.8

assuming again the validity of the existence theorem in a suitable m-dimensional domain. Moreover, let us assume that $x = a$ is a critical point for our system. We shall introduce for it the concept of stability according to Lyapunov, or L-stability. We say that the critical point, or equilibrium solution, $x_i(t) = a_i$ is stable if, for each number $\varepsilon > 0$, we can find a number $\delta > 0$ such that:

i)   any solution of (1.A.32), which belongs to the neighbourhood of a with radius δ for some $t = t_1$, be defined for any $t_1 \leq t < \infty$;

ii)   a solution satisfying (i) remains in the neighbourhood of a with radius ε for any $t > t_1$.

If, furthermore, each solution satisfying (i) and (ii) is such that

iii) the limit for $t \to \infty$ of $x_i(t)$ is a_i, $\forall i$, then the point a is said to be *asymptotically stable*.

If we now apply these definitions to the linear system (1.A.27) and to the considered phase portraits, we can immediately verify that:

1) we have asymptotical stability only when the two roots are negative real or complex conjugate with negative real part,
2) the centre (imaginary roots) is stable but not asymptotically stable.
3) in all other cases, we have instability.

In the case of a general non-linear system, as a rule the solutions are not available, so we cannot directly study their behaviour as for the linear system with constant coefficients (1.A.27). However, Lyapunov has shown that, in most cases, we can get knowledge of the nature of a critical point of a general system (1.A.32) merely by referring to the system obtained from it by linearization. More precisely,

1) if all the roots λ_i of the characteristic equation pertinent to the linear system (1.A.19) have a negative real part, then the equilibrium solution $x_i = 0$ of the system (1.A.32) is *asymptotically stable*, as if the system consisted only of the linearized part;
2) if among the roots of the characteristic equation for the linear system there is at least one λ with a positive real part, then the solution $x_i = 0$ of the system (1.A.32) is *unstable*, whatever its non-linear part may be;
3) if the characteristic equation of the linear system has no roots with a positive real part, but has some imaginary or null root, then we have the so-called *critical cases* and the stability (or instability) will depend on the non-linear part of the system (1.A.32). Needless to say, if the system (1.A.32) had a critical point with coordinates $\neq 0$, one performs a suitable translation of the coordinate axes.

Before closing, we wish to give, concisely, some definitions to outline how the concepts of dynamical systems can be set in abstract form.

First of all, the phase space will be, in general, a differential manifold[6] $(\mathcal{M})$. We have already associated the phase flow to the motion of phase points along phase curves. More formally, we can now introduce the pair $(\mathcal{M}, \{\Phi_t\})$ given by the manifold $\mathcal{M}$ and by a one-parameter group of transformations of $\mathcal{M}$ onto itself. The generic group element will be given by a transformation depending on the parameter t:

$$\Phi_t : \mathcal{M} \to \mathcal{M}, \quad x \to \Phi_t\, x, \quad x,\, \Phi_t\, x \in \mathcal{M} \qquad (\forall\, t \in \mathbb{R}). \qquad (1.A.33)$$

For Φ_t, the group properties hold:

$$\bar{\Phi}_{t+s} = \Phi_t \circ \Phi_s,$$
$$\Phi_0 = \text{identity transformation}, \qquad (1.A.34)$$
$$\Phi_{-t} = \text{inverse trasformation},$$

where $\circ$ means composition of the two transformations.

---

[6] See, for instance, B. F. Schutz: *Geometrical Methods of Mathematical Physics* (Cambridge University Press, Cambridge, 1980).

If, given $x \in \mathcal{M}$, we consider the mapping

$$\Phi : \mathbb{R} \to \mathcal{M}, \qquad \Phi(t) = \Phi_t \, x, \qquad\qquad (1.\text{A}.35)$$

we shall call Φ the *motion* of x due to the phase flow $(\mathcal{M}, \{\Phi_t\})$. The *phase curve* will be the image of the mapping $\Phi : \mathbb{R} \to \mathcal{M}$, i.e. a subset of the phase space $\mathcal{M}$. Moreover, a *critical or equilibrium point* of a flow $(\mathcal{M}, \{\Phi_t\})$ will be a point $x \in \mathcal{M}$ which is, itself, a phase curve:

$$\Phi_t \, x = x, \qquad (\forall \, t \in \mathbb{R}).$$

Finally, recalling that a *diffeomorphism* is a mapping $f : \mathcal{M} \to \mathcal{N}$ that is bijective and differentiable together with $f^{-1} : \mathcal{N} \to \mathcal{M}$, we say that $\{\Phi_t\}$ is a *one-parameter group of diffeomorphisms* of $\mathcal{M}$. In fact $\{\Phi_t\}$ is a mapping

$$\Phi : \mathbb{R} \times \mathcal{M} \to \mathcal{M}, \qquad \Phi(t, x) = \Phi_t \, x, \qquad (t \in \mathbb{R}, \quad x \in \mathcal{M})$$

such that:

a) Φ is a differentiable mapping
b) $\forall \, t \in \mathbb{R}, \quad \Phi : \mathcal{M} \to \mathcal{M}$
c) the family $\{\Phi_t, \, t \in \mathbb{R}\}$ is a one-parameter group of transformations of $\mathcal{M}$.

1.3 The n-Dimensional Oscillator

As we have said, the system (1.A.3) represents the harmonic oscillator of unit frequency; if we consider a frequency $\omega_1 \neq 1$, the Newtonian equation of motion will be

$$\ddot{x}_1 = -\omega_1^2 \, x_1. \qquad\qquad (1.\text{A}.36\text{a})$$

If we put, differently from before, $\dot{x}_1 = \omega_1 \, v_1$, the corresponding system[7] will be

$$\dot{x}_1 = \omega_1 \, v_1, \qquad \dot{v}_1 = -\omega_1 \, x_1, \qquad\qquad (1.\text{A}.36\text{b})$$

and the phase space will be the $x_1 v_1$ plane.

Let us now consider in the Cartesian $x_1 x_2$ plane a second linear oscillator whose equation is

$$\ddot{x}_2 = -\omega_2^2 \, x_2, \qquad\qquad (1.\text{A}.37\text{a})$$

and the corresponding system

$$\dot{x}_2 = \omega_2 \, v_2, \qquad \dot{v}_2 = -\omega_2 \, x_2. \qquad\qquad (1.\text{A}.37\text{b})$$

By the composition of the two motions given by (1.A.36a) and (1.A.37a) we obtain a new physical system: a *planar oscillator*. This can be thought as consisting of a point of unit mass subjected to two orthogonal oscillations. The corresponding dynamical system is

---

[7] Obviously, v_1 is proportional to the velocity, but is not a "true" velocity from the dimensional point of view.

$$\dot{x}_1 = \omega_1\, v_1, \qquad \dot{v}_1 = -\omega_1\, x_1,$$
$$\dot{x}_2 = \omega_2\, v_2, \qquad \dot{v}_2 = -\omega_2\, x_2, \tag{1.A.38}$$

the phase space is 4-dimensional and the phase curves are subsets of this 4-dimensional space; their projections on the $x_1 x_2$ plane give the trajectories (orbits) of the moving point. Different orbits may intersect, whereas the phase curves, as we know, may not. As in (1.A.5a), the first integral can be found immediately (corresponding to the conservation of the total energy of the planar oscillator):

$$I = \frac{1}{2}\,\omega_1^2\,\left(v_1^2 + x_1^2\right) + \frac{1}{2}\,\omega_2^2\,\left(v_2^2 + x_2^2\right) = E. \tag{1.A.39}$$

The 3-dimensional surface defined by (1.A.39) is the energy integral manifold: it is a subset of the 4-dimensional phase space ($\mathbb{R}^4$) and is invariant under the phase flow. On the other hand, having obtained the system (1.A.38) by putting together the two independent systems (1.A.36b) and (1.A.37b), one finds that the solutions, a motion in $\mathbb{R}^4$, will be

$$\begin{aligned}
x_1 &= A_1\,\sin(\omega_1 t + \varphi_1^0),\\
x_2 &= A_2\,\sin(\omega_2 t + \varphi_2^0),\\
v_1 &= A_1\,\cos(\omega_1 t + \varphi_1^0),\\
v_2 &= A_2\,\cos(\omega_2 t + \varphi_2^0),
\end{aligned} \tag{1.A.40}$$

and the space $\mathbb{R}^4$ results from the direct sum of two invariant planes: $\mathbb{R}^4 = \mathbb{R}^2_{x_1,v_1} + \mathbb{R}^2_{x_2,v_2}$. The phase curves are given by the circles on these two planes

$$\begin{aligned}
S^1_{x_1,v_1} &= \{(x_1, v_1) \in \mathbb{R}^2; \quad x_1^2 + v_1^2 = A_1^2\},\\
S^1_{x_2,v_2} &= \{(x_2, v_2) \in \mathbb{R}^2; \quad x_2^2 + v_2^2 = A_2^2\},
\end{aligned}$$

and by the points given by $A_1 = 0$, $A_2 = 0$.

The phase flow consists of rotations of angles $\omega_1 t$, $\omega_2 t$. Every phase curve of the system (1.A.38) belongs to the direct product of the phase curves in the planes $\mathbb{R}^2_{x_1,v_1}$, $\mathbb{R}^2_{x_2,v_2}$. The direct product of two circles will be

$$T^2 = S^1 \times S^1 = \left\{(x_1, v_1, x_2, v_2) \in \mathbb{R}^4 : x_1^2 + v_1^2 = A_1^2, x_2^2 + v_2^2 = A_2^2\right\},$$

i.e a *two-dimensional torus*. As is known, T^2 may be imagined to be obtained by making a circle rotate around a coplanar axis which lies outside the circle itself (Fig. 1.9). A point on the surface of the torus corresponds to two angular coordinates φ_1, φ_2 (mod. 2π), which may be called *longitude* and *latitude*, respectively. The coordinates φ_1, φ_2 define a diffeomorphism between the surface of the torus and the direct product of two circles (T^2). One can also imagine "building" the two-dimensional torus by starting from a rectangle and "gluing" first two opposite sides, obtaining a cylinder, and then, by suitable deformations, the remaining two sides (already transformed into circles). Therefore, in our case, the square $0 \leq \varphi_1 \leq 2\pi$, $0 \leq \varphi_2 \leq 2\pi$ in the $\varphi_1\varphi_2$ plane can be considered a map of the torus T^2. The pair of points $(\varphi_1,0)$, $(\varphi_1,2\pi)$ and $(0,\varphi_2)$, $(2\pi,\varphi_2)$ are identified in T^2.

The flow of the system (1.A.38) leaves the torus $T^2 \subset \mathbb{R}^4$ invariant. The phase curves of (1.A.38) lie on the surface T^2. Now, if φ_1 is the polar angle in the plane $\mathbb{R}^2_{x_1,v_1}$ (considered positive in going from the positive v_1 axis to x_1) and the same happens for φ_2 in the x_2v_2 plane, from (1.A.40) we have

$$\begin{aligned}\varphi_1 &= \omega_1 t + \varphi_1^0,\\ \varphi_2 &= \omega_2 t + \varphi_2^0.\end{aligned} \quad (\text{mod.}\,2\pi) \qquad (1.A.41a)$$

The two (1.A.41a) are the equations of the phase curves. On the surface of T^2 they represent a helix, whereas in the "map", that is, in the $\varphi_1\varphi_2$ plane, we have the straight lines

$$\varphi_2 = \frac{\omega_2}{\omega_1}\,\varphi_1 + \left(\varphi_2^0 - \frac{\omega_2}{\omega_1}\,\varphi_1^0\right) \quad (\text{mod.}\,2\pi). \qquad (1.A.41b)$$

Now the following proposition holds:

If ω_1 and ω_2 are rationally dependent,[8] every phase curve (1.A.41a) on the torus is closed; if, on the contrary, ω_1 and ω_2 are rationally independent, every phase curve is everywhere dense[9] on the torus T^2.

To the proof we shall premise the following lemma:

If the circle S^1 is made to rotate on itself by an angle α not commensurable with 2π, the images of each point due to repeated application of the rotation α (mod. 2π) constitute a set everywhere dense on the circle.

To demonstrate this, let us split up the circle into k arcs corresponding to angles of amplitude $2\pi/k$. If we perform $k+1$ applications of the rotation α, because of the well-known principle of the Dirichlet cells we shall have two images of the same point of the circle in one arc, these images being different since $\alpha/2\pi$ is not rational. Let us assume that these two images correspond to the angles $\varphi + p\alpha$ and $\varphi + q\alpha$ (φ is the initial reference angle), with $p > q$, to fix the ideas. Let us put $s = p - q$; then the angle $s\alpha$ differs from a multiple of 2π by an amount smaller than $2\pi/k$. Therefore, if we consider the sequence φ, $\varphi + s\alpha$, $\varphi + 2s\alpha$, $\varphi + 3s\alpha$, ... (mod. 2π), any two consecutive terms differ by the same quantity, which is smaller than $2\pi/k$. Then, if we fix an arbitrary $\varepsilon > 0$, a k will always exist such that $2\pi/k < \varepsilon$, and, as a consequence, in every neighbourhood of a point of the circle of amplitude ε, there will always be terms of the sequence $\varphi + Ns\alpha$ (mod. 2π). Of course, what has been demonstrated above holds since α is not commensurable with

---

[8] Two numbers ω_1 and ω_2 are rationally dependent if two integers $k_1, k_2 \neq 0$ exist such that $k_1\omega_1 + k_2\omega_2 = 0$, and rationally independent if this happens only for $k_1 = k_2 = 0$.

[9] A set A is everywhere dense in B if, in an arbitrary small neighbourhood of any point of B, there is at least a point of A.

2π; otherwise an integer n would exist such that a rotation by an angle $n\alpha$ should be equal to an integer number m of rotations of 2π, and therefore the image of every point would be superimposed on the "preceding" images every m complete turns.

Let us go back, now, to the proposition concerning the phase curves.

a) Let ω_1 and ω_2 be rationally dependent, i.e. $k_1\omega_1 + k_2\omega_2 = 0$ with k_1 and k_2 non-zero integers. Then it is sufficient to define the period T as given by

$$T = \frac{2\pi}{\omega_1}\, k_2 = -\frac{2\pi}{\omega_2}\, k_1.$$

One immediately checks that this makes the curve (1.A.41a) closed on the torus.

b) Now, let ω_1 and ω_2 be rationally independent, i.e. ω_2/ω_1 is irrational. By referring to the map of T^2 in the $\varphi_1\varphi_2$ plane, let us consider the sequence given by the points in which the straight lines (1.A.41b) intersect the axis φ_2 (mod. 2π). These points will have as coordinate φ_2

$$\varphi_{2,k} = \varphi_2^0 + 2\pi\, k\, \frac{\omega_2}{\omega_1} \quad (\text{mod. } 2\pi).$$

From the above lemma, it follows that the set given by these points is everywhere dense in the segment $0 \le \varphi_2 \le 2\pi$. As a consequence, the set of the segments belonging to the above straight lines is everywhere dense in the square $0 \le \varphi_1 \le 2\pi$, $0 \le \varphi_2 \le 2\pi$, and moreover the corresponding phase curve is everywhere dense on T^2.

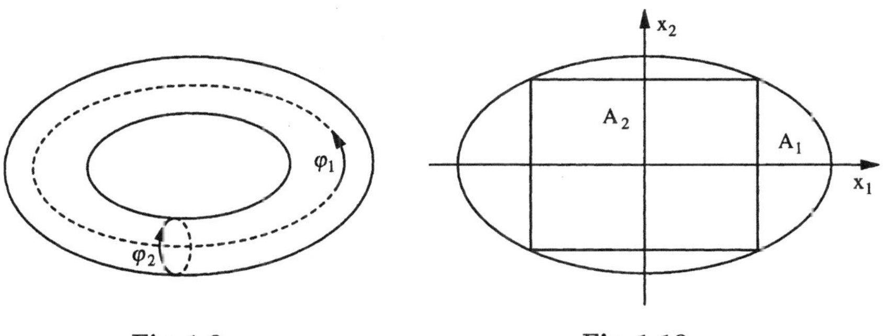

Fig. 1.9   **Fig. 1.10**

Note that is possible to show that the above case can be generalized to n independent harmonic oscillations. Therefore one can demonstrate that the corresponding phase curves lie on an n-dimensional torus

$$T^n = \underbrace{S^1 \times S^1 \times S^1 \times \cdots S^1}_{n \text{ times}} = \big\{(\varphi_1,\, \varphi_2,\, \ldots,\, \varphi_n) \quad \text{mod. } 2\pi\big\}$$

and satisfy the n independent equations

$$\dot{\varphi} = \omega_1, \ \dot{\varphi}_2 = \omega_2, \ldots, \ \dot{\varphi}_n = \omega_n.$$

Moreover, if the frequencies $\omega_1, \omega_2, \ldots, \omega_n$ are rationally dependent, every phase curve is closed; in the opposite case every phase curve is everywhere dense on the torus T^n. For the planar oscillator one can give a graphical representation by exploiting the energy integral. The two oscillatory motions being independent, besides (1.A.39) we also have

$$I_1 = \frac{1}{2}\,\omega_1^2\,\left(v_1^2 + x_1^2\right) = E_1 = \text{const.}$$
$$I_2 = \frac{1}{2}\,\omega_2^2\,\left(v_2^2 + x_2^2\right) = E_2 = \text{const.} \tag{1.A.42}$$

From (1.A.42), since kinetic energies must always be greater than 0, we have

$$|x_1| \leq \sqrt{\frac{2E_1}{\omega_1^2}} = A_1, \qquad |x_2| \leq \sqrt{\frac{2E_2}{\omega_2^2}} = A_2.$$

Thence, the orbit of the point submitted to the two orthogonal oscillations is confined in the rectangle $|x_1| \leq A_1$, $|x_2| \leq A_2$. For the same reason, from (1.A.39), we have

$$\omega_1^2 x_1^2 + \omega_2^2 x_2^2 \leq 2E = 2I,$$

i.e. the motion in the $x_1 x_2$ plane is confined to the ellipse

$$x_1^2 \frac{\omega_1^2}{2E} + x_2^2 \frac{\omega_2^2}{2E} = 1,$$

with $E_1 + E_2 = E$ (see Fig. 1.10).

It is easy to check that the rectangle turns out to be inscribed within the ellipse. We have seen in (1.A.40) that

$$x_1 = A_1 \sin\left(\omega_1 t + \varphi_1^0\right),$$
$$x_2 = A_2 \sin\left(\omega_2 t + \varphi_2^0\right).$$

Of course, the determination of the actual orbit must be done by composing the two independent oscillations of amplitude A_1 and A_2 and frequencies ω_1 and ω_2 respectively. The task is not simple and depends, obviously, on the values of the ratio ω_1/ω_2 and of the phase shift $\varphi_2^0 - \varphi_1^0$. In practice, one can use the following strategy.

A sinusoid with period $2\pi A_1(\omega_1/\omega_2)$ and amplitude A_2 is drawn on a strip of width $2A_2$ (see Fig. 1.11). The strip is wrapped round a cylinder of height $2A_2$ and diameter $2A_1$. The orbit in the $x_1 x_2$ plane is obtained by projecting the sinusoid on this plane. The various orbits are called *Lissajous figures*. One can show, in the same way as for the phase curves, that, if ω_1/ω_2 is a rational number, the trajectory is a closed algebraic curve and, if ω_1/ω_2 is irrational, the trajectory fills the rectangle such that it is everywhere dense.

Let us examine the simplest cases with rational ω_1/ω_2:

1) $\omega_1 = \omega_2$ (isotropic oscillator): The curve on the cylinder is an ellipse. The projection of the ellipse on the $x_1 x_2$ plane depends on the choice made for the directrix of the cylinder (which is projected in the axis x_2), i.e. on the phase shift $\varphi_2^0 - \varphi_1^0$. For $\varphi_2^0 = \varphi_1^0$, we obtain the diagonal of the rectangle; for small values of $\varphi_2^0 - \varphi_1^0$, the ellipse squeezed on the diagonal; for $\varphi_2^0 - \varphi_1^0 = \pi/2$, the ellipse with principal axes on x_1 and x_2, etc. (Fig. 1.12).

2) $\omega_2 = 2\omega_1$: See Fig. 1.13, in which $\varphi_2^0 - \varphi_1^0$ ranges from $-\pi/2$ to 0.

3) $2\omega_2 = 3\omega_1$: In Fig. 1.14, $\varphi_2^0 - \varphi_1^0$ ranges from $-\pi/4$ to 0. Of course, the diagonal in case (1) and the parabola in case (2) (the first curve in Fig. 1.13) are *closed* curves, i.e. described twice.

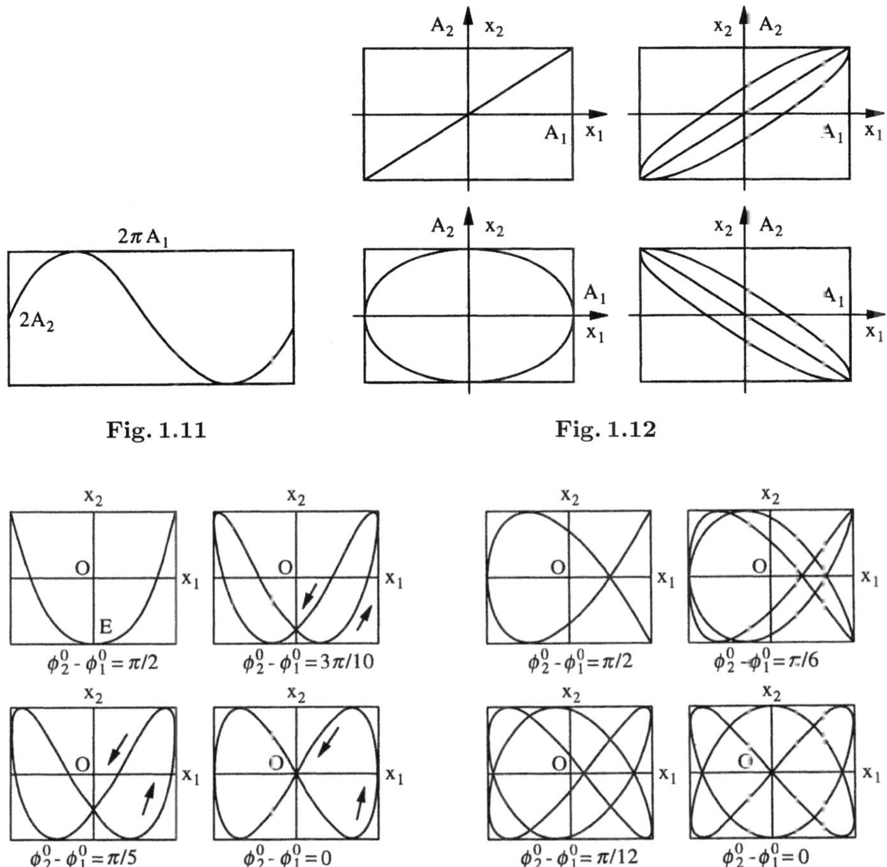

B. *Lagrangian Dynamics*

1.4 Lagrange's Equations

Given a physical system consisting of N mass points, its configuration will be described at any time by means of $3N$ coordinates, for instance Cartesian coordinates x_i $(i = 1, 2, \ldots, 3N)$, with the convention that

$$r_1 \equiv (x_1,\ x_2,\ x_3),$$
$$r_2 \equiv (x_4,\ x_5,\ x_6),$$
$$\vdots$$
$$r_N \equiv (x_{3N-2},\ x_{3N-1},\ x_{3N});$$

r_j, $(j = 1, 2, \ldots, N)$, being the distance vector of the generic point. In general, among the $3N$ coordinates, a certain number p, with $p < 3N$, of kinematical conditions (constraints) may exist; when they are finite relations, the system is called holonomic; when they are not integrable differential relations, the system is non-holonomic. In view of the applications in which we shall be interested, we limit ourselves to considering only holonomic systems, even if the results of this section can be extended to the non-holonomic[10] ones.

The relations among coordinates will be of the type

$$f_m\left(x_1, x_2, \ldots, x_{3N}; t\right) = 0, \qquad m = 1, 2, \ldots, p. \tag{1.B.1}$$

In (1.B.1) the case in which the relations among coordinates may depend explicitly on time is also taken into account. When p relations like (1.B.1) hold, obviously the system will actually depend on only $n = 3N - p$ coordinates; usually one expresses the x_i in terms of n parameters, or generalized coordinates, or Lagrangian coordinates, which are traditionally indicated by q_k $(k = 1, 2, \ldots, n)$:

$$x_i = x_i\left(q_1, q_2, \ldots, q_n; t\right), \qquad i = 1, 2, \ldots, 3N. \tag{1.B.2}$$

The number n of parameters necessary (and sufficient) to completely characterize the configuration of the system is the number of "degrees of freedom" of the system itself. The time derivatives of the q_k, written as $\dot{q}_k$, will be the generalized velocities; it must be kept in mind that, in general, they are not velocities in the strict sense. As, at any time, the configuration of the system is specified by the n-tuple $(q_1, q_2, \ldots, q_n)$, one can make the evolution of the system correspond to the motion of a point of coordinates $(q_1, q_2, \ldots, q_n)$, in an n-dimensional space Q (*configuration space*).

---

[10]See H. Goldstein: *Classical Mechanics*, 2nd ed. (Addison-Wesley, 1980), p. 45.

When the system consists of N mass points, the total kinetic energy will be

$$\mathcal{T} = \frac{1}{2} \sum_{i=1}^{N} m_i v_i^2 = \frac{1}{2} \sum_{k=1}^{3N} m_k \left(\frac{dx_k}{dt} \right)^2, \qquad (1.B.3)$$

thus defining, in a natural way, the metric:[11]

$$ds^2 = 2\mathcal{T}\, dt^2 = \sum_{k=1}^{3N} \left(\sqrt{m_k}\, dx_k \right)^2.$$

In the case of N free mass points (i.e. without constraints), we can take the $\sqrt{m_k}\, x_k$ as rectangular Cartesian coordinates, and therefore the $3N$-dimensional configuration space has a Euclidean structure: any transformation to curvilinear coordinates cannot change its real nature. The squared line element would take the form

$$ds^2 = \sum_{i,k=1}^{3N} g_{ik}\, dq_i\, dq_k, \qquad (1.B.4)$$

with a generic Riemannian appearance, but the geometry would remain Euclidean. If, on the other hand, p relations of the type (1.B.1) exist, things change substantially. Geometrically, each kinematical condition represents an hypersurface in $3N$-dimensional configuration space. The intersection of all hypersurfaces determines a $(3N - p)$-dimensional subspace in which the representative point is constrained to remain. This subspace is no longer Euclidean, but in general Riemannian.

One can also make the choice of expressing x_1, x_2, ..., x_{3N} as functions of $n = 3N - p$ independent parameters q_1, q_2, ..., q_n from the start and then to replace their differentials dx_1, dx_2, ..., dx_{3N} in the expression of ds^2. In this case one gets

$$ds^2 = \sum_{i,k=1}^{n} a_{ik}\, dq_i\, dq_k, \qquad (1.B.5)$$

where the coefficients a_{ik} are functions of the q's. Now the line element has a true Riemannian structure, not only because the q's are curvilinear coordinates, but because the geometry of the (n-dimensional) configuration space does not retain the Euclidean structure of the original $3N$-dimensional space. In the further case of non-holonomic constraints, one cannot proceed in this way and must therefore consider the $3N$-dimensional space with the addition of the constraints. In this case, we no longer have a definite subspace of the $3N$-dimensional space, because the kinematical conditions prescribe certain pencils of directions, but these directions do not envelope any surface.

---

[11]See (A.6) in the Appendix and the following discussion.

If we return now, to Cartesian coordinates and consider a system of N mass points in which forces can be derivable from a work function (or potential) $U(x_1, x_2, \ldots, x_{3N}) \equiv U(\boldsymbol{x})$, the kinetic energy is still expressed by the second equation of (1.B.3) and we shall call the function

$$\mathcal{L} = \mathcal{L}\left(x_1, x_2, \ldots, x_{3N}; \dot{x}_1, \dot{x}_2, \ldots, \dot{x}_{3N}\right) = \mathcal{T} + U. \tag{1.B.6}$$

the kinetic potential, or Lagrangian function, or just *Lagrangian* for short.

In the case proposed above as U is not dependent on time, the system is conservative and we also have

$$U = -V, \qquad \mathcal{L} = \mathcal{T} - V, \tag{1.B.7}$$

where V is the potential energy. In the most general case, we may meet a potential function depending on time and thence a Lagrangian depending explicitly on time. Therefore, we shall often write $\mathcal{L} = \mathcal{L}\left(\boldsymbol{x}, \dot{\boldsymbol{x}}, t\right)$ to emphasize that the results obtained are valid also in the most general case. It can now be seen immediately that the Newtonian equations of motion can be written

$$\frac{\partial \mathcal{L}}{\partial x_i} - \frac{d}{dt}\frac{\partial \mathcal{L}}{\partial \dot{x}_i} = 0, \qquad (i = 1, 2, \ldots, 3N). \tag{1.B.8a}$$

In fact (1.B.8a), with the substitution of (1.B.6), gives

$$m_i \ddot{x}_i = \frac{\partial U}{\partial x_i}, \qquad (i = 1, 2, \ldots, 3N). \tag{1.B.8b}$$

Equations (1.B.8a), universally known as Lagrange's equations, contain more than may appear at first sight: i.e. they are not merely another form in which, in the case considered, Newtonian equations of motion can be written. To further highlight the general significance of Lagrange's equations, let us now show how they can be deduced from a variational principle, *Hamilton's principle*. This is an integral principle which describes the motion of those mechanical systems in which all forces (except those deriving from constraints) can be derived from a single generalized potential function depending on coordinates, velocities and time. Let us assume, therefore, that for our system (1.B.1) and (1.B.2) hold, i.e. the system has n degrees of freedom and his configuration is described by n generalized coordinates $q_1, q_2, \ldots, q_n$. The principle can be stated as follows: the motion of the system in the arbitrary time interval t_1, t_2 is such as to make the action integral

$$I = \int_{t_1}^{t_2} \mathcal{L}\left(q_1, q_2, \ldots, q_n; \dot{q}_1, \dot{q}_2, \ldots \dot{q}_n; t\right) dt \tag{1.B.9}$$

stationary for independent variations of the q_k's, which we indicate by δq_k, with only the prescribed condition that they vanish at the times t_1 and t_2. That is,

$$\delta I = 0, \qquad \text{with} \qquad \delta q_k|_{t=t_1} = \delta q_k|_{t=t_2} = 0, \ \forall k. \tag{1.B.10}$$

By applying the standard methods of the calculus of variations, (1.B.9) and (1.B.10) immediately give

$$\frac{\partial \mathcal{L}}{\partial x_k} - \frac{d}{dt}\frac{\partial \mathcal{L}}{\partial \dot{x}_k} = 0, \qquad (k = 1, 2, \dots , n). \tag{1.B.11}$$

An elementary check can be made in the case of only one degree of freedom: $\mathcal{L} = \mathcal{L}(q, \dot{q}, t)$ and then

$$I = \int_{t_1}^{t_2} \mathcal{L}(q, \dot{q}, t)\, dt.$$

One has to determine a curve $q = q(t)$ such that the variation of the integral I, when passing from evaluating it along the curve to evaluating the same integral along any neighbouring curve, vanishes. The curve is assumed to join two fixed points (q_1, t_1), (q_2, t_2): the integral is evaluated between these two points. Therefore one must look for the necessary and sufficient condition in order that the integral I be stationary:

$$\begin{aligned}
\delta \int_{t_1}^{t_2} \mathcal{L}(q, \dot{q}, t)dt &= \int_{t_1}^{t_2} \left[\frac{\partial \mathcal{L}}{\partial q}\delta q + \frac{\partial \mathcal{L}}{\partial \dot{q}}\,\delta(\dot{q}) \right] dt \\
&= \int_{t_1}^{t_2} \left[\frac{\partial \mathcal{L}}{\partial q}\delta q + \frac{\partial \mathcal{L}}{\partial \dot{q}}\frac{d}{dt}(\delta q) \right] dt \\
&= \int_{t_1}^{t_2} \left[\frac{\partial \mathcal{L}}{\partial q}\delta q + \frac{d}{dt}\left(\frac{\partial \mathcal{L}}{\partial \dot{q}}\,\delta q \right) - \frac{d}{dt}\left(\frac{\partial \mathcal{L}}{\partial \dot{q}}\,\delta q \right) \right] dt \\
&= \frac{\partial \mathcal{L}}{\partial \dot{q}}\,\delta q \Big|_{t_1}^{t_2} + \int_{t_1}^{t_2} \left[\frac{\partial \mathcal{L}}{\partial q} - \frac{d}{dt}\left(\frac{\partial \mathcal{L}}{\partial \dot{q}} \right) \right] \delta q\, dt.
\end{aligned}$$

The first term in the last row is zero, since the end points of the curve are fixed, and thence $\delta q = 0$ for $t = t_1, t_2$. Moreover, as δq is arbitrary along the curve,

$$\delta I = 0 \qquad \Longleftrightarrow \qquad \frac{\partial \mathcal{L}}{\partial q} - \frac{d}{dt}\frac{\partial \mathcal{L}}{\partial \dot{q}} = 0.$$

The procedure is then extended to the case of n degrees of freedom. What we have seen assures us, therefore, that Hamilton's principle is a sufficient condition for obtaining the equations of motion; one can also show that it is a necessary condition, by deriving Hamilton's principle from the Lagrange's equations.[12]

In deriving Lagrange's equations from Hamilton's principle, a point of fundamental importance is raised: the non-uniqueness of the Lagrangian of a system. In fact, if in (1.B.11) we replace $\mathcal{L}$ by $\mathcal{L} + \dot{F}$, $\dot{F}$ being the time derivative of a function $F = F(q_1, q_2, \dots , q_n; \dot{q}_1, \dot{q}_2, \dots , \dot{q}_n; t)$, the content of the equations of motion does not change because the definite integral relates

---

[12]See, for instance, E. T. Whittaker: *A Treatise on the Analytical Dynamics of Particles and Rigid Bodies* (Cambridge University Press, 1937) p. 245 f.

to the values of F at the end points where the q's do not vary. Conversely, one can show that, if $\mathcal{L}$ and $\mathcal{L}'$ are two Lagrangians which give the same equations of motion, it must hold that[13]

$$\mathcal{L}' = \mathcal{L} + \dot{F}.$$

Let us now show one of the fundamental features of Lagrange's equations, that is their invariance in form under any transformation of generalized coordinates (and therefore also of reference systems). In fact, let us assume that our system has n degrees of freedom and is then described by n generalized coordinates. Let them be $q_1, q_2, \ldots, q_n$, and (1.B.11) the equations of motion. If we define a transformation (endowed with all the properties required by later developments) to n new generalized coordinates $y_1, y_2, \ldots, y_n$,

$$q_i = q_i(y_1, y_2, \ldots, y_n), \qquad i = 1, 2, \ldots, n, \tag{1.B.12}$$

we shall have for the Lagrangian of the system

$$\mathcal{L}\left[q_i(y_j),\ \dot{q}_i(y_j, \dot{y}_j),\ t\right] = \widetilde{\mathcal{L}}(y_i, \dot{y}_i, t). \tag{1.B.13}$$

Moreover

$$\frac{\partial \widetilde{\mathcal{L}}}{\partial y_k} = \frac{\partial \mathcal{L}}{\partial q_i}\frac{\partial q_i}{\partial y_k} + \frac{\partial \mathcal{L}}{\partial \dot{q}_i}\frac{\partial \dot{q}_i}{\partial y_k} \qquad (k = 1, 2, \ldots, n),$$

where, for the sake of brevity, sum over repeated indices is implied. By inverting the order of derivation in the second term of the right-hand side, we get

$$\frac{\partial \widetilde{\mathcal{L}}}{\partial y_k} = \frac{\partial \mathcal{L}}{\partial q_i}\frac{\partial q_i}{\partial y_k} + \frac{\partial \mathcal{L}}{\partial \dot{q}_i}\frac{d}{dt}\frac{\partial q_i}{\partial y_k} \qquad (k = 1, 2, \ldots, n).$$

Since q_i does not depend on the $\dot{y}_k$'s, we also have

$$\frac{\partial \widetilde{\mathcal{L}}}{\partial \dot{y}_k} = \frac{\partial \mathcal{L}}{\partial \dot{q}_i}\frac{\partial \dot{q}_i}{\partial \dot{y}_k},$$

where it must be remembered that

$$\dot{q}_i = \frac{\partial q_i}{\partial y_k}\,\dot{y}_k,$$

and then

$$\frac{\partial \dot{q}_i}{\partial \dot{y}_k} = \frac{\partial q_i}{\partial y_k}$$

($\dot{y}_k$ being one particular among the $\dot{y}$'s). We can then collect together the results obtained in the form

---

[13]See, for instance, S. N. Rasband: *Dynamics* (Krieger, 1991) p. 56.

$$\frac{\partial \widetilde{\mathcal{L}}}{\partial y_k} - \frac{d}{dt}\frac{\partial \widetilde{\mathcal{L}}}{\partial \dot{y}_k} = \frac{\partial \mathcal{L}}{\partial q_i}\frac{\partial q_i}{\partial y_k} + \frac{\partial \mathcal{L}}{\partial \dot{q}_i}\frac{d}{dt}\left(\frac{\partial q_i}{\partial y_k}\right) - \frac{d}{dt}\left(\frac{\partial \mathcal{L}}{\partial \dot{q}_i}\frac{\partial q_i}{\partial y_k}\right)$$

$$= \left[\frac{\partial \mathcal{L}}{\partial q_i} - \frac{d}{dt}\frac{\partial \mathcal{L}}{\partial \dot{q}_i}\right]\frac{\partial q_i}{\partial y_k},$$

where the sum over i and $k = 1, 2, \ldots, n$ is implied.

Since, for (1.B.11), every term in the square bracket vanishes,

$$\frac{\partial \widetilde{\mathcal{L}}}{\delta y_k} - \frac{d}{dt}\frac{\partial \widetilde{\mathcal{L}}}{\partial \dot{y}_k} = 0 \qquad (k = 1, 2, \ldots, n).$$

Therefore, (1.B.11) have a much more general significance than would appear from the form (1.B.8a): Lagrange's equations are *invariant with respect to an arbitrary coordinate transformation*. Afterwards $\mathcal{L}(\boldsymbol{q}, \dot{\boldsymbol{q}}, t)$ will always be assumed to be C^2, i.e. a function with continuous derivatives up to the second order in all its $2n - 1$ arguments.

The system (1.B.11) can also be written by developing the time derivative:

$$\frac{\partial \mathcal{L}}{\partial q_i} - \frac{\partial^2 \mathcal{L}}{\partial \dot{q}_i \partial q_k}\dot{q}_k - \frac{\partial^2 \mathcal{L}}{\partial \dot{q}_i \partial \dot{q}_k}\ddot{q}_k - \frac{\partial^2 \mathcal{L}}{\partial \dot{q}_i \partial t} = 0 \qquad (i = 1, 2, \ldots, n);$$

therefore, so that the system can be written in normal form for the q's as unknown functions, the (Hessian) determinant given by the derivates

$$\frac{\partial^2 \mathcal{L}}{\partial \dot{q}_i \partial \dot{q}_k}$$

must always be different from zero. We assume that this is always true.

1.5 Ignorable Variables
and Integration of Lagrange's Equations

Lagrange's equations (1.B.11) are a system of n second-order differential equations in the unknown functions $q_k(t)$, and therefore, like the Newtonian equations, they correspond to a differential system of order $2n$. Whereas general methods of integration of Lagrange's equations do not exist, it may be, however, that the structure of the Lagrangian allows the existence of constants of the motion and thus the possibility of lowering the order of the system. This happens, for example, when the Lagrangian does not contain one (or more) of the q_k's, though containing the corresponding generalized velocities $\dot{q}_k$. In the case in which $\mathcal{L}$ does not depend on q_k, $\partial \mathcal{L}/\partial q_k = 0$, and then, from (1.B.11),

$$\frac{\partial \mathcal{L}}{\partial \dot{q}_k} \equiv p_k = \text{const.} \tag{1.B.14}$$

The quantity p_k, defined in (1.B.14), is called the momentum conjugate to the generalized coordinate q_k. We have already pointed out that the $\dot{q}_k$'s, in general, do not have the dimensions of velocities; analogously, the p_k's may not have the dimensions of momenta. When $\mathcal{L}$ does not depend on q_k one says that q_k is an *ignorable*, or *cyclic*, variable. Therefore (1.B.14) states that, when q_k is an ignorable variable, the corresponding conjugate momentum is a constant of the motion. Of course, it may happen that the Lagrangian turns out not to depend on one or more coordinates when a particular type of coordinate has been chosen; as an example, if we write the Lagrangian of a particle in a central field of force, there are no ignorable variables if we employ Cartesian coordinates, while the thing is evident in polar coordinates. This example suggests the existence of a connection between the absence of certain variables and the symmetries of the problem: the most convenient choice of the variables is that reflecting the symmetries of the physical problem.

Let us now see how one can proceed in the case in which k ($0 < k < n$) ignorable variables exist. Let us assume, for convenience, that they are the first k coordinates; if not, one could always rearrange things in such a way as to return to this case. Therefore we have

$$\frac{\partial \mathcal{L}}{\partial \dot{q}_i} = p_i = c_i, \qquad i = 1, 2, \dots, k, \qquad (1.\text{B}.15)$$

where the c_i's are constants whose values depend only on the initial conditions. It can be verified that, in any case, (1.B.15) are linear in $\dot{q}_1, \dot{q}_2, \dots, \dot{q}_k$ and therefore the first k Lagrangian velocities can be expressed as functions of $c_1, c_2, \dots, c_k$; $\dot{q}_{k+1}, \dot{q}_{k+2}, \dots, \dot{q}_n$; $q_{k+1}, q_{k+2}, \dots, q_n$ that are linear with respect to the k c_i's and the $n - k$ $\dot{q}$'s:

$$\dot{q}_i = \dot{q}_i(c_1, c_2, \dots, c_k; \; \dot{q}_{k+1}, \dot{q}_{k+2}, \dots, \dot{q}_n; \; q_{k+1}, q_{k+2}, \dots, q_n). \qquad (1.\text{B}.16)$$

Substituting (1.B.16) in the Lagrangian, one obviously obtains a function which depends only on $n - k$ coordinates and $n - k$ generalized velocities. We have, at this point, $n - k$ of the old equations which contain only the $n - k$ non-ignorable variables and corresponding velocities (besides, of course, the k constants c_i) and the k equations (1.B.16). If it were possible to integrate the $n - k$ second-order differential equations and obtain the $q_{k+1}, q_{k+2}, \dots, q_n$ as functions of t, of the c_i's and of the initial conditions, it would be sufficient to put them into (1.B.16) and obtain the $q_1, q_2, \dots, q_k$ through quadrature.

Once it was taken for granted that, in general, the integration of the $n - k$ differential equations is not possible, it was found more convenient (and meaningful) to proceed in a slightly different manner. One transforms the reduced system (of the $n - k$ differential equations) in a true Lagrangian system corresponding to $n - k$ degrees of freedom. This system, and (1.B.15) rewritten in a suitable form, completely characterize the problem. To do this, one defines the *reduced Lagrangian* (Routh function):

$$\mathcal{R} = \mathcal{L} - \sum_{i=1}^{k} \dot{q}_i \frac{\partial \mathcal{L}}{\partial \dot{q}_i} = \mathcal{L} - \sum_{i=1}^{k} c_i \dot{q}_i, \tag{1.B.17}$$

which, by substituting (1.B.16) into it, becomes $\mathcal{R} = \mathcal{R}(q_r; \dot{q}_r\ c_i, t)$ with $i = 1, 2, \ldots, k$ and $r = k+1, k+2, \ldots, n$. Let us show now that this function really represents the Lagrangian of a new system with $n-k$ degrees of freedom with coordinates $q_{k+1}, q_{k+2}, \ldots, q_n$, i.e. the following system holds:

$$\frac{\partial \mathcal{R}}{\partial q_r} - \frac{d}{dt}\frac{\partial \mathcal{R}}{\partial \dot{q}_r} = 0, \qquad r = k + 1, k + 2, \ldots, n. \tag{1.B.18}$$

In fact, if we vary $\mathcal{R}$ starting from the definition (1.B.17), i.e. varying also with respect to the c_i's, we get

$$\begin{aligned}
\delta\mathcal{R} &= \sum_{r=k+1}^{n} \frac{\partial \mathcal{L}}{\partial q_r}\delta q_r + \sum_{r=1}^{n} \frac{\partial \mathcal{L}}{\partial \dot{q}_r}\delta\dot{q}_r - \sum_{r=1}^{k} \frac{\partial \mathcal{L}}{\partial \dot{q}_r}\delta\dot{q}_r - \sum_{r=1}^{k} \dot{q}_r\delta c_r \\
&= \sum_{r=k+1}^{n} \left(\frac{\partial \mathcal{L}}{\partial q_r}\delta q_r + \frac{\partial \mathcal{L}}{\partial \dot{q}_r}\delta\dot{q}_r \right) - \sum_{r=1}^{k} \dot{q}_r\delta c_r.
\end{aligned} \tag{1.B.19}$$

From (1.B.19), we then have

$$\frac{\partial \mathcal{R}}{\partial q_r} = \frac{\partial \mathcal{L}}{\partial q_r}, \qquad \frac{\partial \mathcal{R}}{\partial \dot{q}_r} = \frac{\partial \mathcal{L}}{\partial \dot{q}_r}, \qquad r = k + 1, k + 2, \ldots, n \tag{1.B.20}$$

and

$$\dot{q}_i = -\frac{\partial \mathcal{R}}{\partial c_i}, \qquad i = 1, 2, \ldots, k. \tag{1.B.21}$$

From (1.B.20), owing to the Lagrange equations for $q_{k+1}, q_{k+2}, \ldots, q_n$, we immediately have (1.B.18). Therefore the initial system with n degrees of freedom and k ignorable coordinates is transformed into a system with $n - k$ degrees of freedom and the Lagrangian $\mathcal{R}$ with the addition of equations (1.B.21) to determine, through quadrature, the k ignorable coordinates as functions of the time and the initial conditions. Let us see, in the case of only one ignorable coordinate, some interesting features shown by the reduced system. Before doing this, let us consider which structures are possible for the total kinetic energy of the system. If our system is subjected to holonomic constraints depending also on the time, from (1.B.2) we have

$$\mathcal{T} = \frac{1}{2}\sum_{i=1}^{3N} m_i \dot{x}_i^2 = \mathcal{T}_0(q) + \mathcal{T}_1(q, \dot{q}) + \mathcal{T}_2(q, \dot{q}),$$

where $\mathcal{T}_0$ is dependent only on the generalized coordinates, $\mathcal{T}_1$ is linear in the generalized velocities and $\mathcal{T}_2$ is quadratic in the $\dot{q}$'s. On the other hand, when $x_i = x_i(q_1, q_2, \ldots, q_n)$, $i = 1, 2, \ldots, 3N$, that is, the holonomic constraints do not depend on time.

$$T = \frac{1}{2} \sum_{i=1}^{3N} m_i \dot{x}_i^2 = T_2(q, \dot{q},) = \frac{1}{2} \sum_{l,m=1}^{n} a_{lm} \dot{q}_l \dot{q}_m,$$

with

$$a_{lm} = a_{lm}(q) = \sum_{i=1}^{3N} m_i \frac{\partial x_i}{\partial q_l} \frac{\partial x_i}{\partial q_m} = a_{ml}.$$

Systems of the last type are called *natural systems*; as a consequence, the Lagrangian of the natural systems will be of the form

$$\mathcal{L} = \frac{1}{2} \sum_{i,k=1}^{n} a_{ik} \dot{q}_i \dot{q}_k + U. \tag{1.B.22}$$

Let us consider now a natural system with only one ignorable coordinate, e.g., q_1. Then

$$T = \frac{1}{2} a_{11} \dot{q}_1^2 + \sum_{i=2}^{n} a_{1i} \dot{q}_1 \dot{q}_i + \frac{1}{2} \sum_{i,k=2}^{n} a_{ik} \dot{q}_i \dot{q}_k$$

and

$$p_1 = \frac{\partial T}{\partial \dot{q}_1} = a_{11} \dot{q}_1 + \sum_{i=2}^{n} a_{1i} \dot{q}_i = c_1.$$

The Routh function, therefore, will become

$$\mathcal{R} = \mathcal{L} - c_1 \dot{q}_1 = \frac{1}{2} \sum_{i,k=2}^{n} a_{ik} \dot{q}_i \dot{q}_k - \frac{1}{2} a_{11} \dot{q}_1^2 + U. \tag{1.B.23a}$$

Now, the velocity $\dot{q}_1$ must be replaced in $\mathcal{R}$ with $(c_1 - \sum_{i=2}^{n} a_{1i} \dot{q}_i)/a_{11}$. This will give

$$\mathcal{R} = \frac{1}{2} \sum_{i,k=2}^{n} a_{ik} \dot{q}_i \dot{q}_k - \frac{1}{2} \frac{\left(\sum_{i=2}^{n} a_{1i} \dot{q}_i\right)^2}{a_{11}} + \frac{c_1}{a_{11}} \sum_{i=2}^{n} a_{1i} \dot{q}_i - \frac{1}{2} \frac{c_1^2}{a_{11}} + U. \tag{1.B.23b}$$

Then the Routh function (the reduced Lagrangian) contains a part that is quadratic in the velocities, $n-1$ terms linear in the velocities and a term not depending on the velocities. We can consider the last term as an *apparent potential* function and combine it with U to give

$$U_{\text{eff}} = U - \frac{1}{2} \frac{c_1^2}{a_{11}}. \tag{1.B.24}$$

As to the terms that are linear in the velocities, they are called *gyroscopic terms* and are responsible for the fact that the new system with $n-1$ degrees of freedom is no longer a natural system.

Let us go back again to the result shown at the beginning of this section, i.e. to the fact that if a coordinate q_i is ignorable, then the corresponding conjugate momentum is a constant of the motion. We have already recalled the example of a particle moving in a central potential $U = U(r)$. After having proved that the motion is planar, one can choose polar coordinates in the plane of motion: (r, ϑ). The Lagrangian of a unit mass is then

$$\mathcal{L} = \frac{1}{2}(\dot{r}^2 + r^2\dot{\vartheta}^2) + U(r). \tag{1.3.25}$$

The coordinate ϑ is ignorable and, consequently, the conjugate momentum (angular momentum)

$$p_\vartheta = r^2\dot{\vartheta} \tag{1.3.26}$$

is a constant of the motion. Obviously, a lot of examples of this type, well known in classical mechanics, can be given; however, there is a case which, while giving rise to a first integral of paramount importance, seems not to be included in the above description. Let us consider, in fact, a system described by a Lagrangian not explicitly containing the time

$$\mathcal{L} = \mathcal{L}(q_1, q_2, \ldots, q_n; \dot{q}_1, \dot{q}_2, \ldots, \dot{q}_n). \tag{1.3.27}$$

Thus

$$\frac{d\mathcal{L}}{dt} = \frac{\partial \mathcal{L}}{\partial q_i}\dot{q}_i + \frac{\partial \mathcal{L}}{\partial \dot{q}_i}\ddot{q}_i$$

which can be rewritten, owing to (1.B.11), as

$$\frac{d}{dt}\left(\frac{\partial \mathcal{L}}{\partial \dot{q}_i}\dot{q}_i - \mathcal{L}\right) = 0,$$

which is equivalent to saying

$$\frac{\partial \mathcal{L}}{\partial \dot{q}_i}\dot{q}_i - \mathcal{L} = p_i\dot{q}_i - \mathcal{L} = \mathcal{H} = \text{const.} \tag{1.3.28}$$

It is easy to check that, for a natural system, $p_i\dot{q}_i$ (where the sum over i is implied) equals two times the kinetic energy, and therefore

$$\mathcal{H} = E = (\text{total energy}) = \text{const.} \tag{1.3.29}$$

The function $\mathcal{H} = p_i\dot{q}_i - \mathcal{L}$ introduced in (1.B.28) is called Hamilton's function, or the *Hamiltonian* of the system; we refer the reader to Part C for all related matter; here we limit ourselves to pointing out the coincidence of $\mathcal{H}$ with the total energy for a natural system and the fact that the absence of t from the Lagrangian entails the existence of a constant of the motion (the Hamiltonian). However, according to the definition given above, t is not an ignorable coordinate, being not a coordinate but the independent variable. We can circumvent this lack of symmetry between the coordinates q_i $(i = 1, 2, \ldots, n)$ and the time t, by considering $t = q_0$ as a coordinate and all the $n - 1$ coordinates defined in this way as functions of a suitable parameter w. In this case the action integral (1.B.12) will become

$$I = \int_{t_1}^{t_2} \mathcal{L}(q_1, q_2, \ldots, q_n; \dot{q}_1, \dot{q}_2, \ldots, \dot{q}_n)\,dt$$

$$= \int_{w_1}^{w_2} \mathcal{L}\left(q_1, q_2, \ldots, q_n; \frac{q_1'}{t'}, \frac{q_2'}{t'}, \ldots, \frac{q_n'}{t'}\right) t'\,dw, \tag{1.B.30}$$

where the prime means derivation with respect to the parameter w. The Lagrangian is given now by $\mathcal{L}t'$ and does not contain the coordinate $t = q_0$ (it contains, instead, $q_0' = t'$). The conjugate momentum $p_0 = p_t$ will therefore be

$$p_0 = \frac{\partial(\mathcal{L}t')}{\partial t'} = \mathcal{L} + t'\frac{\partial \mathcal{L}}{\partial t'} = \mathcal{L} - t'\frac{\partial \mathcal{L}}{\partial \dot{q}_i}\frac{q_i'}{t'^2} = \mathcal{L} - \frac{\partial \mathcal{L}}{\partial \dot{q}_i}\frac{q_i'}{t'} = \mathcal{L} - \frac{\partial \mathcal{L}}{\partial \dot{q}_i}\dot{q}_i.$$

Therefore, $p_0 = p_t = -\mathcal{H}$. In this new formulation, one replaces the configuration space Q by a new space with $n+1$ dimensions, the space of events[14] defined as $\mathbb{R} \times Q$ where the coordinates are $t(= q_0), q_1, q_2, \ldots, q_n$. At this point, having already said that the existence or otherwise of ignorable coordinates is in correspondence with the generalized coordinates one has chosen, it is proper to ask oneself if it is possible to give a rule for: (1) deciding if a system, through a choice of suitable coordinates, admits ignorable variables; (2) determining such coordinates. Unfortunately, neither of these points can, in general, be satisfied. It is only possible[15] to decide if a system admits $n-1$ ignorable variables, but not to determine the maximum number of ignorable variables which a given system admits. The fundamental problem remains of determining the appropriate coordinate system. We can introduce the problem in this way. If in the Lagrangian $\mathcal{L}(\boldsymbol{q}, \dot{\boldsymbol{q}}, t)$ no one of the q_i's is ignorable, the obvious consequence is that no generalized momentum is constant. However, a transformation to a new system of generalized coordinates may exist, which we shall denote by $Q_1, Q_2, \ldots, Q_n$, such that $\mathcal{L}(\boldsymbol{Q}, \dot{\boldsymbol{Q}}, t)$ (we go on using the same symbol $\mathcal{L}$ for the sake of simplicity) does not contain one of the Q_i's. Let Q_l be this coordinate. Then

$$P_l = \frac{\partial \mathcal{L}}{\partial \dot{Q}_l} = c_l = \text{const.} \tag{1.B.31}$$

In terms of the old coordinates q,

$$P_l = \frac{\partial \mathcal{L}}{\partial \dot{Q}_l} = \frac{\partial \mathcal{L}}{\partial \dot{q}_k}\frac{\partial \dot{q}_k}{\partial \dot{Q}_l}, \tag{1.B.32}$$

since $\mathcal{L}$ depends on the $\dot{Q}$'s only through the $\dot{q}$'s. Moreover,

$$\dot{q}_k = \frac{\partial q_k}{\partial Q_l}\dot{Q}_l + \frac{\partial q_k}{\partial t},$$

---

[14]See J. L. Synge: Classical Dynamics, *Handbuch der Physik,* **III**, 1 (Springer, 1960), p. 105 ff.

[15]See J. L. Synge: On the geometry of the dynamics, *Philos. Transactions (A)* **226**, 31–106 (1927).

and then $\partial \dot{q}_k / \partial \dot{Q}_l = \partial q_k / \partial Q_l$. We have already used this property of "effacement of the point" in obtaining (1.B.13). We therefore get

$$\frac{\partial \mathcal{L}}{\partial \dot{q}_k} \frac{\partial q_k}{\partial Q_l} = c_l = \text{const.} \tag{1.B.33}$$

The problem is that of finding a coordinate transformation

$$q_i = q_i(Q_1, Q_2, \dots, Q_n, t), \qquad i = 1, 2, \dots, n \tag{1.B.34}$$

such to satisfy (1.B.33). A transformation like (1.B.34) is called a *point transformation*. Rather than devoting oneself to pursuing single transformations, it is more fruitful to study families of transformations, and, as we shall see in the next section, to consider more general transformations which also involve generalized velocities.

1.6 Noether's Theorem

The connection between the existence of conserved quantities and the behaviour of a system under transformations of the variables has been studied in a completely general way and the result is contained in a celebrated theorem enunciated by Emmy Noether in 1918.[16] Traditionally, Noether's theorem for discrete systems is given considering only point transformations (see Sect. 1.5). In this way, while the connections are highlighted between the symmetries of space (or space-time) and the existence of constants of motion such as energy and angular momentum, other quantities, which we know (by other means) to be conserved but not corresponding to purely point transformations, are neglected. On the other hand, in the last two decades we have seen an increasing tendency to present Noether's theorem in a more general way involving also generalized velocities.[17] In the following we choose this last approach. Then we have to handle transformations which work in the space $(t, \boldsymbol{q}, \dot{\boldsymbol{q}})$. We have seen in the preceding section how to pass from the configuration space Q to $\mathbb{R} \times Q$. Since Q is, in the language of differential geometry, a differentiable manifold, the space of q's and $\dot{q}$'s in the same language is the tangent bundle of Q and is denoted by TQ. Therefore we shall work in $\mathbb{R} \times TQ$ and shall consider transformations depending on a real parameter ε:

$$\bar{t} = \varphi(t, \boldsymbol{q}, \dot{\boldsymbol{q}}, \varepsilon), \qquad \bar{q}_k = \psi_k(t, \boldsymbol{q}, \dot{\boldsymbol{q}}, \varepsilon), \qquad k = 1, 2, \dots, n, \tag{1.B.35}$$

---

[16]E. Noether: *Nachr. Akad. Wiss. Göttingen, Math. Phys.* **KL. II**, 235–257 (1918). English translation in *Transport Theory and Statis. Physics* **1**, 186–207 (1971).

[17]See W. Sarlet, F. Cantrijn: Generalizations of Noether's theorem in classical mechanics, *Siam Review* **23**, 467–494 (1981).

with the understanding that the transformation of the $\dot{q}$'s is not independent, but comes from (1.B.35).

Usually, the transformations which one encounters in applications make a group; here we consider only one-parameter Lie groups; nevertheless the results we shall explain do not require explicit use of the concept of group. The functions φ and ψ_k must be at least C^2 in each of the $2n+2$ arguments and the parameter ε will be such that $\varepsilon \in I \subset \mathbb{R}$, where I includes the origin as interior point. Moreover,

$$\varphi\left(t,\boldsymbol{q},\dot{\boldsymbol{q}},0\right) = t, \qquad \psi_k\left(t,\boldsymbol{q},\dot{\boldsymbol{q}},0\right) = q_k. \tag{1.B.36}$$

Owing to the hypotheses, we can expand (1.B.35) in Taylor series in the neighbourhood of $\varepsilon = 0$, taking into account only first-order terms in ε:

$$\bar{t} = t + \tau\left(t,\boldsymbol{q},\dot{\boldsymbol{q}}\right)\varepsilon + \mathrm{o}(\varepsilon), \qquad \bar{q}_k = q_k + \xi_k\left(t,\boldsymbol{q},\dot{\boldsymbol{q}}\right)\varepsilon + \mathrm{o}(\varepsilon). \tag{1.B.37a}$$

The principal linear parts, τ and ξ_k, of $\bar{t}$ and $\bar{q}_k$ with respect to ε in $\varepsilon = 0$ are called *infinitesimal generators* of the transformations φ and ψ_k; they are

$$\tau\left(t,\boldsymbol{q},\dot{\boldsymbol{q}}\right) = \frac{\partial\varphi}{\partial\varepsilon}\left(t,\boldsymbol{q},\dot{\boldsymbol{q}},0\right); \qquad \xi_k\left(t,\boldsymbol{q},\dot{\boldsymbol{q}}\right) = \frac{\partial\psi_k}{\partial\varepsilon}\left(t,\boldsymbol{q},\dot{\boldsymbol{q}},0\right).$$

The transformations defined by (1.B.37a) are the infinitesimal transformations associated with (1.B.35). From now on we shall therefore consider the transformations

$$\bar{t} = t + \tau\left(t,\boldsymbol{q},\dot{\boldsymbol{q}}\right)\varepsilon \qquad \bar{q}_k = q_k + \xi_k\left(t,\boldsymbol{q},\dot{\boldsymbol{q}}\right)\varepsilon, \tag{1.B.37b}$$

from which we shall also obtain the variations of the $\dot{q}$'s; the derivatives $\dot{\tau}$ and $\dot{\xi}_k$ will be evaluated on arbitrary curves $t \to \boldsymbol{q}\left(t\right)$, with $t \in [a,b]$ and $[a,b] \subset \mathbb{R}$. Through (1.B.37b), a curve $t \to \boldsymbol{q}\left(t\right)$ is transformed to $\bar{t} \to \bar{\boldsymbol{q}}\left(\bar{t}\right)$; to first order in ε, we have

$$\frac{d\bar{q}_i}{d\bar{t}} = \frac{d\bar{q}_i}{dt}\frac{dt}{d\bar{t}} = \frac{\dot{q}_i + \varepsilon\dot{\xi}_i}{1 + \varepsilon\dot{\tau}} \sim \dot{q}_i + \varepsilon\left(\dot{\xi}_i - \dot{q}_i\dot{\tau}\right). \tag{1.B.38}$$

We shall say that the infinitesimal transformation (1.B.37b) leaves the action integral invariant up to *gauge* terms if a function $f = f(t,\boldsymbol{q},\dot{\boldsymbol{q}})$ exists such that, for any differentiable curve $t \to \boldsymbol{q}\left(t\right)$, one has

$$\int_{\bar{t}_1}^{\bar{t}_2} \mathcal{L}\left[\bar{t},\bar{\boldsymbol{q}}(\bar{t}),\frac{d}{d\bar{t}}\bar{\boldsymbol{q}}(\bar{t})\right] d\bar{t} = \int_{t_1}^{t_2} \mathcal{L}\left(t,\boldsymbol{q}(t),\dot{\boldsymbol{q}}(t)\right) dt + \varepsilon\int_{t_1}^{t_2}\frac{df}{dt}dt + \mathrm{o}(\varepsilon),$$
$$\tag{1.B.39}$$

with $[t_1,t_2] \subset [a,b]$. Transforming the first integral back to the interval $[t_1,t_2]$, (1.B.39) becomes equivalent to

$$\mathcal{L}\left[\bar{t},\bar{\boldsymbol{q}}(\bar{t}),\frac{d}{d\bar{t}}\bar{\boldsymbol{q}}\left(\bar{t}\right)\right]\frac{d\bar{t}}{dt} = \mathcal{L}\left(t,\boldsymbol{q}(t),\dot{\boldsymbol{q}}(t)\right) + \varepsilon\frac{df\left(t,\boldsymbol{q},\dot{\boldsymbol{q}}\right)}{dt} + \mathrm{o}(\varepsilon), \tag{1.B.40a}$$

which must hold for a whole family of curves $t \to \boldsymbol{q}\left(t\right)$ and therefore must be an identity in $t,\boldsymbol{q},\dot{\boldsymbol{q}}$. Developing and retaining only first-order terms:

$$\mathcal{L}\left[\bar{t}, \overline{\boldsymbol{q}}(\bar{t}), \frac{d}{d\bar{t}}\,\overline{\boldsymbol{q}}\,(\bar{t})\right](1 + \varepsilon\dot{\tau}) = \mathcal{L}\,(t, \boldsymbol{q}(t), \dot{\boldsymbol{q}}(t)) + \varepsilon\,\frac{df}{dt},$$

and also

$$\frac{\partial\mathcal{L}}{\partial t}\,\delta t + \frac{\partial\mathcal{L}}{\partial q_i}\,\delta q_i + \frac{\partial\mathcal{L}}{\partial \dot{q}_i}\,\delta\dot{q}_i + \mathcal{L}\,\varepsilon\dot{\tau} = \varepsilon\,\frac{df}{dt}.$$

From (1.B.38) one has $\delta\dot{q}_i = \varepsilon\,(\dot{\xi}_i - \dot{q}_i\dot{\tau})$. Moreover, taking into account that

$$\dot{\tau} = \frac{\partial\tau}{\partial t} + \frac{\partial\tau}{\partial q_i}\,\dot{q}_i + \frac{\partial\tau}{\partial \dot{q}_i}\,\ddot{q}_i, \qquad \dot{\xi}_i = \frac{\partial\xi_i}{\partial t} + \frac{\partial\xi_i}{\partial q_j}\,\dot{q}_j + \frac{\partial\xi_i}{\partial \dot{q}_j}\,\ddot{q}_j,$$

we have

$$\frac{\partial\mathcal{L}}{\partial t}\tau + \frac{\partial\mathcal{L}}{\partial q_i}\xi_i + \frac{\partial\mathcal{L}}{\partial \dot{q}_i}\left[\frac{\partial\xi_i}{\partial t} + \frac{\partial\xi_i}{\partial q_j}\dot{q}_j + \frac{\partial\xi_i}{\partial \dot{q}_j}\ddot{q}_j - \dot{q}_i\left(\frac{\partial\tau}{\partial t} + \frac{\partial\tau}{\partial q_j}\dot{q}_j + \frac{\partial\tau}{\partial \dot{q}_j}\ddot{q}_j\right)\right]$$

$$+\mathcal{L}\left(\frac{\partial\tau}{\partial t} + \frac{\partial\tau}{\partial q_i}\dot{q}_i + \frac{\partial\tau}{\partial \dot{q}_i}\ddot{q}_i\right) = \frac{\partial f}{\partial t} + \frac{\partial f}{\partial q_i}\dot{q}_i + \frac{\partial f}{\partial \dot{q}_i}\ddot{q}_i. \tag{1.B.40b}$$

Now, (1.B.40b) is an identity in $t, q, \dot{q}$ and $\ddot{q}$, which means that the coefficients of $\ddot{q}_i$ have to vanish separately, yielding the $n+1$ equations:

$$\mathcal{L}\frac{\partial\tau}{\partial \dot{q}_i} + \frac{\partial\mathcal{L}}{\partial \dot{q}_j}\left(\frac{\partial\xi_j}{\partial \dot{q}_i} - \frac{\partial\tau}{\partial \dot{q}_i}\dot{q}_j\right) = \frac{\partial f}{\partial \dot{q}_i}, \qquad i = 1, 2, \ldots, n, \tag{1.B.41}$$

$$\frac{\partial\mathcal{L}}{\partial t}\tau + \frac{\partial\mathcal{L}}{\partial q_i}\xi_i + \frac{\partial\mathcal{L}}{\partial \dot{q}_i}\left[\frac{\partial\xi_i}{\partial t} + \frac{\partial\xi_i}{\partial q_j}\dot{q}_j - \dot{q}_i\left(\frac{\partial\tau}{\partial t} + \frac{\partial\tau}{\partial q_j}\dot{q}_j\right)\right]$$

$$+\mathcal{L}\left(\frac{\partial\tau}{\partial t} + \frac{\partial\tau}{\partial q_i}\dot{q}_i\right) = \frac{\partial f}{\partial t} + \frac{\partial f}{\partial q_i}\dot{q}_i. \tag{1.B.42}$$

(1.B.41) and (1.B.42) are given by $n+1$ partial differential equations, linear in the $n+1$ unknown functions τ and ξ^i, and represent the necessary and sufficient conditions for the action integral to be invariant up to *gauge* terms under the infinitesimal transformations with generators τ and ξ. Equations (1.B.41) and (1.B.42) are called generalized Killing equations.[18] Let us see now what the consequence is for the system of the invariance of the action integral under the transformations we have considered. Go back to (1.B.40b) and assume that, in it, the $q_i(t)$'s are solutions of the equations of motion, i.e. satisfy the Lagrange equations with the Lagrangian $\mathcal{L}$.

---

[18]It is easy to show that, when $\mathcal{L} = \frac{1}{2}g_{kl}\dot{q}^k\dot{q}^l$ (a system of free mass points) and one considers a transformation $\bar{t} = t$, $\bar{q}^k = q^k + \xi^k(q)\,\varepsilon$, (1.B.41) and (1.B.42) are replaced by $g_{ik}(\partial\xi^k/\partial q^l) + g_{kl}(\partial\xi^k/\partial q^i) = 0$, which are just the Killing equations. See, for instance, L.P. Eisenhart: *Continuous Groups of Transformations* (Princeton University Press, 1933) p. 208.

By using the identities

$$\frac{\partial \mathcal{L}}{\partial t}\tau = \frac{d\mathcal{L}}{dt}\tau - \frac{\partial \mathcal{L}}{\partial q_i}\dot{q}_i\tau - \frac{\partial \mathcal{L}}{\partial \dot{q}_i}\ddot{q}_i\tau,$$

$$\frac{\partial \mathcal{L}}{\partial \dot{q}_i}\frac{d\xi_i}{dt} = \frac{d}{dt}\left(\frac{\partial \mathcal{L}}{\partial \dot{q}_i}\xi_i\right) - \frac{d}{dt}\left(\frac{\partial \mathcal{L}}{\partial \dot{q}_i}\right)\xi_i,$$

$$\frac{\partial \mathcal{L}}{\partial \dot{q}_i}\dot{q}_i\frac{d\tau}{dt} + \frac{\partial \mathcal{L}}{\partial \dot{q}_i}\ddot{q}_i\tau = \frac{d}{dt}\left(\frac{\partial \mathcal{L}}{\partial \dot{q}_i}\dot{q}_i\tau\right) - \frac{d}{dt}\left(\frac{\partial \mathcal{L}}{\partial \dot{q}_i}\right)\dot{q}_i\tau,$$

one obtains

$$\frac{d\mathcal{L}}{dt}\tau - \frac{\partial \mathcal{L}}{\partial q_i}\dot{q}_i\tau - \frac{\partial \mathcal{L}}{\partial \dot{q}_i}\ddot{q}_i\tau + \frac{\partial \mathcal{L}}{\partial q_i}\xi_i + \frac{d}{dt}\left(\frac{\partial \mathcal{L}}{\partial \dot{q}_i}\xi_i\right) - \frac{d}{dt}\frac{\partial \mathcal{L}}{\partial \dot{q}_i}\xi_i$$

$$+\frac{\partial \mathcal{L}}{\partial \dot{q}_i}\ddot{q}_i\tau - \frac{d}{dt}\left(\frac{\partial \mathcal{L}}{\partial \dot{q}_i}\dot{q}_i\tau\right) + \frac{d}{dt}\frac{\partial \mathcal{L}}{\partial \dot{q}_i}\dot{q}_i\tau + \mathcal{L}\frac{d\tau}{dt}$$

$$= \frac{d}{dt}\left(\mathcal{L}\tau\right) + \frac{d}{dt}\frac{\partial \mathcal{L}}{\partial q_i}(\xi_i - \dot{q}_i\tau) + \frac{d}{dt}\left[\frac{\partial \mathcal{L}}{\partial \dot{q}_i}(\xi_i - \dot{q}_i\tau)\right] - \frac{d}{dt}\frac{\partial \mathcal{L}}{\partial q_i}(\xi_i - \dot{q}_i\tau) = \frac{df}{dt}.$$

That is, finally

$$\frac{d}{dt}\left[\mathcal{L}\tau + \frac{\partial \mathcal{L}}{\partial \dot{q}_i}(\xi_i - \dot{q}_i\tau) - f\right] = 0. \tag{1.B.43}$$

Therefore, if we have a system with Lagrangian $\mathcal{L}$, the invariance up to *gauge* terms of the action integral under the transformation with generators τ and ξ_i has as a consequence that the quantity in square brackets in (1.B.43) is an integral of the motion.

This conclusion constitutes the Noether's theorem. Conversely, owing to what we have already demonstrated, we can say that a necessary and sufficient condition for a transformation to be *Noetherian*, i.e., given a system with Lagrangian $\mathcal{L}$, the quantity in the square brackets in (1.B.43) be a constant of motion, is that its generators satisfy the generalized Killing equations (1.B.41) and (1.B.42). So far as these equations are concerned, it is clear that they can be used in two ways which are, so to speak, complementary. If we start with a given Lagrangian $\mathcal{L}$ and a transformation with generators τ and ξ_i, (1.B.41) and (1.B.42) enable us to check if the transformation is Noetherian and then, from (1.B.43), to obtain the corresponding constant of motion. Conversely, if a transformation is given with known generators τ and ξ_i, (1.B.41) and (1.B.42) can be used to single out the set of Lagrangians enjoying the property of invariance under the given transformation. In the first case, it may also happen that one has the Lagrangian $\mathcal{L}$ and wants to determine all the possible Noetherian transformations. This is the case in which we are interested. It being impossible, for obvious reasons, to determine a complete integral of (1.B.41) and (1.B.42), in this case one will be obliged to make systematic attempts starting from the simplest and obvious cases. This will be done in the next section, where we consider a particularly meaningful system, the *n-dimensional oscillator*.

Before we pass on to the applications, it is interesting to devote our attention to the possibility of extending Noether's theorem also to non-conservative systems and at last to the inverse of the theorem itself. When a system is acted on by non-conservative forces of generalized components Q_i ($i = 1, 2, \ldots, n$), the Lagrangian equations[19] become

$$\frac{d}{dt}\frac{\partial \mathcal{L}}{\partial \dot{q}_i} - \frac{\partial \mathcal{L}}{\partial q_i} = Q_i, \tag{1.B.44a}$$

where $\mathcal{L}$ represents the Lagrangian corresponding to the conservative forces. Let us write (1.B.44a) in the form

$$L_{(i)} \equiv \frac{d}{dt}\frac{\partial \mathcal{L}}{\partial \dot{q}_i} - \frac{\partial \mathcal{L}}{\partial q_i} - Q_i = 0, \tag{1.B.44b}$$

and consider infinitesimal transformations

$$\bar{t} = t + \varepsilon \tau\left(t, \boldsymbol{q}, \dot{\boldsymbol{q}}\right), \quad \bar{q}_i = q_i + \varepsilon \xi_i\left(t, \boldsymbol{q}, \dot{\boldsymbol{q}}\right),$$
$$\frac{d\bar{q}_i}{d\bar{t}} = \frac{dq_i}{dt} + \varepsilon\left[\dot{\xi}_i - \dot{q}_i\dot{\tau} + \varphi_i\left(t, \boldsymbol{q}, \dot{\boldsymbol{q}}\right)\right], \tag{1.B.45}$$

where, unlike in (1.B.39), the transformations of the generalized velocities are not determined by those of the coordinates and of the time, but depend on n functions φ_i (which we assume to be C^2) associated with the presence of the non-conservative forces. Let us now assume that transformations (1.B.45) leave the action invariant up to *gauge* terms, that is, as in (1.B.40a),

$$\mathcal{L}\left(\bar{\boldsymbol{q}}(\bar{t}), \frac{d\bar{\boldsymbol{q}}}{d\bar{t}}, \bar{t}\right)\frac{d\bar{t}}{dt} - \mathcal{L}\left(\boldsymbol{q}, \dot{\boldsymbol{q}}, t\right) = \varepsilon\frac{df\left(t, \boldsymbol{q}, \dot{\boldsymbol{q}}\right)}{dt} + \mathrm{o}(\varepsilon)$$

holds. Expanding this in the usual manner and retaining only first order terms, we now have

$$\frac{\partial \mathcal{L}}{\partial q_i}\xi_i + \frac{\partial \mathcal{L}}{\partial \dot{q}_i}\left(\dot{\xi}_i - \dot{q}_i\dot{\tau} + \varphi_i\right) + \frac{\partial \mathcal{L}}{\partial t}\tau + \mathcal{L}\dot{\tau} = \frac{df}{dt}, \tag{1.B.46a}$$

which can be rewritten

$$\frac{d}{dt}\left[\mathcal{L}\tau + (\xi_i - \dot{q}_i\tau)\frac{\partial \mathcal{L}}{\partial \dot{q}_i}\right] + \frac{\partial \mathcal{L}}{\partial \dot{q}_i}\varphi_i - Q_i(\xi_i - \dot{q}_i\tau) - (\xi_i - \dot{q}_i\tau)L_{(i)} = \frac{df}{dt}. \tag{1.B.46b}$$

Following on from these premises, we can now state the extended Noether's theorem:

If the action integral is invariant up to gauge terms under the transformations (1.B.45) and if, moreover, the identity

---

[19]See, for instance, H. Goldstein: *Classical Mechanics*, 2nd edn. (Addison-Wesley, 1980) Chap. I.

$$\frac{\partial \mathcal{L}}{\partial \dot{q}_i}\, \varphi_i = Q_i\left(\xi_i - \dot{q}_i\, \tau\right) \tag{1.B.47}$$

holds, then the first integral

$$I = \mathcal{L}\tau + \left(\xi_i - \dot{q}_i\tau\right)\frac{\partial \mathcal{L}}{\partial \dot{q}_i} - f \tag{1.B.48}$$

exists.

The proof is quite obvious; in fact, it is sufficient to substitute (1.B.44b) and (1.B.47) in (1.B.46b). The relation providing the first integral, as can be seen, retains the form (1.B.43). By substituting (1.B.47) in (1.B.46a), we have

$$\mathcal{L}\dot{\tau} + \frac{\partial \mathcal{L}}{\partial t}\tau + \frac{\partial \mathcal{L}}{\partial q_i}\xi_i + \frac{\partial \mathcal{L}}{\partial \dot{q}_i}\left(\dot{\xi}_i - \dot{q}_i\,\dot{\tau}\right) + Q_i\left(\xi_i - \dot{q}_i\tau\right) = \frac{df}{dt}$$

and, by developing the total derivatives,

$$\mathcal{L}\left(\frac{\partial \tau}{\partial t} + \frac{\partial \tau}{\partial q_i}\dot{q}_i + \frac{\partial \tau}{\partial \dot{q}_i}\ddot{q}_i\right) + \frac{\partial \mathcal{L}}{\partial t}\tau + \frac{\partial \mathcal{L}}{\partial q_i}\xi_i$$

$$+\frac{\partial \mathcal{L}}{\partial \dot{q}_i}\left(\frac{\partial \xi_i}{\partial t} + \frac{\partial \xi_i}{\partial q_j}\dot{q}_j + \frac{\partial \xi_i}{\partial \dot{q}_j}\ddot{q}_j - \dot{q}_i\frac{\partial \tau}{\partial t} - \dot{q}_i\frac{\partial \tau}{\partial q_j}\dot{q}_j - \dot{q}_i\frac{\partial \tau}{\partial \dot{q}_j}\ddot{q}_j\right)$$

$$+Q_i\left(\xi_i - \dot{q}_i\tau\right) = \frac{\partial f}{\partial t} + \frac{\partial f}{\partial q_i}\dot{q}_i + \frac{\partial f}{\partial \dot{q}_i}\ddot{q}_i.$$

Since it must hold for any $q, \dot{q}, \ddot{q}$, it will be equivalent to the system of $n+1$ partial differential equations:

$$\mathcal{L}\frac{\partial \tau}{\partial \dot{q}_i} + \frac{\partial \mathcal{L}}{\partial \dot{q}_j}\left(\frac{\partial \xi_j}{\partial \dot{q}_i} - \frac{\partial \tau}{\partial \dot{q}_i}\dot{q}_j\right) = \frac{\partial f}{\partial \dot{q}_i} \qquad (i = 1, 2, \dots, n), \tag{1.B.49}$$

$$\frac{\partial \mathcal{L}}{\partial t}\tau + \frac{\partial \mathcal{L}}{\partial q_i}\xi_i + \frac{\partial \mathcal{L}}{\partial \dot{q}_i}\left[\frac{\partial \xi_i}{\partial t} + \frac{\partial \xi_i}{\partial q_j}\dot{q}_j - \dot{q}_i\left(\frac{\partial \tau}{\partial t} + \frac{\partial \tau}{\partial q_j}\dot{q}_j\right)\right]$$

$$+\mathcal{L}\left(\frac{\partial \tau}{\partial t} + \frac{\partial \tau}{\partial q_i}\dot{q}_i\right) + Q_i\left(\xi_i - \dot{q}_i\tau\right) = \frac{\partial f}{\partial t} + \frac{\partial f}{\partial q_i}\dot{q}_i. \tag{1.B.50}$$

Equations (1.B.49), (1.B.50) are the generalized Killing equations for the non-conservative systems. As can be seen, they differ from (1.B.41) and (1.B.42) only for the presence of the term $Q_i(\xi_i - \dot{q}_i\tau)$ in the $(n+1)$th equation. Also in this case, the Killing equations represent necessary and sufficient condition for the invariance of the action up to *gauge* terms and therefore, owing to Noether's theorem, for the existence of a first integral.

Returning to the case of conservative systems, let us see if Noether's theorem has an inverse and, if so, under what conditions. Let us consider again the expression for the first integral (1.B.48). If we differentiate it with respect to $\dot{q}_j$ and substitute (1.B.41) in the obtained expression, we get $\partial I/\partial \dot{q}_j = H_{ij}\left(\xi_i - \dot{q}_i\tau\right)$, where

$$|H_{ij}| = \left| \frac{\partial^2 \mathcal{L}}{\partial \dot{q}_i \partial \dot{q}_j} \right|$$

is the Hessian determinant we have already assumed to be nonvanishing. If we call J^{ki} the elements of the inverse determinant, that is, $H_{ij} J^{ki} = \delta_j^k$, we finally have

$$\xi_i = J_{ij} \frac{\partial I}{\partial \dot{q}_j} + \dot{q}_i \tau. \tag{1.B.51}$$

In most cases concerning our applications, the Lagrangian contains the $\dot{q}$'s through $\frac{1}{2} \delta_{ij} \dot{q}_i \dot{q}_j$; therefore $J_{ij} = \delta_{ij}$ and (1.B.51) becomes

$$\xi_i = \frac{\partial I}{\partial \dot{q}_i} + \dot{q}_i \, \tau.$$

If, now, we substitute (1.B.51) in (1.B.48), we shall have for τ

$$\tau = \mathcal{L}^{-1} \left(I + f - J_{ij} \frac{\partial I}{\partial \dot{q}_j} \frac{\partial \mathcal{L}}{\partial \dot{q}_i} \right). \tag{1.B.52}$$

At this point, we can state the inverse of Noether's theorem:

> To any constant of motion $I = I(\boldsymbol{q}, \dot{\boldsymbol{q}}, t)$ for the dynamical system described by the Lagrangian $\mathcal{L}$ there correspond infinitesimal transformations (1.B.51), (1.B.52) which leave the action integral invariant up to "gauge" terms, in the sense stated in (1.B.40b), for all the solutions $t \to \boldsymbol{q}(t)$ of the equations of motion.

As the function $f = f(\boldsymbol{q}, \dot{\boldsymbol{q}}, t)$, in (1.B.52) is quite arbitrary, it also follows that the function $\tau = \tau(\boldsymbol{q}, \dot{\boldsymbol{q}}, t)$ is arbitrary, and thus (1.B.51) does not define a coordinate transformation but a family of coordinate transformations. This consequence, which a priori seems to introduce a troublesome arbitrariness, as a matter of fact enables us to "simplify" the transformation corresponding to a given first integral in the most convenient way. To this end, we state an easily applied proposition for real cases.

Let us assume we have a first integral $I = \overline{I}(\boldsymbol{q}, \dot{\boldsymbol{q}}, t)$ relative to the system described by the Lagrangian $\mathcal{L}$ and, moreover, this integral, through Noether's theorem, corresponds to the transformation defined by

$$\tau \equiv 0, \qquad \xi_i = \overline{\xi}_i (\boldsymbol{q}, \dot{\boldsymbol{q}}, t), \qquad i = 1, 2, \ldots, n, \tag{1.B.53}$$

with $f = \overline{f}(\boldsymbol{q}, \dot{\boldsymbol{q}}, t)$; then, as a consequence, an infinity of transformations like

$$\xi_i = \overline{\xi}_i + \tau \dot{q}_i, \qquad \tau = \tau(\boldsymbol{q}, \dot{\boldsymbol{q}}, t), \qquad i = 1, 2, \ldots, n \tag{1.B.54}$$

are defined, with $f = \overline{f} + \mathcal{L}\tau$ (τ is a quite arbitrary function), corresponding to some integral $\overline{I}$. We omit the proof, which consists of a simple check. It is also easy to verify that the inverse theorem and the above proposition can be extended to non-conservative systems.

1.7 An Application of Noether's Theorem: The n-Dimensional Oscillator

We have studied the n-dimensional oscillator (both isotropic and anisotropic) in the framework of Newtonian dynamics. This system is particularly important; in a certain sense it is the unique meaningful model of an n-dimensional physical system which is integrable for any n. In any case it is of great importance as a term of comparison.

Before continuing with the study of the oscillator, we should note that, when in (1.B.48) one has $f \equiv 0$, one speaks of *absolute invariance* and of symmetry of the Lagrangian, and when $f \neq 0$, of invariance up to *gauge* terms and *quasi-symmetry*. For the oscillator we shall have both cases. Since in the forthcoming applications we shall have to handle Lagrangians not explicitly dependent on time, $\mathcal{L} = \mathcal{L}(\boldsymbol{q}, \dot{\boldsymbol{q}})$, we shall limit ourselves to transformations with

$$\xi_i = \xi_i(\boldsymbol{q}, \dot{\boldsymbol{q}}), \qquad \tau = \tau(\boldsymbol{q}, \dot{\boldsymbol{q}}), \qquad f = f(\boldsymbol{q}, \dot{\boldsymbol{q}}). \tag{1.B.55}$$

Consequently, the generalized Killing equations become

$$\xi_i \frac{\partial \mathcal{L}}{\partial q_i} + \frac{\partial \mathcal{L}}{\partial \dot{q}_i} \left[\frac{\partial \xi_i}{\partial q_j} \dot{q}_j - \dot{q}_i \dot{q}_j \frac{\partial \tau}{\partial q_j} \right] + \mathcal{L} \frac{\partial \tau}{\partial q_i} \dot{q}_i = \frac{\partial f}{\partial q_i} \dot{q}_i, \tag{1.B.56}$$

$$\mathcal{L} \frac{\partial \tau}{\partial \dot{q}_i} + \frac{\partial \mathcal{L}}{\partial \dot{q}_j} \left[\frac{\partial \xi_j}{\partial \dot{q}_i} - \dot{q}_j \frac{\partial \tau}{\partial \dot{q}_i} \right] = \frac{\partial f}{\partial \dot{q}_i} \qquad i, j = 1, 2, \ldots, n. \tag{1.B.57}$$

The n-dimensional Isotropic Oscillator

Consider first an n-dimensional isotropic oscillator (of frequency ω and unit mass). The Lagrangian will be

$$\mathcal{L} = \frac{1}{2} \dot{q}_i \dot{q}_i - \frac{1}{2} \omega^2 q_i q_i, \tag{1.B.58}$$

and the equations of motion

$$\ddot{q}_i + \omega^2 q_i = 0 \qquad i = 1, 2, \ldots, n. \tag{1.B.59}$$

As we have already emphasized, we cannot substitute (1.B.58) in (1.B.56) and (1.B.57) and try to solve the $n + 1$ generalized Killing equations, to obtain all the possible solutions ξ_i and τ. Nevertheless, we can equally well use equations (1.B.57) and (1.B.56) for a systematic exploration of the first integrals corresponding to the "simplest" transformations.

a) Then let us begin by considering a transformation which is a pure "time translation", and of the simplest type, i.e. $\xi_i \equiv 0$, $\forall i$, and $\tau = $ const. One can immediately check that, in this case, (1.B.56) and (1.B.57) are identically satisfied with $f \equiv 0$; τ being an arbitrary constant, we shall choose it equal to -1 for the sake of simplicity. Then $\bar{t} = t - \varepsilon$, $\bar{q}_i = q_i$. From (1.B.48) the corresponding first integral is

$$-\mathcal{L} + \frac{\partial \mathcal{L}}{\partial \dot{q}_i}(\dot{q}_i) = \dot{q}_i \dot{q}_i - \mathcal{L} = E = \text{const}, \qquad (1.B.60)$$

where E is the total energy of the oscillator. Therefore, the conservation of energy corresponds to the invariance of the Lagrangian under time translations. This is a more formal way of saying that the time is an ignorable coordinate.

b) If we consider an infinitesimal space translation along the direction given by a vector $\boldsymbol{e}$ (arbitrary unit vector), then

$$\delta q_i = \varepsilon e_i, \quad \xi_i = e_i. \qquad (1.B.61)$$

It is easy to check that the transformation could be Noetherian only if

$$q_i e_i = 0, \qquad (1.B.62)$$

that is $\boldsymbol{q} \cdot \boldsymbol{e} = 0$. It not being possible, in general, to satisfy this condition, the transformation (1.B.61) cannot be Noetherian and so the Lagrangian of the oscillator is not invariant under space translations.

c) Another simple case may be that of a space transformation depending on the q's: the simplest one is when the dependence is linear, i.e.:

$$\xi_i = \omega_{ij}\, q_j, \qquad \tau \equiv 0, \qquad (1.B.63)$$

with a matrix $\boldsymbol{\omega}$ having constant elements. Let us see, by substituting (1.B.63) in (1.B.56) and (1.B.57), what are the conditions, if any, for the matrix $\boldsymbol{\omega}$. From (1.B.56), with $f \equiv 0$, one obtains

$$\omega_{ij}(\dot{q}_i \dot{q}_j - \omega^2\, q_i q_j) = 0. \qquad (1.B.64)$$

Therefore the equation is identically satisfied under the condition that $\omega_{ij} = -\omega_{ji}$, i.e. $\boldsymbol{\omega} = -\boldsymbol{\omega}^T$. Then the transformation is such that $\bar{q}_i = q_i + \varepsilon\, \omega_{ij} q_j$, $\bar{t} = t$, so that $\xi_i = \omega_{ij} q_j$, $\omega_{ij} = -\omega_{ji}$ and the corresponding first integral is

$$\omega_{ij}\, \dot{q}_i q_j = \text{const}. \qquad (1.B.65)$$

In an n-dimensional space, $\delta q_i = \varepsilon\, \omega_{ij}\, q_j$ with $\omega_{ij} = -\omega_{ji}$ represents an infinitesimal rotation: since there are $n(n-1)/2$ linearly independent anti-symmetrical $n \times n$ matrices, (1.B.65) represents a set of $n(n-1)/2$ linearly independent first integrals. In the case $n = 3$, (1.B.65) yields $\boldsymbol{r} \times \dot{\boldsymbol{r}} = \text{const}$, that is, the conservation of angular momentum.

d) Finally let us consider the possibility of a spatial transformation depending on velocities:

$$\tau \equiv 0, \qquad \xi_i = a_{ij}\, \dot{q}_j, \qquad (1.B.66)$$

where $\boldsymbol{a}$ is a matrix with constant elements. By substituting (1.B.66) in (1.B.56) and (1.B.57), we obtain

$$-\omega^2\, a_{ij}\, q_i\, \dot{q}_j = \frac{\partial f}{\partial q_i}\, \dot{q}_i, \qquad a_{ji}\, \dot{q}_j = \frac{\partial f}{\partial \dot{q}_i}, \qquad i = 1, 2, \ldots, n. \qquad (1.\text{B}.67)$$

We have (1.B.67) identically satisfied if we assume

$$f\,(q, \dot{q}) = \frac{1}{2}\, a_{ij}\, \dot{q}_i\, \dot{q}_j - \frac{1}{2}\, \omega^2 a_{ij}\, q_i\, q_j, \qquad (1.\text{B}.68\text{a})$$

with $a_{ij} = a_{ji}$, i.e. $\boldsymbol{a} = \boldsymbol{a}^T$. Then the transformation which turns out to be Noetherian is

$$\bar{t} = t, \quad \bar{q}_i = q_i + \varepsilon\xi_i, \qquad \xi_i = a_{ij}\dot{q}_j, \; a_{ij} = a_{ji}. \qquad (1.\text{B}.69)$$

The corresponding integral is $a_{ij}\, \dot{q}_i\dot{q}_j - f$; that is

$$\frac{1}{2}\, a_{ij}\, \left(\dot{q}_i\dot{q}_j + \omega^2 q_i q_j\right) = \text{const.} \qquad (1.\text{B}.70)$$

Before entering into a discussion about the physical meaning of (1.B.70) let us try to understand what type of transformation (1.B.69) represent. Let us refer to TQ, i.e. to the space of the q's and $\dot{q}$'s. Remembering that from (1.B.38) one has $\delta\dot{q}_i = \varepsilon(\dot{\xi}_i - \dot{q}_i\dot{\tau})$, we have in TQ that $\delta\, q_i = \varepsilon\, a_{ij}\, \dot{q}_j$, $\delta\, \dot{q}_i = \varepsilon\, a_{ij}\, \ddot{q}_j$. From the equations of motion (1.B.59), $\delta\, \dot{q}_i = -\varepsilon\, \omega^2\, a_{ij}\, q_j$. If, as in Sect. 1.3, we perform a change of scale defined by

$$\dot{q}_i = \omega\, v_i, \qquad (1.\text{B}.71)$$

we obtain

$$\delta\, q_i = \varepsilon\, \omega\, a_{ij}\, v_j, \qquad \delta\, v_i = -\varepsilon\, \omega\, a_{ij}q_j. \qquad (1.\text{B}.72)$$

Therefore the Noetherian transformation (1.B.69) can be interpreted as a rotation of the two subspaces of the q_i's and v_i's as a whole around an axis orthogonal to TQ. The first integral (1.B.70), therefore, corresponds to a rotational symmetry of the space of the q's and $\dot{q}$'s: it is a symmetry which was not displayed in the traditional versions of Noether's theorem and was considered a "hidden symmetry". Let us consider now (1.B.70) from the point of view of its physical meaning. First of all, it must be pointed out that, since $n(n+1)/2$ is the number of components of a second-order symmetric tensor and since also the a_{ij}'s are arbitrary constants, (1.B.70) states the existence of the $n(n+1)/2$ first integrals:

$$\frac{1}{2}\, \left(\dot{q}_i\dot{q}_j + \omega^2 q_i q_j\right) = A_{ij} = \text{const}; \qquad A_{ij} = A_{ji}. \qquad (1.\text{B}.73)$$

For a better understanding of the implications of (1.B.73), it is convenient to consider the already familiar case $n = 2$ (planar oscillator). In this case, A_{ij} will have three independent components:

$$A_{11} = \frac{1}{2}(\dot{q}_1)^2 + \frac{1}{2}\omega^2(q_1)^2, \; A_{22} = \frac{1}{2}(\dot{q}_2)^2 + \frac{1}{2}\omega^2(q_2)^2,$$

$$A_{12} = A_{21} = \frac{1}{2}\,\left(\dot{q}_1\dot{q}_2 + \omega^2 q_1 q_2\right).$$

We recognize immediately that $A_{11} = E_1$ and $A_{22} = E_2$, where E_1 and E_2 are the energies of the two oscillations; moreover, the trace of the tensor A_{ij} is $A = A_{11} + A_{22} = E$, the total energy, and the determinant

$$|A| = A_{11} A_{22} - A_{12} A_{21} = \frac{1}{4} \omega^2 \left(q_1 \dot{q}_2 - q_2 \dot{q}_1 \right)^2 = \frac{1}{4} \omega^2 L^2,$$

where L is the angular momentum. As to $A_{12} = A_{21}$, if we start from the solutions (which we rewrite in a slightly different way from those in Sect. 1.3)

$$q_1 = c \cos \omega \left(t - t_0 \right), \qquad q_2 = b \cos \left[\omega \left(t - t_0 \right) + \delta \right], \tag{1.B.74}$$

where δ is the initial phase difference and t_0 the initial time, we have

$$A_{12} = \frac{1}{2} \left(\dot{q}_1 \, \dot{q}_2 + \omega^2 \, q_1 \, q_2 \right) = \frac{1}{2} \, \omega^2 \, a b \cos \delta. \tag{1.B.75}$$

Therefore A_{12} is connected to the initial phase difference; we have already learned in Sect. 1.3 that it is just this phase difference which determines the orientation of the orbit: this is the meaning of A_{12}. We also learned that for $\delta = \pi/2$ we have an ellipse with the axes coinciding with the directions of the q_1 and q_2 axes: in this case $A_{12} = 0$ and the tensor A_{ij} is reduced to diagonal form with q_1 and q_2 as principal axes.

At this point, let us see what relations exist among the integrals we have found, since they surely exist in excess. In fact, from (1.B.74) it is evident that we may have at most three independent first integrals (corresponding to the constants a, b and δ), apart from t_0 (a constant always present in the case of autonomous systems and connected with the arbitrariness of the origin of times); we have met five integrals: total energy, angular momentum and the three independent components of A_{ij}. Thence two relations among them must exist: one is obviously given by $A_{11} + A_{22} = E$ and the other, which we have already seen when we evaluated the determinant of A, can be rewritten as

$$E^2 - \frac{1}{2} \omega^2 L^2 = \left(A_{11} \right)^2 + \left(A_{22} \right)^2 + 2 \left(A_{12} \right)^2. \tag{1.B.76}$$

The Anisotropic Oscillator

We already know from Sect. 1.3 that, when the frequencies are different and not commensurable, the system consists of n uncoupled oscillations without any mutual relation: therefore no first integral related to the system as a whole exists, apart from the total energy owing to the additivity property of the energy itself. Therefore we have to study only the case of commensurable frequencies. In this case an element exists which is common to all oscillations: the period; in fact, as we have seen, when the frequencies are commensurable, a period exists (a multiple of every single period through a different coefficient) which is common to the whole system. In other words, the anisotropic oscillator with commensurable frequencies is a periodic, and not a multiperiodic, system. We expect that this fact entails the existence of first integrals (one or more) concerning the system as a whole.

The Lagrangian will be

$$\mathcal{L} = \frac{1}{2}\dot{q}_i\,\dot{q}_i - \frac{1}{2}\sum_{i=1}^{n}\omega_i^2 q_i^2. \tag{1.B.77}$$

If we investigate if there are the same integrals of the isotropic case, we realize immediately that the existence of different frequencies makes transformation (1.B.63) no longer Noetherian ((1.B.56) cannot be satisfied) and then the angular momentum is not conserved. Moreover, (1.B.66) in its general form is also no longer Noetherian. Only the diagonal elements remain and the transformation which still holds is

$$\xi_i = a_{ii}\,\dot{q}_i \quad \text{(no sum over } i\text{)}, \tag{1.B.78}$$

and

$$f(\boldsymbol{q}, \dot{\boldsymbol{q}}) = \sum_{i=1}^{n}\frac{1}{2}\,a_{ii}\left(\dot{q}_i^2 - \omega_i^2 q_i^2\right). \tag{1.B.68b}$$

Consequently, we have the integrals A_{ii} (namely the energies of the individual modes of oscillation) but not the non-diagonal ones A_{ij}. On the other hand it is easy enough to realize that the isotropy of the space TQ which led us to the interpretation provided by (1.B.72) is now removed owing to the existence of different frequencies, which prevents us from arriving at (1.B.72) through (1.B.71) (the different frequencies are "mixed" performing the transformation). Therefore we shall now have to investigate the nature of transformations, different from (1.B.66), corresponding to the integrals which certainly exist (we know, in fact, that closed trajectories correspond to commensurable frequencies). For the sake of simplicity, let us limit ourselves to the two-dimensional case (planar anisotropic oscillator). Then

$$\mathcal{L} = \frac{1}{2}\left(\dot{q}_1^2 + \dot{q}_2^2\right) - \frac{1}{2}\left(\omega_1^2\,q_1^2 + \omega_2^2\,q_2^2\right). \tag{1.B.79}$$

With ω_1 and ω_2 being commensurable we can, without loss of generality, put $\omega_2 = n\omega_1 = n\omega$, with $n > 0$. Let us start from the parametric equations of the trajectory:

$$q_1 = a\cos\omega\,(t - t_0), \qquad q_2 = b\cos[n\omega\,(t - t_0) + \delta], \tag{1.B.80}$$

As $E_1 = \frac{1}{2}\omega^2 a^2$, $E_2 = \frac{1}{2}n^2\omega^2 b^2$ are the two energies, we have only to look for a third integral related to the phase difference δ. Differentiating (1.B.80) one gets

$$\dot{q}_1 = -\omega\,a\sin\omega\,(t - t_0), \qquad \dot{q}_2 = -n\omega\,b\sin[n\omega\,(t - t_0) + \delta]. \tag{1.B.81}$$

From (1.B.80), one has

$$\delta = \arccos\frac{q_2}{b} - n\arccos\frac{q_1}{a}, \tag{1.B.82}$$

and, using (1.B.81),

$$\cos \delta = \frac{q_2}{b} \cos \left(n \arccos \frac{q_1}{a} \right) - \frac{\dot{q}_2}{n\omega b} \sin \left(n \arccos \frac{q_1}{a} \right). \tag{1.B.83}$$

When n is an integer, (1.B.83) can be rewritten as

$$\cos \delta = \frac{q_2}{b} T_n \left(\frac{q_1}{a} \right) + \frac{\dot{q}_1 \dot{q}_2}{n^2 \omega^2 ab} T'_n \left(\frac{q_1}{a} \right), \tag{1.B.84}$$

where $T_n(x) = \cos(n \arccos x)$ is the Chebyshev polynomial of the first kind and degree n and $T'_n(x)$ its derivative with respect to the argument x. It is easy now to check, even if calculations are a little cumbersome, that to every integer n a first integral corresponds:

$$I_n = (2E_1)^{n/2}(2E_2)^{1/2} \cos \delta. \tag{1.B.85}$$

For $n = 1$, one obtains $2A_{12}$, where A_{12} is the integral (1.B.75). Then (1.B.85) and (1.B.84) enable us, for any integer n, to express the corresponding first integral. From a formal point of view, we can also consider I_n as a scalar product $\boldsymbol{A} \cdot \boldsymbol{B}$ between two vectors of moduli $|\boldsymbol{A}|^2 = (2E_1)^n$ and $|\boldsymbol{B}|^2 = 2E_2$. That is,

$$A_1^2 + A_2^2 = (2E_1)^n, \qquad B_1 = \dot{q}_2, \qquad B_2 = n\,\omega\,q_2.$$

For $n = 2$, we have

$$I_2 = (2E_1)\sqrt{2E_2} \cos \delta = (2\omega\, q_1\, \dot{q}_1)\, \dot{q}_2 + (\omega^2 q_1^2 - \dot{q}_1^2)\, 2\,\omega q_2, \tag{1.B.86}$$

and, for $n = 3$,

$$I_3 = (2E_1)^{3/2}\sqrt{2E_2} \cos \delta = (3\omega^2 q_1^2 \dot{q}_1 - \dot{q}_1^3)\, \dot{q}_2 + (\omega^3 q_1^3 - 3\omega q_1 \dot{q}_1^2)\, 3\omega q_2. \tag{1.B.87}$$

It remains, now, to consider the case of n given by a non-integer rational number. One can check that (1.B.85) also holds in this case, but it is no longer possible to express $\cos \delta$ in a compact form as in (1.B.84). In the simplest case, $n = 3/2$, by means of the formulae of elementary trigonometry one obtains

$$I_{3/2} = (2E_1)^{3/4}(2E_2)^{1/2} \cos \delta = \dot{q}_2 A_1 + \frac{3}{2}\, \omega\, q_2 A_2, \tag{1.B.88}$$

with

$$A_{1,2} = \mp \sqrt{\frac{1}{2}(\omega^3 a^3 \mp \omega^3 q_1^3 \pm 3\omega q_1 \dot{q}_1^2)}.$$

Obviously, $A_1^2 + A_2^2 = \omega^3 a^3 = (2E_1)^{3/2}$.

We recall that, for $n = 2$ and $n = 3/2$, the trajectories for various values of δ have been given in Sect. 1.3. To sum up, we can say that, for the anisotropic case also the third integral exists and is given by a one-valued function of q and $\dot{q}$ (the integral contains $\cos \delta$ and not simply δ). We have now to go back again to Noether's theorem and to realize what are the transformations corresponding to the integrals we have found. From examples (1.B.86) and (1.B.87) it is evident that now the integral proportional to $\cos\delta$ is no longer, like A_{12} in the isotropic case, a second-degree polynomial in the variables q_1, q_2, $\dot{q}_1$, $\dot{q}_2$ but on the contrary a polynomial whose degree is increasing with n $(n = \omega_2/\omega_1)$; the situation becomes even more complicated when n is not an integer because of the appearance of roots, as one can see from the example (1.B.88). Therefore, we shall no longer have any single transformation like (1.B.66), but different transformations corresponding to different values of n (and of increasing complexity with increasing n). In the simplest case $(n = 2)$, we are still able to proceed tentatively in search of the correct transformation. The structure exhibited by integral (1.B.86) suggests trying a transformation like

$$\xi_i = A_{ijk}\dot{q}_j q_k, \qquad \tau \equiv 0, \qquad i,j,k = 1,2, \qquad (1.B.89)$$

where the components of A_{ijk} are some constants. In this case (1.B.56) and (1.B.57) become

$$\xi_i \frac{\partial \mathcal{L}}{\partial q_i} + \frac{\partial \mathcal{L}}{\partial \dot{q}_i}\frac{\partial \xi_i}{\partial q^l}\dot{q}^l = \frac{\partial f}{\partial q_i}\dot{q}_i, \qquad (1.B.90)$$

$$\frac{\partial \mathcal{L}}{\partial \dot{q}^l}\frac{\partial \xi^l}{\partial \dot{q}_i} = \frac{\partial f}{\partial \dot{q}_i}. \qquad (1.B.91)$$

Substituting (1.B.89) in (1.B.90) and taking into account that now $\mathcal{L} = \frac{1}{2}(\dot{q}_1^2 + \dot{q}_2^2) - \frac{1}{2}\omega^2(q_1^2 + 4q_2^2)$, one obtains (for the sake of brevity we omit the calculations) for f the expression

$$f = A\left(q_1\dot{q}_1\dot{q}_2 - q_2\dot{q}_1^2 - \omega^2 q_1^2 q_2\right) \qquad (1.B.92)$$

and for the components of A_{ijk} the values

$$A_{121} = A_{211} = A, \qquad A_{112} = -2A,$$
$$A_{111} = A_{221} = A_{122} = A_{222} = A_{212} \equiv 0, \qquad (1.B.93)$$

where A is an arbitrary constant which can be put equal to 1. The wanted integral will then be

$$I = \frac{\partial \mathcal{L}}{\partial \dot{q}_i}\xi_i - f = \dot{q}_1(-2q_2\dot{q}_1 + q_1\dot{q}_2) + \dot{q}_2 q_1\dot{q}_1$$
$$- (q_1\dot{q}_1\dot{q}_2 - q_2\dot{q}_1^2 - \omega^2 q_1^2 q_2) = -\dot{q}_1^2 q_2 + q_1\dot{q}_1\dot{q}_2 + \omega^2 q_1^2 q_2. \qquad (1.B.94)$$

We can check immediately, by comparing with (1.B.86), that $I = I_2/2\omega$ and that the integral in (1.B.94) corresponds to the transformation generated by $\xi_1 = -2q_2\dot{q}_1 + q_1\dot{q}_2$, $\xi_2 = q_1\dot{q}_1$, $\tau \equiv 0$. From (1.B.89) and (1.B.90) it is evident that the transformations we have to handle now are no longer identifiable as rotations, translations with constant coefficients, etc., but are more complex. We can rewrite the generators in the form

$$\xi_1 = -\dot{q}_1 q_2 + (q_1\dot{q}_2 - q_2\dot{q}_1)\,, \qquad \xi_2 = \dot{q}_1 q_1\,, \qquad (1.\text{B}.95)$$

and interpret ξ_1 and ξ_2 as generators of a (differential) rotation in the $q_1 q_2$ plane (with coefficients given by $\dot{q}_i$) and of a translation along q_1 of an amount given by the angular momentum $q_1\dot{q}_2 - q_2\dot{q}_1$ (which in this case is not constant). Anyhow they are no longer "simple" transformations involving the whole space of the q's and $\dot{q}$'s. Nevertheless, the method we have adopted has enabled us all the same to obtain integral (1.B.94), which is a polynomial (in the q's and $\dot{q}$'s) of a higher degree than A_{12}.

The success achieved so far, however, does not mean that the method works in every case. When the Lagrangian of the system has a relatively simple expression in rectangular coordinates (as in the above case) the method can always be successfully applied. In fact, with rectangular coordinates, the "simplest" transformations are with constant coefficients and the procedure is unambiguous. On the other hand, when the Lagrangian is given in curvilinear coordinates, the transformations (even the simplest ones) do not in general have constant coefficients, and then the choice is to a great extent arbitrary.

1.8 The Principle of Least Action in Jacobi Form

For a conservative system, Hamilton's principle assumes a particularly meaningful form due to Jacobi.

In the sixth of his celebrated lectures on dynamics,[20] Jacobi proposed eliminating from Hamilton's principle any feature which could make one think of a *metaphysical cause*, not considering himself satisfied with the formulation given by Lagrange and Hamilton, which, in his opinion, still keeping the presence of the time variable, could allow a finalistic interpretation. In the Jacobi formulation, as we shall see, the principle is, so to speak, geometrized.

Let us then consider a conservative natural system (see Sect. 1.5). The action integral will be

$$I = \int_{t_1}^{t_2} L(q_1, q_2, \ldots, q_n; \dot{q}_1, \dot{q}_2, \ldots, \dot{q}_n)\, dt$$

$$= \int_{t_1}^{t_2} \left(\sum_{k=1}^{n} p_k \dot{q}_k - \mathcal{H} \right) dt. \qquad (1.\text{B}.96\text{a})$$

As the system is conservative,

$$\mathcal{H} = \mathcal{H}(\boldsymbol{q}, \boldsymbol{p}) = E = \text{const}, \qquad (1.\text{B}.97)$$

and therefore (1.B.96a) will become

---

[20] K. G. J. Jacobi: *Vorlesungen Über Dynamik*, 2nd rev. edn. (Reiner, Berlin, 1884), pp. 43–51. (Reprinted by Chelsea, New York, 1969).

$$I = \int_{t_1}^{t_2} \left(\sum_{k=1}^{n} p_k dq_k - \mathcal{H}dt \right) = \int_{t_1}^{t_2} \sum_{k=1}^{n} p_k dq_k - (t_2 - t_1)E. \qquad (1.B.96b)$$

If we now consider any two neighbouring curves Γ_1 and Γ_2 with the same end points and with tangents satisfying (1.B.97), we have

$$I(\Gamma_1) = \int_{\Gamma_1} \sum_{k=1}^{n} p_k dq_k + (t_1 - t_2)\, E,$$

$$I(\Gamma_2) = \int_{\Gamma_2} \sum_{k=1}^{n} p_k dq_k + (t_1 - t_2)\, E,$$

and then

$$\delta I = I(\Gamma_2) - I(\Gamma_1) = \delta \int \sum_{k=1}^{n} p_k dq_k.$$

On the other hand, owing to the Hamilton principle, we have $\delta I = 0$ and thus

$$\delta \int \sum_{k=1}^{n} p_k dq_k = 0. \qquad (1.B.98)$$

The obtained result states that, in the Q space, the trajectories of the system satisfy the variational equation (1.B.98) with supplementary condition (1.B.97). It is understood that the limits of integration are fixed in Q.

Let us now define the reduced action S^* (Maupertuis action):

$$S^* = \int \sum_{k} p_k dq_k,$$

where p_k and q_k satisfy (1.B.97). Recalling the definition $p_k = \partial \mathcal{L}/\partial \dot{q}_k$, we get

$$S^* = \int \sum_{k} \frac{\partial \mathcal{L}}{\partial \dot{q}_k} \dot{q}_k \, dt.$$

Recall also that for a natural system (see Sect. 1.5),

$$\mathcal{T} = \frac{1}{2} a_{ik} \dot{q}_i \dot{q}_k.$$

Moreover, $V = V(q)$ is the potential energy and $\mathcal{L} = \mathcal{T} - V(q)$ the Lagrangian. Accordingly, we have

$$\sum_{k} \frac{\partial \mathcal{L}}{\partial \dot{q}_k} \dot{q}_k = 2\,\mathcal{T}.$$

Therefore the reduced action becomes

$$S^* = 2 \int \mathcal{T} \, dt, \qquad (1.B.99)$$

and the variational principle assumes the form

$$\delta S^* = \delta \int 2\mathcal{T}\, dt = 0, \tag{1.B.100}$$

with $\mathcal{H} - E = 0$ and, consequently, $\mathcal{T} = E - V$.

Now, following Jacobi, it remains to eliminate the time. Let us refer again to the (kinematical) line element in the configuration space introduced in Sect. 1.4. From $ds^2 = 2\mathcal{T}dt^2$, we have

$$dt = \frac{ds}{\sqrt{2\mathcal{T}}} \tag{1.B.101}$$

and therefore

$$2\mathcal{T}\,dt = \sqrt{2\mathcal{T}}\,ds = \sqrt{2(E-V)}\,ds.$$

Neglecting the factor $\sqrt{2}$, we can rewrite principle (1.B.100) in the form

$$\delta \int \sqrt{E-V}\,ds = 0. \tag{1.B.102}$$

It must be remembered that ds in (1.B.102) is, as is obvious, the square root of a quadratic differential form and not the exact differential of some function.

Finally, if we define the action line element[21] ds_1 by

$$ds_1^2 = (E-V)\,ds^2, \tag{1.B.103}$$

the variational principle will be completely geometrized and will take the form

$$\delta \int ds_1 = 0. \tag{1.B.104}$$

This means that the solution of the problem of the motion of a conservative natural system leads to the search for the geodesic of a Riemannian manifold whose metric is given by (1.B.103). The time no longer appears in this formulation; (1.B.104) determines the path of the representative point in the configuration space, not the motion in time.

For this part of the problem, one will have to integrate (1.B.101). Moreover, it must be said that the equilibrium states of the system are excluded from this formulation; in fact, in this case $E = V$, and then the metric (1.B.103) becomes singular. When the system consists of only one particle moving under a potential $U = -V$, the line element ds is simply the line element of the ordinary three-dimensional space in arbitrary curvilinear coordinates. With regard to (1.B.103), we can further remark that the metric ds_1 is obtained from ds by multiplying by a factor depending only on the q's

---

[21]The phrases "kinematic line element" and "action line element" are due to J. L. Synge, see *Handbuch der Physik*, **III**, 1 (Springer, 1960) and the paper already quoted.

and not $\dot{q}$'s, i.e. only on the "position" and not on the "direction". Therefore we are in the presence of a conformal transformation and thence in passing from one metric to another, the angles are conserved. As to the length of any arc of curve, it decreases more and more for kinetic energies tending to zero, untill vanishing for $\mathcal{T} = 0$; in fact, we have already remarked that, for $E = V$, the metric ds_1 becomes singular.

Lastly, a comment regarding the term "least action": we have preserved it in homage to tradition, but more correctly one must say *stationary* action, since it is required that the action integral be stationary and not necessarily minimum. The same, from the geometrical point of view, is true for the geodesics. As Jacobi[22] had already pointed out, if we consider, for instance, the geodesics on the sphere, the geodesic consisting of an arc of a great circle joining two points A and C corresponding to an angular distance greater than π is not the shortest distance. Therefore we can speak of a minimum only when we are comparing two neighbouring curves. We take from Jacobi Fig. 1.15. Let B be the diametrically opposed point to a given point A. The great circles joining A and B obviously have all the same length; then, if $A\alpha BC$ lies on a great circle, we have

$$A\alpha BC = A\beta B + BC = A\beta + \beta B + BC.$$

If β is a point close to B, let us draw the arc of a great circle joining β with C. Then:

$$\beta C < \beta B + BC.$$

Thus $A\beta + \beta C$ is smaller than the arc of great circle $A\alpha BC$.

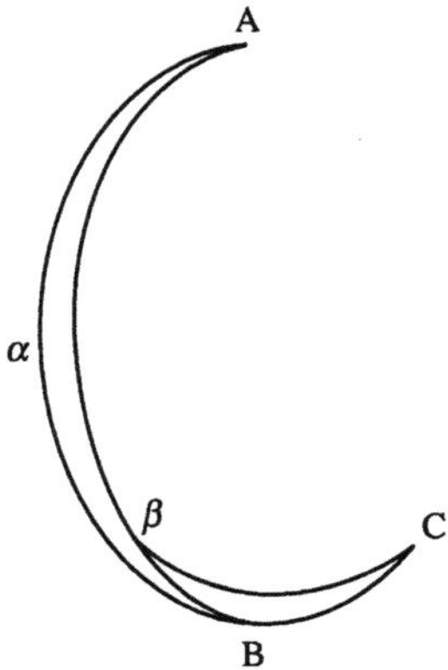

Fig. 1.15

---

[22]See Jacobi, op. cit., p. 46.

C. Hamiltonian Dynamics and Hamilton–Jacobi Theory

1.9 The Canonical Equations

As we have seen, in the Lagrangian formulation generalized coordinates $q_1, q_2, \ldots, q_n$, have been introduced, for a system having n degrees of freedom, and their time derivatives enter explicitly in the Lagrangian besides the q's themselves. Nevertheless, the equations of motion are n differential equations of the second order in the unknown functions $q_i(t)$. The further step of passing to $2n$ first-order differential equations was taken in the Hamiltonian formulation of the dynamics. We have already met in (1.B.14) the definition of the momentum conjugate to the coordinate q_i,

$$p_i = \frac{\partial \mathcal{L}}{\partial \dot{q}_i} \qquad (i = 1, 2, \ldots, n), \tag{1.C.1}$$

and in (1.B.28), the definition of the Hamiltonian,

$$\mathcal{H} = p_i \, \dot{q}_i - \mathcal{L}. \tag{1.C.2a}$$

In (1.B.28) $\mathcal{H}$ appeared as a constant of motion when in the Lagrangian the time was not explicitly contained; therefore the Hamiltonian also did not contain the time explicitly. Now, on the other hand, we consider (1.C.2a) a general definition with $\mathcal{H}$ and $\mathcal{L}$ also containing, for all eventualities, the time in an explicit way. Relation (1.C.2a) which contains $\mathcal{H}$ and $\mathcal{L}$ in a, so to speak, symmetrical role, represents a *Legendre transformation*. If, for the sake of simplicity, we limit ourselves to only two variables, a Legendre transformation transforms a function $f(x, y)$ into another one $g(x, z)$ with $z = \partial f / \partial y$ and such that $\partial g / \partial z = y$. In our case $f(x, y) = \mathcal{L}(q, \dot{q})$ and $g(x, z) = \mathcal{H}(q, p)$.

Now, differentiating (1.C.2a), we have

$$d\mathcal{H} = \dot{q}_k \, dp_k + p_k \, d\dot{q}_k - \frac{\partial \mathcal{L}}{\partial q_k} dq_k - \frac{\partial \mathcal{L}}{\partial \dot{q}_k} d\dot{q}_k - \frac{\partial \mathcal{L}}{\partial t} dt.$$

Owing to (1.C.1), the coefficients of $d\dot{q}_k$ mutually cancel and then

$$\frac{\partial \mathcal{H}}{\partial q_k} = -\frac{\partial \mathcal{L}}{\partial q_k}, \qquad \frac{\partial \mathcal{H}}{\partial p_k} = \dot{q}_k.$$

Exploiting again Lagrange's equations and (1.C.1), one obtains $\partial \mathcal{H}/\partial q_k = -\dot{p}_k$, and then the system

$$\dot{q}_k = \frac{\partial \mathcal{H}}{\partial p_k}, \qquad \dot{p}_k = -\frac{\partial \mathcal{H}}{\partial q_k} \qquad (k = 1, 2, \ldots, n). \tag{1.C.3}$$

Equations (1.C.3) are called *Hamilton's equations*, or *canonical equations*. Summarizing: the function $\mathcal{H}$ is, in general, a function of the q_k p_k and the time; system (1.C.3) consists of $2n$ first-order equations; the conjugate variables q_k, p_k do not necessarily correspond to coordinates and components of momenta, but may be completely general. Every time it is possible to characterize the state of a system by means of n generalized coordinates q_k and n conjugate momenta, given by (1.C.1), obeying (1.C.3), one says that the system is a Hamiltonian dynamical system: the particular structure of (1.C.3) distinguishes it among the systems of $2n$ first-order differential equations.

Hamilton's equations can also be obtained directly from the variational principle, as we did for Lagrange's ones. In fact, owing to the duality of the Legendre transformations, we can start from the Hamiltonian to obtain the Lagrangian

$$\mathcal{L} = p_i\,\dot{q}_i - \mathcal{H} \tag{1.C.2b}$$

and write down the action integral in the form

$$I = \int_{t_1}^{t_2} (p_i\dot{q}_i - \mathcal{H})dt. \tag{1.C.4}$$

Since the second term in (1.C.2b) does not contain the derivatives of the p_k's, these ones do not even appear in the integrand of the variational principle and therefore the variation of I has the same conditions we have already fixed for the q_k's. The variational problem, now, has $2n$ variables and the resulting equations are

$$\frac{d}{dt}\frac{\partial\mathcal{L}}{\partial\dot{q}_i} - \frac{\partial\mathcal{L}}{\partial q_i} \equiv \frac{dp_i}{dt} + \frac{\partial\mathcal{H}}{\partial q_i} = 0,$$
$$\frac{d}{dt}\frac{\partial\mathcal{L}}{\partial\dot{p}_i} - \frac{\partial\mathcal{L}}{\partial p_i} \equiv 0 - \dot{q}_i + \frac{\partial\mathcal{H}}{\partial p_i} = 0, \tag{1.C.5}$$

which are just the canonical equations (1.C.3) obtained directly from the variational principle.

The Hamiltonian system (1.C.3), owing to its peculiar structure, can be written in a form, called symplectic, which employs the matrix formalism; this form will be useful in applications. For a system with n degrees of freedom, one defines a column vector $\boldsymbol{z}$, with $2n$ components, such that

$$z_i = q_i, \qquad z_{i+n} = p_i \qquad (i = 1, 2, \ldots, n) \tag{1.C.6}$$

and the columm vector $\partial\mathcal{H}/\partial\boldsymbol{z}$, also with $2n$ components, such that

$$\left(\frac{\partial\mathcal{H}}{\partial\boldsymbol{z}}\right)_i = \frac{\partial\mathcal{H}}{\partial q_i}, \qquad \left(\frac{\partial\mathcal{H}}{\partial\boldsymbol{z}}\right)_{i+n} = \frac{\partial\mathcal{H}}{\partial p_i}. \tag{1.C.7}$$

Lastly, one defines the $2n \times 2n$ square matrix

$$\boldsymbol{J} = \begin{pmatrix} \mathbf{0} & \mathbf{1} \\ -\mathbf{1} & \mathbf{0} \end{pmatrix}, \tag{1.C.8}$$

where $\mathbf{0}$ is the $n \times n$ matrix with vanishing elements and $\mathbf{1}$ is the $n \times n$ identity matrix. Then it is easy to verify that system (1.C.3) can be compactified in the form

$$\dot{z} = J \frac{\partial \mathcal{H}}{\partial z}. \tag{1.C.9}$$

Moreover the matrix J has the following properties:

$$J^2 = -\mathbf{1}, \tag{1.C.10}$$

$$J^T J = \mathbf{1} \quad \Longrightarrow \quad J^T = -J = J^{-1}, \tag{1.C.11}$$

$$|J| = 1. \tag{1.C.12}$$

Returning to the usual formalism, if we evaluate the total derivative $d\mathcal{H}/dt$, we obtain

$$\frac{d\mathcal{H}}{dt} = \frac{\partial \mathcal{H}}{\partial q_k} \dot{q}_k + \frac{\partial \mathcal{H}}{\partial p_k} \dot{p}_k + \frac{\partial \mathcal{H}}{\partial t}$$

and, because of (1.C.3),

$$\frac{d\mathcal{H}}{dt} = \frac{\partial \mathcal{H}}{\partial t}. \tag{1.C.13}$$

When $\mathcal{H} = \mathcal{H}(\boldsymbol{q}, \boldsymbol{p})$, that is, the Hamiltonian does not depend explicitly on time, (1.C.13) gives $d\mathcal{H}/dt = 0$ and then

$$\mathcal{H}(q_1, q_2, \ldots, q_n; p_1, p_2, \ldots, p_n) = \text{const.} \tag{1.C.14}$$

Obviously, (1.C.14) is a first integral of system (1.C.3) and the result is not unexpected, since we have already obtained that $\mathcal{H} = \text{const.}$ when $\mathcal{L}$ does not contain the time: in this case, the mere inspection of (1.C.2a) provides (1.C.14). Nevertheless, it is important, for the forthcoming applications, to clarify the implications of the first integral (1.C.14). To do this, it is necessary to deduce what type of function (of the q's and p's) the Hamiltonian may be.

As regards the kinetic energy $\mathcal{T}$, we have already said in Sect. 1.5 that for natural systems it is a quadratic form homogeneous in the $\dot{q}$'s, indicated by $\mathcal{T}_2$, whereas for non-natural systems it may have also terms linear in the $\dot{q}$'s or terms not containing the $\dot{q}$'s at all. It happens that a system is, for instance, a natural one when considered in an inertial reference frame whereas it becomes non-natural if we pass to a rotating frame. Therefore, in general, $\mathcal{T} = \mathcal{T}_0 + \mathcal{T}_1 + \mathcal{T}_2$, where $\mathcal{T}_n$ means a homogeneous function of degree n in the $\dot{q}$'s (which may also depend on the q's). Owing to the Euler theorem,

$$n \mathcal{T}_n = \sum_k \frac{\partial \mathcal{T}_n}{\partial \dot{q}_k} \dot{q}_k,$$

and then

$$\sum_k p_k \dot{q}_k = \sum_k \frac{\partial \mathcal{L}}{\partial \dot{q}_k} \dot{q}_k = \sum_k \frac{\partial \mathcal{T}}{\partial \dot{q}_k} \dot{q}_k = \sum_k \frac{\partial (\mathcal{T}_0 + \mathcal{T}_1 + \mathcal{T}_2)}{\partial \dot{q}_k} \dot{q}_k = \mathcal{T}_1 + 2\mathcal{T}_2.$$

From (1.C.2a) and (1.B.7), since V depends only on the q's,

$$\mathcal{H} = \mathcal{T}_1 + 2\mathcal{T}_2 - (\mathcal{T}_0 + \mathcal{T}_1 + \mathcal{T}_2) + V = -\mathcal{T}_0 + \mathcal{T}_2 + V.$$

For natural systems, $\mathcal{T}_0 = \mathcal{T}_1 = 0$ and then $\mathcal{T} = \mathcal{T}_2$. In an inertial reference frame, the system is natural, and therefore

$$\mathcal{H} = \mathcal{T} + V, \tag{1.C.15}$$

that is, the Hamiltonian coincides with the total energy and (1.C.14) represents the energy integral.

In a noninertial reference frame, with $\mathcal{T}_0$ and $\mathcal{T}_1$ different from zero, if $\mathcal{H}$ does not depend explicitly on time, the first integral (1.C.14) still exists, but it does not coincide any more with the energy integral. Let us look, as an example, at the case of a coordinate system $\Omega\,(\xi, \eta, \zeta)$ rotating with angular velocity ω with respect to an inertial system $O\,(x, y, z)$, the origins Ω and O and the axes ζ and z coinciding. The two coordinate systems are connected by the relations

$$x = \xi \cos \omega t - \eta \sin \omega t,$$
$$y = \xi \sin \omega t + \eta \cos \omega t,$$
$$z = \zeta,$$

and therefore the kinetic energy in $\Omega(\xi, \eta, \zeta)$ will be[23]

$$\mathcal{T} = \sum \frac{1}{2} m \left[\omega^2(\xi^2 + \eta^2) + 2\omega(\xi\dot{\eta} - \eta\dot{\xi}) + \dot{\xi}^2 + \dot{\eta}^2 + \dot{\zeta}^2 \right].$$

The momenta conjugate to ξ, η, ζ will be

$$P_\xi = m(\dot{\xi} - \omega\eta),$$
$$P_\eta = m(\dot{\eta} + \omega\xi),$$
$$P_\zeta = m\dot{\zeta}$$

so that

$$\mathcal{T} = \sum \frac{1}{2m} (P_\xi^2 + P_\eta^2 + P_\zeta^2).$$

For $\mathcal{H} = p_i\dot{q}_i - \mathcal{T} + V$,

$$\mathcal{H} = \sum \left[\dot{\xi}P_\xi + \dot{\eta}P_\eta + \dot{\zeta}P_\zeta \right] - \mathcal{T} + V = \omega \sum (\eta P_\xi - \xi P_\eta) + \mathcal{T} + V.$$

If V has a rotational symmetry around the z axis, $\mathcal{H}$ will not explicitly contain the time, and therefore $\mathcal{H} = $ const. This is called the *Jacobi integral*.

One can see immediately that it is not the energy integral; in fact,

$$E = \mathcal{T} + V = \sum \frac{1}{2m} (P_\xi^2 + P_\eta^2 + P_\zeta^2) + V.$$

---

[23]From now on, to simplify the exposition, we omit the indices labelling the individual mass points.

But, in this case, the energy is also conserved, and then $E - \mathcal{H} = \text{const.}$ as well. This provides the conservation of angular momentum; in fact,

$$E - \mathcal{H} = \omega \sum (\xi P_\eta - \eta P_\xi) = \omega \sum m\left(\xi\dot\eta - \eta\dot\xi\right) + \omega^2 \sum m\left(\xi^2 + \eta^2\right)$$

and, transforming this to the inertial system, we get

$$E - \mathcal{H} = \omega \sum m\left(x\dot y - y\dot x\right).$$

1.10 The Integral Invariants – Liouville's Theorem

As we have seen, the Hamiltonian dynamics introduces, besides the n generalized coordinates q_i, the n conjugate momenta p_i as independent variables. Going on to use geometrical language, we therefore have a $2n$-dimensional space (*phase space*). From the point of view of differential geometry, every vector $\boldsymbol{p}$ related to a point $\boldsymbol{q}$ of the manifold Q (*configuration space*) is an element of T^*Q_q (*cotangent space* in $\boldsymbol{q}$), this one being the dual space of TQ_q (*tangent space* in $\boldsymbol{q}$), which consists of all real linear functions on TQ_q. In fact, as we know, $\boldsymbol{p} = \partial\mathcal{L}/\partial\dot{\boldsymbol{q}} = \boldsymbol{p}\,(\boldsymbol{q},\,\dot{\boldsymbol{q}})$. The union of all cotangent spaces is called *cotangent fibre bundle* and indicated by T^*Q. Then, the $2n$-dimensional phase space with coordinates (q_i, p_i) is, in geometrical language, the cotangent fibre bundle of the manifold Q (configuration space). By analogy with what we have done in Lagrangian dynamics, where we introduced the space of events $\mathbb{R} \times Q$, here we shall introduce the space $\mathbb{R} \times T^*Q$, called by Cartan the *state space*.[24] In this $2n + 1$-dimensional space, the problem of motion is completely geometrized and the complete solution of the Hamilton equations is represented by a family of infinite curves which fill without intersecting the space itself. We shall now deal with a problem introduced by Poincaré[25] and later generalized by Cartan:[26] the problem of the existence of integral invariants. These are quantities, represented by means of integrals which stay constant during the motion and assume a particular significance if one looks at the motion of the phase points in the state space.

To introduce the subject, we shall first consider a completely general system of equations, later moving on to the case that is of interest to us: that of a Hamiltonian system.

---

[24]E. Cartan: *Leçons sur les invariants integraux* (Hermann, Paris, 1922), p. 4.

[25]H. Poincaré: Sur le problème des trois corps et les equations de la dynamique, *Acta Mathematica* **13**, 1–271 (1890).

[26]E. Cartan: op. cit.

The Integral Invariants

Let us consider, in all its generality, an autonomous system (the conclusions will be the same in the case of a non-autonomous system – our choice is made only with the aim of simplifying the presentation) of the form (1.A.32), where the functions $X_i(\boldsymbol{x})$ are C^1 in a given m-dimensional domain. The solutions will be C^2 functions

$$x_i = \varphi_i\,(t;\, x_1^0,\, x_2^0,\, \ldots\,,\, x_m^0). \tag{1.C.16}$$

In (1.C.16), the constants x_1^0, x_2^0, $\ldots$, x_m^0 indicate the values of x_1, x_2, $\ldots$, x_m corresponding to a fixed (initial) value of t, which can even be assumed equal to zero. System (1.A.32) can be considered to be the equations describing the motion of a point (with coordinates x_1, x_2, $\ldots$, x_m) in an m-dimensional space. If we consider all the points occupying an r-dimensional region $(r \leq m)$ Ω_0 at $t = 0$, the same points at a time t will occupy a region Ω_t, still r-dimensional. If the r-dimensional integral of a given function F on the region Ω_0 retains the same value for any t, i.e.

$$\int_{\Omega_0} F\,d\Omega = \int_{\Omega_t} F\,d\Omega, \quad \forall t, \tag{1.C.17}$$

we say that the integral considered is an *integral invariant* of the system (1.A.32) and r is the *order* of the invariant. Let us begin from the simplest case: $r = 1$. As we have seen in Sect. 1.2, (1.C.16) can also be interpreted as the definition of a transformation (depending on the parameter t) from $\boldsymbol{x}^0$ to $\boldsymbol{x}$:

$$\boldsymbol{x} = \Phi_t\,\boldsymbol{x}^0. \tag{1.C.18}$$

The operator Φ_t transforms the point $\boldsymbol{x}^0$, at $t = 0$, into the point $\boldsymbol{x}$, at the time t. Let us assume that the Jacobian of the transformation,

$$J = \frac{\partial(\varphi_1,\, \varphi_2,\, \ldots\,,\, \varphi_m)}{\partial(x_1^0,\, x_2^0,\, \ldots\,,\, x_m^0)}, \tag{1.C.19}$$

does not vanish for the values of $\boldsymbol{x}^0$ and t in which we are interested. Let us now consider, not a single initial point $\boldsymbol{x}_0$, but the points of a whole curve, γ_0. The transformation defined by Φ_t will transform every point of γ_0 into a corresponding point of a new curve: γ_t. In this way, we have a tube of trajectories which has the curve γ_0 as generatrix; every point of γ_0, as a consequence of the transformation Φ_t, will move like a particle of a fluid: at the time t the particles that were on γ_0 at the time $t = 0$ will occupy the points of the curve γ_t (see Fig. 1.16).

Let us now consider a vector field $\boldsymbol{v} = \boldsymbol{v}(\boldsymbol{x})$ whose components $v_i(\boldsymbol{x})$ are functions of class C^1 defined in the domain in which (1.A.32) are defined. If the relation

$$\int_{\gamma_0} v_1 dx_1 + v_2 dx_2 + \ldots + v_m dx_m = \int_{\gamma_t} v_1 dx_1 + v_2 dx_2 + \ldots + v_m dx_m \tag{1.C.20}$$

holds true for any curve γ_0 in the domain considered and for all values of t, then we say that the integral

$$I_1 = \int_{\gamma_t} \boldsymbol{v} \cdot d\boldsymbol{x} \qquad (1.C.21a)$$

is a linear integral invariant, or a first-order invariant, for the system (1.A.32). When (1.C.20) is valid for any curve, open or closed, the invariant is said to be *absolute*; when (1.C.20) holds only for closed curves, the invariant is said to be *relative*. We can parametrize the curve γ_t: let α be the parameter and 0 and 1 the values of α corresponding to the end points of the curve. In this way, to every point of γ_t there corresponds a value of α and vice versa. Moreover, on the surface of the tube of trajectories, if we vary t (for fixed α), we obtain the trajectory of a particle of the fluid; if we vary α (for fixed t) we obtain the simultaneous positions of all particles. Therefore the integral I_1 can be written

$$I_1 = \int_{\gamma_t} \boldsymbol{v} \cdot d\boldsymbol{x} = \int_{\gamma_t} v_i dx_i = \int_0^1 v_i \frac{\partial x_i}{\partial \alpha} d\alpha. \qquad (1.C.22)$$

Starting from (1.C.22) it is possible to show[27] that the necessary and sufficient condition for the validity of (1.C.20) is given by

$$v_j X_j \Big|_{\alpha=0}^{\alpha=1} + \int_{\gamma_t} \left(\frac{\partial v_i}{\partial x_j} - \frac{\partial v_j}{\partial x_i} \right) X_j \, dx_i = 0. \qquad (1.C.23)$$

Obviously, when the curve is closed, the first term vanishes and the resulting relation can be expressed by considering that the differential form

$$\left(\frac{\partial v_i}{\partial x_j} - \frac{\partial v_j}{\partial x_i} \right) X_j \, dx_i$$

must be the exact differential of a given function of x_1, x_2, ... , x_m. Before we consider the application of all this to Hamiltonian systems, it is worth looking at the other extreme case of integral invariants: when the order r of the invariant coincides with the dimensions of the space, i.e. with the order m of the system. Given a function $M = M(x_1, x_2, \ldots, x_m)$, of class C^1 and defined in the same domain of the $X_i(x_1, x_2, \ldots, x_m)$, let us define the integral

$$I_m = \int_{E_t} M \, dx_1 \, dx_2 \, \ldots \, dx_m \qquad (1.C.24)$$

over an m-dimensional region E_t with finite volume, which is the region occupied by a given whole of points at the time t. If E_0 is the region occupied by the same whole of points at $t = 0$, in order to have I_m invariant it must hold that

---

[27]See, for instance, L. A. Pars: *A Treatise on Analytical Dynamics* (Heinemann, London, 1964) Sect. 21 6.

$$\int_{E_0} M\,J\,dx_1^0\,dx_2^0\,\ldots\,dx_m^0 = \int_{E_t} M\,dx_1\,dx_2\,\ldots\,dx_m, \quad \forall t, \qquad (1.C.25)$$

which can also be rewritten as

$$\frac{dI_m}{dt} = \dot{I}_m = \int_{E_0} \frac{d}{dt}\,(M\,J)\,dx_1^0\,dx_2^0\,\ldots\,dx_m^0 = 0.$$

Therefore, the necessary and sufficient condition for I_m being an integral invariant of order m for the system (1.A.32) is given by

$$\frac{d}{dt}\,(MJ) = 0. \qquad (1.C.26)$$

By applying the Liouville lemma,[28] which states that

$$\dot{J} = \left(\frac{\partial X_1}{\partial x_1} + \frac{\partial X_2}{\partial x_2} + \ldots + \frac{\partial X_m}{\partial x_m}\right) J, \qquad (1.C.27)$$

we can rewrite condition (1.C.26) in the form

$$\sum_{i=1}^{m} \frac{\partial}{\partial x_i}\,(MX_i) = \frac{dM}{dt} + M\sum_{i=1}^{m}\frac{\partial X_i}{\partial x_i} = 0, \qquad (1.C.28)$$

(where $J \neq 0$ according to our hypotheses). The functions M satisfying (1.C.28) were called by Jacobi *multipliers* for the system (1.A.32).

The Liouville Theorem

Let us see, now, how the theory of integral invariants can be applied to Hamiltonian systems, that is, when the system (1.A.32) coincides with the system (1.C.3) of Sect. 1.9. In this case, according to (1.C.6–9),

$$X_i = \frac{\partial \mathcal{H}}{\partial x_{i+n}} = \frac{\partial \mathcal{H}}{\partial p_i}, \qquad X_{i+n} = -\frac{\partial \mathcal{H}}{\partial x_i} = -\frac{\partial \mathcal{H}}{\partial q_i},$$

where now $i = 1, 2, \ldots n = m/2$. Immediately we have

$$\sum_{k=1}^{m} \frac{\partial X_k}{\partial x_k} = \sum_{i=1}^{n} \left(\frac{\partial^2 \mathcal{H}}{\partial p_i \partial q_i} - \frac{\partial^2 \mathcal{H}}{\partial q_i \partial p_i}\right) = 0,$$

and therefore, from (1.C.28), $M = $ const, and in particular $M = 1$ is a multiplier that gives $I_m = $ const. The choice $M = 1$ makes $I_m = I_{2n}$ be the volume of a given region of phase space, and this region (defined as that containing a given number of points) keeps its volume invariant (yet changing, obviously, its geometrical shape). Going back again to the analogy with a fluid, we see that the phase space behaves like an incompressible fluid. Of course, the above conclusion may also hold true for non-Hamiltonian systems:

---

[28]See Pars, op. cit., Sect. 21.7.

Every time

$$\sum_{i=1}^{m} \frac{\partial X_i}{\partial x_i} = 0,$$

that is the vector field $\mathbf{X}$ has vanishing divergence, the phase volume is conserved. (Liouville's Theorem).

If we reconsider the linear invariants (still dealing with Hamiltonian systems as above) and the condition (1.C.23) in the case of a closed curve, we see for

$$v_i = p_i, \qquad v_{n+i} \equiv 0, \tag{1.C.29}$$

that the integral (1.C.21a) becomes

$$I_1 = \int_{\gamma_t} \mathbf{p} \cdot d\mathbf{q} = \int_{\gamma_t} \sum_{i=1}^{n} p_i dq_i, \tag{1.C.21b}$$

where the sum is written explicitly to emphasize that i goes from 1 to n only. The integrand of (1.C.23), by substitution of (1.C.29), becomes

$$\sum_{i=1}^{n} \left(\frac{\partial v_i}{\partial x_j} - \frac{\partial v_j}{\partial x_i} \right) X_j dx_i + \sum_{i=1}^{n} \left(\frac{\partial v_{n+i}}{\partial x_j} - \frac{\partial v_j}{\partial x_{n+i}} \right) X_j dx_{n+i}$$

$$= \sum_{i=1}^{n} \left(-\frac{\partial \mathcal{H}}{\partial q_i} \right) dq_i + \sum_{i=1}^{n} \left(-\frac{\partial \mathcal{H}}{\partial p_i} \right) dp_i = -d\,\mathcal{H}(q,p),$$

that is, the exact differential of a function of the q's and p's $(-\mathcal{H})$. The integral (1.C.21b) is therefore a relative linear invariant of any Hamiltonian system: it is the well-known *Poincaré linear invariant*.

The Cartan Integral Invariant

A generalization of the Poincaré invariant is due to Cartan, who introduced the quantity

$$\omega_\delta = p_i \delta q_i - \mathcal{H}\delta t. \tag{1.C.30}$$

Let us now consider the action integral (1.C.4) and more general variations than those considered before, that is, variations also implying changes at the extremes, for instance with the generalized coordinates $q_1, q_2, \ldots, q_n$, functions, besides the time t, of a parameter α. To any variation $\delta\alpha$ of α there will correspond a variation of the action[29] integral:

$$\delta I = \delta \int_{t_1}^{t_2} (p_i \dot{q}_i - \mathcal{H})\, dt = \omega_\delta \Big|_1^2 + \int_{t_1}^{t_2} \left\{ \left[\dot{q}_i - \frac{\partial \mathcal{H}}{\partial p_i} \right] \delta p_i - \left[\dot{p}_i + \frac{\partial \mathcal{H}}{\partial q_i} \right] \delta q_i \right\} dt. \tag{1.C.31}$$

---

[29]Cartan, op. cit.

Let us suppose, now, that we are considering a family of actual trajectories (i.e. corresponding to real motions of the system) depending on the parameter α and that we limit ourselves to the interval t_1, t_2 (also dependent on α). The variation of the action integral along these trajectories, owing to (1.C.31), will be

$$\delta I = \omega_\delta\big|_{t=t_2} - \omega_\delta\big|_{t=t_1}.$$

If we refer to a closed tube of trajectories, i.e. to a family of trajectories closed in on itself (see Fig. 1.17), with every trajectory limited to a time interval (t_1, t_2), the total variation of the action integral when one goes back again to the initial trajectory (going along a closed curve around the tube) obviously vanishes. Therefore, if we integrate over α, we have

$$\int \omega_\delta\big|_1 = \int \omega_\delta\big|_2,$$

and then

$$\int \omega_\delta = \int (p_i \delta q_i - \mathcal{H}\delta t) = \text{const.} \tag{1.C.32}$$

That is, given any tube of actual trajectories, the integral $\int \omega_\delta$ evaluated along a closed curve winding round the tube does not depend on the curve itself but only on the tube. $\int \omega_\delta$ is called the *Cartan integral invariant*. If, in particular, we consider a closed curve consisting of points corresponding to the same t (simultaneous states), then $\int \omega_\delta = \int p_i \delta q_i = \text{const.}$

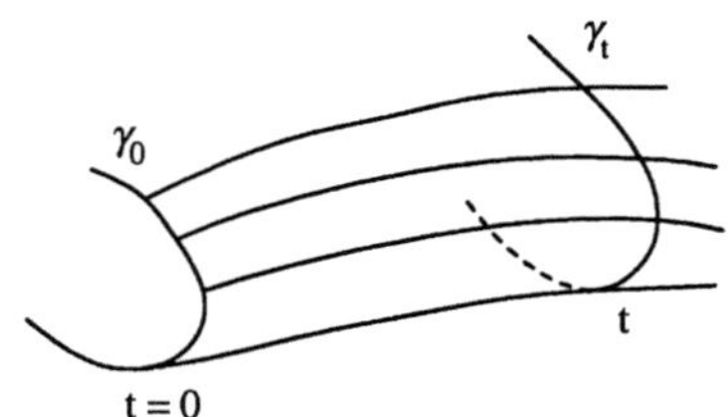

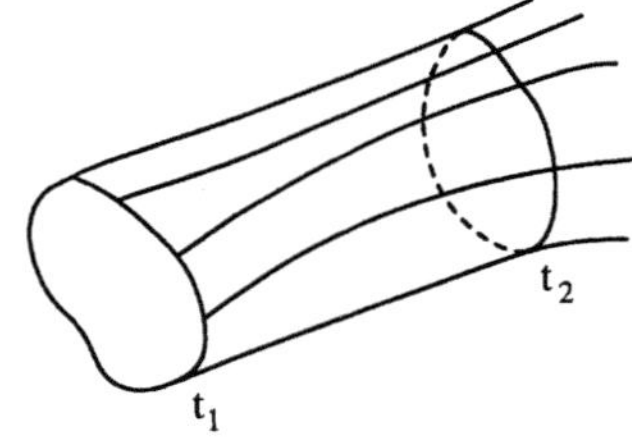

Fig. 1.16   Fig. 1.17

In this way, we recover the Poincaré integral invariant, which always retains the same value if the closed curve along which it is evaluated is made to slide along the tube (obviously, always corresponding to simultaneous states). As we have seen, a Hamiltonian system admits the Cartan invariant (1.C.32) as its integral invariant. It can be conversely shown,[30] that if a system admits the integral invariant (1.C.32), it is a Hamiltonian system. In Sect. 1.5, we showed that if one considers the time t as a Lagrangian coordinate ($t = q_0$),

---

[30]Cartan, op. cit.

the corresponding conjugate momentum is $p_0 = -\mathcal{H}$. Therefore, in this new phase space[31] with $(2n+2)$ dimensions,

$$\omega_\delta = p_\alpha \delta q_\alpha \qquad (\alpha = 0, 1, 2, \ldots, n). \qquad (1.C.33)$$

In this *extended phase space* (the name is due to Lanczos)[32] the relation between the Cartan's and Poincaré's invariants becomes more transparent.

The Poincaré Recurrence Theorem

The integral invariants are important, as we have seen, when one considers instead of one particular trajectory, a whole family of possible trajectories. Their use, as we have recalled, was introduced by Poincaré in his celebrated paper on the three-body problem; in that paper there also appeared for the first time the theorem which is today called *recurrence theorem*. The argument which is decisive in the proof of this theorem is provided by the property of the phase space which has already been codified in Liouville's theorem, i.e. the property of behaving like an incompressible fluid; in other words, the phase flow preserves the volume. Let us assume, now, that the motion of our system is confined to a bounded region of the phase space: let D be this region. Poincaré's theorem states that if Φ_t is a transformation of D into itself, that is continuous, bijective and volume preserving, then any neighbourhood U of every point of D contains a point x which comes back again to U after repeated applications of the transformation Φ_t. Let t_0 be the initial instant of time and τ an arbitrary interval $t - t_0 > 0$; then consider the subsequent transformations Φ_τ, $\Phi_{2\tau}$, $\ldots$, $\Phi_{m\tau}$, which transform the neighbourhood U to $\Phi_\tau U$, $\Phi_{2\tau} U$, $\ldots$, $\Phi_{m\tau} U$, etc. Then $\Phi_{n\tau} x \in U$ for some integer n. The theorem also entails a "stronger" version of itself. That is, not only does it happen that for almost every x in U at least one element of the sequence $\Phi_\tau x$, $\Phi_{2\tau} x$, $\ldots$ belongs to U, but also that, for almost every x in U, there are infinite values of n for which $\Phi_{n\tau} x \in U$. A rigorous demonstration of Poincaré's theorem needs the concepts of measure theory; we therefore refer the reader to specialized[33] textbooks. Of course, in such a context, instead of volume preserving transformations, one speaks of measure-preserving transformations (*Lebesgue measure*).

It should be remarked, at this point, that this theorem which appears to belong "naturally" to the subjects we have dealt with so far, is in reality of a completely different nature and can be considered the starting point of a new way of studying dynamical systems. From the "classical" point of view, the problem is that of explicitly determining the configuration of the system at

---

[31]Called "space of the states and energy" by J. L. Synge (op. cit.).

[32]C. Lanczos, *The Variational Principles of Mechanics* (Toronto University Press, 1970; Dover, 1986).

[33]See, for instance, P. R. Halmos: *Lectures On Ergodic Theory* (Chelsea, New York, 1956) p. 10.

the time t, as a function of the time and the initial conditions at $t = 0$. This, as we know, is almost never possible; Poincaré's theorem, therefore, renounces any claim to deal with individual trajectories and instead considers the global properties of all trajectories.

The Use of Integral Invariants

Let us now see what relations exist between the integral invariants and the first integrals for the system (1.A.32). Let us first examine the linear invariants and the necessary and sufficient condition (1.C.23). If we have a first integral of the autonomous system (1.A.32) given by

$$F(x_1, x_2, \ldots, x_m) = \text{const}, \tag{1.C.34}$$

putting $v_i = \partial F / \partial x_i$ and substituting in (1.C.23), we obtain

$$\dot{F}\Big|_{\alpha=0}^{\alpha=1} + \int_\gamma \left(\frac{\partial^2 F}{\partial x_i \partial x_j} - \frac{\partial^2 F}{\partial x_j \partial x_i} \right) X_j dx_i = 0.$$

That is, whatever $F = \text{const.}$ and of class C^2 satisfies (1.C.23) identically: if (1.C.34) is a first integral of our system, then $\int (\partial F / \partial x_i) dx_i$ is a linear invariant for the same system. Conversely, if $\int (\partial U / \partial x_r) dx_r$ (with $U = U(x_1, x_2, \ldots, x_m)$ and C^2) is a linear integral invariant, then still using (1.C.23),

$$\frac{\partial U}{\partial x_r} X_r \Big|_{\alpha=0}^{\alpha=1} + \int_\gamma \left(\frac{\partial^2 U}{\partial x_i \partial x_j} - \frac{\partial^2 U}{\partial x_j \partial x_i} \right) X_j dx_i = 0,$$

from which $(\partial U / \partial x_r) X_r$ is a function not depending on the x's. Therefore $dU/dt = K = \text{const.}$ and $dU/dt - K = 0$ immediately provides the first integral $U - Kt = \text{const.}$

Let us now consider the integral invariants of order m and the condition (1.C.28). If we know two of these invariants for the system and the related multipliers M_1 and M_2, substituting in (1.C.28), we have

$$\frac{dM_1}{dt} + M_1 \mathrm{div}\, \boldsymbol{X} = 0, \qquad \frac{dM_2}{dt} + M_2 \mathrm{div}\, \boldsymbol{X} = 0,$$

from which

$$\frac{1}{M_2} \frac{dM_2}{dt} - \frac{1}{M_1} \frac{dM_1}{dt} = 0$$

and also

$$\frac{M_1}{M_2} \left(\frac{1}{M_1} \frac{dM_1}{dt} - \frac{1}{M_2} \frac{dM_2}{dt} \right) = \frac{d}{dt} \left(\frac{M_1}{M_2} \right) = 0.$$

Therefore, if M_1 and M_2 are multipliers for the autonomous system (1.A.32), then M_1 / M_2 is a first integral for the same system.

1.11 Poisson Brackets and Poisson's Theorem – The Generation of New Integrals

Let us now return to Hamilton's equations and suppose that they have a first integral $F = F(q_1, q_2, \ldots, q_n; p_1, p_2 \ldots, p_n; t)$, also depending on time, to consider the most general case. Hence

$$\frac{dF}{dt} = \frac{\partial F}{\partial t} + \frac{\partial F}{\partial q_k}\dot{q}_k + \frac{\partial F}{\partial p_k}\dot{p}_k = \frac{\partial F}{\partial t} + \left(\frac{\partial F}{\partial q_k}\frac{\partial \mathcal{H}}{\partial p_k} - \frac{\partial F}{\partial p_k}\frac{\partial \mathcal{H}}{\partial q_k}\right) = 0. \quad (1.\text{C}.35\text{a})$$

The expression contained in brackets, which is the sum of n second-order Jacobian determinants, is called the *Poisson bracket* and enjoys some remarkable properties, which we shall now study. We rewrite (1.C.35a) as

$$\frac{dF}{dt} = \frac{\partial F}{\partial t} + (F, \mathcal{H}) = 0, \quad (1.\text{C}.35\text{b})$$

having put

$$(F, \mathcal{H}) = \sum_{k=1}^{n} \frac{\partial(F, \mathcal{H})}{\partial(q_k, p_k)}.$$

In general, if u and v are two functions (which we shall assume are C^2) of the q's and p's and of t, then

$$(u, v) = \sum_{k=1}^{n} \frac{\partial(u, v)}{\partial(q_k, p_k)} = \left(\frac{\partial u}{\partial q_k}\frac{\partial v}{\partial p_k} - \frac{\partial u}{\partial p_k}\frac{\partial v}{\partial q_k}\right). \quad (1.\text{C}.36\text{a})$$

The time derivative of any function, evaluated along a solution of the system, will be written, for the Hamiltonian systems, as in (1.C.35b), and Hamilton's equations themselves will become

$$\dot{q}_k = (q_k, \mathcal{H}), \qquad \dot{p}_k = (p_k, \mathcal{H}). \quad (1.\text{C}.37)$$

Owing to the definitions we have introduced in (1.C.7) and (1.C.8), (1.C.36a) can be written as

$$(u, v) = \left(\frac{\partial u}{\partial z}\right) J \left(\frac{\partial v}{\partial z}\right)^T. \quad (1.\text{C }36\text{b})$$

The following properties are easily checked:

$$(u, v) = -(v, u), \quad (u, u) = 0 : \text{antisymmetry};$$
$$(c, u) = 0, \qquad\qquad \text{if } c = \text{const.};$$
$$\frac{\partial}{\partial t}(u, v) = \left(\frac{\partial u}{\partial t}, v\right) + \left(u, \frac{\partial v}{\partial t}\right); \qquad\qquad (1.\text{C}.38)$$

but checking the so-called *Jacobi identity* is a little more arduous

$$(u, (v, w)) + (v, (w, u)) + (w, (u, v)) = 0, \quad (1.\text{C}.39)$$

where u, v, w are three functions of class C^2.

Owing to (1.C.35b), if $F_1(\boldsymbol{q}, \boldsymbol{p}, t)$ and $F_2(\boldsymbol{q}, \boldsymbol{p}, t)$ are two first integrals of Hamilton's equations,

$$\frac{\partial F_1}{\partial t} + (F_1, \mathcal{H}) = 0, \qquad \frac{\partial F_2}{\partial t} + (F_2, \mathcal{H}) = 0. \tag{1.C.40}$$

Let us now evaluate

$$\frac{\partial}{\partial t}(F_1, F_2) = \left(\frac{\partial F_1}{\partial t}, F_2\right) + \left(F_1, \frac{\partial F_2}{\partial t}\right)$$
$$= -\big((F_1, \mathcal{H}), F_2\big) - \big(F_1, (F_2, \mathcal{H})\big),$$

where we have used (1.C.40). Then, by using (1.C.38),

$$\frac{\partial}{\partial t}(F_1, F_2) + \big((F_1, F_2), \mathcal{H}\big) = \big(F_2, (F_1, \mathcal{H})\big) + \big(F_1, (\mathcal{H}, F_2)\big) + \big(\mathcal{H}, (F_2, F_1)\big) = 0,$$

owing to the Jacobi identity (1.C.39). Therefore we have obtained

$$\frac{\partial}{\partial t}(F_1, F_2) + \big((F_1, F_2), \mathcal{H}\big) = 0;$$

that is, if F_1 and F_2 are two first integrals, their Poisson bracket (F_1, F_2) is also a first integral (*Poisson's theorem*). At first sight, Poisson's theorem appears to be a concrete possibility of generating new first integrals starting from two that are already known and of generating them in such a number as to allow the integration of the system. As a matter of fact, all is not so rosy as one might think; first of all, since the Poisson bracket with the Hamiltonian of any first integral not containing the time vanishes, in this case we need two first integrals different from the total energy and mutually independent, and we are seldom able to satisfy such a condition. Secondly, in the case of an integral depending on time, $(F, \mathcal{H}) = -\partial F/\partial t$ will be the new integral, but for autonomous systems this result can be obtained directly; obviously the successive derivatives $\partial^2 F/\partial t^2$, $\partial^3 F/\partial t^3$, ... are also first integrals; however, they are not necessarily mutually independent.

Nevertheless, sometimes it happens that new integrals that do not depend on those already found can be generated. As an example of this, let us consider the planar oscillator which has already been studied. With $p_1 = \dot{q}_1$, $p_2 = \dot{q}_2$, the Hamiltonian of the isotropic planar oscillator will be

$$\mathcal{H} = \frac{1}{2}\left(p_1^2 + p_2^2\right) + \frac{1}{2}\omega^2\left(q_1^2 + q_2^2\right). \tag{1.C.41}$$

The energies of the two modes being separately conserved, we can choose their difference

$$B = \frac{1}{2}\left(p_1^2 + \omega^2 q_1^2\right) - \frac{1}{2}\left(p_2^2 + \omega^2 q_2^2\right) \tag{1.C.42}$$

as an integral independent of $\mathcal{H}$. Moreover, owing to the rotational symmetry, the angular momentum immediately gives another integral independent of B and $\mathcal{H}$:

$$L = q_1 p_2 - q_2 p_1. \tag{1.C.43}$$

Their Poisson bracket will be

$$C = (L, B) = 2\,p_1 p_2 + 2\,\omega^2 q_1 q_2 = 4\,A_{12} = 2\,\omega^2 ab\,\cos\hat{c},$$

where we refer to (1.B.75) for the definition of A_{12}. Therefore C is really a new integral, independent of the preceding ones. If we go on to search for new integrals, we find

$$(C, B) = 4\,\omega^2(q_2 p_1 - q_1 p_2) = -4\,\omega^2 L,$$
$$(C, L) = 2\big[(p_1^2 - p_2^2) + \omega^2(q_1^2 - q_2^2)\big] = 4\,B;$$

that is, no further new integral can be found in this way.

For Hamiltonian systems, the lowering of the order by means of first integrals goes in the same way as for Lagrangian systems: the reduced system is of the same type (Hamiltonian) as the original system. Therefore, rather than repeat that type of procedure, we shall deal with a separate case, i.e. the reduction which can be obtained by means of the energy integral. The procedure has been described as the *elimination of the time* and is important in the n-body problem in celestial mechanics. We have seen in the preceeding section, dealing with the Cartan invariant, how the differential form

$$p_1\,dq_1 + p_2\,dq_2 + \ldots + p_n\,dq_n - \mathcal{H}\,dt \tag{1.C.44}$$

is associated with a Hamiltonian system and in what way the one determines the other.

If we have an autonomous system, $\mathcal{H}$ will not contain the time explicitly, and then $\mathcal{H} = h = \text{const.}$, or also (with $p_0 = -\mathcal{H}$),

$$h + p_0 = 0. \tag{1.C.45}$$

Solving (1.C.45) with regard to p_1, we can write

$$K\,(p_2, p_3, \ldots, p_n;\ q_1, q_2, \ldots, q_n;\ p_0) + p_1 = 0. \tag{1.C.46}$$

The differential form associated with the system is

$$p_1\,dq_1 + p_2\,dq_2 + \ldots + p_n\,dq_n + p_0\,dt,$$

with the $2n + 2$ variables

$$p_1, p_2, \ldots, p_n;\ q_1, q_2, \ldots, q_n;\ p_0,\ t$$

satisfying (1.C.46). Therefore we can also write

$$p_2 dq_2 + p_3 dq_3 + \ldots + p_n dq_n + p_0 dt - K(p_2, p_3, \ldots, p_n;\ q_1, q_2, \ldots, q_n;\ p_0)dq_1$$

and consider

$$p_2, \ldots, p_n;\ q_1, q_2, \ldots, q_n;\ p_0,\ t$$

as the $2n + 1$ variables But the system corresponding to this form will be

$$\frac{dq_i}{dq_1} = \frac{\partial \mathcal{K}}{\partial p_i}, \qquad \frac{dp_i}{dq_1} = -\frac{\partial \mathcal{K}}{\partial q_i},$$
$$\frac{dt}{dq_1} = \frac{\partial \mathcal{K}}{\partial p_0}, \qquad \frac{dp_0}{dq_1} = -\frac{\partial \mathcal{K}}{\partial t} \equiv 0 \qquad (i = 2, 3, \dots, n). \tag{1.C.47}$$

The last two equations are separate from the other $2n - 2$, since these do not contain t, and p_0 is constant. Therefore the outcome is that, from a Hamiltonian system with n degrees of freedom, we have passed, eliminating time, to another one with $n - 1$ degrees of freedom:

$$\frac{dq_i}{dq_1} = \frac{\partial \mathcal{K}}{\partial p_i}, \qquad \frac{dp_i}{dq_1} = -\frac{\partial \mathcal{K}}{\partial q_i} \qquad (i = 2, 3, \dots, n). \tag{1.C.48}$$

So that the procedure is advantageous, $\mathcal{K}$ must be a one-valued function for the correspondence between the values of q_1 (which plays the role of a parameter) and the points of the trajectory in the configuration space to be one to one. Clearly, the system (1.C.48) is a system which determines the trajectory but does not tell us when the moving point is at a certain position on the trajectory itself. This must be deduced from the last two equations of (1.C.47).

1.12 Canonical Transformations

Obviously, the integration of the $2n$ canonical equations (1.C.3) represents the central problem of the Hamiltonian theory; since general rules do not exist which can be applied to integrate (1.C.3), it is convenient to investigate if first integrals exist (which enable us to lower the order of the system) or to proceed in performing variable transformations which make the system simpler and then easier to be integrated. It is clear that the second issue implies the first, as we have already seen speaking about transformations of variables which make one or more variables ignorable. An extreme case is when $\mathcal{H}$ depends only on half of the canonical variables (and all of the same type) and not on the time; suppose that the variables in question are the p_k's: $\mathcal{H} = \mathcal{H}(p_1, p_2, \dots, p_n)$. We have immediately

$$\dot{p}_i = -\frac{\partial \mathcal{H}}{\partial q_i} = 0, \qquad \dot{q}_i = \frac{\partial \mathcal{H}}{\partial p_i} = \text{const.} = \omega_i,$$

that is,

$$q_i = \omega_i\, t + \delta_i, \qquad p_i = \text{const.} = \alpha_i, \qquad \forall i,$$

where α_i and δ_i are integration constants. Therefore, in this way, (1.C.3) are immediately integrated. We shall deal with this case, which is of great importance in applications, in a subsequent section. If we now consider, in general, the problem of "simplifying" the system (1.C.3) by means of a variable transformation, the first question is of what type must the transformation be to it be meaningful. It seems quite natural to impose the condition that the new variables must also be canonical so that one can write new equations in place of (1.C.3) which in turn are canonical.

According to a convention introduced by Whittaker, we shall call Q_i, P_i the new canonical variables and $\mathcal{K}$ the new Hamiltonian:

$$
\begin{aligned}
Q_i &= Q_i(q_1, q_2, \ldots, q_n;\ p_1, p_2, \ldots, p_n;\ t), \\
P_i &= P_i(q_1, q_2, \ldots, q_n;\ p_1, p_2, \ldots, p_n;\ t).
\end{aligned}
\tag{1.C.49}
$$

The condition which must be satisfied, so that Q_i, P_i may be canonical variables, is obtained by again imposing the stationarity of the action integral (1.C.4), but with the integrand now given by

$$
\sum_{k=1}^{n} P_k \dot{Q}_k - \mathcal{K}
\tag{1.C.50}
$$

instead of (1.C.2b). A necessary and sufficient condition for having this is that the two integrands (1.C.2b) and (1.C.50) differ from one another by the total time derivative of a function W depending on $2n$ among the new and old variables and the time.[54] In fact, in this way, dW will be an exact differential, its integral will depend only on the end points and not on the path of integration, and if $W = W(q_1, q_2, \ldots, q_n;\ Q_1, Q_2, \ldots, Q_n;\ t)$ its variation will vanish at the end points. Therefore, one must look for an arbitrary function $W = W(q_1, q_2, \ldots, q_n;\ Q_1, Q_2, \ldots, Q_n;\ t)$ which satisfies the condition

$$
\sum_k p_k\,\dot{q}_k - \mathcal{H}(q_i, p_i, t) = \sum_k P_k\,\dot{Q}_k - \mathcal{K}(Q_i, P_i, t) + \frac{dW(q_i, Q_i, t)}{dt},
\tag{1.C.51}
$$

that is,

$$
\begin{aligned}
\sum_k p_k\,dq_k - \mathcal{H}(q_i, p_i, t)\,dt &= \sum_k P_k\,dQ_k - \mathcal{K}(Q_i, P_i, t)dt + dW(q_i, Q_i, t) \\
&= \sum_k P_k\,dQ_k - \mathcal{K}(Q_k, P_k, t)dt + \sum_k \left(\frac{\partial W}{\partial q_k}\,dq_k + \frac{\partial W}{\partial Q_k}\,dQ_k \right) - \frac{\partial W}{\partial t}dt.
\end{aligned}
$$

For (1.C.51) to be satisfied, it must be the case that:

$$
p_k = \frac{\partial W(q_i, Q_i, t)}{\partial q_k}, \quad P_k = -\frac{\partial W(q_i, Q_i, t)}{\partial Q_k}, \quad \mathcal{K} = \mathcal{H} + \frac{\partial W(q_i, Q_i, t)}{\partial t}.
\tag{1.C.52}
$$

By inverting the second group of equations in (1.C.52), we can obtain $q_k = q_k(Q_i, P_i, t)$, and, by substituting in the first group, $p_k = p_k(Q_i, P_i, t)$;

---

[54] As a rule, the function W must depend on $(4n + 1)$ variables, i.e. on q, p, Q, P and the time; since (1.C.49) provide $2n$ relations among them, the actual dependence is on $2n$ canonical variables and the time.

therefore (1.C.52) constitute the $2n$ equations of transformation. For the sketched procedure to be possible, the P_k must satisfy the conditions required by the implicit function theorem, i.e.

$$\left| \frac{\partial^2 W(q_i, Q_i, t)}{\partial Q_k\, \partial q_i} \right| \neq 0. \tag{1.C.53}$$

Obviously, this condition also guarantees the existence of the inverse transformation. In fact, starting now from the first of (1.C.52) we can obtain the Q_i's as $Q_i(q_k, p_k, t)$, with the Jacobian (1.C.53) $\neq 0$, and, by substituting in the second one, have the P_i's as $P_i(q_k, p_k, t)$. Moreover, it should be remarked that $W = W(q_i, Q_i, t)$ is not defined on a set belonging to the phase space but on a set of the Cartesian product of the two configuration spaces, $\mathbb{R}_q^n \times \mathbb{R}_Q^n$. As a further remark, note that (1.C.51) or (1.C.52) does not define the most general canonical transformation; in fact, the most general form is obtained with the left-hand side given by $c\left(\sum_k p_k\, dq_k - \mathcal{H}\, dt\right)$, where c is an arbitrary constant. Usually, only the case $c = 1$ is taken into account. The function W is called the generating function of the canonical transformation: in fact, it is enough to know only this function (instead of $2n$ functions $q_i = q_i(Q_k, P_k, t)$ and $p_i = p_i(Q_k, P_k, t)$) to be able to write by means of (1.C.52) the new canonical equations.

The condition (1.C.51) refers to the most general canonical transformation in which the new coordinates and the new conjugate momenta depend, not only on the old canonical variables, but also on the time; if we consider it for a fixed value $\bar{t}$ of the variable t, it will be

$$\sum_k p_k\, dq_k = \sum_k P_k\, dQ_k + dW(q, Q, \bar{t}). \tag{1.C.54}$$

Equation (1.C.54) is therefore the condition which must be satisfied for a transformation independent of time to be canonical. Conversely, if we consider (1.C.54) for various given values of the parameter $\bar{t}$ (and so for every value of $\bar{t}$ the transformation is canonical) and consider the function $\mathcal{K}$ defined by

$$\mathcal{K} = \mathcal{H} + \frac{\partial W}{\partial t}, \tag{1.C.55}$$

then, summing (1.C.54) and (1.C.55), we again get the condition (1.C.51). This shows that the necessary and sufficient condition for a transformation depending on time to be canonical is that all the transformations not depending on time, obtained by (1.C.51) by replacing t by an arbitrary $\bar{t}$, are canonical. Therefore, to set up rules for deciding if a given transformation is canonical, we can limit ourselves to considering transformations not explicitly dependent on time. Therefore, it is enough to verify if the difference $\sum_i p_i\, dq_i - \sum_i P_i\, dQ_i$ is an exact differential. In other words, the necessary

and sufficient condition for (1.C.49) to represent[35] a canonical transformation consists in the invariance of the integral $\oint \sum_i p_i \, dq_i$ evaluated along any
closed curve, that is,

$$\oint \sum_i p_i \, dq_i = \oint \sum_i P_i \, dQ_i = \text{const.} \qquad (1.C.56)$$

The question now is to find for (1.C.56) equivalent formulations which are
more directly applicable in actual cases. If we consider the integral $\oint \sum_i p_i \, dq_i$
along any closed curve in the $2n$-dimensional phase space, we have, obviously,

$$\oint \sum_i p_i \, dq_i = \sum_{i=1}^{n} \oint p_i \, dq_i.$$

Consider then, for any given i, the line integral $\oint p_i \, dq_i$ along the closed
curve resulting from the projection on the $q_i p_i$ plane of the curve in the $2n$-
dimensional phase space. It represents, as we know, the oriented area bounded
by the curve itself and therefore, if we introduce a system of orthogonal
curvilinear coordinates (u_i, v_i) in the $q_i p_i$ plane, it will also be

$$\oint p_i \, dq_i = \int_{S_i} dp_i \, dq_i = \int_{S_i} \left(\frac{\partial q_i}{\partial u_i} \frac{\partial p_i}{\partial v_i} - \frac{\partial p_i}{\partial u_i} \frac{\partial q_i}{\partial v_i} \right) du_i \, dv_i,$$

where the term within the brackets is the Jacobian determinant of the transformation $(q_i, p_i) \to (u_i, v_i)$. Denoting by δ the increment that the canonical
variables undergo when only v varies and by d that due to the variation of
u only, the last integrand can also be written $(\delta p_i \, dq_i - dp_i \, \delta q_i)$ and then in
addition

$$\sum_{i=1}^{n} \left(\frac{\partial q_i}{\partial u_i} \frac{\partial p_i}{\partial v_i} - \frac{\partial p_i}{\partial u_i} \frac{\partial q_i}{\partial v_i} \right) du_i \, dv_i = \sum_{i=1}^{n} (\delta p_i \, dq_i - dp_i \, \delta q_i). \qquad (1.C.57)$$

The right-hand side in (1.C.57) represents a bilinear differential form to which
we shall very soon return. By repeating our procedure backwards, the integral
becomes

$$\oint_\Gamma \sum_i p_i dq_i = \sum_i \oint p_i dq_i = \sum_i \int_{S_i} \left(\frac{\partial q_i}{\partial u_i} \frac{\partial p_i}{\partial v_i} - \frac{\partial p_i}{\partial u_i} \frac{\partial q_i}{\partial v_i} \right) du_i \, dv_i$$

$$= \int_S \sum_i \left(\frac{\partial q_i}{\partial u_i} \frac{\partial p_i}{\partial v_i} - \frac{\partial p_i}{\partial u_i} \frac{\partial q_i}{\partial v_i} \right) du_i \, dv_i.$$

$$(1.C.58)$$

The integral in the left-hand side must remain invariant under the transformation $(q, p) \to (Q, P)$, whatever the closed curve Γ (and then the surface

---

[35]These arguments have a local validity, i.e. "in small", and leave the topology of
   the phase space out of consideration.

S having Γ as boundary) may be; from this the invariance of the integrand follows and therefore also the invariance of

$$[u_i, v_i] = \sum_{i=1}^{n} \left(\frac{\partial q_i}{\partial u_i} \frac{\partial p_i}{\partial v_i} - \frac{\partial p_i}{\partial u_i} \frac{\partial q_i}{\partial v_i} \right), \tag{1.C.59}$$

where we have introduced a new symbol called the *Lagrange bracket* of u_i and v_i, which enjoys the property

$$[u_i, \, v_i] = -[u_i, \, v_i]. \tag{1.C.60}$$

From what was said above, it follows that this is equivalent to the invariance of the bilinear differential form

$$\sum_{i=1}^{n} (\delta p_i \, dq_i - dp_i \, \delta q_i). \tag{1.C.61}$$

In Lagrangian dynamics, as we have seen, it appeared quite natural to define, by means of the kinetic energy, a distance and then the ds^2 of a Riemannian metric in the configuration space. Equation (1.C.61) shows that, in the phase space of Hamiltonian dynamics, the invariant quantity is no longer the square of a distance, but an area.

For establishing if a transformation is canonical or not, therefore we have now the rule:

The transformations from the variables (q_i, p_i) to (Q_i, P_i) are canonical if they leave the Lagrange bracket (1.C.59) *invariant, whatever the dependence of the q_i's, p_i's, on u and v may be.*

As the parameters u and v are quite arbitrary, we can choose for them any pair Q_i, Q_k or Q_i, P_k or P_i, P_k of the new variables and then construct the Lagrange brackets. If we do the same in the new variables, all the Q_i's and P_i's in this case being mutually independent, we obtain

$$[Q_i, Q_k] = 0, \qquad [P_i, P_k] = 0, \qquad [Q_i, P_k] = \delta_{ik}. \tag{1.C.62}$$

Therefore (1.C.62) provides a necessary and sufficient condition for a transformation $(q_i, p_i) \rightarrow (Q_i, P_i)$ to be canonical. When the canonical transformation depends explicitly on the time, the conditions (1.C.62) must hold true for any t.

Now, exploiting the matrix form of the canonical equations and the properties of the matrix $\boldsymbol{J}$ we have already brought in (see Sect. 1.9), we will obtain another equivalent condition to characterize a canonical transformation. From (1.C.6–8), it is evident that

$$[u, v]_z = \left(\frac{\partial \boldsymbol{z}}{\partial u} \right)^T \boldsymbol{J} \left(\frac{\partial \boldsymbol{z}}{\partial v} \right), \tag{1.C.63}$$

where we have denoted by $[u, v]_z$ the Lagrange bracket related to the old variables. Denoting by $[u, v]_Z = (\partial Z/\partial u)^T J (\partial Z/\partial v)$ that related to the new ones, we must have, owing to the invariance,

$$\left(\frac{\partial z}{\partial u}\right)^T J \left(\frac{\partial z}{\partial v}\right) = \left(\frac{\partial Z}{\partial u}\right)^T J \left(\frac{\partial Z}{\partial v}\right).$$

But if we call $M = (\partial Z/\partial z)$ the Jacobian matrix of the transformation, then

$$\left(\frac{\partial Z}{\partial u}\right) = M \left(\frac{\partial z}{\partial u}\right), \qquad \left(\frac{\partial Z}{\partial v}\right) = M \left(\frac{\partial z}{\partial v}\right).$$

Therefore

$$\left(\frac{\partial Z}{\partial u}\right)^T J \left(\frac{\partial Z}{\partial v}\right) = \left(\frac{\partial z}{\partial u}\right)^T M^T J M \left(\frac{\partial z}{\partial v}\right),$$

and, hence, the invariance is verified if the Jacobian matrix obeys the condition

$$M^T J M = J. \tag{1.C.64}$$

A matrix satisfying (1.C.64) is called *symplectic*. Consequently, we can express the necessary and sufficient condition for a transformation to be a canonical one if and only if the Jacobian matrix of the transformation is symplectic. Reconsider now the definition (1.C.36b) of the Poisson brackets and exploit the matrix formalism to see what the relationship is between (u, v) and the Lagrange bracket $[u, v]$. To do this, it is convenient to introduce two $m \times m$, with $m = 2n$, square matrices:

$$[\boldsymbol{u}, \boldsymbol{u}] = \begin{pmatrix} [u_1, u_1] & \cdots & [u_1, u_m] \\ [u_2, u_1] & \cdots & [u_2, u_m] \\ \vdots & \vdots & \vdots \\ [u_m, u_1] & \cdots & [u_m, u_m] \end{pmatrix}, \ (\boldsymbol{u}, \boldsymbol{u}) = \begin{pmatrix} (u_1, u_1) & \cdots & (u_1, u_n) \\ (u_2, u_1) & \cdots & (u_2, u_n) \\ \vdots & \vdots & \vdots \\ (u_m, u_1) & \cdots & (u_m, u_m) \end{pmatrix}$$

which we call the Lagrange matrix and Poisson matrix respectively and which have as a generic element, whose position in the array is labelled by i and j, the corresponding bracket between u_i and u_j. The $2n$ functions $u_1, u_2, \ldots, u_m$ must be considered arbitrary functions of $z_1, z_2, \ldots, z_m$. If we evaluate the product $[\boldsymbol{u}, \boldsymbol{u}] (\boldsymbol{u}, \boldsymbol{u})^T$, we obtain

$$[\boldsymbol{u}, \boldsymbol{u}](\boldsymbol{u}, \boldsymbol{u})^T = \left(\frac{\partial z}{\partial u}\right)^T J \left(\frac{\partial z}{\partial u}\right) \left(\frac{\partial u}{\partial z}\right)^T J^T \left(\frac{\partial u}{\partial z}\right)$$

$$= \left(\frac{\partial z}{\partial u}\right)^T J J^T \left(\frac{\partial u}{\partial z}\right) = 1, \tag{1.C.65a}$$

where we have used (1.C.11) and the properties of the matrix product. Equation (1.C.65a) can be rewritten as

$$\sum_{k=1}^{m} [u_i, u_k]\,(u_j, u_k) = \delta_{ij} \qquad (i, j = 1, 2, \ldots, m = 2n). \qquad (1.C.65b)$$

Therefore, between the Poisson and Lagrange brackets a reciprocity relationship exists: this means that the invariance of Lagrange brackets entails the invariance of Poisson brackets and vice versa. That is, (1.C.62) has as a consequence that

$$(Q_i, Q_k) = 0, \qquad (P_i, P_k) = 0, \qquad (Q_i, P_k) = \delta_{ik} \qquad (1.C.66)$$

and vice versa.

1.13 Generating Functions – Infinitesimal Canonical Transformations

In (1.C.51) and (1.C.52) in Sect. 1.12 we have chosen for the function W the form $W = W(q_i, Q_i, t)$, that is we have chosen the q_i's and Q_i's (among the $4n$ variables available) as the $2n$ independent variables; canonical transformations of this type are called *free canonical transformations*, thus emphasizing that the Q_i's can be fixed *independently* of the q_i's. It is clear that a transformation like the identical transformation $Q_i = q_i$ ($i = 1, \ldots, n$) could not be included among the transformations of this type; therefore the problem exists of listing all the possible classes of generating functions. Since there are four groups of variables, (q_i, p_i, Q_i, P_i), we want to pair, in such a way to have in each pair a "new" group and an "old" one, there are also four possibilities:

$$W_1(q_i, Q_i, t), \quad W_2(q_i, P_i, t), \quad W_3(p_i, Q_i, t), \quad W_4(p_i, P_i, t). \qquad (1.C.67)$$

The generating function so far employed in (1.C.51–53) in Sect. 1.12 is $W_1(q_i, Q_i, t)$. If we start again from (1.C.51), we can write

$$\sum_k p_k dq_k - \mathcal{H}dt = \sum_k P_k dQ_k - \mathcal{K}dt + dW_1$$

$$= -\sum_k Q_k dP_k - \mathcal{K}dt + d\left(W_1 + \sum_k Q_k P_k\right),$$

from which

$$\sum_k p_k dq_k + \sum_k Q_k dP_k + (\mathcal{K} - \mathcal{H})dt = d\left(W_1 + \sum_k Q_k P_k\right).$$

If we now define

$$W_2(q_i, P_i, t) = W_1(q_i, Q_i, t) + \sum_k Q_k P_k, \qquad (1.C.68)$$

we have

$$\sum_k p_k dq_k + \sum_k Q_k dP_k + (\mathcal{K} - \mathcal{H})dt$$
$$= \sum_k \frac{\partial W_2}{\partial q_k} dq_k + \sum_k \frac{\partial W_2}{\partial P_k} dP_k + \sum_k \frac{\partial W_2}{\partial t} dt, \tag{1 C.69}$$

and then

$$p_k = \frac{\partial W_2}{\partial q_k}, \qquad Q_k = \frac{\partial W_2}{\partial P_k}, \qquad \mathcal{K} = \mathcal{H} + \frac{\partial W_2}{\partial t}. \tag{1.C.70}$$

Equation (1.C.68) is a Legendre transformation (see the definition in Sect. 1.9) which, together with (1.C.70), defines the canonical transformation generated by $W_2(q_i, P_i, t)$. The above-mentioned identical transformation belongs to this class. In fact, if we consider

$$W_2(q_i, P_i, t) = \sum_i q_i\, P_i, \tag{1.C.71}$$

from (1.C.70) we have

$$p_i = P_i, \qquad Q_i = q_i, \qquad \mathcal{K} = \mathcal{H}.$$

By means of analogous Legendre transformations, we can obtain

$$W_3(p_i, Q_i, t) = W_1(q_i, Q_i, t) - \sum_k p_k q_k,$$
$$W_4(p_i, P_i, t) = W_1(q_i, Q_i, t) + \sum_k Q_k P_k - \sum_k p_k q_k. \tag{1.C.72}$$

We summarize in the table below relations (1.C.52), (1.C.70) and those regarding W_3 and W_4 defined in (1.C.72). The three determinants, analogous to (1.C.53), will never vanish.

Generating functions	q_i	p_i	Q_i	P_i	$\mathcal{K} - \mathcal{H}$
$W_1(q_i, Q_i, t)$		$\frac{\partial W_1}{\partial q_i}$		$-\frac{\partial W_1}{\partial Q_i}$	$\frac{\partial W_1}{\partial t}$
$W_2(q_i, P_i, t)$		$\frac{\partial W_2}{\partial q_i}$	$\frac{\partial W_2}{\partial P_i}$		$\frac{\partial W_2}{\partial t}$
$W_3(p_i, Q_i, t)$	$-\frac{\partial W_3}{\partial p_i}$			$-\frac{\partial W_3}{\partial Q_i}$	$\frac{\partial W_3}{\partial t}$
$W_4(p_i, P_i, t)$	$-\frac{\partial W_4}{\partial p_i}$		$\frac{\partial W_4}{\partial P_i}$		$\frac{\partial W_4}{\partial t}$

It is evident that it is possible to generate the identical transformation also by $W_3(p_i, Q_i, t) = -\sum_k p_k Q_k$, from which, in fact, one obtains

$$q_k = Q_k, \qquad P_k = p_k, \qquad \mathcal{K} = \mathcal{H}.$$

Other bilinear generating functions are

$$W_1 = \sum_k q_k Q_k, \qquad W_4 = \sum_k p_k P_k,$$

which exchange the q's for the p's in an asymmetric way:

$$Q_i = p_i, \qquad P_i = -q_i.$$

A worthy particular case is when the transformation is merely a "point transformation", i.e. when the new coordinates Q_i are functions only of the q_i's, without the presence of the conjugate momenta: this can be obtained by means of transformations of the type $W_2 = \sum_{k=1}^{n} f_k(q_j, t) P_k$, which give $Q_i = f_i(q_j, t)$.

Infinitesimal Canonical Transformations

Consider now a transformation "close" to the identical one, i.e. which differs from it by an infinitesimal amount, in the sense that the variation of the q_i's and the p_i's is infinitely small. Its generating function will then differ infinitely little from the one of the identical transformation.

Therefore we shall write it in the form

$$\sum_i q_i P_i + \varepsilon \overline{W}_2(q_i, P_i), \tag{1.C.73}$$

where ε is a first-order infinitesimal and $\overline{W}_2$ is a differentiable function. By applying (1.C.70), we obtain

$$p_i = P_i + \varepsilon \frac{\partial \overline{W}_2}{\partial q_i}, \qquad Q_i = q_i + \varepsilon \frac{\partial \overline{W}_2}{\partial P_i}. \tag{1.C.74}$$

Equations (1.C.74) define an "infinitesimal canonical transformation" depending on the parameter ε; neglecting second-order terms, from these equations we also obtain

$$Q_i = q_i + \varepsilon \frac{\partial \overline{W}_2}{\partial p_i}, \qquad P_i = p_i - \varepsilon \frac{\partial \overline{W}_2}{\partial q_i}. \tag{1.C.75}$$

If we now rewrite (1.C.75) as

$$Q_i - q_i = \delta q_i = \varepsilon \frac{\partial \overline{W}_2}{\partial p_i}, \qquad P_i - p_i = \delta p_i = -\varepsilon \frac{\partial \overline{W}_2}{\partial q_i}$$

and consider the case in which $\varepsilon = \delta t$ and $\overline{W}_2 = \mathcal{H}$, we have

$$\delta q_i = \delta t \frac{\partial \mathcal{H}}{\partial p_i}, \qquad \delta p_i = -\delta t \frac{\partial \mathcal{H}}{\partial q_i};$$

that is, we recover the canonical equations written in terms of differentials. This means that the infinitesimal canonical transformation having $\sum q_i P_i + \delta t \mathcal{H}$ as a generating function (and therefore the infinitesimal time δt as a parameter) is given by the evolution of the system in the infinitesimal time δt.

It can be shown that, by performing several infinitesimal canonical transformations, one again has a canonical transformation and hence, at last, the conclusion that a finite motion of our system consists of a sequence of an infinite number of infinitesimal transformations generated by the Hamiltonian.

1.14 The Extended Phase Space

When defining the phase space for a system with n degrees of freedom, one considers the conjugate canonical variables (q_k, p_k) as dynamical variables and the time t as an independent variable. However, sometimes it is convenient (and in certain cases necessary on principle, as for instance in the theory of relativity) to consider also the variable t as a dynamical variable, and then one describes the evolution of the system by means of the variation of a parameter w, instead of the time. We have already treated, in part, the problem by introducing in Lagrangian dynamics the space of events (see Sect. 1.5) and the extended phase space for Hamiltonian dynamics (Sect. 1.10). We want to discuss just the extended phase space here in more detail. As we have seen, if the time t were considered a coordinate q_0, its conjugate momentum would be given by $p_0 = -\mathcal{H}$, and one would have to handle a $2n+2$-dimensional space: the extended phase space. Therefore, the coordinates would be given by $q_0 = t$, q_1, q_2, ..., q_n and the momenta by $p_0 = -\mathcal{H}$, p_1, p_2, ..., p_n. For greater clarity, it is convenient to use different symbols:

$$x_0 = t, \qquad x_i = q_i, \qquad y_0 = -\mathcal{H}, \qquad y_i = p_i \tag{1.C.76}$$

and Greek letters for the indexes running from zero to n; in this way the extended phase space will have coordinates x_ϱ (with $\varrho = 0, 1, 2, \ldots, n$) and momenta y_ϱ (with $\varrho = 0, 1, 2, \ldots, n$).

The action integral becomes

$$I = \int_{w_1}^{w_2} \mathcal{L}\left(x_0, x_1, \ldots, x_n; \frac{x_1'}{x_0'}, \frac{x_2'}{x_0'}, \ldots, \frac{x_n'}{x_0'}\right) x_0' \, dw,$$

where we have replaced the dependence on t by the dependence on a parameter w: $t = x_0 = x_0(w)$, $x_i = x_i(t(w))$, etc. (see (1.B.30) for the method used). We have, in this case, as Lagrangian

$$\Lambda = \mathcal{L} x_0' = \mathcal{L}\left(x_0, x_1, \ldots, x_n; \frac{x_1'}{x_0'}, \frac{x_2'}{x_0'}, \ldots, \frac{x_n'}{x_0'}\right) x_0'.$$

This is a homogeneous function of first degree in the $n+1$ variables $x_0', x_1', \ldots, x_n'$ and, owing to Euler's theorem, we also have

$$\Lambda = \sum_{\varrho=0}^{n} \frac{\partial \Lambda}{\partial x'_\varrho}\, x'_\varrho = \sum_{\varrho=0}^{n} y_\varrho\, x'_\varrho.$$

As a consequence, the corresponding Hamiltonian, which we shall call $\mathcal{K}$, will be equal to zero:

$$\mathcal{K} = \sum_{\varrho=0}^{n} y_\varrho\, x'_\varrho - \Lambda \equiv 0.$$

The action integral, therefore, will be reduced to the completely symmetric form:

$$I = \int_{w_1}^{w_2} \Lambda dw = \int_{w_1}^{w_2} \left(\sum_{\varrho=0}^{n} y_\varrho x'_\varrho - \mathcal{K} \right) dw = \int_{w_1}^{w_2} \sum_{\varrho=0}^{n} y_\varrho x'_\varrho dw, \qquad (1.\mathrm{C}.77)$$

which does not contain any Hamiltonian. In the integral (1.C.77), which must be considered evaluated along a curve parametrized by w, the y_ϱ's must be given in conformity with $y_0 + \mathcal{H}\left(x_0, x_1, \ldots, x_n;\ y_1, y_2, \ldots, y_n\right) = 0$ (from (1.C.76)), which it may be convenient to replace by the more general *equation of energy*,[36]

$$\Gamma\left(x_\varrho, y_\varrho\right) = 0, \qquad (1.\mathrm{C}.78)$$

instead of making one of the variables explicit. Then (1.C.78) is a relationship among the $2n + 2$ variables $x_0, x_1, \ldots, x_n,\ y_0, y_1, \ldots y_n$ and, if we want to deduce the equations of motion in the extended phase space from a variation of (1.C.77), in the same way as is done by (1.C.4), (1.C.5), we must take into account the condition (1.C.78). Therefore, using the method of Lagrange multipliers, we shall perform the variation of the integral

$$\bar{I} = \int_{w_1}^{w_2} \left(\sum_{\varrho=0}^{n} y_\varrho x'_\varrho - \lambda \Gamma \right) dw.$$

The multiplier λ will be a function of w which, with a suitable choice of w itself, can be put equal to 1. We have therefore to perform the variation of

$$\bar{I} = \int_{w_1}^{w_2} \left(\sum_{\varrho=0}^{n} y_\varrho\, x'_\varrho - \Gamma \right) dw,$$

which yields

$$\delta \bar{I} = \int_{w_1}^{w_2} \sum_{\varrho=0}^{n} \left(y_\varrho \delta x'_\varrho + x'_\varrho \delta y_\varrho - \frac{\partial \Gamma}{\partial x_\varrho} \delta x_\varrho - \frac{\partial \Gamma}{\partial y_\varrho} \delta y_\varrho \right) dw = \sum_{\varrho=0}^{n} y_\varrho \delta x_\varrho \Big|_{w_1}^{w_2}$$

$$+ \int_{w_1}^{w_2} \sum_{\varrho=0}^{n} \left[\left(\frac{dx_\varrho}{dw} \delta y_\varrho - \frac{dy_\varrho}{dw} \delta x_\varrho \right) - \left(\frac{\partial \Gamma}{\partial x_\varrho} \delta x_\varrho + \frac{\partial \Gamma}{\partial y_\varrho} \delta y_\varrho \right) \right] dw = 0,$$

---

[36]See J. L. Synge: *Handbuch der Physik*, **III**, 1 (Springer, 1960), Sects. 67, 68.

upon integrating by parts in the usual way. By imposing the condition that the end points of the curve remain fixed, we finally have, from Hamilton's principle for the extended phase space, the canonical equations

$$\frac{dx_\varrho}{dw} = \frac{\partial \Gamma}{\partial y_\varrho}, \qquad \frac{dy_\varrho}{dw} = -\frac{\partial \Gamma}{\partial x_\varrho} \qquad (\varrho = 0, 1, 2, \ldots, n), \qquad (1.C.79a)$$

where Γ plays the role of Hamiltonian; we call it the *extended Hamiltonian*. If we make the particular choice

$$\Gamma = p_0 + \mathcal{H}(q_0, q_1, \ldots, q_n; \, p_1, \ldots, p_n),$$

(1.C.79a) give us

$$\frac{dq_i}{dw} = \frac{\partial \Gamma}{\partial p_i} = \frac{\partial \mathcal{H}}{\partial p_i}, \qquad \frac{dp_i}{dw} = -\frac{\partial \Gamma}{\partial q_i} = -\frac{\partial \mathcal{H}}{\partial q_i}, \qquad (1.C.79b)$$

which are the usual canonic equations, and

$$\frac{dq_0}{dw} = \frac{\partial \Gamma}{\partial p_0} = 1 \to t = q_0 = w + \text{const.},$$

$$\frac{dp_0}{dw} = -\frac{\partial \Gamma}{\partial q_0} = -\frac{\partial \mathcal{H}}{\partial q_0} \to \frac{d\mathcal{H}}{dt} = \frac{\partial \mathcal{H}}{\partial t}, \qquad (1.C.79c)$$

which coincides with (1.C.13). This means that the choice made for the multiplier λ corresponds to having chosen $w = q_0$ (the arbitrary constant can be put to zero) and that now (1.C.13) must be integrant part of the equations of motions in parametric form, with p_0 one of the independent variables. Through (1.C.79b), therefore, we have verified the complete equivalence of the two formulations, in both ordinary phase space and extended phase space. It is worth remarking that since Γ is always independent of the parameter w, in extended phase space every system is a conservative system. As to canonical transformations in extended phase space, the starting point will of course be the variational principle: if one passes to new variables X_ϱ, Y_ϱ,

$$\sum_\varrho \left[y_\varrho \, dx_\varrho - Y_\varrho \, dX_\varrho \right] = dW \qquad (1.C.80)$$

will be the fundamental relation for the generating function of the transformations in the $2n + 2$-dimensional extended phase space. The function W, which a priori could depend on $4n + 5$ variables $(x_\varrho, y_\varrho, X_\varrho, Y_\varrho, w)$, owing to the $2n + 2$ relations among the canonical variables, will depend only on $2n + 3$ independent variables, and therefore there will still be four possible classes of generating functions. The table of the generating functions, already established for the $2n$-dimensional phase space, can still be used replacing t by w and taking into account now that $i \to \varrho = 0, 1, 2, \ldots, n$. A different and particularly important case occurs when (in the extended phase space) one wants to transform only the parameter t (as a function of coordinates) leaving the canonical variables unchanged. We shall dwell upon a transformation of this type in more detail since, besides the interest of the subject *per se*, it is also important for applications.

When varying $\bar{I}$, we chose $\lambda = 1$ and then $w = t$; it is now obvious that if we want to transform the variable t, we have to take into account that the parameter w will also change. But a change in w entails a change in the condition (1.C.78), to which the choice for w is strictly tied. In fact, the elementary action is given by

$$dI = \sum_{\varrho=0}^{n} y_\varrho dx_\varrho = \sum_{\varrho=0}^{n} y_\varrho \frac{dx_\varrho}{dw}\, dw = \sum_{\varrho=0}^{n} y_\varrho \frac{d\Gamma}{dy_\varrho}\, dw,$$

and this determines dw. The same relationship among the $2n + 2$ variables x_ϱ, y_ϱ, however, can be expressed by means of different equations and, if we pass from $\Gamma = 0$ to $\Gamma^\star = 0$, both expressing the same relationship, the corresponding parameters will satisfy

$$\frac{dw^\star}{dw} = \frac{d\Gamma}{d\Gamma^\star}.$$

Let us verify this directly. Suppose then we have a system with n degrees of freedom described by a Hamiltonian that does not depend on time. In extended phase space, we have

$$\dot{q}_k = \frac{\partial \Gamma}{\partial p_k}, \qquad \dot{p}_k = -\frac{\partial \Gamma}{\partial q_k}, \tag{1.C.81}$$

with $\Gamma = p_0 + \mathcal{H}\left(q_1, q_2, \dots, q_n;\, p_1, \dots, p_n\right)$ and $w = t$. If we now want to pass from the variable t to T by means of

$$dt = f\left(q_1, q_2, \dots, q_n\right) dT, \tag{1.C.82}$$

with f an arbitrary function of q_k's (having the required properties of regularity), as a consequence T will be equal to the new parameter $w^\star$ and

$$\Gamma^\star = f\left(q_1, q_2, \dots, q_n\right) \Gamma \tag{1.C.83}$$

as well. In fact, the corresponding Hamiltonian system will be

$$\begin{aligned}
\frac{dq_0}{dT} &= \frac{\partial \Gamma^\star}{\partial p_0}, & \frac{dp_0}{dT} &= -\frac{\partial \Gamma^\star}{\partial q_0}, \\[2mm]
\frac{dq_i}{dT} &= \frac{\partial \Gamma^\star}{\partial p_i}, & \frac{dp_i}{dT} &= -\frac{\partial \Gamma^\star}{\partial q_i}, & (i = 1, 2, \dots, n)
\end{aligned} \tag{1.C.84}$$

with $\Gamma^\star = f\Gamma = f\left[p_0 + \mathcal{H}\left(q_1, q_2, \dots, q_n;\, p_1, p_2, \dots, p_n\right)\right]$. The first equation will give $dq_0/dT = f$, and then (1.C.82), and the second one

$$\frac{dp_0}{dT} = f\frac{dp_0}{dt} = 0 \rightarrow p_0 = \text{const.}$$

From the remaining equations, we obtain

$$\frac{dq_i}{dT} = f\frac{dq_i}{dt} = f\frac{\partial \Gamma}{\partial p_i} \rightarrow \frac{dq_i}{dt} = \frac{\partial \mathcal{H}}{\partial p_i},$$

$$\frac{dp_i}{dT} = f\frac{dp_i}{dt} = -\frac{\partial(f\Gamma)}{\partial q_i} = -\frac{\partial f}{\partial q_i}(p_0 + \mathcal{H}) - f\frac{\partial \mathcal{H}}{\partial q_i} = -f\frac{\partial \mathcal{H}}{\partial q_i},$$

that is

$$\frac{dp_i}{dt} = -\frac{\partial \mathcal{H}}{\partial q_i},$$

since $p_0 + \mathcal{H} = 0$ along a solution. Therefore the system (1.C.84) is completely
equivalent to the system (1.C.81), as asserted. The considered transformation,
then, has the effect of transforming the time according to (1.C.82) and the
Hamiltonian according to (1.C.83), leaving the canonical variables invariant.
It must be stressed that what we have stated has to be thought as *operating*
in extended phase space, i.e. the transformation (1.C.82) cannot be carried
out in *ordinary* phase space leaving the canonical character of the equations
of motion invariant. To be clearer, let us demonstrate this assertion in detail.
Suppose then we apply (1.C.82) in (1.C.3). Then $\dot{q}_i = q_i'/f$, $\dot{p}_i = p_i'/f$, where
we have denoted by the prime the derivative with respect to T, and then

$$q_i' = f(q_1, \ldots, q_n)\frac{\partial \mathcal{H}}{\partial p_i}, \qquad p_i' = -f(q_1, \ldots, q_n)\frac{\partial \mathcal{H}}{\partial q_i}. \tag{1.C.85}$$

For (1.C.85) to remain canonical equations, a Hamiltonian $\widetilde{\mathcal{H}}$ must exist such
that

$$q_i' = f\frac{\partial \mathcal{H}}{\partial p_i} = \frac{\partial \widetilde{\mathcal{H}}}{\partial p_i}, \qquad p_i' = -f\frac{\partial \mathcal{H}}{\partial q_i} = -\frac{\partial \widetilde{\mathcal{H}}}{\partial q_i}.$$

Differentiating, we get

$$\frac{\partial}{\partial q_j}\left(f\frac{\partial \mathcal{H}}{\partial p_i}\right) = \frac{\partial^2 \widetilde{\mathcal{H}}}{\partial q_j \partial p_i} = \frac{\partial^2 \widetilde{\mathcal{H}}}{\partial p_i \partial q_j} = \frac{\partial}{\partial p_i}\left(f\frac{\partial \mathcal{H}}{\partial q_j}\right)$$

and then

$$\frac{\partial f}{\partial q_j}\frac{\partial \mathcal{H}}{\partial p_i} = \frac{\partial f}{\partial p_i}\frac{\partial \mathcal{H}}{\partial q_j}.$$

However, owing to (1.C.82), one has

$$\frac{\partial f}{\partial p_i} = 0, \quad \forall i;$$

moreover, since it must be the case that (at least for some i) $\partial \mathcal{H}/\partial p_i \neq 0$,
it is also necessary that $\partial f/\partial q_i = 0$. Therefore (1.C.82), in ordinary phase
space, is admitted only for $f = $ const; and indeed one can show the same
thing starting from a function $f(p_1, \ldots, p_n)$.

1.15 The Hamilton–Jacobi Equation and the Problem of Separability

Given a system with n degrees of freedom, let $\mathcal{H}(q_1, q_2, \ldots, q_n; p_1, p_2, \ldots, p_n; t)$ be its Hamiltonian (which we assume, in the most general case, also depends on time) and (1.C.3) the corresponding canonical equations. Furthermore, let $W_1(q_i, Q_i, t)$ be the generating function of a canonical transformation into the new variables Q_i, P_i. Then

$$p_i = \frac{\partial W_1}{\partial q_i}, \qquad P_i = -\frac{\partial W_1}{\partial Q_i}, \qquad \mathcal{K} = \mathcal{H} + \frac{\partial W_1}{\partial t}. \qquad (1.C.86)$$

Now, if the new Hamiltonian $\mathcal{K}$ is zero, the new set of canonical variables will consist of constants and the system will have been transformed, so to speak, into an equilibrium state. In fact, from $\mathcal{K}(Q_i, P_i) = 0$, one obtains $\dot{Q}_i = 0$ so that $Q_i = \text{const.}$, and $\dot{P}_i = 0$ so that $P_i = \text{const.}$ We call $\overline{S}(q_i, Q_i, t)$ the generating function W_1 which gives this result. Therefore

$$\mathcal{H}(q_i, p_i, t) + \frac{\partial \overline{S}}{\partial t} = \mathcal{K} = 0; \qquad (1.C.87)$$

that is, the generating function $\overline{S}$ will have to satisfy the partial differential equation (Hamilton–Jacobi equation)

$$\mathcal{H}\left(q_i, \frac{\partial \overline{S}}{\partial q_i}, t\right) + \frac{\partial \overline{S}}{\partial t} = 0. \qquad (1.C.88)$$

By means of (1.C.88), the problem, which at the beginning required the solution of $2n$ differential equations, is changed into that of the solution of only one partial differential equation. Despite appearances, however, this does not lead to a simplification, since (in general) to solve a partial differential equation is one of the most complicated problems of analysis. Moreover, the problems that become explicitly solvable by means of (1.C.88) are, for the most part, the same as those that are solvable in another way. Therefore, (1.C.88) does not represent, so to speak, an easing of matters from the technical point of view in the solution of problems; instead it consists of a method which, besides inspiring a deeper understanding of the dynamical systems, turns out to be particularly suitable for the study of theories such as celestial mechanics, quantum mechanics, etc. Let us now take care of the solution of (1.C.88). As is known, in the case of a first-order partial differential equation, a solution which contains as many arbitrary constants as are the independent variables is called a complete integral; since in (1.C.88) the unknown function $\overline{S}$ always appears through its derivatives, a solution is determined up to an additive constant which can be omitted, being uninfluential in the canonical transformation. Furthermore, we recall that $\overline{S}$, being the generating function of a canonical transformation, must satisfy the condition (1.C.53). The method of solution of (1.C.88) will therefore consist of the following steps:

1) Obtaining a solution $\overline{S}$ containing n essential integration constants $\alpha_1, \ldots, \alpha_n$. Since one must have $\overline{S} = \overline{S}(q_1, \ldots, q_n; Q_1, \ldots, Q_n, t)$ the constants will be the new canonical coordinates: $Q_i = \alpha_i$.
2) According to the second equation of (1.C.86), putting $-\partial \overline{S}/\partial \alpha_i = P_i = \beta_i$, where β_i are n new constants. In fact, since $\mathcal{K} = 0$, then also $\dot{P}_i = -\partial \mathcal{K}/\partial Q_i = 0$ and $P_i = \mathrm{const}$.
3) Inverting these relations to obtain the $q_i = f_i(\alpha_1, \alpha_2, \ldots, \alpha_n; \beta_1, \beta_2, \ldots, \beta_n; t)$ which, together with $p_i = \partial \overline{S}/\partial q_i$, will provide the initial canonical variables as functions of the time t and $2n$ integration constants α_i, β_i.

In this way we have fulfilled the aim: to have the general solution of the $2n$ differential equations (1.C.3). Obviously, the complete integrals of (1.C.88) which generate a canonical transformation that satisfies the required conditions, i.e. $Q_i = \alpha_i$, $P_i = \beta_i$ constant, and $\mathcal{K} = 0$, are infinite in number. However, one among them is particularly interesting and suitable for clarifying the physical meaning of the transformation itself: that one in which the double n-tuple of the α_i, β_i consists of the set of the initial values of the q_i, p_i, at a time t_0, taken as the origin of time, in which the Hamiltonian is zero.

That is, one considers the motion of the system as consisting of a canonical transformation. In fact, if we evaluate $d\overline{S}/dt$, we obtain

$$\frac{d\overline{S}}{dt} = \sum_{i=1}^{n} \frac{\partial \overline{S}}{\partial q_i} \dot{q}_i + \sum_{i=1}^{n} \frac{\partial \overline{S}}{\partial Q_i} \dot{Q}_i + \frac{\partial \overline{S}}{\partial t} = \sum_{i=1}^{n} p_i \dot{q}_i - \mathcal{H}$$

and from this

$$d\overline{S} = \sum_{i=1}^{n} p_i \, dq_i - \mathcal{H} dt, \qquad \overline{S} = \int_{t_0}^{t} \left[\sum_{i=1}^{n} p_i \dot{q}_i - \mathcal{H} \right] dt = \int_{t_0}^{t} \mathcal{L} dt,$$

that is, the action of the system. If the Hamiltonian $\mathcal{H}$ does not explicitly contain the time (this is nearly always the case in the problems of celestial mechanics) one obtains a considerable simplification. In this case we shall define a generating function

$$\overline{S} = S^{\star}(q_1, q_2, \ldots, q_n; \alpha_1, \ldots, \alpha_n) - \alpha_1 t, \tag{1.C.89}$$

where $S^{\star}$ is independent of time. By substituting (1.C.89) in (1.C.88), we obtain

$$\mathcal{H}\left(q_i, \frac{\partial \overline{S}}{\partial q_i} \right) + \frac{\partial \overline{S}}{\partial t} = \mathcal{H}\left(q_i, \frac{\partial S^{\star}}{\partial q_i} \right) - \alpha_1 = 0, \tag{1.C.90}$$

where it is evident that one of the constants, α_1, turns out to be equal to h (the constant corresponding to the energy integral). Furthermore

$$\beta_i = -\frac{\partial \overline{S}}{\partial \alpha_i} = -\frac{\partial S^{\star}}{\partial \alpha_i} \qquad (i \neq 1),$$

$$\beta_1 = -\frac{\partial \overline{S}}{\partial \alpha_1} = -\frac{\partial S^{\star}}{\partial \alpha_1} + t = -\frac{\partial S^{\star}}{\partial h} + t, \qquad p_i = \frac{\partial S^{\star}}{\partial q_i}.$$

Equation (1.C.90) is sometimes called the stationary Hamilton–Jacobi equation or, directly, the Hamilton–Jacobi equation *tout court* it being mistaken for (1.C.88). The function $S^\star$ in it is also called the *reduced action*. A calculation, analogous to that performed above for (1.C.88), now gives

$$S^\star = \int_{t_0}^{t} \sum_{i=1}^{n} p_i \, \dot{q}_i \, dt = 2 \int_{t_0}^{t_1} \mathcal{T} dt,$$

that is, the *Maupertuis action* (1.B.99). Regarding the solution of (1.C.90), as in the case of (1.C.88), general methods which can be applied do not exist. However, there is a particular class of problems, separable problems, for which it is possible to find a complete solution: in these cases the partial differential equation in n variables splits into n ordinary differential equations in one variable; they are then immediately integrable. The method of separation consists in the following: one tries to solve the partial differential equation by putting $S^\star$ equal to a sum of functions, each depending on a single variable:

$$S^\star = S_1^\star (q_1) + S_2^\star (q_2) + \ldots + S_n^\star (q_n).$$

There is also the case of partial separability, that is, the case in which only one, or more (but not all), coordinates are separable. For each of them,

$$S^\star = S_i^\star (q_i) + S^\star (q_1, q_2, \ldots, q_{i-1}, q_{i+1}, \ldots, q_n).$$

This resumes the old argument of ignorable coordinates, and consequently it is quite natural to ask oneself in what way it is possible a priori to know if a given problem is separable (completely separable) or not. From a general point of view, the question is still unsolved, but for a certain class of problems we have general rules stated in a theorem due to Stäckel (1893–1895) and further developed by Darboux, Levi-Civita, Burgatti, Dall'Acqua, Eisenhart, etc. Referring the reader to Chap. 5 for a detailed study of the results concerning the motion of a mass point in a given potential, here we limit ourselves to mentioning the Stäckel conditions. If one considers a natural system ($\mathcal{T} = \mathcal{T}_2$) and a system of orthogonal curvilinear coordinates, then

$$\mathcal{T} = \frac{1}{2} \sum_{i=1}^{n} c_i p_i^2 = \frac{1}{2} \sum_{i=1}^{n} \frac{1}{c_i} \dot{q}_i^2.$$

Moreover, one assumes that the coefficients c_i and the potential U are functions of class C^1 of $q_1, q_2, \ldots, q_n$ in a domain of the configuration space Q and $c_i > 0, \forall i$. The Stäckel theorem states that the system is separable if and only if a regular $n \times n$ matrix exists (with element u_{ij} depending only on the q_i's) together with a column vector w_j (with its components depending only on the q_i's) such that

$$\sum_{i=1}^{n} c_i u_{ij} = \delta_1^j, \qquad \sum_{i=1}^{n} c_i w_i = U \qquad (j = 1, 2, \ldots, n). \qquad (1.C.91)$$

As a consequence, if (v_{ij}) is the inverse matrix of (u_{ij}), one has

$$c_i = v_{1i}, \qquad U = \sum_{i=1}^{n} v_{1i} w_i. \tag{1.C.92}$$

In actual cases, one can directly verify the separability hypothesis by substituting (1.C.91) in (1.C.90). If the operation meets with success, every conjugate momentum becomes

$$p_k = \frac{\partial S^\star}{\partial q_k} = \frac{dS_k^\star(q_k)}{dq_k},$$

that is, a function of the sole coordinate q_k, and the partial differential equation splits into n ordinary differential equations. The dynamical system with n degrees of freedom can then be considered a superposition of n systems with only one degree of freedom. However, this result must not be interpreted from the dynamical point of view, because the actual equations of motion are contained in

$$-\frac{\partial S^\star}{\partial \alpha_i} = \beta_i, \qquad \frac{\partial S^\star}{\partial h} = t - \beta_1, \tag{1.C.93}$$

which are not separated, since certain α_i and also h in general will be present in more than one of the $S_i^\star$'s. Furthermore it must be emphasized that the separable nature of a problem is not so inherent in the "physical" features of the dynamical system considered as in the chosen system of canonical coordinates. It may happen that some problems are separable only in one system of coordinates and some other problems (for instance, the Kepler problem) are separable in more than one system of coordinates. Unfortunately, rules do not exist to find the "right" coordinate system. Let us suppose we have to handle a separable system and we have determined a complete integral of (1.C.90),

$$S^\star = S^\star(q_1, \dots, q_n; \alpha_1, \alpha_2, \dots, \alpha_n), \tag{1.C.94}$$

where $\alpha_1 = h$. Let us perform a point transformation on the constants α_i:

$$\begin{aligned}
\alpha_1 &= h = f_1(\gamma_1, \gamma_2, \dots, \gamma_n), \\
\alpha_2 &= f_2(\gamma_1, \gamma_2, \dots, \gamma_n), \\
&\vdots \qquad \vdots \\
\alpha_n &= f_n(\gamma_1, \gamma_2, \dots, \gamma_n),
\end{aligned} \tag{1.C.95}$$

where we intend such a transformation to be one to one and differentiable to the extent required. If we substitute (1.C.95) in (1.C.94), then

$$S^\star = S^\star\left[q_1, q_2, \dots, q_n; f_1(\gamma_1, \dots, \gamma_n), \dots, f_n(\gamma_1, \dots, \gamma_n)\right].$$

By an abuse of notation, we shall go on to use the symbol $S^\star$ for the sake of simplicity and write

$$S^\star = S^\star(q_1, \dots, q_n; \gamma_1, \dots, \gamma_n). \tag{1.C.96}$$

Now,

$$\mathcal{H}\left[q_i, \frac{\partial S^\star(q_i,\gamma_i)}{\partial q_i}\right] = \alpha_1\Bigg|_{\alpha_i=f_i(\gamma_j)} = f_1(\gamma_j)$$

and $S^\star(q_1,\dots,q_n;\gamma_1,\dots,\gamma_n)$ will be a complete integral of (1.C.90) of the type

$$\overline{S}(q_i,\gamma_i,t) = S^\star(q_i,\gamma_i) - f_1(\gamma_i)t.$$

As before in the case of $\alpha_1 = h$, $\alpha_2,\dots,\alpha_n$, the $\gamma_1,\gamma_2,\dots,\gamma_n$ can also be considered the new canonical coordinates $Q_1,\dots,Q_n$: that is, we have a canonical transformation (for the sake of simplicity we still call $S^\star$ the generating function) from the q_i,p_i to the Q_i,P_i (as we shall see only one half of them will be constant) which transforms the Hamiltonian $\mathcal{H}$ into

$$\widetilde{\mathcal{H}} = f_1(\gamma_1,\gamma_2,\dots,\gamma_n) = f_1(Q_1,Q_2,\dots,Q_n).$$

Instead of (1.C.90), we now have

$$\widetilde{\mathcal{H}}\left(q_i, \frac{\partial S^\star}{\partial q_i}\right) = h(Q_i) \qquad (i = 1,2,\dots,n). \tag{1.C.97}$$

A stationary Hamilton–Jacobi equation of the nature of (1.C.97), that is with the Hamiltonian depending only on one group of canonical variables (assumed to be constants), can also be obtained with the momenta P_i in place of the coordinates Q_i, i.e. with $\mathcal{H} = \mathcal{H}(P_i)$. In fact, it is sufficient to apply to the initial canonical variables (q_i,p_i) a transformation generated by $S(q_i,P_i) = S^\star(q_i,Q_i) + \sum_{i=1}^{n} Q_i P_i$ (which is of the W_2 type), demanding the constancy of the P_i's and proceeding as above. One ends up with

$$\mathcal{H}\left(q_i, \frac{\partial S}{\partial q_i}\right) = h(P_i). \tag{1.C.98}$$

The two equations (1.C.97) and (1.C.98) are completely equivalent; one will employ the one or the other according to the convenience of having to handle a generating function of type W_1 or W_2. Incidentally, we point out that the same operation can obviously also be performed for the Hamilton–Jacobi equation (1.C.88), by introducing a function $\widetilde{S} = \widetilde{S}(q_1,P_i,t)$ of type W_2 instead of $\overline{S} = \overline{S}(q_1,Q_i,t)$.

Looking at (1.C.97) and (1.C.98) it might appear that the problem has been complicated in comparison with (1.C.90), going from a Hamiltonian equal to only one constant to another one which (in general) will depend on n constants. As a matter of fact it is not like that because, since $\mathcal{H}$ depends on only one group of variables (for (1.C.98) the P_i's), the system remains explicitly integrable. In fact, from (1.C.98),

$$\dot{P}_i = 0 \iff P_i = \text{const.}; \qquad \dot{Q}_i = \frac{\partial \mathcal{H}}{\partial P_i} = \text{const.} = \gamma_i \iff Q_i = \gamma_i t + \delta_i,$$

$$\tag{1.C.99}$$

the P_i's being constant and then the canonical equations completely integrated. The advantage of a solution like (1.C.99), in comparison with the preceeding case with $\alpha_1 = h$, is the following: if the initial separation by means of the constants $\alpha_1 = h, \alpha_2, \ldots, \alpha_n$ induces one to represent the system by means of variables devoid of direct physical meaning, now, on the contrary, one has the possibility through $\mathcal{H} = \mathcal{H}(P_i)$ of introducing variables which better reflect its intrinsic physical features.

Moreover, it often happens that one wants to use the solution of a relatively simple problem (for instance, the unperturbed motion of two bodies) as the starting point for the solution of a more complex problem (for instance, perturbation theory). Therefore, it is necessary to choose the variables in such a way that the solution of the simplest problem is expressed in the simplest possible way.

Application to the Case of the n-Dimensional Oscillator

Let us see now, as an example of the application of the Hamilton–Jacobi theory, the case of the oscillator. Consider at first a generic anisotropic planar oscillator in rectangular coordinates. The Hamiltonian (for unit mass) will be

$$\mathcal{H} = \frac{1}{2}(p_1^2 + p_2^2) + \frac{1}{2}(\omega_1^2 q_1^2 + \omega_2^2 q_2^2) \tag{1.C.100}$$

and the Hamilton–Jacobi equation (1.C.90) is

$$\frac{1}{2}\left[\left(\frac{\partial S^\star}{\partial q_1}\right)^2 + \left(\frac{\partial S^\star}{\partial q_2}\right)^2\right] + \frac{1}{2}(\omega_1^2 q_1^2 + \omega_2^2 q_2^2) = \alpha_1. \tag{1.C.101}$$

We can put

$$S^\star(q_1, q_2) = S_1^\star(q_1) + S_2^\star(q_2) \tag{1.C.102}$$

and obtain

$$\left[\frac{1}{2}\left(\frac{dS_1^\star}{dq_1}\right)^2 + \frac{1}{2}\omega_1^2 q_1^2\right] + \left[\frac{1}{2}\left(\frac{dS_2^\star}{dq_2}\right)^2 + \frac{1}{2}\omega_2^2 q_2^2\right] = \alpha_1. \tag{1.C.101a}$$

(1.C.101a) is satisfied by

$$\frac{1}{2}\left(\frac{dS_1^\star}{dq_1}\right)^2 + \frac{1}{2}\omega_1^2 q_1^2 = \alpha_2,$$

$$\frac{1}{2}\left(\frac{dS_2^\star}{dq_2}\right)^2 + \frac{1}{2}\omega_2^2 q_2^2 = \alpha_1 - \alpha_2, \tag{1.C.103}$$

α_2 being a new constant. Since

$$\left(\frac{dS_1^\star}{dq_1}\right)^2, \quad \left(\frac{dS_2^\star}{dq_2}\right)^2 \geq 0,$$

from (1.C.103) one has

$$-\frac{\sqrt{2\alpha_2}}{\omega_1} \le q_1 \le \frac{\sqrt{2\alpha_2}}{\omega_1}; \quad -\frac{\sqrt{2(\alpha_1 - \alpha_2)}}{\omega_2} \le q_2 \le \frac{\sqrt{2(\alpha_1 - \alpha_2)}}{\omega_2}. \quad (1.C.104)$$

From (1.C.102) and (1.C.103)

$$S^\star(q_1, q_2) = \int \sqrt{2\alpha_2 - \omega_1^2 q_1^2}\, dq_1 + \int \sqrt{2(\alpha_1 - \alpha_2) - \omega_2^2 q_2^2}\, dq_2 \quad (1.C.105)$$

and from (1.C.93)

$$\frac{\partial S^\star}{\partial \alpha_1} = \int \frac{dq_2}{\sqrt{2(\alpha_1 - \alpha_2) - \omega_2^2 q_2^2}} = -\beta_1 + t = t - t_0, \quad (1.C.106)$$

$$\frac{\partial S^\star}{\partial \alpha_2} = \int \frac{dq_1}{\sqrt{2\alpha_2 - \omega_1^2 q_1^2}} - \int \frac{dq_2}{\sqrt{2(\alpha_1 - \alpha_2) - \omega_2^2 q_2^2}} = -\beta_2. \quad (1.C.107)$$

Finally, from (1.C.106) and (1.C.107)

$$t - t_0 = \frac{1}{\omega_2} \arcsin\left(\frac{\omega_2 q_2}{\sqrt{2(\alpha_1 - \alpha_2)}} \right)$$

and

$$\beta_2 = -\frac{1}{\omega_1} \arcsin\left(\frac{\omega_1 q_1}{\sqrt{2\alpha_2}} \right) + \frac{1}{\omega_2} \arcsin\left(\frac{\omega_2 q_2}{\sqrt{2(\alpha_1 - \alpha_2)}} \right),$$

from which

$$q_1 = \frac{\sqrt{2\alpha_2}}{\omega_1} \sin[\omega_1(t - t_0) - \omega_1 \beta_2],$$

$$q_2 = \frac{\sqrt{2(\alpha_1 - \alpha_2)}}{\omega_2} \sin[\omega_2(t - t_0)]. \quad (1.C.108)$$

The solutions (1.C.108) are the same as those already obtained in the frameworks of Newtonian and Lagrangian dynamics, and obviously $\alpha_1 - \alpha_2$ and α_2 represent the energies of the two modes of oscillation. It is clear from our procedure that the same method can be applied for an n-dimensional oscillator, with n however large. This is no longer the case if we give up the rectangular coordinates. Only two cases exist of separability in non-rectangular coordinates: the *isotropic* planar oscillator in polar coordinates and the *anisotropic* planar oscillator with $\omega_2 = 2\omega_1 = 2\omega$ in parabolic coordinates. We shall briefly discuss both of them, in view of future applications. For the isotropic oscillator, in polar coordinates

$$\mathcal{H} = \frac{1}{2}\left(p_r^2 + \frac{1}{r^2} p_\vartheta^2 \right) + \frac{1}{2}\omega^2 r^2, \quad (1.C.109)$$

$$\frac{1}{2}\left[\left(\frac{\partial S^\star}{\partial r} \right)^2 + \frac{1}{r^2}\left(\frac{\partial S^\star}{\partial \vartheta} \right)^2 \right] + \frac{1}{2}\omega^2 r^2 = \alpha_1, \quad (1.C.110)$$

$$S^\star(r, \vartheta) = S_r^\star(r) + S_\vartheta^\star(\vartheta),$$

from which

$$\frac{dS^\star_\vartheta}{d\vartheta} = \alpha_\vartheta, \qquad \frac{1}{2}\left[\left(\frac{dS^\star_r}{dr}\right)^2 + \frac{\alpha^2_\vartheta}{r^2}\right] + \frac{1}{2}\omega^2 r^2 = \alpha_1,$$

and then

$$S^\star(r,\vartheta) = \alpha_\vartheta \cdot \vartheta + \int \sqrt{2\alpha_1 - \omega^2 r^2 - \frac{\alpha^2_\vartheta}{r^2}}\, dr. \tag{1.C.111}$$

From (1.C.111), one obtains

$$\frac{\partial S^\star}{\partial \alpha_1} = t - t_0 = \int \frac{dr}{\sqrt{2\alpha_1 - \omega^2 r^2 - \frac{\alpha^2_\vartheta}{r^2}}} = -\frac{1}{2\omega} \arcsin \frac{(\alpha_1 - \omega^2 r^2)}{\sqrt{\alpha^2_1 - \omega^2 \alpha^2_\vartheta}},$$

$$\frac{\partial S^\star}{\partial \alpha_\vartheta} = -\beta = \vartheta - \alpha_\vartheta \int \frac{dr}{r^2\sqrt{2\alpha_1 - \omega^2 r^2 - \frac{\alpha^2_\vartheta}{r^2}}} = \vartheta - \frac{1}{2}\arcsin \frac{\left(\alpha_1 - \frac{\alpha^2_\vartheta}{r^2}\right)}{\sqrt{\alpha^2_1 - \omega^2 \alpha^2_\vartheta}}$$

and the equation of the trajectory becomes

$$r^2 = \frac{\alpha^2_\vartheta}{\alpha_1 - \sqrt{\alpha^2_1 - \omega^2 \alpha^2_\vartheta}\, \sin[2(\vartheta + \beta)]}. \tag{1.C.112}$$

The constants α_1 and $\alpha_\vartheta = p_\vartheta$ are respectively the total energy and the angular momentum.

For the anisotropic oscillator with $\omega_2 = 2\omega_1 = 2\omega$, in parabolic coordinates (see Sect. 5.4) with λ and μ defined by

$$q_1 = \sqrt{\lambda\mu}, \qquad q_2 = \frac{1}{2}(\lambda - \mu), \qquad \lambda, \mu > 0, \tag{1.C.113}$$

the Hamiltonian (1.C.100) becomes

$$\mathcal{H} = 2\left(\frac{\lambda p^2_\lambda + \mu p^2_\mu}{\lambda + \mu}\right) + \frac{1}{2}\omega^2 \frac{(\lambda^3 + \mu^3)}{\lambda + \mu}, \tag{1.C.114}$$

with

$$p_\lambda = \frac{1}{4}\dot\lambda\frac{(\lambda + \mu)}{\lambda}, \qquad p_\mu = \frac{1}{4}\dot\mu\frac{(\lambda + \mu)}{\mu}. \tag{1.C.115}$$

The Hamilton–Jacobi equation is then

$$\frac{2}{(\lambda + \mu)}\left[\lambda\left(\frac{\partial S^\star}{\partial \lambda}\right)^2 + \mu\left(\frac{\partial S^\star}{\partial \mu}\right)^2\right] + \frac{1}{2}\omega^2 \frac{(\lambda^3 + \mu^3)}{\lambda + \mu} = \alpha_1 \tag{1.C.116}$$

and can be satisfied by putting

$$2\lambda\left(\frac{dS^\star_\lambda}{d\lambda}\right)^2 + \frac{1}{2}\omega^2\lambda^3 - \alpha_1\lambda = \alpha_2, \qquad 2\mu\left(\frac{dS^\star_\mu}{d\mu}\right)^2 + \frac{1}{2}\omega^2\mu^3 - \alpha_1\mu = -\alpha_2,$$

where α_2 is a new constant. By subtracting the two equations, we have for α_2

$$\alpha_2 = \lambda p_\lambda^2 - \mu p_\mu^2 + \frac{1}{4}\,\omega^2(\lambda^3 - \mu^3) - \frac{1}{2}\,\alpha_1(\lambda - \mu), \qquad (1.C.117)$$

where we have used

$$\frac{\partial S^\star}{\partial \lambda} = p_\lambda, \qquad \frac{\partial S^\star}{\partial \mu} = p_\mu.$$

The constant α_2 has been introduced as an integration constant, without any particular expedient; but if now, using (1.C.113) and (1.C.115), we rewrite (1.C.117) in rectangular coordinates, we realize that α_2 coincides with the first integral (1.B.94). Therefore, the introduction of parabolic coordinates has enabled us to obtain, in this case, the first integral in an automatic way, without any particular device. We shall see in Chap. 2 an analogous situation concerning the Laplace–Runge–Lenz vector for the Kepler problem.

1.16 Action–Angle Variables

We have already pointed out in (1.C.99) that, when the Hamiltonian of a conservative system depends only on n dynamical variables (and all of the same type: n coordinates or n momenta) and they are constants of the motion, then the system is explicitly integrable. Now we will deal with a method that, when applicable, enables us to reach this condition. This method, concerning the bounded motions of conservative Hamiltonian systems, which was originally introduced in a problem of celestial mechanics (the motion of the Moon), was subsequently applied in several branches of physics, beginning with early quantum mechanics. For the sake of simplicity, we shall start from a system with only one degree of freedom. In this case the bounded motions of a conservative Hamiltonian system are periodic in time, and this can occur in two ways:

1) different values of q correspond to different configurations of the system and both p and q are periodic functions of the time (and consequently of any variable being a linear function of time): after a period τ, both p and q again take the initial values (*libration*);
2) every time q increases by a given constant amount q_0, the same configuration of the system is repeated: after a period τ, p again take the same value, whereas q is increased by q_0 (*rotation*).

Figures 1.18 and 1.19 give the phase curves for libration and rotation respectively. Figure 1.18 represents the phase curve of the oscillator (see below).

It may occur that the same system can have both types of motion; the typical case is provided by the pendulum: for a small amplitude one has *libration*, whereas if the energy is sufficient for a rotation in the vertical plane one has *rotation*. In Fig. 1.21, are shown the phase curves of the pendulum. The closed curves (libration) are separated from the open ones (rotation) by a dashed line (separatrix) which is the *boundary line* of regions with phase curves having different behaviours. Later on we shall discuss the case of the pendulum in more detail. In general, one has libration when the system moves between two states of vanishing kinetic energy; in the case of rotation, it is always possible to choose the variables in such a way that q is an angle and to normalize them so that $q_0 = 2\pi$. Let us suppose we have found a solution of the Hamilton–Jacobi equation (1.C.98) for our system ($i = 1$, only one degree of freedom) and let $\alpha = P$ and $\beta = Q$ be the new canonical variables. Then β will be a linear function of the time: $\beta = \gamma(t - t_0)$. Therefore we can write in the two cases

$$\text{libration:} \quad q\,(\beta + \gamma\tau) = q\,(\beta),$$
$$\text{rotation:} \quad q\,(\beta + \gamma\tau) = q\,(\beta) + 2\pi.$$

Now we perform a point transformation from α and β into two new variables J and ϑ so that $\vartheta = 2\pi\beta/\gamma\tau$. This can be accomplished by means of a generating function

$$\widetilde{S} = 2\pi\frac{J\beta}{\gamma\tau},$$

from which it also follows that $\alpha = 2\pi J/\gamma\tau$ and then $J = \alpha\gamma\tau/2\pi$. In the new variables ϑ, we have

$$\text{libration:} \quad q\,(\vartheta + 2\pi) = q\,(\vartheta),$$
$$\text{rotation:} \quad q\,(\vartheta + 2\pi) = q\,(\vartheta) + 2\pi. \tag{1.C.118}$$

The comprehensive canonical transformation, from (q,p) into (J,ϑ), is still a canonical transformation of the type generated by the S which occurs in (1.C.98) since the one generated by $\widetilde{S}$ was a point transformation, and then J is proportional to α without depending on β; and this means that the Hamiltonian depends on J only. Therefore, we shall continue, for the sake of simplicity, also to use the symbol S for the generating function of the transformation from (q,p) into (J,ϑ). Then

$$p = \frac{\partial S}{\partial q}, \qquad \vartheta = \frac{\partial S}{\partial J}. \tag{1.C.119}$$

Through this transformation, p, as well as q, will be a periodic function of ϑ with period 2π; the Hamiltonian $\mathcal{H}$ will depend on J and the equations of motion will be

$$J = \text{const.}, \qquad \dot\vartheta = \frac{d\mathcal{H}}{dJ} = \text{const.} = \omega \qquad \Rightarrow \vartheta = \omega t + \delta, \tag{1.C.120}$$

where ω represents the *angular frequency* of the motion, while the period (in time) is given by $2\pi/\omega$. From (1.C.119), it follows that

$$\frac{\partial \vartheta}{\partial q} = \frac{\partial}{\partial q}\left(\frac{\partial S}{\partial J}\right) = \frac{\partial}{\partial J}\left(\frac{\partial S}{\partial q}\right) = \frac{\partial p}{\partial J},$$

and then

$$\frac{\partial}{\partial J}\oint pdq = \oint \frac{\partial p}{\partial J}\,dq = \oint d\vartheta = 2\pi,$$

where $\oint$ means integration over a whole period (in the case of libration a motion there and back for q; in the case of rotation an increase of 2π). From

$$\frac{\partial}{\partial J}\oint pdq = 2\pi,$$

we then have (considering the integration constant to be equal to zero):

$$J = \frac{1}{2\pi}\oint pdq. \tag{1.C.121}$$

Comparing (1.C.121) with the first of equations (1.C.119), one has that $2\pi J$ equals the variation undergone by S in a whole period. Referring to Figs. 1.18 and 1.19, $2\pi J$ will be the measure of the area bounded by the phase curve in the first case and of the area bounded by the curve, the q axis and the two ordinates at a distance $q_0 = 2\pi$ in the second case. Therefore $2\pi J$ is exactly the action integral over a period: J has the dimensions of angular momentum and ϑ of an angle; they are examples of quantities called *action–angle variables*. In the plane J, ϑ the phase curves become straight lines $J = \text{const.}$ parallel to the ϑ axis, and then Figs. 1.18 and 1.19 have their equivalent in Fig. 1.20. Since, in our case (only one degree of freedom), the generating function $S^\star$ is given by $S^\star = S - \vartheta J$, its variation in one period (in which ϑ varies by 2π and S by $2\pi J$) will be $\Delta S^\star = \Delta S - 2\pi J = 0$. Therefore $S^\star$ is a periodic function of ϑ with period 2π. This follows from the determination (1.C.121) of J. If we had considered $S^\star$ as our starting point, we should of course have imposed the condition of periodicity to attain the same determination of J. If we make use of $S^\star = S^\star(q, \vartheta)$ as the generating function of the transformation $(q, p) \to (\vartheta, J)$, we have

$$p = \frac{\partial S^\star(q, \vartheta)}{\partial q}, \qquad J = -\frac{\partial S^\star(q, \vartheta)}{\partial \vartheta}. \tag{1.C.122}$$

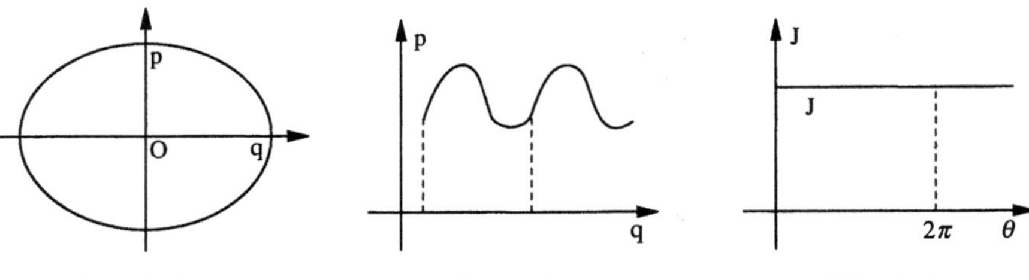

Fig. 1.18   Fig. 1.19   Fig. 1.20

To sum up, we can go from the variables (q, p) to (ϑ, J) by means of a canonical transformation generated either by $S(q, J)$ or by $S^\star(q, \vartheta)$. Every time the new momentum J is a constant of motion and the new coordinate ϑ is a linear function of the time, we shall say that J is an *action variable* and ϑ an *angle variable*.

The Linear Oscillator

As an example of the application of the method of action–angle variables, consider first the linear oscillator. From the Hamiltonian

$$\mathcal{H} = \frac{p^2}{2m} + \frac{1}{2}\, m\, \omega^2 q^2 = \text{const.} = h, \tag{1.C.123}$$

we have

$$p = \pm\sqrt{2\, m\, [h - \frac{m}{2}\, \omega^2 q^2]}, \quad J = \frac{1}{2\pi} \oint p\, dq = \frac{1}{\pi} \int_{-\bar{q}}^{+\bar{q}} \sqrt{2m(h - \frac{m}{2}\omega^2 q^2)}\, dq,$$

where $\pm\bar{q}$ are the roots of $m\,\omega^2\bar{q}^2 = 2h$, that is, the coordinates of the points where the motion is inverted. Thus $J = h/\omega$, and then the Hamiltonian in action–angle variables will be

$$\mathcal{H} = \omega J, \tag{1.C.124}$$

with the frequency ω given by

$$\omega = \frac{\partial \mathcal{H}}{\partial J}. \tag{1.C.125}$$

Equating the two expressions for the Hamiltonian (1.C.123) and (1.C.124), we obtain

$$p = \sqrt{2m\omega J - (m\omega q)^2}$$

and, since $\partial\vartheta/\partial q = \partial p/\partial J$, also

$$\vartheta = \int_0^q \frac{\partial p}{\partial J}\, dq = \int_0^q dq \sqrt{\frac{m\omega}{2J - m\omega q^2}} = \arcsin\left(q\sqrt{\frac{m\omega}{2J}}\right),$$

from which

$$q = \sqrt{\frac{2J}{m\omega}}\, \sin\vartheta, \qquad p = \sqrt{2m\omega J}\, \cos\vartheta. \tag{1.C.126}$$

The relations between (q, p) and (ϑ, J) have been obtained, in this case, without resorting to S or $S^\star$, which are given by

$$S = \int_0^q pdq = \int_0^q dq\sqrt{2m\omega J - (m\omega q)^2}$$

$$= J\arcsin\left(q\sqrt{\frac{m\omega}{2J}}\right) + \frac{1}{2}q\sqrt{2m\omega J - m^2\omega^2 q^2}$$

$$S^\star = S(q,J) - \vartheta J = S - J\arcsin\left(q\sqrt{\frac{m\omega}{2J}}\right) = \frac{1}{2}q\sqrt{2m\omega J - m^2\omega^2 q^2}$$

$$= \frac{1}{2}q\sqrt{\frac{m^2\omega^2 q^2}{\sin^2\vartheta} - m^2\omega^2 q^2} = \frac{1}{2}m\omega q^2 \cot\vartheta.$$

$$(1.C.127)$$

It is easy to check, using the first of equations (1.C.127) and of equations (1.C.126), that the variation ΔS corresponding to $\Delta\vartheta = 2\pi$ is $2\pi J$. Equations (1.C.126), since J is constant, represent the parametric equations of the phase curve of Fig. 1.18. As we have seen in Sect. 1.3, by means of a scale transformation, we obtain as the phase curve a circle which is just a one-dimensional torus; in this case, using the variables q and $p/m\omega$, the radius of the circle will be $\sqrt{2J/m\omega}$.

The Pendulum

As a further example of an application of the method of action–angle variables to systems with one degree of freedom, let us consider the pendulum. Not only is it important *per se*, as a mechanical system, but also the structure of its Hamiltonian is the same as that present in most non-linear resonant problems of celestial mechanics. The pendulum can be thought of as consisting of a mass point P (of mass m) suspended in O (Fig. 1.22) by means of a weightless string of length l. The angle q is measured from the downward vertical line through O; the point is subjected only to gravity, whose acceleration is g. If we set, conventionally, the zero of the potential energy to that corresponding to $q = \pi/2$, the Hamiltonian will be

$$\mathcal{H} = \frac{P^2}{2G} - mgl\cos q = h = \text{const.,} \qquad (1.C.128a)$$

where $G = ml^2$ is the moment of inertia and $P = G\omega = G\dot{q}$ the angular momentum. In (1.C.128a), the potential energy $V(q) = -mgl\cos q$ is a periodic function of period 2π. By choosing units such that $G = ml^2 = 1$ and putting $b = g/l$, the Hamiltonian (1.C.128a) can be rewritten as

$$\mathcal{H} = \frac{P^2}{2} - b\cos q = h. \qquad (1.C.128b)$$

The quantity b is positive and, for $-b < h < b$, one has libration with q varying between the values

$$q' = \arccos\left(-h/b\right) \quad \text{and} \quad q'' = -\arccos\left(-h/b\right); \qquad (1.C.129)$$

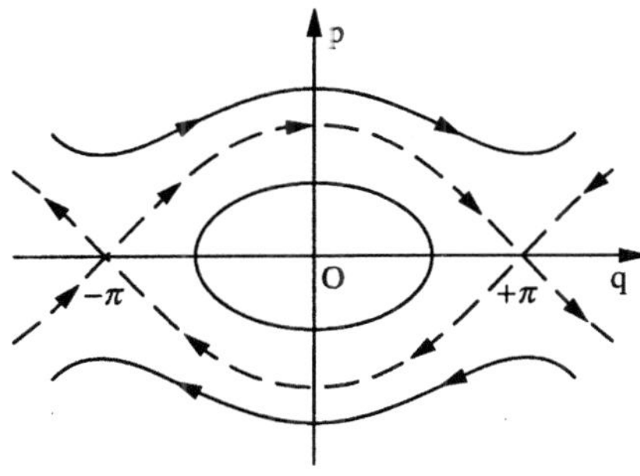
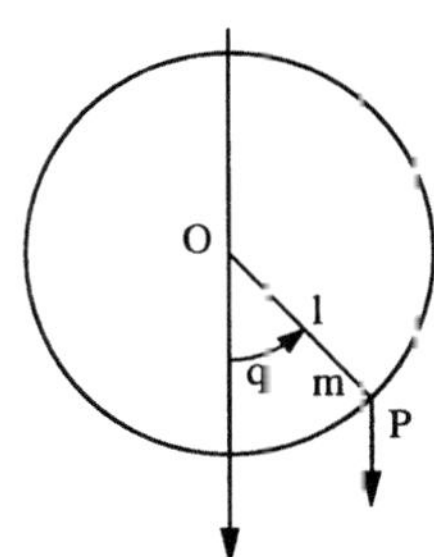

Fig. 1.21   Fig. 1.22

in Fig. 1.21 the curve is an ellipse. For $h > b$, one has rotation; the upper curve corresponds to positive P and the lower curve to negative P.

At this point, it is natural for us to consider as phase space the cylindrical surface obtained by identifying the lines $q = -\pi$ and $q = +\pi$. In the limiting case $h = b$, the phase curve is given by the *separatrix* (the dashed line in the figure); in this case the pendulum attains asymptotically the position $q = \pi$ in an infinite time. In fact, from (1.C.128b), in the units we have chosen,

$$b = \frac{1}{2}\dot{q}^2 - b\cos q,$$

and then

$$dt = \frac{dq}{\sqrt{2b(1 + \cos q)}};$$

integrating this, one will obtain $t \to \infty$, for $q \to \pi$. Let us see, now, in what way one passes to the action–angle variables in the cases of libration and rotation. The Hamilton–Jacobi equation will be

$$\mathcal{H}\left(q, \frac{\partial S}{\partial q}\right) = h(J), \tag{1.C.130}$$

with

$$p = \frac{\partial S}{\partial q}, \qquad \vartheta = \frac{\partial S}{\partial J}, \qquad J = \frac{1}{2\pi}\oint p\,dq, \qquad S = S(q, J).$$

From (1.C.128b)

$$J = \frac{1}{2\pi}\oint \sqrt{2(h + b\cos q)}\,dq. \tag{1.C.131}$$

As is known, (1.C.131) is not evaluable by means of elementary functions: through suitable substitutions we arrive at an elliptic integral. In the libration case, we put

$$\sin\frac{q}{2} = \sqrt{\frac{h + b}{2b}}\sin x.$$

From the definition, x varies from 0 to 2π during a complete libration, and also

$$\alpha = \sqrt{\frac{h+b}{2b}} < 1.$$

We obtain

$$\cos q = 1 - \frac{h+b}{b}\sin^2 x,$$

and from (1.C.128a)

$$p = \sqrt{2(h + b\cos q)} = \sqrt{2(h+b)}\cos x = 2\sqrt{b}\,\alpha\cos x.$$

As

$$dq = \frac{2\alpha\cos x\,dx}{\sqrt{1 - \alpha^2\sin^2 x}},$$

we obtain

$$\begin{aligned}
S = \int_0^q p\,dq &= 4\sqrt{b}\int_0^x \frac{\alpha^2\cos^2 x\,dx}{\sqrt{1 - \alpha^2\sin^2 x}}\\
&= 4\sqrt{b}\left[\int_0^x \sqrt{1 - \alpha^2\sin^2 x}\,dx + (\alpha^2 - 1)\int_0^x \frac{dx}{\sqrt{1 - \alpha^2\sin^2 x}}\right] \qquad (1.C.132)\\
&= 4\sqrt{b}\left[E(x,\alpha^2) + (\alpha^2 - 1)F(x,\alpha^2)\right],
\end{aligned}$$

where the following elliptic integrals have been introduced:[37]

$$E(x,\alpha^2) = \int_0^x \sqrt{1 - \alpha^2\sin^2 u}\,du, \qquad F(x,\alpha^2) = \int_0^x \frac{du}{\sqrt{1 - \alpha^2\sin^2 u}}.$$

In the same way, we can calculate J, this time by means of the complete elliptic integrals

$$\begin{aligned}
J &= \frac{1}{2\pi}\int_0^{2\pi} 2\sqrt{b}\,\alpha\cos x\frac{2\alpha}{\sqrt{1 - \alpha^2\sin^2 x}}\cos x\,dx\\
&= 4\left\{\frac{1}{2\pi}4\sqrt{b}\left[\int_0^{\pi/2}\sqrt{1 - \alpha^2\sin^2 x}\,dx + (\alpha^2 - 1)\int_0^{\frac{\pi}{2}}\frac{dx}{\sqrt{1 - \alpha^2\sin^2 x}}\right]\right\}\\
&= \frac{8}{\pi}\sqrt{b}\left[E(\alpha^2) + (\alpha^2 - 1)K(\alpha^2)\right], \qquad\qquad (1.C.133)
\end{aligned}$$

where

$$E(\alpha^2) = E(x = \pi/2,\,\alpha^2), \qquad K(\alpha^2) = F(x = \pi/2,\,\alpha^2).$$

It remains now to evaluate

$$\vartheta = \frac{\partial S}{\partial J} = \frac{\partial S}{\partial\alpha}\frac{\partial\alpha}{\partial J} = \frac{\partial S}{\partial\alpha}\left(\frac{dJ}{d\alpha}\right)^{-1}.$$

---

[37]See, for instance, E. T. Whittaker, G. N. Watson: *A Course of Modern Analysis* (Cambridge University Press, 1927), pp. 512 ff.

Applying the known properties of elliptic integrals,

$$\frac{\partial[\alpha\,F(x,\alpha^2)]}{\partial\alpha} = \frac{E(x,\alpha^2)}{1-\alpha^2} - \frac{\alpha^2}{1-\alpha^2}\frac{\sin x\,\cos x}{\sqrt{1-\alpha^2\sin^2 x}},$$

$$\frac{\partial E(x,\alpha^2)}{\partial\alpha} = \frac{E(x,\alpha^2)-F(x,\alpha^2)}{\alpha},$$

$$\frac{d[\alpha\,K(\alpha^2)]}{d\alpha} = \frac{E(\alpha^2)}{1-\alpha^2}, \qquad \frac{dE(\alpha^2)}{d\alpha} = \frac{E(\alpha^2)-K(\alpha^2)}{\alpha}$$

and considering $S = S(q,\alpha)$ as

$$4\sqrt{b}\left[E\left(\arcsin\left(\frac{1}{\alpha}\sin\frac{q}{2}\right),\alpha^2\right) + (\alpha^2-1)F\left(\arcsin\left(\frac{1}{\alpha}\sin\frac{q}{2}\right),\alpha^2\right)\right],$$

one obtains

$$\frac{\partial S}{\partial\alpha} = 4\sqrt{b}\,\alpha F(x,\alpha^2).$$

Analogously

$$\frac{dJ}{d\alpha} = \frac{8}{\pi}\sqrt{b}\frac{d}{d\alpha}\left[E(\alpha^2)+(\alpha^2-1)K(\alpha^2)\right] = \frac{8}{\pi}\sqrt{b}\,\alpha K(\alpha^2).$$

And finally

$$\vartheta = \frac{\partial S}{\partial\alpha}\left(\frac{dJ}{d\alpha}\right)^{-1} = \frac{\pi}{2K(\alpha^2)}F(x,\alpha^2). \tag{1.C.134}$$

By substituting the definition of x in (1.C.132–134), we obtain the wanted expressions for S, J, ϑ. Regarding the Hamiltonian, it is understood that it is defined implicitly by (1.C.133) through

$$\alpha^2 = \frac{\mathcal{H}(J)+b}{2b}. \tag{1.C.135}$$

In the case of rotation, $h > b$, so that $\alpha > 1$. We should make substitutions different from the previous ones, so as always to have real square roots. From (1.C.128b)

$$h + b = \frac{1}{2}p^2 + b(1-\cos q),$$

$$p = \sqrt{2[(h+b)-b(1-\cos q)]} = \sqrt{2(h+b)}\sqrt{1-\frac{1}{\alpha^2}\sin^2\frac{q}{2}},$$

with $1/\alpha^2 < 1$. Then,

$$J = \frac{1}{2\pi}\int_0^{2\pi} p\,dq = \frac{4}{\pi}\sqrt{\frac{(h+b)}{2}}\int_0^{\frac{\pi}{2}}\sqrt{1-\frac{1}{\alpha^2}\sin^2 x}\,dx,$$

where we have put $x = q/2$. Therefore

$$J = \frac{4}{\pi}\sqrt{\frac{(h+b)}{2}}\,E\!\left(\frac{1}{\alpha^2}\right). \tag{1.C.136a}$$

Through an analogous procedure and exploiting the recalled properties of the elliptic integrals, we also get

$$\vartheta = \frac{\pi}{K\left(1/\alpha^2\right)}F\!\left(\frac{q}{2}, \frac{1}{\alpha^2}\right). \tag{1.C.136b}$$

It is interesting enough to dwell upon the evaluation of the frequency. We have already said that the Hamiltonian will result only through an implicit definition; however, this does not forbid us from obtaining the frequency explicitly. In fact, differentiating (1.C.135) we shall get

$$\omega = \frac{d\mathcal{H}}{dJ} = 4\alpha b\left(\frac{dJ}{d\alpha}\right)^{-1} = \frac{\pi}{2}\frac{\sqrt{b}}{K(\alpha^2)}$$

in the case of libration. Analogously, for the rotation case

$$\omega = \frac{\pi}{K\left(1/\alpha^2\right)}\sqrt{\frac{h+b}{2}}.$$

If now, in (1.C.128b), we expand $\cos q$ around the equilibrium position $q = 0$ and take into account only the first non-constant term (this is equivalent to considering only small oscillations and then a *linearized* pendulum), the equation of motion will be $\ddot{q} + bq = 0$. Therefore $\sqrt{b} = \omega_0$ is the frequency of the linearized pendulum: if we call ω_L and ω_R the frequencies of libration and rotation respectively, we obtain

$$\frac{\omega_L}{\omega_0} = \frac{\pi}{2}\frac{1}{K(\alpha^2)}, \qquad \frac{\omega_R}{\omega_0} = \frac{\pi\alpha}{K\left(1/\alpha^2\right)}.$$

For $h \to b$, both libration and rotation tend to the separatrix, and then, by using the asymptotic expansion for K, we have the two frequencies tending logarithmically to zero when $\alpha \to 1$:

$$\lim_{\alpha\to 1}\frac{\omega_L}{\omega_0} = \frac{\pi}{2\ln\left[\frac{4}{\sqrt{1-\alpha^2}}\right]}, \qquad \lim_{\alpha\to 1}\frac{\omega_R}{\omega_0} = \frac{\pi}{\ln\left[\frac{4}{\sqrt{\alpha^2-1}}\right]}.$$

1.17 Separable Multiperiodic Systems – Uniqueness of the Action–Angle Variables

Before extending the results obtained to systems with more than one degree of freedom, we introduce (following Born)[38] some concepts and definitions regarding multiperiodic functions.

---

[38]M. Born: *The Mechanics of the Atom* (G. Bell and Sons, London, 1927).

Definition. A function $F(x_1, \ldots, x_n; y_1, \ldots)$ is periodic in the variables $x_1, x_2, \ldots, x_n$ with a period $\boldsymbol{\tau}$ with components $\tau_1, \tau_2, \ldots, \tau_n$ if it identically holds that

$$F(x_1 + \tau_1,\, x_2 + \tau_2,\, \ldots, x_n + \tau_n;\, y_1, \ldots) = F(x_1, x_2, \ldots, x_n;\, y_1,\, \ldots).$$

If $x_1, x_2, \ldots, x_n$ are considered in an n-dimensional space, a period $\boldsymbol{\tau}$ is a vector in this space: $\boldsymbol{\tau} \equiv (\tau_1, \tau_2, \ldots, \tau_n)$.

Proposition 1. If $\boldsymbol{\tau}$ is a period for the function F, $m\,\boldsymbol{\tau}$ (m positive or negative integer) is also a period for F.

Proposition 2. If $\boldsymbol{\tau}^{(1)}$ and $\boldsymbol{\tau}^{(2)}$ are periods for F, $\boldsymbol{\tau}^{(1)} + \boldsymbol{\tau}^{(2)}$ is also a period for F.

Proposition 3. If a function F has several periods $\boldsymbol{\tau}^{(1)}, \boldsymbol{\tau}^{(2)}, \ldots, \boldsymbol{\tau}^{(f)}$, then any linear combination of them $\sum_{k=1}^{f} c_k \boldsymbol{\tau}^{(k)}$, with coefficients given by positive or negative integers, is also a period. Proposition 3 obviously comes from 1 and 2.

Now we impose the following condition: The function F does not have arbitrary small periods (that is, given $\varepsilon > 0$, no period $\boldsymbol{\tau}$ exists with $|\boldsymbol{\tau}| < \varepsilon$). The consequence of this is that every period $\boldsymbol{\tau}$ is an integer multiple of $\boldsymbol{\tau}_1$ ($\boldsymbol{\tau} = m\boldsymbol{\tau}_1$), where $\boldsymbol{\tau}_1$ is the smallest among the periods of the type $\lambda\boldsymbol{\tau}$. In fact, from the imposed condition, we have that if $\boldsymbol{\tau}$ and $\lambda\boldsymbol{\tau}$ are two parallel periods of the function F, then they must be commensurable, i.e. λ must be rational. If this were not the case (since, owing to a well-known theorem of number theory due to Kronecker, if λ is not rational one can always find two integers m and m' such as to make $(m + m'\lambda)$ arbitrarily small) one should have (by using 3) an arbitrarily small period $(m + m'\lambda)\,\boldsymbol{\tau}$. If we now call p/q the number λ (with p, q prime numbers), we have that $\boldsymbol{\tau}/q$ is a period for F; in fact, in this case, again resorting to a theorem of number theory, we know that two integers m and m' must exist for which $mq + m'p = 1$, and then $m + m'\lambda = m + m'p/q = 1/q$. By hypothesis, we must then get a period $\boldsymbol{\tau}/q_1$, which is the smallest among all the periods of the form $\boldsymbol{\tau}/q$. Therefore we can express every period whose vector has a certain direction as an integer multiple of a given minimum period. One can show that all possible periods constitute an f-dimensional lattice: that is, f periods $\boldsymbol{\tau}^{(1)}, \boldsymbol{\tau}^{(2)}, \ldots, \boldsymbol{\tau}^{(f)}$ exist such that for any other period one has $\boldsymbol{\tau} = \sum_{k=1}^{f} c_k \boldsymbol{\tau}^{(k)}$, with integer c_k, uniquely determined.

We shall consider the case where the number f of periods coincides with the number of variables n. Therefore any one of the periods of the function F will be given by a linear combination of the type $\boldsymbol{\tau} = \sum_{k=1}^{n} c_k \boldsymbol{\tau}^{(k)}$. We shall then say that the function $F(x_1, x_2, \ldots, x_n; y_1, \ldots)$, periodic in the n variables $x_1, x_2, \ldots, x_n$, has a fundamental system of periods $\boldsymbol{\tau}^{(1)}, \boldsymbol{\tau}^{(2)}, \ldots, \boldsymbol{\tau}^{(n)}$. If, now, we perform a transformation in order to obtain a new fundamental system of periods, this transformation, besides having integer coefficients, must also be unimodular, that is, with determinant equal to ± 1. In fact, if $(\tau^{(k)})$ and $(\sigma^{(k)})$ are two fundamental systems of periods $(k = 1, 2, \ldots, n)$, it must be the case that $\tau^{(k)} = \sum_i n_{ik}\sigma^{(i)}$, $\sigma^{(k)} = \sum_i m_{ik}\tau^{(i)}$, where $n = (n_{ij})$ and $m = (m_{ij})$ are two $n \times n$ matrices with integer elements, and then also

$$\tau^{(k)} = \sum_i \sum_j n_{ik}\, m_{ji}\, \tau^{(j)}.$$

Consequently $mn = 1$, and then $\det m \det n = 1$. As $\det n$ and $\det m$ are integer, the only solution is $\det m \det n = \pm 1$.

We summarize the above by means of the following propositions:

Proposition 4. For any function $F(x_1, x_2, \ldots, x_n; y_1, \ldots)$ periodic in the $x_1, x_2, \ldots, x_n$ variables a fundamental system of periods does exist.

Proposition 5. All the fundamental systems of periods of a function are connected through linear transformations with integer coefficients and determinant equal to ± 1.

If now, in place of the variables $x_1, x_2, \ldots, x_n$, we introduce in the n-dimensional space a new system of coordinates $\nu_1, \nu_2, \ldots, \nu_n$, with axes parallel to the vectors of a fundamental system of periods, then the function F expressed as a function of the variables $\nu_1, \nu_2, \ldots, \nu_n$ has the fundamental system of periods

$$\tau^{(1)} = (1, 0, \ldots, 0),$$
$$\tau^{(2)} = (0, 1, \ldots, 0),$$
$$\vdots = \quad \vdots$$
$$\tau^{(n)} = (0, 0, \ldots, 1).$$

By means of a suitable scale transformation one can obtain

$$\tau^{(1)} = (2\pi, 0, \ldots, 0),$$
$$\tau^{(2)} = (0, 2\pi, \ldots, 0),$$
$$\vdots = \quad \vdots$$
$$\tau^{(n)} = (0, 0, \ldots, 2\pi).$$

In this case, one says that the $F(\nu_1, \nu_2, \ldots, \nu_n; \ldots)$ *has fundamental period* 2π.

In conclusion we can state the further proposition:

Proposition 6. By means of a linear transformation of the variables in which a function is periodic, one can transform the function itself into another one having fundamental period 2π.

Finally one can demonstrate[39] the following:

Proposition 7. All the sets of variables in which a function has fundamental period 2π are connected through transformations of the type

$$\nu_1 = \sum_k c_{1k}\overline{\nu}_k + \psi_1(\overline{\nu}_1, \overline{\nu}_2, \dots, \overline{\nu}_n; y_1, \dots),$$

$$\nu_2 = \sum_k c_{2k}\overline{\nu}_k + \psi_2(\overline{\nu}_1, \overline{\nu}_2, \dots, \overline{\nu}_n; y_1, \dots),$$

$$\vdots = \vdots$$

$$\nu_n = \sum_k c_{nk}\overline{\nu}_k + \psi_n(\overline{\nu}_1, \overline{\nu}_2, \dots, \overline{\nu}_n; y_1, \dots),$$

where the c_{ik} are integer coefficients, positive or negative, whose determinant has modulus 1, and the ψ_i are periodic in the $\overline{\nu}_i$ with period 2π. The function F, being a periodic function in the variables $\nu_1, \nu_2, \dots, \nu_n$, can be expressed as an n variables Fourier series:

$$F(\nu_1, \nu_2, \dots, \nu_n) = \sum_{\lambda_1=-\infty}^{+\infty} \cdots \sum_{\lambda_n=-\infty}^{+\infty} c_{\lambda_1,\lambda_2,\dots,\lambda_n} e^{i(\lambda_1\nu_1 + \lambda_2\nu_2 - \dots + \lambda_n\nu_n)},$$

where, if $F(\nu_1, \dots, \nu_n)$ is real, $c_{\lambda_1,\dots,\lambda_n}$ $c_{\lambda_{-1},\dots,\lambda_{-n}}$ are complex conjugate quantities. In vector language, the above relation can be rewritten as

$$F(\boldsymbol{\nu}) = \sum_{\lambda} c_{\lambda} e^{i(\boldsymbol{\lambda}\cdot\boldsymbol{\nu})},$$

with the coefficients given by

$$c_{\lambda} = \int F(\boldsymbol{\nu}) e^{-i(\boldsymbol{\lambda}\cdot\boldsymbol{\nu})} d\boldsymbol{\nu},$$

where the vector $\boldsymbol{\lambda}$ is a vector with n components which span all the positive and negative integer values from $-\infty$ to $+\infty$ and $0 \le \nu_k \le 2\pi$, $\forall k$.

Now the task is to exploit what was said above to extend the method of action–angle variables to multiperiodic separable systems, i.e. to systems for which the Hamilton–Jacobi equation is separable and in each partial plane (q_i, p_i) one has either a closed curve (libration) or a periodic function q_i (rotation). Obviously, the simplest case occurs when the Hamiltonian of the system (with n degrees of freedom) consists of a sum of n terms, each of them containing only a pair of conjugate variables q_i, p_i:

$$\mathcal{H} = \mathcal{H}_1(q_1, p_1) + \mathcal{H}_2(q_2, p_2) + \dots + \mathcal{H}_n(q_n, p_n). \tag{1.C.137}$$

---

[39] For the proof, see M. Born: op. cit., pp. 74–75.

The Hamilton–Jacobi equation (1.C.98), Sect. 1.15, is solved by separating the variables, by putting

$$\mathcal{H}_i\left(q_i, \frac{\partial S_i}{\partial q_i}\right) = h_i, \quad \text{with} \quad h_1 + h_2 + \ldots + h_n = h;$$

therefore the motion corresponds to that of a system resulting from the superposition of n independent systems with one degree of freedom each. One can generalize the method of the preceeding section, by defining the actions

$$J_i = \frac{1}{2\pi} \oint p_i dq_i \tag{1.C.138}$$

and expressing the functions S_i by means of q_i and J_i. Then

$$\vartheta_i = \frac{\partial S_i}{\partial J_i}. \tag{1.C.139}$$

The classic example in this case is that given by the superposition of n linear oscillators; considering unit mass and $q_1 = x_1, q_2 = x_2, \ldots, q_n = x_n$; $p_1 = \dot{x}_1, \ldots, p_n = \dot{x}_n$ (i.e. using the Cartesian components as canonical variables) we can refer to what we have written in Sect. 1.3. For $n = 2$,

$$\mathcal{H} = \frac{1}{2}\left(p_1^2 + p_2^2\right) + \frac{1}{2}\left(\omega_1^2 q_1^2 + \omega_2^2 q_2^2\right), \tag{1.C.140}$$

and from (1.C.126), Sect. 1.16,

$$q_1 = \sqrt{\frac{2J_1}{\omega_1}} \sin \vartheta_1, \qquad q_2 = \sqrt{\frac{2J_2}{\omega_2}} \sin \vartheta_2,$$
$$p_1 = \sqrt{2\omega_1 J_1} \cos \vartheta_1, \qquad p_2 = \sqrt{2\omega_2 J_2} \cos \vartheta_2, \tag{1.C.141}$$

with

$$\vartheta_1 = \omega_1 t + \delta_1, \qquad \vartheta_2 = \omega_2 t + \delta_2. \tag{1.C.142}$$

The Hamiltonian will become

$$\mathcal{H} = \omega_1 J_1 + \omega_2 J_2. \tag{1.C.143}$$

As was shown in Sect. 1.3, (compare (1.C.142) with (1.A.41a)), the phase curves wind round tori and are closed if ω_1 and ω_2 are rationally dependent and everywhere dense in the opposite case. For $n > 2$, the conclusions are obviously the same; only the possibility of a graphic representation fails. Returning to the general problem, it may happen that the Hamiltonian does not result a sum of n independent terms as in (1.C.137), but that the Hamilton–Jacobi equation (1.C.98) can be solved by separation of variables as well, that is, by putting

$$S = S_1(q_1) + S_2(q_2) + \ldots + S_n(q_n). \tag{1.C.144}$$

In this case $p_k = \partial S_k / \partial q_k$ depends only on q_k.

For any pair q_k, p_k one has the situation[40] considered in the preceding section; that is, one has either *libration* or *rotation*. Every action integral of the form (1.C.138) is constant: therefore we shall assume that the actions J_i are constant momenta. The Hamiltonian will be a function of only the J_i, and S will be expressed as a function of the q_i and J_i; the angle variables ϑ_i, conjugate of the J_i, will be given by

$$\vartheta_i = \frac{\partial S}{\partial J_i} = \sum_l \frac{\partial S_l}{\partial J_i}. \tag{1.C.145}$$

Let us now show that the variables J_i, ϑ_i introduced in this way behave as ϑ and J did in the case of only one degree of freedom, that is, the q_i are multiperiodic functions of the ϑ_i with the fundamental system of periods

$$\begin{aligned}
&(2\pi, 0, \ldots, 0), \\
&(0, 2\pi, \ldots, 0), \\
&\qquad \vdots \\
&(0, 0, \ldots, 2\pi).
\end{aligned} \tag{1.C.146}$$

Let us evaluate the variation of ϑ_k during a period (of libration or rotation) of variation of q_k, keeping the other coordinates fixed. If we call Δ_k the increment one has in making only q_k vary then

$$\Delta_k \vartheta_i = \oint (\partial \vartheta_i / \partial q_k) dq_k.$$

Partially differentiating (1.C.145), we get

$$\frac{\partial \vartheta_i}{\partial q_k} = \sum_l \frac{\partial^2 S_l}{\partial J_i \partial q_k} = \frac{\partial}{\partial J_i} \sum_l \frac{\partial S_l}{\partial q_k} = \frac{\partial}{\partial J_i} \frac{\partial S_k}{\partial q_k},$$

and integrating,

$$\Delta_k \vartheta_i = \oint \frac{\partial}{\partial J_i} \frac{\partial S_k}{\partial q_k} dq_k = \frac{\partial}{\partial J_i} \oint \frac{\partial S_k}{\partial q_k} dq_k = 2\pi \frac{\partial}{\partial J_i} J_k = 2\pi \delta_{ik}, \tag{1.C.147}$$

with $\delta_{ik} = 0$ for $i \neq k$ and $= 1$ for $i = k$.

Vice versa, if we consider the $q_k(\vartheta_1, \vartheta_2, \ldots, \vartheta_n)$ and increase ϑ_i by 2π, keeping the other ϑ fixed, one has that q_i performs a complete libration (or rotation); the other q may depend on ϑ_i as well, but go back to the initial value without having run a whole cycle (otherwise the corresponding ϑ should vary by 2π). In this way it turns out that the q are multiperiodic functions of the ϑ with the fundamental system of periods (1.C.146). Every q_k can then be expressed by means of the Fourier series

---

[40]This refers to the projections onto each phase plane (q_i, p_i) of the trajectory in the $2n$-dimensional phase space.

$$q_k = \sum_\lambda c_\lambda^{(k)} e^{i(\boldsymbol{\lambda}\cdot\boldsymbol{\vartheta})}. \tag{1.C.148}$$

Since ϑ are functions of time, they can be obtained from the canonical equations, and, as in (1.C.142),

$$\vartheta_k = \omega_k t + \delta_k, \quad \text{with} \quad \omega_k = \frac{\partial \mathcal{H}}{\partial J_k} \quad (k = 1, 2, \dots, n). \tag{1.C.149}$$

From (1.C.148) and (1.C.149) one then has

$$q_k = \sum_\lambda c_\lambda^{(k)} e^{i[(\boldsymbol{\lambda}\cdot\boldsymbol{\omega})t + (\boldsymbol{\lambda}\cdot\boldsymbol{\delta})]}, \tag{1.C.150}$$

with

$$\boldsymbol{\lambda}\cdot\boldsymbol{\omega} = \lambda_1\omega_1 + \lambda_2\omega_2 + \dots + \lambda_n\omega_n,$$
$$\boldsymbol{\lambda}\cdot\boldsymbol{\delta} = \lambda_1\delta_1 + \lambda_2\delta_2 + \dots + \lambda_n\delta_n.$$

Here also, for every $\boldsymbol{J}$, the phase curves will wind round the corresponding n-dimensional torus: if the ω_i are rationally dependent, the curves will be closed; in the opposite case the curves will be everywhere dense on each torus. In the first case $\omega_k = n_k\omega$ (where the n_k are integer), $\forall k$, and then one has a unique frequency, and the motion is periodic in the real sense of the word (the system is also said to be *completely degenerate*). In the second case the motion is said to be *conditionally periodic*. As to the function $S^\star$, in this case given by

$$S^\star = S - \sum_k \vartheta_k J_k, \tag{1.C.151}$$

since S increases by $2\pi J_k$ every time the coordinate q_k runs a complete cycle (the other q_i remaining fixed), it turns out that it is a multiperiodic function with fundamental period 2π. In fact, one has from (1.C.151) that, while S increases by $2\pi J_k$, ϑ_k increases by 2π and the other ϑ_i remain unchanged: then $\Delta S^\star = 0$ on every cycle. As usual, one can consider $S^\star$ as generating a canonical transformation from (q_i, p_i) into (ϑ_i, J_i): $S^\star = S^\star(q_i, \vartheta_i)$. Then

$$J_k = -\frac{\partial S^\star(q_i, \vartheta_i)}{\partial \vartheta_k}, \qquad p_k = \frac{\partial S^\star(q_i, \vartheta_i)}{\partial q_k}. \tag{1.C.152}$$

Now one can ask oneself if the integrals (1.C.138) which define the action–angle variables are fully determined by the set of variables (q_k, p_k) in which the Hamilton–Jacobi equation is separable or not. It can be shown that the answer is different according to whether one has rationally independent frequencies (nondegenerate system) or rationally dependent ones (degenerate system). In the former case one sees that the J_k are uniquely determined by the set (q_k, p_k) which makes the Hamilton-Jacobi equation separable; in the latter (degenerate system) this is no longer true: different sets of variables can exist, giving rise to separability, connected by transformations which are not only point transformations but more generally canonical transformations.

Stating the problem in more general terms, one can reach some conclusions which we shall explain without proof.[41]

Let a Hamiltonian system be given whose function $\mathcal{H} = \mathcal{H}(q_1, q_2, \ldots, q_n;\ p_1, p_2, \ldots, p_n)$ is independent of time and such that one can introduce new variables ϑ_k, J_k, by means of the canonical transformation generated by $S(q_1, \ldots, q_n, J_1, \ldots, J_r)$:

$$p_k = \frac{\partial S}{\partial q_k}, \qquad \vartheta_k = \frac{\partial S}{\partial J_k}.$$

Moreover, let us assume that the following conditions hold true:

a) The configuration of the system is periodic with regard to the ϑ_k, with a fundamental period 2π.
b) The Hamiltonian $\mathcal{H}(q_i, p_i)$ is transformed into a function $h(J_k)$ depending only on the J_k.
c) The function $S^\star$ defined by (1.C.151) and generating the transformation given by (1.C.152) is periodic in the ϑ_k with a period 2π.

Now one has that, if (a), (b), and (c) hold true and if among the $\omega_k = \partial h/\partial J_k$ no relation of rational dependence exists, then the J_k are uniquely determined up to a linear transformation with integer coefficients having unit determinant. In the case in which, besides the validity of (a), (b), and (c), relations of rational dependence also exist among the ω_k (degenerate systems), it is always possible (by means of a transformation) to make a certain number f of the n frequencies be not commensurable (rationally independent), whereas the other $n - f$ vanish. Having reached this condition, again the transformed J_k are uniquely determined, obviously up to the linear transformation with integer coefficients of the former case.

1.18 Integrals in Involution – Liouville's Theorem for Integrable Systems

Given a set of n functions $f_1, f_2, \ldots, f_n$ depending on the q_i, the p_i and on the time t, they are said to be in involution, or to constitute an involution system, if all their Poisson brackets vanish identically, that is if

$$(f_i, f_j) \equiv 0, \qquad \forall\, i, j = 1, 2, \ldots, n. \tag{1.C.153}$$

Now we will demonstrate that, when the Jacobian

$$J = \frac{\partial(f_1, f_2, \ldots, f_n)}{\partial(p_1, p_2, \ldots, p_n)} \neq 0,$$

---

[41]See M. Born: op. cit., Sect. 15.

if we consider the functions $f_1, f_2, \ldots, f_n$ as n new coordinates $Q_1, \ldots, Q_n$, (1.C.153) in addition guarantees the existence of n functions $P_1, P_2, \ldots, P_n$ such as to constitute together with the Q_i a new set of canonical variables.

Therefore let us set

$$f_i(\boldsymbol{q}, \boldsymbol{p}, t) = Q_i, \qquad i = 1, 2, \ldots, n. \tag{1.C.154}$$

In a neighbourhood of $(\boldsymbol{p}, \boldsymbol{q})$, since $J \neq 0$, we can solve (1.C.154), thus obtaining the p_i:

$$p_i = \varphi_i(\boldsymbol{q}, \boldsymbol{Q}, t). \tag{1.C.155}$$

If we now replace the p_i by the φ_i in (1.C.154), we again obtain n identities in the $\boldsymbol{q}, \boldsymbol{Q}, t$:

$$f_i(\boldsymbol{q}, \boldsymbol{\varphi}, t) - Q_i = 0, \qquad i = 1, 2, \ldots, n. \tag{1.C.156}$$

By deriving (1.C.156) with respect to the q_i, one has

$$\frac{\partial f_i}{\partial q_j} + \frac{\partial f_i}{\partial p_l} \frac{\partial \varphi_l}{\partial q_j} = 0. \tag{1.C.157}$$

Let us now multiply (1.C.157) by $\partial f_k / \partial p_j$ and sum over j from 1 to n:

$$\frac{\partial f_i}{\partial q_j} \frac{\partial f_k}{\partial p_j} + \frac{\partial f_i}{\partial p_l} \frac{\partial \varphi_l}{\partial q_j} \frac{\partial f_k}{\partial p_j} = 0. \tag{1.C.158}$$

If now in (1.C.158) we interchange i with k and subtract the equations obtained from equations (1.C.158) themselves, we obtain, taking into account (1.C.153),

$$\frac{\partial f_i}{\partial p_l} \frac{\partial f_k}{\partial p_j} \frac{\partial \varphi_l}{\partial q_j} - \frac{\partial f_k}{\partial p_l} \frac{\partial f_i}{\partial p_j} \frac{\partial \varphi_l}{\partial q_j} = 0,$$

which, interchanging the dummy indices in the second term, can be rewritten as

$$\frac{\partial f_i}{\partial p_l} \frac{\partial f_k}{\partial p_j} \left(\frac{\partial \varphi_l}{\partial q_j} - \frac{\partial \varphi_j}{\partial q_l} \right) = 0. \tag{1.C.159}$$

Equations (1.C.159) constitute a system of n^2 homogeneous equations in the

$$\left(\frac{\partial \varphi_l}{\partial q_j} - \frac{\partial \varphi_j}{\partial q_l} \right)$$

and the determinant of the coefficients is certainly different from zero, being given by J^2; the only solution is then

$$\frac{\partial \varphi_l}{\partial q_j} = \frac{\partial \varphi_j}{\partial q_l}, \qquad \forall\, j, l = 1, 2, \ldots, n. \tag{1.C.160}$$

Consequently, a function must exist (we call it $\overline{S}$ in view of a forthcoming application) such that

$$p_i = \varphi_i(\boldsymbol{q}, \boldsymbol{Q}, t) = \frac{\partial \overline{S}(\boldsymbol{q}, \boldsymbol{Q}, t)}{\partial q_i}. \tag{1.C.161}$$

If, in addition to (1.C.161), we also take

$$P_i = -\frac{\partial \overline{S}(\boldsymbol{q}, \boldsymbol{Q}, t)}{\partial Q_i}, \tag{1.C.162}$$

it is easy to check that (1.C.161) and (1.C.162) define a canonical transformation $(\boldsymbol{q}, \boldsymbol{p}) \to (Q, P)$ generated by $\overline{S} = \overline{S}(\boldsymbol{q}, \boldsymbol{Q}, t)$. When the n functions $f_1, f_2, \ldots, f_n$ are first integrals for our system, we can also, according to what we have just demonstrated, consider them as new coordinates, constants of the motion:

$$f_i(\boldsymbol{q}, \boldsymbol{p}, t) = \alpha_i, \qquad i = 1, 2, \ldots, n. \tag{1.C.163}$$

Because of what we have seen in the preceding sections, the system is completely integrable. Therefore, when we have a system of n (independent) first integrals in involution, the Hamiltonian system is completely integrable But, as a matter of fact, we can obtain an even more meaningful result: *if the n first integrals (1.C.163) are in involution, then the corresponding momenta (1.C.162) are also constant and we have all the possible $2n$ independent first integrals and therefore the complete solution of the problem.*

   This result is due to Liouville.[42] To demonstrate it, let us evaluate the derivatives $d\varphi_i/dt$ along the solutions of the equations of motion

$$\frac{d\varphi_i}{dt} = \frac{dp_i}{dt} = -\frac{\partial \mathcal{H}(\boldsymbol{q}, \boldsymbol{p}, t)}{\partial q_i} = \frac{\partial \varphi_i}{\partial t} + \frac{\partial \varphi_i}{\partial q_j}\frac{dq_j}{dt} = \frac{\partial \varphi_i}{\partial t} + \frac{\partial \varphi_j}{\partial q_i}\frac{dq_j}{dt},$$

where we have used (1.C.160). Therefore, in addition,

$$-\frac{\partial \mathcal{H}(\boldsymbol{q}, \boldsymbol{p}, t)}{\partial q_i} = \frac{\partial \varphi_i}{\partial t} + \frac{\partial \varphi_j}{\partial q_i}\frac{\partial \mathcal{H}}{\partial p_j},$$

that is,

$$\frac{\partial \varphi_i}{\partial t} = -\frac{\partial \mathcal{H}}{\partial q_i} - \frac{\partial \mathcal{H}}{\partial p_j}\frac{\partial \varphi_j}{\partial q_i} = -\frac{\partial \widetilde{\mathcal{H}}}{\partial q_i}, \tag{1.C.164}$$

where we have written as $\widetilde{\mathcal{H}} = \widetilde{\mathcal{H}}(\boldsymbol{q}, \boldsymbol{\alpha}, t)$ the Hamiltonian expressed as a function of the q_i, α_i and t, through the φ_i. The existence of (1.C.160) and (1.C.164) means the existence of a function $\overline{S}(\boldsymbol{q}, \boldsymbol{\alpha}, t)$ whose exact differential is given by

$$\varphi_1 dq_1 + \varphi_2 dq_2 + \ldots + \varphi_n dq_n - \widetilde{\mathcal{H}}dt = d\overline{S}. \tag{1.C.165}$$

If, finally, we consider the canonical transformation generated by $\overline{S}$ according to (1.C.161) and (1.C.162), the new Hamiltonian will be $\mathcal{K} = \widetilde{\mathcal{H}} + \partial \overline{S}/\partial t$; but, on the other hand, (1.C.165) gives

$$\widetilde{\mathcal{H}} + \frac{\partial \overline{S}}{\partial t} = 0, \tag{1.C.166}$$

---

[42] J. Liouville: Note sur les équations de la Dynamique, *J. Math. et Appl.* **XX** 137–138 (1855).

and therefore $\mathcal{K} = 0$ and the new momenta P_i are constants of the motion. In this case, (1.C.162) will be written

$$P_i = \beta_i = -\frac{\partial \overline{S}}{\partial \alpha_i}, \qquad i = 1, 2, \ldots, n, \qquad (1.C.167)$$

and the constants β_i are the wanted integrals, also in involution being canonical momenta. In turn, $\overline{S}$ will be a complete integral of the Hamilton–Jacobi equation:

$$\mathcal{H}\left(q, \frac{\partial \overline{S}}{\partial q}, t\right) + \frac{\partial \overline{S}}{\partial t} = 0. \qquad (1.C.168)$$

When the integrals (1.C.163) are independent of t, the demonstration becomes even simpler and in place of (1.C.168) one will have the stationary Hamilton-Jacobi equation. The existence of n first integrals in involution for a system with n degrees of freedom also has consequences of a geometrical nature (with those geometrical features we have already met but from a different point of view). In fact, the existence of n integrals $f_i(\boldsymbol{q}, \boldsymbol{p})$ implies that every trajectory of the system is confined on an n-dimensional manifold M defined by the $f_i(\boldsymbol{q}, \boldsymbol{p}) = \alpha_i = \text{const.}$, $i = 1, 2, \ldots, n$. Since we are considering *bounded* motions, that is, motions occurring in a finite part of phase space, M must be a compact manifold. If, now, we define the n vector fields (with $2n$ components) in phase space,

$$\boldsymbol{V}_i \equiv \left(\frac{\partial f_i}{\partial \boldsymbol{p}}, -\frac{\partial f_i}{\partial \boldsymbol{q}}\right), \qquad (1.C.169)$$

on every M, defined by fixing an n-tuple of constants α_i for the $f_i(\boldsymbol{q}, \boldsymbol{p})$, the $\boldsymbol{V}_i$ are regular and linearly independent functions, such as the f_i.

Moreover, the n linearly independent vectors (1.C.169) are tangential to the manifold M. On the other hand, we know from topology (Poincaré–Hopf theorem) that an n-dimensional manifold for which it is possible to build up n linearly independent vectors, tangential to the manifold itself and nowhere vanishing, has the topology of an n-dimensional torus.

Therefore, to every n-tuple of constants for the f_i, there corresponds an n-dimensional torus T^n: these tori are called *invariant tori*, because any orbit originating on one of them remains there indefinitely. The existence of these tori in phase space enables us to define the action–angle variables in a, so to say, intrinsic way. That is, now the angles ϑ_i will be the angles on the torus and the actions I_i will be the n constants of motion α_i which define the torus itself. For a completely integrable system, the transformation into action–angle variables is complete, i.e. we can imagine the phase space completely *filled* with tori; each trajectory will lie entirely on one torus or another according to the initial conditions and, as we have said, will remain there for ever.

Lie's Theorem

Let us suppose now that, for a Hamiltonian system with n degrees of freedom, we know f independent first integrals in involution with $f < n$. One can demonstrate that in this case Lie's theorem[43] holds: *If the independent first integrals in involution are also solvable for the p_k $(k = 1, 2, \ldots, f)$, as in (1.C.155), then the order of the system can be reduced to $2(n - f)$.*

1.19 Lax's Method – The Painlevé Property

Until now we have developed the discussion based upon the integrability of Hamiltonian systems in such a way that one might feel entitled to consider the terms *integrability* and *separability* synonymous. In fact, in all cases, the problem has been brought back to the search for a suitable set of variables which enabled us to separate the various degrees of freedom and then reduce the system to quadratures. The search itself for the first integrals picked out, in successful cases, the set of variables in which the system was integrable. However, there exist cases which do not come into the picture sketched above. The most famous example is the so-called *Toda lattice*.[44] The physical system consists of n particles, moving on a straight line, acted upon by exponentially decreasing forces. As $x_1, x_2, \ldots, x_n$ are the coordinates of the n mass points on the straight line, the equations of motion are

$$\ddot{x}_j = \frac{\partial U}{\partial x_j} \qquad (j = 1, 2, \ldots, n), \qquad (1.C.170)$$

with

$$U = -\sum_{k=1}^{n} e^{(x_k - x_{k+1})} \qquad (k = 1, 2, \ldots, n) \qquad (1.C.171)$$

and $x_{n+1} = x_1$ (periodicity condition). To integrate the problem, one can now apply a method, first described by Lax[45] in a quite different context (the search for solitonic solutions of the Korteweg – de Vries equation; then continuous systems and infinite degrees of freedom). Immediately afterwards it was shown that the method could also be applied to discrete systems and in particular to the Toda lattice. Let us see now, briefly, how this method works (in the case of the Toda lattice). The basic statement is that, if the equations of motion can be transformed in a matrix equation of the form

---

[43]See S. Lie: Begründung einer Invarianten theorie der Berührungs transformationen. *Math. Annalen* **VIII**, 215–303 (1875).

[44]M. Toda: Waves in nonlinear lattice, *Prog. Theor. Phys. Suppl.* **15**, 174–200 (1970).

[45]P. D. Lax: Integrals of nonlinear equations of evolution and solitary waves, *Comm. on Pure and Appl. Math.* **XXI**, 467–490 (1968).

$$\dot{L} = BL - LB, \tag{1.C.172}$$

where the square matrices B and L have elements that are (in general complex) functions of $x_i, \dot{x}_i$, then the eigenvalues of the matrix L are first integrals and therefore do not vary with time. A matrix which, while varying as a function of a parameter (in this case the time t) keeps its eigenvalue spectrum unchanged, is said to undergo an *isospectral deformation*.

Equation (1.C.172) is the equation which regulates this deformation and coincides with the equations of motion of the system. Let us now try to demonstrate the assertion; that is, if we find two matrices L and B for which the equations of motion can be written in the form (1.C.172), then the eigenvalues of L are independent of time. For this it is sufficient to demonstrate that the various forms of $L(t)$ as t varies remain similar. Hence, one must find a nonsingular and differentiable matrix $U(t)$ for which

$$\frac{d}{dt}\left[U(t)^{-1}\,L(t)\,U(t)\right] = 0. \tag{1.C.173}$$

As $U(t)$ is nonsingular, $U(t)\,U^{-1}(t) = 1$.

Taking the derivative with respect to t, we get

$$\frac{dU}{dt}\,U^{-1} + U\,\frac{dU^{-1}}{dt} = 0, \tag{1.C.174}$$

from which

$$\frac{dU^{-1}}{dt} = -U^{-1}\,\frac{dU}{dt}U^{-1}. \tag{1.C.175}$$

Consequently, (1.C.173) will become

$$\begin{aligned}
\frac{d}{dt}\left(U^{-1}LU\right) &= -U^{-1}\frac{dU}{dt}U^{-1}LU + U^{-1}\frac{dL}{dt}U + U^{-1}L\frac{dU}{dt} \\
&= U^{-1}\left[-\frac{dU}{dt}U^{-1}L + \frac{dL}{dt} + L\frac{dU}{dt}U^{-1}\right]U \\
&= U^{-1}\left[-\frac{dU}{dt}U^{-1}L + BL - LB + L\frac{dU}{dt}U^{-1}\right]U,
\end{aligned} \tag{1.C.176}$$

because of (1.C.172). Let us now put

$$\frac{dU}{dt}\,U^{-1} = B(t). \tag{1.C.177}$$

If the matrix $B(t)$ is known, $U(t)$ can be obtained as a solution of the equation

$$\frac{dU}{dt} = B(t)\,U(t), \tag{1.C.178}$$

with the initial condition

$$U(0) = 1, \tag{1.C.179}$$

by assuming that the solution exists in a real interval in which B is continuous as a function of t. By substituting (1.C.178) into (1.C.176), one has $\frac{d}{dt}\left(U^{-1}LU\right) = 0$, and then $L(t) = U^{-1}LU = L(0) = \text{const}$. $L(t)$ and $L(0)$ are similar matrices; therefore they have the same spectrum and the same characteristic polynomial. Hence, we have demonstrated that, if the equations of motion can be written in the form (1.C.172), then the eigenvalues of $L(t)$ are first integrals. In practice it is more convenient to make use of the coefficients of the characteristic polynomial $\det(L - \lambda 1)$, rather than the eigenvalues λ roots of the characteristic equation. It remains then to make out which integrals among these, found in this way, are independent. For the Toda lattice, (1.C.170) and (1.C.171), for the matrices $L(t)$ and $B(t)$ has been given the representation[46]

$$L = \begin{pmatrix} b_1 & a_1 & 0 & 0 & \cdots & a_n \\ a_1 & b_2 & a_2 & 0 & \cdots & 0 \\ 0 & a_2 & b_3 & a_3 & \cdots & 0 \\ 0 & 0 & a_3 & b_4 & \cdots & 0 \\ \vdots & \vdots & \vdots & \vdots & \ddots & \vdots \\ a_n & 0 & 0 & 0 & \cdots & b_n \end{pmatrix} \quad B = \begin{pmatrix} 0 & a_1 & 0 & 0 & \cdots & -a_n \\ -a_1 & 0 & a_2 & 0 & \cdots & 0 \\ 0 & -a_2 & 0 & a_3 & \cdots & 0 \\ \vdots & \vdots & \vdots & \vdots & \ddots & \vdots \\ a_n & 0 & 0 & 0 & \cdots & 0 \end{pmatrix}$$

$$(1.C.180)$$

Here: $2a_k = e^{1/2(x_k - x_{k-1})}$, $2b_k = -\dot{x}_k$ and it is easy to check that a_k and b_k satisfy

$$\dot{a}_k = a_k(b_{k+1} - b_k), \quad \dot{b}_k = 2(a_k^2 - a_{k-1}^2). \tag{1.C.181}$$

Moreover it has been shown that the eigenvalues of L, besides being first integrals of (1.C.170), are independent and in involution. As an application, let us look at the case $n = 3$:

$$L = \frac{1}{2}\begin{pmatrix} -\dot{x}_1 & e^{1/2(x_1 - x_2)} & e^{1/2(x_3 - x_1)} \\ e^{1/2(x_1 - x_2)} & -\dot{x}_2 & e^{1/2(x_2 - x_3)} \\ e^{1/2(x_3 - x_1)} & e^{1/2(x_2 - x_3)} & -\dot{x}_3 \end{pmatrix}, \tag{1.C.182}$$

and the characteristic polynomial,

$$\det(L - \lambda 1) = -\left(\frac{1}{2}\dot{x}_1 + \lambda\right)\left(\frac{1}{2}\dot{x}_2 + \lambda\right)\left(\frac{1}{2}\dot{x}_3 + \lambda\right) + \left(\frac{1}{2}\dot{x}_1 + \lambda\right)\frac{1}{4}e^{(x_2 - x_3)}$$

$$+ \frac{1}{4}\left(\frac{1}{2}\dot{x}_3 + \lambda\right)e^{(x_1 - x_2)} + \frac{1}{4} + \frac{1}{4}\left(\frac{1}{2}\dot{x}_2 + \lambda\right)e^{(x_3 - x_1)}$$

$$= -\lambda^3 - \frac{1}{2}\left(\dot{x}_1 + \dot{x}_2 + \dot{x}_3\right)\lambda^2$$

$$+ \frac{1}{4}\left[e^{(x_1 - x_2)} + e^{(x_2 - x_3)} + e^{(x_3 - x_1)} - \dot{x}_1\dot{x}_2 - \dot{x}_2\dot{x}_3 - \dot{x}_3\dot{x}_1\right]\lambda$$

$$+ \frac{1}{8}\left[\dot{x}_1 e^{x_2 - x_3} + \dot{x}_2 e^{x_3 - x_1} + \dot{x}_3 e^{x_1 - x_2} - \dot{x}_1\dot{x}_2\dot{x}_3\right] + \frac{1}{4}.$$

[46]M. Hénon: Integrals of the Toda lattice, *Phys. Rev. B* **9**, 1921–1923 (1974); H. Flaschka: The Toda lattice. II. Existence of integrals, *Phys. Rev. B* **9**, 1924–1925 (1974).

Let us put

$$
\begin{aligned}
P(\lambda) = \lambda^3 &+ \frac{1}{2}\left(\dot{x}_1 + \dot{x}_2 + \dot{x}_3\right)\lambda^2 \\
&- \frac{1}{4}\left[e^{(x_1-x_2)} + e^{(x_2-x_3)} + e^{(x_3-x_1)} - \dot{x}_1\dot{x}_2 - \dot{x}_2\dot{x}_3 - \dot{x}_3\dot{x}_1\right]\lambda \\
&- \frac{1}{8}\left[\dot{x}_1 e^{(x_2-x_3)} + \dot{x}_2 e^{(x_3-x_1)} + \dot{x}_3 e^{(x_1-x_2)} - \dot{x}_1\dot{x}_2\dot{x}_3\right] - \frac{1}{4}.
\end{aligned}
$$

$$(1.\text{C}.183\text{a})$$

If we rewrite this as

$$
P(\lambda) = \lambda^3 + \frac{1}{2}I_1\lambda^2 + \frac{1}{4}I_2\lambda + \frac{1}{8}I_3 - \frac{1}{4},
\tag{1.C.183b}
$$

because of what was said above, I_1, I_2, I_3 will be first integrals.

Recalling that, in this case, the Hamiltonian can be written

$$
\mathcal{H} = \frac{1}{2}\left(\dot{x}_1^2 + \dot{x}_2^2 + \dot{x}_3^2\right) + e^{x_1-x_2} + e^{x_2-x_3} + e^{x_3-x_1},
\tag{1.C.184}
$$

we then have as first integrals $\mathcal{H}, I_1, I_2, I_3$. It is easy to see that I_1 represents the total momentum and the relation

$$
I_2 = \frac{1}{2}I_1^2 - \mathcal{H}
\tag{1.C.185}
$$

holds. It is not possible to interpret I_3 in a similar way, since it is a third-degree polynomial in the momenta. Moreover, it is easy to foresee that, as the number $4n$ of particles increases, the degree of the polynomials in the momenta which constitute the first integrals will also increase. We have already encountered an analogous situation when dealing with the anisotropic planar oscillator in the case of an integer ratio between the two frequencies (for $n = 3$ we had a third-degree polynomial in the momenta). Furthermore, after the introduction of Lax's method for the integration of discrete systems, it was discovered that the method can be applied in almost all problems which had been integrated by means of other methods. A schematization of the problem according to (1.C.172), in some cases, besides providing the first integrals, is also useful in the explicit integration of the equations of motion.

To summarize, the method by separation does not exhaust all the possibilities that a system has to be integrated: separability and integrability are not synonymous. Therefore it is convenient to make a distinction between an integrability "à la Liouville" (the one just obtained by separation) and the integrability obtained in another way.

Before we move on to outline another method of studying the integrability of a system, let us go back to the Toda lattice to point out a very interesting feature. Let us consider again the case $n = 3$, but now put the three particles on a circle (see Fig. 1.23). Now the coordinates will be given by the three angles $\varphi_1, \varphi_2, \varphi_3$ and the conjugate momenta by the corresponding momenta p_1, p_2, p_3 . The Hamiltonian will be

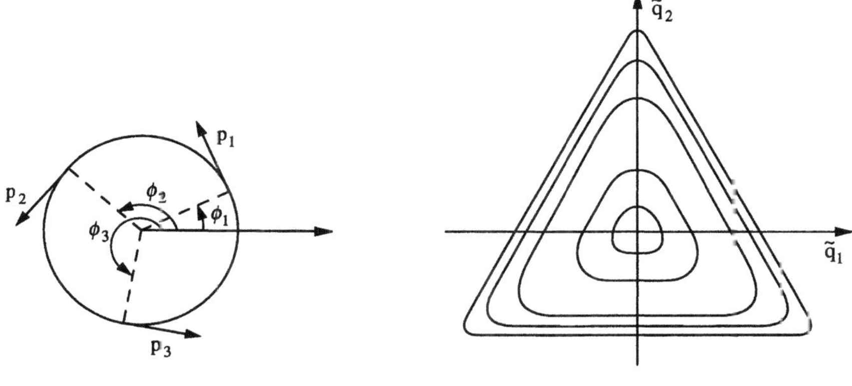

Fig. 1.23   **Fig. 1.24**

$$\mathcal{H} = \frac{1}{2}\left(p_1^2 + p_2^2 - p_3^2\right) + e^{\phi_1 - \phi_2} + e^{\phi_2 - \phi_3} + e^{\phi_3 - \phi_1} - 3, \qquad (1.\text{C}.186)$$

where we have fixed the zero of the potential energy according to a convention which will be advantageous later on.

Now perform a canonical transformation from (φ_i, p_i) to (Φ_i, P_i), generated by a function W_2 given by

$$W_2(\varphi_i, P_i) = P_1\varphi_1 + P_2\varphi_2 + (P_3 - P_1 - P_2)\varphi_3.$$

From

$$p_i = \frac{\partial W_2}{\partial \varphi_i}, \qquad \Phi_i = \frac{\partial W_2}{\partial P_i} \qquad (i = 1, 2, 3), \qquad (1.\text{C}.187)$$

we get

$$p_1 = P_1 \qquad p_2 = P_2, \qquad p_3 = P_3 - P_1 - P_2,$$
$$\Phi_1 = \varphi_1 - \varphi_3, \qquad \Phi_2 = \varphi_2 - \varphi_3, \qquad \Phi_3 = \varphi_3,$$

and then $p_1 + p_2 + p_3 = P_3$. The new Hamiltonian will be

$$\mathcal{K} = \frac{1}{2}\left[2P_1^2 + 2P_2^2 + P_3^2 - 2P_1P_3 - 2P_2P_3 + 2P_1P_2\right]$$
$$+ e^{-\Phi_1} + e^{\Phi_2} + e^{\Phi_1 - \Phi_2} - 3. \qquad (1.\text{C}.188)$$

As Φ_3 is an ignorable variable, P_3 (corresponding to the total momentum in the old variables) will be a first integral. If we now put $P_3 = 0$ (and this is completely justified), (1.C.188) becomes

$$\mathcal{K}' = P_1^2 + P_2^2 + P_1P_2 + e^{-\Phi_1} + e^{\Phi_2} + e^{\Phi_1 - \Phi_2} - 3. \qquad (1.\text{C}.189)$$

From a formal point of view, we now have to handle a system with two degrees of freedom characterized by the canonical variables Φ_1, Φ_2, P_1, P_2 and the Hamiltonian $\mathcal{K}'$. Perform a new canonical transformation generated by

$$\overline{W}_2 = \frac{1}{4\sqrt{3}}\left[(\overline{p}_1 - \sqrt{3}\overline{p}_2)\Phi_1 + (\overline{p}_1 + \sqrt{3}\overline{p}_2)\Phi_2\right] \qquad (1.\text{C}.190)$$

to pass from (Φ_i, P_i) to $(\overline{q}_i, \overline{p}_i)$. We will obtain

$$\overline{q}_i = \frac{\partial \overline{W}_2}{\partial \overline{p}_i}, \qquad P_i = \frac{\partial \overline{W}_2}{\partial \Phi_i}. \qquad (1.\text{C}.191)$$

Therefore

$$P_1 = \frac{1}{4\sqrt{3}}\left(\overline{p}_1 - \sqrt{3}\,\overline{p}_2\right), \qquad \overline{q}_1 = \frac{1}{4\sqrt{3}}\left(\Phi_1 + \Phi_2\right),$$
$$P_2 = \frac{1}{4\sqrt{3}}\left(\overline{p}_1 + \sqrt{3}\,\overline{p}_2\right), \qquad \overline{q}_2 = \frac{1}{4}\left(\Phi_2 - \Phi_1\right), \qquad \overline{K} = K'.$$

If, instead of the transformation generated by (1.C.190), we use the same transformation but with the multiplicative constant $c = 1/(8\sqrt{3})$ (see Sect. 1.12), we obtain

$$\begin{aligned}
\widetilde{q}_1 &= \frac{1}{8\sqrt{3}}\,\overline{q}_1, \qquad \widetilde{p}_1 = \overline{p}_1, \\
\widetilde{q}_2 &= \frac{1}{8\sqrt{3}}\,\overline{q}_2, \qquad \widetilde{p}_2 = \overline{p}_2, \qquad \overline{\overline{K}} = \frac{\overline{K}}{8\sqrt{3}},
\end{aligned} \qquad (1.\text{C}.192)$$

where $\widetilde{q}_i, \widetilde{p}_i$ are the new canonical variables and $\overline{\overline{K}}$ the Hamiltonian. Lastly, let us also transform the time (in extended phase space):

$$\overline{\overline{K}}\,dt = \widetilde{K}\,d\tau, \qquad (1.\text{C}.193)$$

with $\widetilde{K}/\overline{\overline{K}} = dt/d\tau = 1/\sqrt{3}$. The final result is

$$\begin{aligned}
\widetilde{K} = \frac{1}{2}\left(\widetilde{p}_1^2 + \widetilde{p}_2^2\right) + \frac{1}{24}\Big[&\exp\left(2\widetilde{q}_2 + 2\sqrt{3}\,\widetilde{q}_1\right) \\
&+ \exp\left(2\widetilde{q}_2 - 2\sqrt{3}\,\widetilde{q}_1\right) + \exp\left(-4\widetilde{q}_2\right)\Big] - \frac{1}{8}.
\end{aligned} \qquad (1.\text{C}.194)$$

The function $\widetilde{K} = \widetilde{K}(\widetilde{q}_1, \widetilde{q}_2, \widetilde{p}_1, \widetilde{p}_2)$ in (1.C.194) represents the Hamiltonian of a particle of unit mass subjected to a potential whose level curves ($U = $ const.) are represented in Fig. 1.24. The potential grows, going from the inside to the outside. The system with Hamiltonian $\widetilde{K}$ is obviously integrable, and we have performed canonical transformations.

If we now expand the Hamiltonian $\widetilde{K}$ in series of $\widetilde{q}_1$ and $\widetilde{q}_2$ and retain terms up to the third degree, we obtain

$$\widetilde{K}' = \frac{1}{2}\left(\widetilde{p}_1^2 + \widetilde{p}_2^2 + \widetilde{q}_1^2 + \widetilde{q}_2^2\right) + \widetilde{q}_1^2\,\widetilde{q}_2 - \frac{1}{3}\,\widetilde{q}_2^3. \qquad (1.\text{C}.195)$$

Equation (1.C.195) is the Hamiltonian of a famous non-integrable system (Hénon–Heiles),[47] which will be dealt with in Volume 2. The surprising fact is that what turns out to be non-integrable is the truncated expansion instead of the series itself: that is, the opposite of what one would normally expect occurs. It is evident, at this point, that the term -3 in (1.C.188) has been introduced to make the truncated series expansion give exactly the Hénon–Heiles Hamiltonian.

Let us rewrite this Hamiltonian in the most general form:

$$\mathcal{H} = \frac{1}{2}\left(p_x^2 - p_y^2 + x^2 + y^2\right) + Dx^2y - \frac{1}{3}Cy^3. \tag{1.C.196}$$

The corresponding Newtonian equations of motion are

$$\ddot{x} = -x - 2D\,xy, \qquad \ddot{y} = -y - Dx^2 + Cy^2. \tag{1.C.197}$$

They constitute a system, in normal form, of two non-linear second order equations. To such a system one can apply[48] a further criterion (usually called the *Painlevé property*) to find for what values of the coefficients D and C (besides, obviously, $D = C = 0$) it is integrable. Since the subject is technically complex, for the moment we confine ourselves only to an outline. First of all, one must no longer study the equations in the real field but in the complex field: that is, one considers the time t (and thus also x and y in this case) as a complex quantity; then the various functions involved must be analytical functions, that is, representable by means of power series. The theory of differential equations in the complex domain[49] says that if we are dealing with linear equations in normal form, the solutions can have singularities only at the points at which the coefficients of the equations themselves are singular. Therefore, we have *fixed* singularities, that is, ones not depending on the initial conditions. If, on the contrary, we consider non-linear equations, then we may also have *moving* singularities, that is, depending on the initial conditions. For example, the simple equation $y' + y^2 = 0$ has the general solution $y = (x - c)^{-1}$.

One says that y has a *pole* (of the first order) at the point $x = c$, which moves as the arbitrary constant c varies. In order, the singular points may be poles (we have already seen an example), algebraic branching points (e.g. a square root), trascendental branching points (e.g. a logarithm), or essential singularities, like $\exp(-1/t)$, whose expansion in powers of t has negative powers at any order. The criterion mentioned above consists in this: *if a system of non-linear differential equations has, as moving singularities, only simple poles (Painlevé property), then it is integrable.*

---

[47]M. Hénon, C. Heiles: loc. cit. in Footnote 2 in the Preface.

[48]See Y. F. Chang, M. Tabor, J. Weiss: Analytic structure of the Hénon–Heiles Hamiltonian in integrable and non integrable regimes, *J. Math. Phys.* **23**, 531–538 (1982).

[49]See, for instance: E. L. Ince: *Ordinary Differential Equations* (Dover, 1944).

In the last few years, the criterion has also been extended to the presence of algebraic branching points (e.g. for the Hénon–Heiles Hamiltonian); in this case, one speaks of the *weak Painlevé property*. Of course, the Painlevé property constitutes, in a certain sense, a sufficient condition for integrability and not a necessary condition. Up to now we do not even have a fully exhaustive demonstration of the criterion itself; anyhow, one point is well founded: the criterion can certainly be applied when we have Hamiltonians that are algebraically integrable, i.e. when the first integrals are polynomials in the canonical variables.

Historically, the first case in which the integrability of a mechanical system was demonstrated by studying the equations in the complex domain goes back to Sophya Kovalevskaya, who showed the integrability of the problem of the motion under gravity of a body one of whose points is fixed (in the case where the three principal moments of inertia are in the relation $A = B = 2C$).[50]

At this point of our exposition, we can draw a conclusion, even if temporary: given a Hamiltonian system (the same occurs if we are in the Lagrangian or the Newtonian scheme), no *systematic* method exists for determining if it is integrable or not. The methods we have explained until now do not enable us to give a definite answer in every case. We shall see in volume 2, which deals with almost integrable systems, what aid can be obtained from numerical methods to *guess* the existence of first integrals. What can be henceforth anticipated is that the method which has been very successful in dealing with these problems is a *mixed* one, that is a method consisting of numerical investigations whose result indicates where to concentrate the analytical study.

We should not even forget what Poincaré maintained, that is, that a system of differential equations is only more or less integrable.

---

[50]S. Kowalevski: Sur le problème de la rotation d'un corps solide autour d'un point fixe, *Acta Mathematica* **XII**, 177–232 (1888); Sur une propriété du système d'équations différentielles qui définit la rotation d'un corps solide autour d'un point fixe, *Acta Mathematica* **XIV**, 81–93 (1890).

Chapter 2

The Two-Body Problem

The two-body problem is the only case of the N-body problem that one can solve completely; therefore it has been investigated in all its details and there are innumerable treatments of it even at an elementary level with special regard to its application to the solar system. But the "solution" of this problem also has certain features which deserve to be discussed both on their own and as a basis from which to begin studying more complex problems.

This chapter is devoted mainly to this end, reducing to a minimum the explanation of traditional subjects that the reader can find in any one of the available textbooks on celestial mechanics.

2.1 The Two-Body Problem and Kepler's Three Laws

As we noted in the introduction, from Kepler's three laws it is possible to derive Newton's law of universal gravitation:

> *Between any two bodies[1] of masses m_1 and m_2 at a distance r from each other there exist attractive forces $\boldsymbol{F}_{12}$ and $\boldsymbol{F}_{21}$ directed from one body to the other. The two forces have equal magnitude which is directly proportional to the product of the masses and inversely proportional to the square of the distance.*

The law of universal gravitation can thus be written

$$F_{12} = F_{21} = G\,\frac{m_1\,m_2}{r^2},$$

where the universal constant G depends only on the chosen system of units.

Let us now consider, in a given inertial reference system, two bodies of mass m_1 and m_2 respectively; $\boldsymbol{r}_1$ and $\boldsymbol{r}_2$ are their distance vectors from the origin of the system. Moreover, to preserve greater generality, rather than to explicitly assume the dependence on the inverse square, let us assume only a generic dependence on r, leaving undetermined the form of the function. The accelerations[2] for the two bodies will be respectively

$$\ddot{\boldsymbol{r}}_1 = -G\,m_2\,f(r)\,\frac{\boldsymbol{r}_1 - \boldsymbol{r}_2}{r}, \qquad \ddot{\boldsymbol{r}}_2 = -G\,m_1\,f(r)\,\frac{\boldsymbol{r}_2 - \boldsymbol{r}_1}{r}, \qquad (2.1)$$

where $r = |\boldsymbol{r}_2 - \boldsymbol{r}_1|$. System (2.1) is of order 12 and thus the solution, in general, will depend on 12 constants. If it were possible to obtain $\boldsymbol{r}_1$ and $\boldsymbol{r}_2$ explicitly as functions of time, we could obtain at any instant of time the position of each body and then the orbit described by it in any given interval of time. But we can adopt another point of view. If we multiply the first of equations (2.1) by m_1 and the second by m_2 and then sum them, we obtain

$$m_1\,\ddot{\boldsymbol{r}}_1 + m_2\,\ddot{\boldsymbol{r}}_2 = 0,$$

and, by defining

$$\boldsymbol{r}_{\text{c.m.}} = \frac{m_1\boldsymbol{r}_1 + m_2\boldsymbol{r}_2}{m_1 + m_2}, \qquad (2.2)$$

the distance vector of the barycentre of the two bodies, $\ddot{\boldsymbol{r}}_{\text{c.m.}} = 0$. By integrating, we immediately get

$$\boldsymbol{r}_{\text{c.m.}} = \boldsymbol{a}\,t + \boldsymbol{b}, \qquad (2.3)$$

i.e. the barycentre moves in a straight line with constant velocity ($\boldsymbol{a}$ and $\boldsymbol{b}$ are two arbitrary constant vectors, therefore corresponding to six scalar

---

[1] The bodies must be point masses or spheres and homogeneous in concentric layers.
[2] We consider the equivalence law for the gravitational and inertial mass of a body to be already proved.

constants). Finally, if we subtract the first of equations (2.1) from the second and define $\boldsymbol{r} = \boldsymbol{r}_2 - \boldsymbol{r}_1$, we obtain

$$\ddot{\boldsymbol{r}} = -G\left(m_1 + m_2\right) f(r)\,\frac{\boldsymbol{r}}{r}. \tag{2.4}$$

This is the equation of the relative motion and is the same equation we should have for the motion of a particle attracted by a fixed centre with mass $m_1 + m_2$. Therefore each particle moves as if it were attracted by a fixed centre (at the position of the other particle) with mass $M = m_1 + m_2$. The orbit of each particle as seen from the other one is the *relative orbit*. Since the equation of the relative motion is invariant under the exchange $(\boldsymbol{r}, -\boldsymbol{r})$, the relative orbits are geometrically identical.[3] The original problem of the two bodies of the system (2.1), has thus been split into the following two problems:

1) the motion of the barycentre,
2) the motion of a body attracted by a fixed centre.

If one is able to solve problem 2), one automatically obtains the solution of system (2.1).

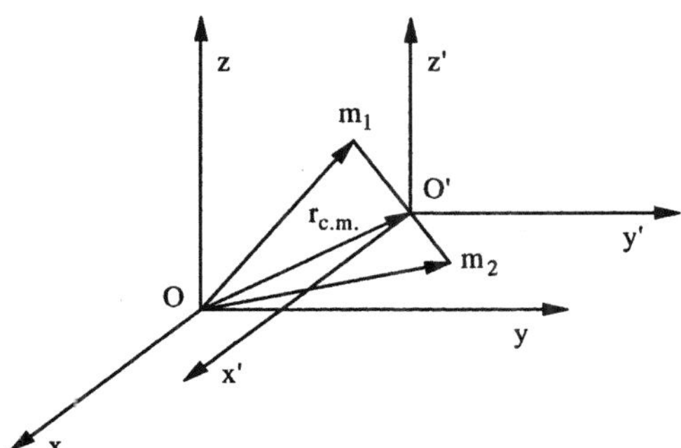

Fig. 2.1

Another point of view is that of referring the motion of the two bodies to the barycentre, that is, of considering the barycentre O' as the origin of a new coordinate system (see Fig. 2.1). Since the motion of the barycentre is rectilinear and uniform, the system with its origin at O' is also an inertial system and therefore the equations of motion will be of the same type. In fact, if we call $\boldsymbol{r}_1' = \boldsymbol{r}_1 - \boldsymbol{r}_{\text{c.m.}}$, $\boldsymbol{r}_2' = \boldsymbol{r}_2 - \boldsymbol{r}_{\text{c.m.}}$ the new coordinates, then

$$m_1 \boldsymbol{r}_1' + m_2 \boldsymbol{r}_2' = 0 \tag{2.5}$$

---

[3] "Two bodies attracting each other mutually describe similar figures about their common centre of gravity, and about each other mutually." (Newton: *Principia*, Proposition LVII, Theorem XX, op. cit. p. 164.)

and also $m_1 r_1' = m_2 r_2'$, with $r_1' = |\boldsymbol{r}_1'|$, $r_2' = |\boldsymbol{r}_2'|$. Therefore:

$$r = r_1' + r_2' = \frac{m_1 + m_2}{m_2} r_1' = \frac{m_1 + m_2}{m_1} r_2',$$

and, by substituting in (2.1), we get

$$\ddot{\boldsymbol{r}}_1' = -G\, m_2\, f\left(\frac{M}{m_2} r_1'\right) \boldsymbol{r}_1'/r_1', \qquad \ddot{\boldsymbol{r}}_2' = -G\, m_1\, f\left(\frac{M}{m_1} r_2'\right) \boldsymbol{r}_2'/r_2'. \qquad (2.6)$$

Equations (2.6) are formally independent equations, but actually only one equation is required in force of equation (2.5): there are then always six essential constants. From (2.6) it can be seen that each body moves as if it were attracted by a fixed centre having a suitable mass (for calculating the mass one needs to know the form of the function $f(r)$).

So far we have dwelt on a rather elementary and well-known subject, to show that the two-body problem can be reduced to the one-body problem only because the two bodies are subjected to a force depending only on the mutual distance. In the Lagrangian scheme this is expressed by saying that $\boldsymbol{r}_{\text{c.m.}}$ corresponds to three ignorable coordinates.[4] We can conclude that in the Newtonian theory of gravitation, the two-body problem is completely equivalent to considering the motion of a test particle attracted by a fixed centre. This result, unfortunately, is no longer true in the theory of general relativity, where the problem of *one body* and that of *two bodies* are deeply different. However, when problems regarding the solar system are considered, in most cases one can consider the Sun to be fixed (mass $\sim \infty$), its mass being much greater than any other mass involved.

Kepler's Laws

Let us now consider a particle of mass m subjected to a central force directed from the particle itself to the origin of the coordinate system. The equation of motion will be

$$m\ddot{\boldsymbol{r}} = -m f\left(r\right) \frac{\boldsymbol{r}}{r}. \qquad (2.7a)$$

Later on we shall specialize $f(r)$ to represent Newton's law; in that case it will be $f(r) = \mu/r^2$, with $\mu > 0$; in fact $\mu = GM$. Equation (2.7a) can be rewritten as a system

$$\dot{\boldsymbol{r}} = \boldsymbol{v}, \quad \dot{\boldsymbol{v}} = -f(r) \frac{\boldsymbol{r}}{r}. \qquad (2.7b)$$

We have three degrees of freedom and then a six-dimensional phase space. If we vectorially multiply by $\boldsymbol{r}$ the second equation of (2.7b), we obtain $\boldsymbol{r} \times \dot{\boldsymbol{v}} = 0$ and then

$$\frac{d}{dt}(\boldsymbol{r} \times \boldsymbol{v}) = \boldsymbol{v} \times \boldsymbol{v} + \boldsymbol{r} \times \dot{\boldsymbol{v}} = 0,$$

---

[4] See H. Goldstein: op. cit., p. 71.

that is, the vector $c = r \times v$ (angular momentum of the unit mass) is a constant of the motion. From the definition, it is obvious also that $c \cdot r = 0$. This means that:

a) If $c \neq 0$, since r must be orthogonal to c, the motion occurs in a plane orthogonal to c. In this case, the phase space is reduced to four dimensions.
b) If $c = 0$, since

$$\frac{d}{dt}\left(\frac{r}{r}\right) = \frac{(r \times \dot{r}) \times r}{r^3} = \frac{(r \times v) \times r}{r^3} = \frac{c \times r}{r^3},$$

we also have $r/r = \text{const.}$, therefore the motion occurs on a straight line passing through the origin.

In the first case, if we choose the plane of motion as the xy plane and introduce polar coordinates in this plane, then $r \equiv (r\cos\vartheta,\ r\sin\vartheta,\ 0)$, $c \equiv (0,\ 0,\ c)$, and therefore

$$c = |c| = |r \times v| = |r \times \dot{r}| = r^2\,\dot{\vartheta}. \tag{2.8}$$

Considering that $dS = \frac{1}{2}r^2\,d\vartheta$ is the area swept out on the xy plane by rotation of the angle $d\vartheta$ of the vector r, it turns out that $c = \text{const.}$ represents the *Kepler's second law*.

If we now scalarly multiply the second of equations (2.7b) by v, we have

$$\dot{v} \cdot v = -f(r)\,r^{-1}(r \cdot v) = -f(r)\,r^{-1}r\,\dot{r} = -f(r)\,\frac{dr}{dt},$$

and then $d(v^2/2) = -f(r)\,dr$. By integrating this, we obtain

$$\frac{1}{2}\,v^2 = f_1(r) + h,$$

where h is an integration constant and $f_1(r) = \int_r^a f(x)dx$, where the choice for a must be such as to make the integral converge. When $f(r) = \mu r^{-p}$, $a = \infty$ if $p > 1$, $a = 0$ if $p < 1$, $a = 1$ if $p = 1$. In the Newtonian case, $f(r) = \mu r^{-2}$, $f_1 = \mu r^{-1}$. For a particle of unit mass, $f_1 = U(r)$ is the potential function and $V(r) = -U(r)$ the potential energy. The relation just obtained can be rewritten as

$$\frac{1}{2}\,v^2 + V(r) = h, \tag{2.9a}$$

with $V(r) = -f_1(r)$. Then (2.9a) represents the law of energy conservation and h is the total energy. Expressing v^2 by means of polar coordinates, (2.9a) becomes

$$\frac{1}{2}\,(\dot{r}^2 + r^2\dot{\vartheta}^2) + V(r) = h,$$

and, by using (2.8),

$$\frac{1}{2}\,\dot{r}^2 + \frac{c^2}{2\,r^2} + V(r) = h. \tag{2.9b}$$

We shall denote by $V_{\text{eff}}(r)$ the quantity[5] $c^2/2\,r^2 + V(r)$. We can therefore rewrite the relations obtained in the form

$$r^2\dot\vartheta = c, \qquad \frac{1}{2}\,\dot r^2 + V_{\text{eff}}(r) = h. \tag{2.10}$$

The second of equations (2.10) is the equation of energy conservation of a system with one degree of freedom and potential function $U_{\text{eff}}(r) = -V_{\text{eff}}(r)$. We shall call the new potential, the *effective potential* and the term $c^2/2\,r^2$, the *centrifugal term*.[6]

The case $r = r_0 = $ const. will correspond to a circular orbit and in that case h will coincide with a minimum or a maximum of $V_{\text{eff}}(r)$ (the orbit is said Lagrange–stable or Lagrange–unstable respectively). For $V_{\text{eff}}(r)$ to have an extremum in r_0, it must be the case that

$$\left.\frac{dV_{\text{eff}}(r)}{dr}\right|_{r=r_0} = \left.\left(\frac{dV(r)}{dr} - \frac{c^2}{r^3}\right)\right|_{r=r_0} = 0,$$

that is,

$$\left.\frac{dV(r)}{dr}\right|_{r=r_0} = \frac{c^2}{r_0^3}. \tag{2.11}$$

To have a stable orbit $(V_{\text{eff}}(r_0) = \min V_{\text{eff}}(r))$, the following must hold

$$\left.\frac{d^2 V_{\text{eff}}(r)}{dr^2}\right|_{r=r_0} > 0,$$

that is,

$$\left.\frac{d^2 V(r)}{dr^2}\right|_{r=r_0} + 3\,\frac{c^2}{r_0^4} > 0.$$

By using (2.11), we finally obtain

$$\left.\frac{d^2 V(r)}{dr^2}\right|_{r=r_0} + \frac{3}{r_0}\left.\frac{dV(r)}{dr}\right|_{r=r_0} > 0. \tag{2.12}$$

If the potential is given by a power law, $V(r) = Kr^\alpha$, and with $K\alpha > 0$, (2.12) gives

$$\alpha\,(\alpha - 1)\,K\,r_0^{\alpha-2} + 3\,\alpha\,K\,r_0^{\alpha-2} > 0,$$

that is $\alpha > -2$.

If we now go back again to (2.10), we see from the first equation that ϑ varies monotonically, since $\dot\vartheta$ is always positive and thus ϑ is always increasing. From the second one, we get

$$\dot r = \frac{dr}{dt} = \pm\sqrt{2\,(h - V_{\text{eff}})}$$

---

[5] What we have done corresponds, in the Lagrangian formalism, to the construction of the Routh function in the case of an ignorable variable (ϑ in our case).

[6] In fact, it corresponds to a repulsive force $\propto 1/r^3$.

and, by integrating,[7]

$$\int dt = \int \frac{dr}{\sqrt{2\,h - 2\,V(r) - c^2/r^2}}. \tag{2.13}$$

From the first of equations (2.10),

$$d\vartheta = \frac{c}{r^2}\,dt = \frac{c}{r^2}\,dr\left(\frac{dt}{dr}\right) = \pm\frac{c}{r^2}\,\frac{dr}{\sqrt{2\,(h - V_{\text{eff}})}}.$$

Then, we have, in addition to (2.13),

$$\int d\vartheta = \int \frac{c\,dr}{r^2\sqrt{2\,(h - V(r)) - c^2/r^2}}. \tag{2.14}$$

If the motion occurs in a bounded region, then $r_{\min} \le r \le r_{\max}$ and, corresponding to these values for r, the denominators of the integrals in (2.13) and (2.14) will also vanish, which in this case means that $\dot{r} = 0$ (points of inversion of the motion). In the interval of time in which r varies between $r_{\min}$ and $r_{\max}$ and goes back again to $r_{\min}$, the radius vector corresponding to the moving particle rotates by an angle $\Delta\vartheta$ (remember that $\vartheta(t)$ is an ever increasing function) given by

$$\Delta\vartheta = 2 \int_{r_{\min}}^{r_{\max}} \frac{c\,dr}{r\sqrt{2r^2\,(h - V(r)) - c^2}}. \tag{2.15}$$

A necessary and sufficient condition for the motion to correspond to a closed trajectory is that this angle is commensurable with 2π, i.e. $\Delta\vartheta = 2\pi m/n$ with integer m and n. If this does not occur, the trajectory will "fill" the annulus $r_{\min} \le r \le r_{\max}$ in an everywhere dense manner; this statement can be proved in a manner analogous to the case of the planar oscillator in Sect. 1.3. It is clear that the condition $\Delta\vartheta = 2\pi m/n$ will only exceptionally be fulfilled, while in general it will not. However, two remarkable cases exist which give rise to closed trajectories, while still remaining in the class of potentials $V(r) \propto r^\alpha$; these are the cases $\alpha = 2$, $\alpha = -1$. We defer to Sect. 2.3 the proof of the fact that among all the central forces only the mentioned potentials give rise to closed trajectories (Bertrand's theorem). For now we proceed to a direct verification. A few lines are sufficient to check that the case $\alpha = 2$ (elastic potential) is nothing but the planar oscillator already studied. As to $\alpha = -1$ (Newtonian or Coulombian potential), returning again to our previous definitions, $V(r) = -f_1(r) = -\mu r^{-1}$. Instead of demonstrating that in this case one has $\Delta\vartheta = 2\pi m/n$ (with integer m, n), we shall directly obtain the equation of the orbit.

Let us again consider system (2.7), which, in the Newtonian case, will be

$$\dot{\boldsymbol{r}} = \boldsymbol{v}, \qquad \dot{\boldsymbol{v}} = -\mu\,\frac{\boldsymbol{r}}{r^3}, \tag{2.16}$$

---

[7] See Footnote 2, Sect. 1.1.

with the first integrals

$$\boldsymbol{c} = \boldsymbol{r} \times \boldsymbol{v} = \text{const.}, \qquad h = \frac{1}{2}\dot{r}^2 + \frac{c^2}{2r^2} - \frac{\mu}{r} = \text{const.} \tag{2.17}$$

The energy integral, in the form (2.10), enabled us to arrive at (2.13) but, unlike (1.A.11a), this is integrable in terms of elementary functions but we cannot obtain by inversion an explicit function $r = r(t)$. However, we can obtain an explicit relation between r and ϑ (the equation of the orbit): this will be done by showing that a further vector exists which stays constant during the motion. From the already used relation

$$\frac{d}{dt}\left(\frac{\boldsymbol{r}}{r}\right) = \frac{\boldsymbol{c} \times \boldsymbol{r}}{r^3},$$

by multiplying it by $-\mu$ and by substituting (2.16), we have

$$\mu\frac{d}{dt}\left(\frac{\boldsymbol{r}}{r}\right) = \dot{\boldsymbol{v}} \times \boldsymbol{c}.$$

By integrating this relation, we obtain

$$\mu\left(\boldsymbol{e} + \frac{\boldsymbol{r}}{r}\right) = \boldsymbol{v} \times \boldsymbol{c}, \tag{2.18}$$

where

$$\boldsymbol{e} = \frac{\boldsymbol{v} \times \boldsymbol{c}}{\mu} - \frac{\boldsymbol{r}}{r} = \text{const.} \tag{2.19}$$

The constant vector $\boldsymbol{e}$ is called the Laplace–Runge–Lenz vector. In the case $\boldsymbol{c} = 0$, $\boldsymbol{r}/r = -\boldsymbol{e}$, and then $\boldsymbol{e}$ lies on the line of motion and $|\boldsymbol{e}| = 1$.

When $\boldsymbol{c} \neq 0$, $\boldsymbol{r} \cdot \boldsymbol{c} = 0$, and one sees immediately that $\boldsymbol{e} \cdot \boldsymbol{c} = 0$; then $\boldsymbol{e}$ lies in the plane of motion. By multiplying (2.18) scalarly by $\boldsymbol{r}$, one obtains

$$\mu(\boldsymbol{e} \cdot \boldsymbol{r} + r) = \boldsymbol{r} \cdot \boldsymbol{v} \times \boldsymbol{c} = \boldsymbol{r} \times \boldsymbol{v} \cdot \boldsymbol{c} = \boldsymbol{c} \cdot \boldsymbol{c} = c^2,$$

that is,

$$\boldsymbol{e} \cdot \boldsymbol{r} + r = \frac{c^2}{\mu}. \tag{2.20}$$

If $\boldsymbol{e} = 0$, then $r = c^2/\mu = \text{const.}$ and the motion is circular; from $v^2 = \dot{r}^2 + r^2\dot{\vartheta}^2$, now $\dot{r} = 0$ and $r^2\dot{\vartheta} = c$ and one has for the velocity $v = \text{const.} = \mu/c$ and, from the second of equations (2.17), $h = v^2/2 - \mu/r = -\mu^2/2c^2$. Therefore, for $\boldsymbol{e} = 0$, we have a circular motion with negative energy. In the case $\boldsymbol{e} \neq 0$, let us call ω the angle between $\boldsymbol{e}$ and the x axis and $f = \vartheta - \omega$ (see Fig. 2.2). By substituting in (2.20), one obtains $e\,r\cos f + r = c^2/\mu$, which we rewrite as

$$r = \frac{c^2/\mu}{1 + e\cos f}. \tag{2.21}$$

Equation (2.21) is the equation of the orbit (a conic section with a focus at O). The distance r of the moving point Q is minimum when $\cos f = 1$, $e > 0$.

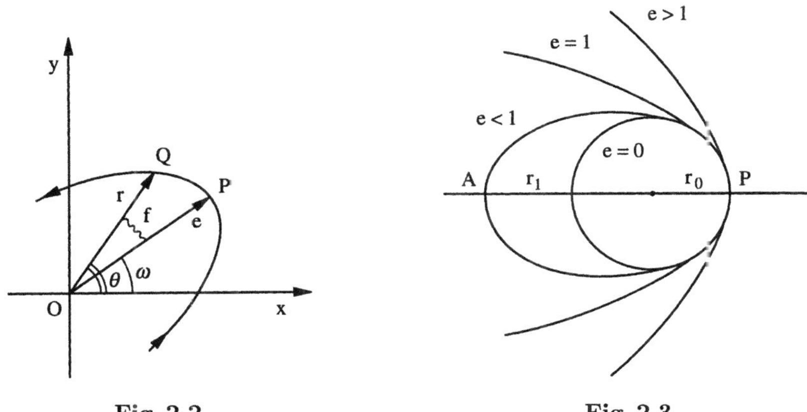

Fig. 2.2   **Fig. 2.3**

The vector e has a length equal to the eccentricity and points towards the *pericentre* (nearest point to the focus). The opposite point, the farthest one from the focus (which, obviously, exists only for closed orbits) is called the *apocentre* and the angle f the *true anomaly*. Pericentre and apocentre are called *apses* and the line joining them the *apse line*. In the apses, being $\dot{r} = 0$, the velocity is orthogonal to the radius vector. From the theory of conic sections, we know that:

1) for $\quad 0 < e < 1 \quad$ the curve is an ellipse,
2) for $\quad e = 1 \quad$ the curve is a parabola,
3) for $\quad e > 1 \quad$ the curve is a hyperbola.

Obviously, the case $e = 0$ is the circular orbit already considered (see Fig. 2.3). In the figure, we have

$$r_0 = \frac{c^2/\mu}{1+e}, \qquad r_1 = \frac{c^2/\mu}{1-e} = r_0 \frac{1+e}{1-e}.$$

As e is a constant vector, ω will also be constant, and then $\dot{f} = \dot{\vartheta}$. Since, as we know, $\dot{\vartheta} > 0$, f will be ever increasing with time and then (2.21) represents *Kepler's first law*.

So far, we have found, for the system (2.16), two vectors and one scalar which are constants of the motion: this means seven scalar constants. As the system is of order six, obviously the seven scalar constants cannot be independent: there must exist at least one relation among them. Actually there are two relations; one has been already seen: $e \cdot c = 0$. To obtain the second one, let us square (2.18):

$$\mu^2 \left(e^2 + \frac{2}{r} e \cdot r + 1 \right) = v^2 c^2$$

and, since

$$v^2 = 2\left(h + \frac{\mu}{r}\right), \qquad \boldsymbol{e} \cdot \boldsymbol{r} = \frac{c^2}{\mu} - r,$$

we get

$$\mu^2(e^2 - 1) = 2\,h\,c^2. \tag{2.22a}$$

We therefore have $7 - 2 = 5$ first integrals and our original system of order 6 can be considered reduced to a quadrature. From (2.22a), we can further infer that:

1) $e < 1$ (elliptic motion)   corresponds to $h < 0$,
2) $e = 1$ (parabolic motion)   corresponds to $h = 0$,
3) $e > 1$ (hyperbolic motion)   corresponds to $h > 0$.

For an ellipse the parameter $(c^2/\mu$ in our case) is given by $(1 - e^2)\,a$, where a is the semi-major axis. From (2.22a), we therefore have

$$c^2 = a\,\mu\,(1 - e^2) \tag{2.22b}$$

and

$$a = -\frac{\mu}{2\,h}$$

(the semi-major axis depends only on the total energy). Moreover, from (2.8) and (2.22b), we have for the area swept out by the radius vector in unit time

$$\frac{dS}{dt} = \frac{1}{2}\,c = \frac{1}{2}\sqrt{\mu\,a(1 - e^2)}.$$

On the other hand the area of the ellipse will be $S = \pi\,a\,b = \pi\,a^2\,\sqrt{1 - e^2}$ (b = semi-minor axis); if we call P the period of the motion and n the mean motion defined by $n = 2\pi/P$, then

$$\frac{dS}{dt} = \frac{S}{P} = \frac{\pi a b}{P} = \frac{n a b}{2} = \frac{1}{2}\,n\,a^2\sqrt{1 - e^2}.$$

Therefore,

$$\frac{1}{2}\sqrt{\mu a\,(1 - e^2)} = \frac{1}{2}\,n a^2\sqrt{1 - e^2},$$

from which

$$n^2 a^3 = \mu.$$

This is the analytic expression of *Kepler's third law*.

For the sake of completeness, we now present another method (Binet's method) for obtaining (2.21). From (2.8), $r^2 d\vartheta = c\,dt$, and then:

$$\frac{d}{dt} = \frac{c}{r^2}\frac{d}{d\vartheta}, \qquad \frac{d^2}{dt^2} = \frac{c}{r^2}\frac{d}{d\vartheta}\left(\frac{c}{r^2}\frac{d}{d\vartheta}\right).$$

For the radial acceleration, which, for a force $-f(r)\frac{\boldsymbol{r}}{r}$, will be given in the plane of motion by $a_r = \ddot{r} - r\,\dot{\vartheta}^2 = -f(r)$, we have

$$\frac{c}{r^2}\frac{d}{d\vartheta}\left(\frac{c}{r^2}\frac{dr}{d\vartheta}\right) - \frac{c^2}{r^3} = -f(r).$$

As

$$\frac{1}{r^2}\frac{dr}{d\vartheta} = -\frac{d}{d\vartheta}\left(\frac{1}{r}\right),$$

and setting $u = 1/r$, we finally get

$$c^2 u^2 \frac{d}{d\vartheta}\left(-\frac{du}{d\vartheta}\right) - c^2 u^3 = -f\left(\frac{1}{u}\right),$$

from which

$$\frac{d^2 u}{d\vartheta^2} + u = \frac{1}{c^2 u^2}\, f\left(\frac{1}{u}\right) \tag{2.23}$$

(the differential equation of the orbit in the variable u). Equation (2.23) is invariant under the exchange $\vartheta \leftrightarrow -\vartheta$: therefore if we assume $\vartheta = 0$ to be the position of one apse we have immediately that the orbit is symmetric with respect to the radius passing through the apse itself. Obviously this applies to all apses. In terms of potential, by using the function $f_1(r) = f_1(1/u)$ introduced at the beginning and exploiting the relation

$$\frac{d}{du} = \frac{dr}{du}\frac{d}{dr} = -\frac{1}{u^2}\frac{d}{dr},$$

we have, in place of (2.23),

$$\frac{d^2 u}{d\vartheta^2} + u = \frac{1}{c^2}\frac{d}{du}\, f_1\left(\frac{1}{u}\right). \tag{2.24}$$

For the Newtonian potential, $f_1(r) = \mu/r = \mu u$, (2.24) becomes

$$\frac{d^2 u}{d\vartheta^2} + u = \frac{\mu}{c^2},$$

which, integrated, gives $u = A\cos(\vartheta - \vartheta_0) + \mu/c^2$ and finally

$$r = \frac{1}{A\cos(\vartheta - \vartheta_0) + \mu/c^2}.$$

If we take $\vartheta_0 = 0$ and put $Ac^2/\mu = e$, we again find (2.21).

Let us pause for a moment to weighing the results we have obtained. Formally our problem was the integration of a system of order six: system (2.7). By leaving the function $f(r)$ undetermined and by using only the general properties of the central potentials, we have reduced the order of the system by four. To go further, determining another first integral, we were obliged to specify the form of $f(r)$ and, in particular, to put $f(r) = \mu/r^2$. This enabled us to attain, in an elegant way, the equation of the orbit without being really obliged to perform an integration. In fact, the introduction of the vector e provides us with the equation of the orbit, directly from the expression of the first integral. That is, (2.20) is a relation $r = r(\vartheta)$ which, by defining the true anomaly, becomes (2.21), the polar equation of a conic. At this point, as the system is reduced to order one, it is by definition integrable and then formally solved. As a matter of fact, things are not so simple: it will not be possible to write $r = r(t)$ and $\vartheta = \vartheta(t)$ explicitly, but we shall be obliged to introduce another variable, connected with r through trigonometric functions, whose expression as a function of time will be obtained only in an approximate form.

2.2 The Laplace–Runge–Lenz Vector

The results of the preceding section, even though obtained very simply and, in the case of the deduction of the equation of the orbit, also with speed and elegance, do not arise from the application of a general procedure but rather from intuitions. In every case the task was in fact to guess by what quantity a certain expression had to be multiplied in order that the time derivative of another expression (the first integral) would be equal to zero. Essentially, one guessed, in each case, the right integrating factor. Now, on the other hand, we want to see how these results can be obtained by applying the formal tools we have explained in the Parts B and C of Chap. 1.

Let us begin with Noether's theorem. To make things easier, let us assume that we have already demonstrated that the motion occurs in a plane and then consider the problem in the plane. The Lagrangian of the Kepler problem is

$$\mathcal{L} = \frac{1}{2}|\dot{\boldsymbol{r}}|^2 + \frac{\mu}{|\boldsymbol{r}|},$$

so that, if $\boldsymbol{r} \equiv (q_1, q_2, 0)$, and $\dot{\boldsymbol{r}} \equiv (\dot{q}_1, \dot{q}_2, 0)$,

$$\mathcal{L} = \frac{1}{2}\left(\dot{q}_1^2 + \dot{q}_2^2\right) + \frac{\mu}{\sqrt{q_1^2 + q_2^2}}, \tag{2.25}$$

for a unit mass. Let us rewrite the generalized Killing equations (1.B.41) and (1.B.42) and the expression for the conserved quantity (1.B.48), confining ourselves to functions that do not depend on time, since the Lagrangian is like this:

$$\xi_i \frac{\partial \mathcal{L}}{\partial q_i} + \frac{\partial \mathcal{L}}{\partial \dot{q}_i}\left[\frac{\partial \xi_i}{\partial q_l}\dot{q}_l - \dot{q}_i\dot{q}_l\frac{\partial \tau}{\partial q_l}\right] + \mathcal{L}\frac{\partial \tau}{\partial q_i}\dot{q}_i = \frac{\partial f}{\partial q_i}\dot{q}_i, \tag{2.26a}$$

$$\mathcal{L}\frac{\partial \tau}{\partial \dot{q}_i} + \frac{\partial \mathcal{L}}{\partial \dot{q}_l}\left[\frac{\partial \xi_l}{\partial \dot{q}_i} - \dot{q}_l\frac{\partial \tau}{\partial \dot{q}_i}\right] = \frac{\partial f}{\partial \dot{q}_i}, \qquad i = 1, 2, \tag{2.26b}$$

$$I = \left(\mathcal{L} - \dot{q}_i\frac{\partial \mathcal{L}}{\partial \dot{q}_i}\right)\tau + \frac{\partial \mathcal{L}}{\partial \dot{q}_i}\xi_i - f. \tag{2.27}$$

It is easy to check that the transformation defined by

$$\tau \equiv -1, \qquad \xi_i \equiv 0 \ (\forall i), \qquad f \equiv 0,$$

corresponds to $\mathcal{H} = \dot{q}_i\,\partial\mathcal{L}/\partial\dot{q}_i - \mathcal{L} = \text{const.}$, i.e to the conservation of energy. For the spatial transformation with constant coefficients

$$\tau \equiv 0, \qquad \xi_i = \omega_{ij}\,q_j,$$

one has from equation (2.26a), with $f \equiv 0$,

$$\omega_{ij}\left[-\mu\,(q_1^2 + q_2^2)^{-3/2}q_i\,q_j + \dot{q}_i\,\dot{q}_j\right] = 0$$

and then $\omega_{ij} = -\omega_{ji}$ ($\omega_{11} = \omega_{22} = 0$, $\omega_{12} = -\omega_{21}$). As ω_{12} is an arbitrary constant, we can put it equal to -1 and obtain at once

$$I = q_1\,\dot{q}_2 - q_2\,\dot{q}_1 = c,$$

that is, the conservation of angular momentum.

If we now try to apply the spatial transformation depending on the velocity,

$$\tau \equiv 0, \qquad \xi_i = a_{ij}\,\dot{q}_j, \qquad (a_{ij}\ \text{const.}\ \ \forall\,i,j),$$

we again obtain, with $f = \mathcal{L}$ and $a_{11} = a_{22} = 1$, $a_{12} = a_{21} \equiv 0$, the conservation of energy (we omit the calculations for the sake of brevity). In this case we also have new evidence that the transformation corresponding to a given first integral is not unique. In fact, if the transformation is

$$\overline{\xi}_1 = \dot{q}_1, \qquad \overline{\xi}_2 = \dot{q}_2, \qquad \overline{f} = \mathcal{L},$$

by applying (1.B.54) with $\tau = -1$, we get

$$\begin{cases} \tau = -1, \\ f = \overline{f} + \mathcal{L}\tau = 0, \\ \xi_1 = \overline{\xi}_1 + \dot{q}_1\tau = 0, \\ \xi_2 = \overline{\xi}_2 + \dot{q}_2\tau = 0, \end{cases}$$

which is just the transformation considered above. If we now want to determine the transformation (or the transformations) corresponding to the conservation of the Laplace–Runge–Lenz vector, in this case the integral no longer being quadratic only in the q and $\dot{q}$, we must refer to a more complicated transformation. As in the case of the anisotropic planar oscillator[8] with $\omega_2 = 2\,\omega_1$ considered in Sect. 1.7, let us study the transformation

$$\xi_i = A_{ijk}\,\dot{q}_j\,q_k, \qquad \tau \equiv 0, \qquad (A_{ijk}\ \text{const.}\ \forall\,i,j,k).$$

Equations (2.26b) give

$$A_{111}\,q_1\,\dot{q}_1 + A_{112}\,q_2\,\dot{q}_1 + A_{211}\,q_1\,\dot{q}_2 + A_{212}\,q_2\,\dot{q}_2 = \frac{\partial f}{\partial \dot{q}_1},$$

$$A_{121}\,q_1\,\dot{q}_1 + A_{122}\,q_2\,\dot{q}_1 + A_{221}\,q_1\,\dot{q}_2 + A_{222}\,q_2\,\dot{q}_2 = \frac{\partial f}{\partial \dot{q}_2},$$

---

[8] The analogies and the interdependencies between Keplerian motion and the planar oscillator will be investigated at various places later in the book

and, by integrating, we get

$$\frac{1}{2}\left(A_{111}\,q_1 + A_{112}\,q_2\right)\dot{q}_1^2 + \left(A_{211}\,q_1 + A_{212}\,q_2\right)\dot{q}_1\dot{q}_2 + g_1\left(q_1, q_2, \dot{q}_2\right) = f,$$

$$\frac{1}{2}\left(A_{221}\,q_1 + A_{222}\,q_2\right)\dot{q}_2^2 + \left(A_{121}q_1 + A_{122}q_2\right)\dot{q}_1\dot{q}_2 + g_2\left(q_1, q_2, \dot{q}_1\right) = f,$$

g_1, g_2 being arbitrary functions. The last two relations can be satisfied by $A_{211} = A_{121}$, $A_{212} = A_{122}$; hence f is of the form

$$\frac{1}{2}\left(A_{111}q_1 + A_{112}q_2\right)\dot{q}_1^2 + \frac{1}{2}\left(A_{221}q_1 + A_{222}q_2\right)\dot{q}_2^2 + \left(A_{211}q_1 + A_{212}q_2\right)\dot{q}_1\dot{q}_2 + g,$$

where $g = g(q_1, q_2)$ is a new function to be determined. By substituting in (2.26a), we have

$$\begin{aligned}
\frac{\partial f}{\partial q_1}\dot{q}_1 + \frac{\partial f}{\partial q_2}\dot{q}_2 &= \frac{\partial g}{\partial q_1}\dot{q}_1 + \frac{\partial g}{\partial q_2}\dot{q}_2 + \frac{1}{2}A_{111}\dot{q}_1^3 + \frac{1}{2}A_{221}\dot{q}_1\dot{q}_2^2 \\
&\quad + A_{211}\dot{q}_1^2\dot{q}_2 + \frac{1}{2}A_{112}\dot{q}_1^2\dot{q}_2 + \frac{1}{2}A_{222}\dot{q}_2^3 + A_{212}\dot{q}_1\dot{q}_2^2 \\
&= \frac{\partial g}{\partial q_1}\dot{q}_1 + \frac{\partial g}{\partial q_2}\dot{q}_2 + \frac{1}{2}A_{111}\dot{q}_1^3 + \frac{1}{2}A_{222}\dot{q}_2^3 \\
&\quad + \left(A_{212} + \frac{1}{2}A_{221}\right)\dot{q}_1\dot{q}_2^2 + \left(A_{211} + \frac{1}{2}A_{112}\right)\dot{q}_1^2\dot{q}_2 \\
&= -\frac{\mu}{r^3}\Big[A_{111}q_1^2\dot{q}_1 + A_{112}q_1q_2\dot{q}_1 + A_{121}q_1^2\dot{q}_2 + A_{122}q_1q_2\dot{q}_2 \\
&\quad + A_{211}q_1q_2\dot{q}_1 + A_{212}q_2^2\dot{q}_1 + A_{221}q_1q_2\dot{q}_2 + A_{222}q_2^2\dot{q}_2\Big] \\
&\quad + A_{111}\dot{q}_1^3 + (A_{121} + A_{112})\dot{q}_1^2\dot{q}_2 + A_{122}\dot{q}_1\dot{q}_2^2 + A_{211}\dot{q}_1^2\dot{q}_2 \\
&\quad + A_{221}\dot{q}_1\dot{q}_2^2 + A_{212}\dot{q}_1\dot{q}_2^2 + A_{222}\dot{q}_2^3.
\end{aligned}$$

For this to be meaningful, it must be the case that

$$A_{111} = A_{222} \equiv 0, \qquad A_{221} = -2A_{122}, \qquad A_{112} = -2A_{121}.$$

Then

$$A_{212} = A_{122} = A, \quad A_{211} = A_{121} = B,$$
$$A_{221} = -2A, \quad A_{112} = -2B, \quad A_{111} = A_{222} \equiv 0,$$

where A and B are to be determined. One obtains

$$\frac{\partial g}{\partial q_1} = \frac{\mu}{r^3}\left(B\,q_1q_2 - A\,q_2^2\right), \qquad \frac{\partial g}{\partial q_2} = \frac{\mu}{r^3}\left(A\,q_1q_2 - B\,q_1^2\right),$$

and then $g = -\frac{\mu}{r}(A\,q_1 + B\,q_2)$, where we have set the integration constant to zero. Finally,

$$f = -B\,q_2\,\dot{q}_1^2 - A\,q_1\,\dot{q}_2^2 + (B\,q_1 + A\,q_2)\,\dot{q}_1\,\dot{q}_2 - \frac{\mu}{r}\left(A\,q_1 + B\,q_2\right).$$

The generators of the transformation will be

$$\xi_1 = A\dot{q}_2q_2 + B\left(q_1\dot{q}_2 - 2q_2\dot{q}_1\right), \qquad \xi_2 = A\left(q_2\dot{q}_1 - 2q_1\dot{q}_2\right) + Bq_1\dot{q}_1.$$

Clearly, one of the two constants can be taken equal to zero, and therefore two choices are possible: $B = 0$, $A \neq 0$ and $A = 0$, $B \neq 0$. Correspondingly, one will have two a priori different first integrals. Let us look at the first case: $B = 0$. Here

$$f = -A\, q_1\, \dot{q}_2^2 + A\, q_2\, \dot{q}_1\, \dot{q}_2 - A\mu\, \frac{q_1}{r}, \qquad \xi_1 = A\, q_2\, \dot{q}_2, \qquad \xi_2 = -A(c - \dot{q}_2\, q_1).$$

The first integral turns out to be $I_1 = A(-c\dot{q}_2 + \mu\, \frac{q_1}{r})$. By choosing $A = -1/\mu$, one obtains

$$I_1 = \frac{c}{\mu}\, \dot{q}_2 - \frac{q_1}{r} = e_1,$$

that is, the component along the q_1 axis of the Laplace–Runge–Lenz vector. The second case, $A = 0$, will give

$$f = -B\, q_2\dot{q}_1^2 + B\, q_1\dot{q}_1\dot{q}_2 - B\mu\, \frac{q_2}{r}, \qquad \xi_1 = B(c - \dot{q}_1 q_2), \qquad \xi_2 = B\, q_1\dot{q}_1$$

and the first integral $I_2 = B(c\dot{q}_1 + \mu\, \frac{q_2}{r})$. By choosing $B = -1/\mu$, one obtains

$$I_2 = -\frac{c}{\mu}\, \dot{q}_1 - \frac{q_2}{r} = e_2,$$

that is, the component along the q_2 axis of the Laplace–Runge–Lenz vector. The conservation of the Laplace–Runge–Lenz vector corresponds then to the transformations having as their generators:

$$\begin{aligned}
\xi_1^1 &= -\frac{1}{\mu}\, \dot{q}_2 q_2, & \xi_1^2 &= \frac{1}{\mu}\, \dot{q}_1 q_2 - \frac{c}{\mu}, \\
\xi_2^1 &= \frac{1}{\mu}\, \dot{q}_2 q_1 + \frac{c}{\mu}, & \xi_2^2 &= -\frac{1}{\mu}\, \dot{q}_1 q_1.
\end{aligned} \qquad (2.29)$$

They are then "rotations" in the $q_1 q_2$ plane (with coefficients depending on velocity) and translations along the axes with constant coefficients.[9] To sum up, through Noether's theorem we have obtained all the first integrals of the Kepler problem. Obviously, concerning the real applicability of the theorem the remarks already made in Sect. 1.7 remain sound. Furthermore, one might object that, after all, the same result has already been obtained in a simpler way, though not in the context of a general procedure, and then the use of a general formalism even becomes unnecessary. The obvious answer is that cases more complex than the simple Kepler problem (actually not so simple) exist and, therefore, general methods help us when intuition cannot. Moreover, in simple cases also, Noether's theorem enables us to penetrate better the nature of the problem.

Let us now move on to apply, still to the problem of Keplerian motion, another general method, the method of integration of the Hamilton–Jacobi

---

[9] In this case, a direct geometric interpretation in the space of the q_i and $\dot{q}_i$, as in the case of the oscillator, does not exist.

equation by separation of the variables. We have already seen, in Sect. 1.15, this method as applied to the anisotropic planar oscillator (with $\omega_2 = 2\omega_1$) in parabolic coordinates. Let us now define, in contrast to that case, the parabolic coordinates λ, ν in the following way[10]

$$q_1 = \lambda^2 - \nu^2, \qquad q_2 = 2\lambda\nu, \qquad r = \sqrt{q_1^2 + q_2^2} = \lambda^2 + \nu^2. \tag{2.30}$$

In this case, the coordinate lines consist of two families of mutually orthogonal parabolas, with the axis coinciding with the q_1 axis.

By substituting (2.30), the Lagrangian (2.25) becomes

$$\mathcal{L} = 2\left(\lambda^2 + \nu^2\right)\left(\dot\lambda^2 + \dot\nu^2\right) + \frac{\mu}{\lambda^2 + \nu^2}. \tag{2.31}$$

From this, one obtains the momenta

$$p_\lambda = 4\left(\lambda^2 + \nu^2\right)\dot\lambda, \qquad p_\nu = 4\left(\lambda^2 + \nu^2\right)\dot\nu \tag{2.32}$$

and the Hamiltonian

$$\mathcal{H} = p_\lambda\dot\lambda + p_\nu\dot\nu - \mathcal{L} = \frac{1}{8}\frac{p_\lambda^2 + p_\nu^2}{\lambda^2 + \nu^2} - \frac{\mu}{\lambda^2 + \nu^2}. \tag{2.33}$$

The Hamilton–Jacobi equation is

$$\frac{1}{8\left(\lambda^2 + \nu^2\right)}\left[\left(\frac{\partial S^\star}{\partial\lambda}\right)^2 + \left(\frac{\partial S^\star}{\partial\nu}\right)^2\right] - \frac{\mu}{\lambda^2 + \nu^2} = \alpha_1, \tag{2.34}$$

where the first integration constant $\alpha_1 = h$ is the total energy. Equation (2.34) can then be written in the form

$$\frac{1}{8}\left[\left(\frac{\partial S^\star}{\partial\lambda}\right)^2 + \left(\frac{\partial S^\star}{\partial\nu}\right)^2\right] - \mu = \alpha_1\lambda^2 + \alpha_1\nu^2$$

and can be satisfied by putting

$$\frac{1}{8}\left(\frac{dS_\lambda^\star}{d\lambda}\right)^2 - \frac{1}{2}\mu - \alpha_1\lambda^2 = \alpha_2, \qquad \frac{1}{8}\left(\frac{dS_\nu^\star}{d\nu}\right)^2 - \frac{1}{2}\mu - \alpha_1\nu^2 = -\alpha_2,$$

where α_2 is a new arbitrary constant. Then we have for α_2

$$\alpha_2 = \frac{1}{8}p_\lambda^2 - \frac{1}{2}\mu - \alpha_1\lambda^2 = -\frac{1}{8}p_\nu^2 + \frac{1}{2}\mu + \alpha_1\nu^2.$$

Taking the half sum of the two different expressions,

$$\alpha_2 = \frac{1}{16}\left(p_\lambda^2 - p_\nu^2\right) + \frac{1}{2}\alpha_1(\nu^2 - \lambda^2),$$

and substituting $\alpha_1 = \mathcal{H}$ (with $\mathcal{H}$ provided by (2.33) as an explicit function of the canonical variables), we finally get

---

[10]The reason for this choice will become clear in Sects. 2.6 and 5.4

$$\alpha_2 = \frac{1}{8} \frac{\nu^2 p_\lambda^2 - \lambda^2 p_\nu^2}{\lambda^2 + \nu^2} - \frac{1}{2} \mu \frac{(\nu^2 - \lambda^2)}{\lambda^2 + \nu^2}. \tag{2.35}$$

If we now substitute (2.32) and (2.30) into (2.35), we find after some calculations that

$$\alpha_2 = -\frac{1}{2} \mu \, e_1.$$

In this case, therefore, the first integral connected to the constancy of e has been obtained in a quite automatic way, by separating the variables in the Hamilton–Jacobi equation. No particular search has been carried out, no *ad hoc* position has been adopted.

Lastly, if we ask ourselves what is the conclusion that can be drawn from the application of Liouville's theorem (Sect. 1.18), the answer is immediate. Going back again to the three-dimensional formulation, if we denote by q_1, q_2, q_3 the components of r and p_1, p_2, p_3 the components of $\dot{r}$, the Hamiltonian for unit mass will be

$$\mathcal{H} = \frac{1}{2}(p_1^2 + p_2^2 + p_3^2) - \mu / \sqrt{q_1^2 + q_2^2 + q_3^2},$$

and the components of the angular momentum

$$c_1 = q_2 \, p_3 - q_3 \, p_2, \qquad c_2 = q_3 \, p_1 - q_1 \, p_3, \qquad c_3 = q_1 \, p_2 - q_2 \, p_1.$$

We have already seen in Sect. 2.1 that $\mathcal{H}$ and c are constants of the motion. It is easy to check that for $c^2 = c_1^2 + c_2^2 + c_3^2$ one has

$$(c^2, \mathcal{H}) = 0, \qquad (c_i, \mathcal{H}) = 0, \qquad (c^2, c_i) = 0, \qquad i = 1, 2, 3;$$

therefore, if we choose any one of the components of c, it represents together with $\mathcal{H}$ and c^2 a tern of independent integrals in involution. According to Lie's theorem, this corresponds to the existence of six first integrals and then the system (of order six) is completely integrable.

2.3 Bertrand's Theorem and Related Questions

Let us now tackle the problem of determining what are all the possible forms of central forces giving rise to closed orbits (i.e. $\Delta\vartheta$ of equation (2.15) commensurable with 2π). The question was solved by J. Bertrand in 1873.[11] To demonstrate Bertrand's theorem, we shall make use of a more restrictive definition of stability than the already mentioned Lagrange-stability. The difference between the two definitions will be made clear later on. Now, we shall adopt the following point of view: all central forces, as we have seen, admit circular orbits (whether stable or unstable); if we slightly perturb a circular

---

[11] J. Bertrand: Théorème relatif au mouvement d'un point attiré vers un centre fixe, *Comptes Rendus*, **LXXVII**, 849–853 (1873).

orbit we can have a perturbed orbit, whether closed or not. It will be closed if the initial circular orbit is stable, open in the opposite case. Therefore closed orbits will be generated by those central forces which admit stable circular orbits.

Starting from the equation of the radial motion,

$$\ddot{r} - r\dot{\vartheta}^2 = -\frac{dV(r)}{dr} \tag{2.36}$$

and using the integral of areas $r^2\dot{\vartheta} = c$, we have

$$\ddot{r} - \frac{c^2}{r^3} = -\frac{dV(r)}{dr}. \tag{2.37}$$

For a circular orbit, $r = \text{const.} = r_0$, $\ddot{r} = 0$ and then (as we have already seen)

$$\frac{c^2}{r_0^3} = \left.\frac{dV(r)}{dr}\right|_{r=r_0}. \tag{2.38}$$

If the circular orbit is perturbed, and writing $x = r - r_0$, the perturbation, one obtains from (2.37)

$$\ddot{x} - c^2(x + r_0)^{-3} = -\left.\frac{dV(r)}{dr}\right|_{r=x+r_0}.$$

By expanding in series in the small quantity x, we get

$$\ddot{x} - \frac{c^2}{r_0^3}\left(1 - 3\frac{x}{r_0} + \ldots\right) = -\left.\frac{dV(r)}{dr}\right|_{r=r_0} - \left.\frac{d^2V(r)}{dr^2}\right|_{r=r_0} x + \ldots \quad .$$

Finally, by using (2.38), we obtain

$$\ddot{x} + \left(3\left.\frac{dV(r)}{dr}\right|_{r=r_0} \cdot \frac{1}{r_0} + \left.\frac{d^2V(r)}{dr^2}\right|_{r=r_0} + \ldots\right)x = 0, \tag{2.39}$$

where we have neglected all the terms of higher order than the first.

Now, if the term in parenthesis in (2.39) is positive, then (2.39) is the equation of a harmonic oscillation, and this means that the perturbation oscillates around r_0: we say that this condition must be satisfied in order that the circular orbit be stable (see Fig. 2.4).

Actually, the condition $3V'(r_0)/r_0 + V''(r_0) > 0$ is the necessary and sufficient condition, as we have already seen, for the existence of a minimum of the potential $V_{\text{eff}}(r)$ in $r = r_0$ (i.e. Lagrange–stability). Now we consider this condition necessary only. For the circular orbit to be stable, it is thus necessary that the term considered is positive and so corresponds to the squared frequency of the oscillations of the perturbation around r_0:

$$\omega^2 = 3\frac{V'(r_0)}{r_0} + V''(r_0) > 0. \tag{2.40}$$

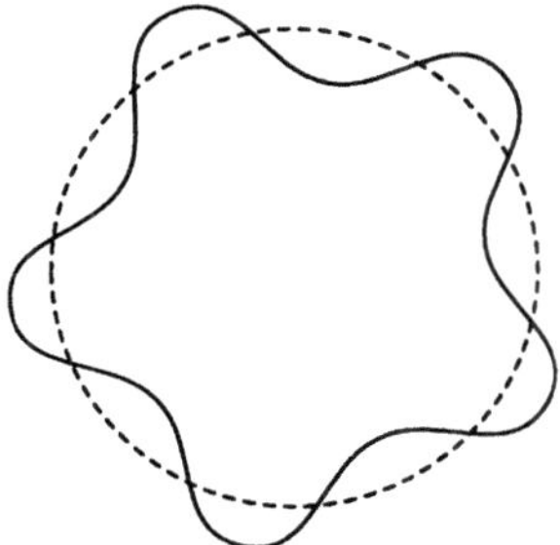

Fig. 2.4

Let us now move on to deal directly with the orbit. For an almost circular orbit, we have seen that r oscillates around the value r_0 (if the orbit is stable). From (2.40), the period of these oscillations is

$$\tau = \frac{2\pi}{\omega} = \frac{2\pi}{\sqrt{V''(r_0) + \frac{3}{r_0}V'(r_0)}}. \tag{2.41}$$

Let us evaluate the apsidal angle, that is the variation of the polar angle ϑ during the time in which r oscillates from a minimum to the next maximum. The time is given, obviously, by $1/2\tau$; to evaluate the angle swept out in this time we shall have to multiply by a mean angular velocity. Since $\dot\vartheta = c/r^2$ and r oscillates around r_0, we shall assume the mean angular velocity to be

$$\overline{\dot\vartheta} = \frac{c}{r_0^2} = \sqrt{\frac{V'(r_0)}{r_0}}$$

because of (2.33). The apsidal angle will therefore be

$$\frac{1}{2}\,\Delta\vartheta \simeq \frac{1}{2}\,\tau\overline{\dot\vartheta} = \frac{\pi}{\sqrt{V''(r_0) + \frac{3}{r_0}V'(r_0)}}\sqrt{\frac{V'(r_0)}{r_0}} = \pi\sqrt{\frac{V'(r_0)}{r_0 V''(r_0) + 3V'(r_0)}}. \tag{2.42}$$

Within certain limits, the apsidal angle will not depend on the particular chosen value r_0. In fact, in the case of closed orbits the number of oscillations must be the same at any distance where closed orbits are permitted; otherwise one would have discontinuity in the number of oscillations when r_0 varies, and corresponding to this discontinuity the orbit could not be closed. Therefore we shall impose[12] the condition that the quantity

$$\frac{V'(r)}{rV''(r) + 3V'(r)}$$

---

[12]We write r for the variable r_0.

be a positive constant. Introducing the constant $\alpha > -2$, we express this condition in the form

$$\frac{V'(r)}{rV''(r) + 3V'(r)} = \frac{1}{(2+\alpha)} > 0. \tag{2.43}$$

This condition is the necessary and sufficient condition we wanted for the stability of the circular orbit. At variance with Lagrange's criterion, which only refers to the behaviour of $V_{\mathrm{eff}}(r)$ (see (2.12)), now the polar angle ϑ is also involved (see (2.42)); therefore the conditions to be fulfilled are more stringent. By integrating (2.43), we have two cases:

$$V(r) = A\ln r \qquad (\text{when } \alpha = 0),$$
$$V(r) = kr^{\alpha} \qquad (\alpha \neq 0,\ > -2),$$

with A and k being integration constants. The logarithmic potential cannot give rise to closed orbits, since in this case $\Delta\vartheta = \pi\sqrt{2}$, and then the apsidal angle is not commensurable with 2π.

We are left with the cases of $V(r) = kr^{\alpha}$ with $-2 < \alpha < 0$ and $\alpha > 0$. To determine what are the allowed values for α in these two cases, let us try to obtain in another way an estimate of the integral coresponding to $\frac{1}{2}\Delta\vartheta$ (Arnold). The method consists in evaluating the integral for a limiting situation, where it is easily evaluable; since, as we have said, it must stay constant when r varies, the value obtained is valid in general. From (2.15), we have

$$\frac{1}{2}\Delta\vartheta = \int_{r_{\min}}^{r_{\max}} \frac{c\,dr}{r\sqrt{2r^2\left(h - V(r)\right) - c^2}}.$$

For the case $\alpha > 0$, let us consider the limit for $r \to \infty : V(r) = kr^{\alpha}$ will tend to ∞ for $r \to \infty$; for the denominator to remain finite, h must also tend to ∞, while the last term will tend to zero. It is evident from this that the constant k must be greater than zero. Let us now make the change of variable $u = c/r$, $du = -(c/r^2)dr$; then

$$\frac{1}{2}\Delta\vartheta = \int_{u_{\min}}^{u_{\max}} \frac{du}{\sqrt{2(h - V_{\mathrm{eff}}(u))}},$$

with $V_{\mathrm{eff}}(u) = V(c/u) + u^2/2$. Let us further set $y = u/u_{\max}$; then

$$\frac{1}{2}\Delta\vartheta = \int_{y_{\min}}^{1} \frac{dy}{\sqrt{2\left(\widetilde{V}_{\mathrm{eff}}(1) - \widetilde{V}_{\mathrm{eff}}(y)\right)}}, \tag{2.44}$$

with

$$\widetilde{V}_{\mathrm{eff}}(y) = \frac{1}{2}y^2 + \frac{1}{u_{\max}^2}V\left(\frac{c}{yu_{\max}}\right),$$

taking into account that, for $r = r_{\min}$,

$$h = \frac{c^2}{2r_{\min}^2} + V(r_{\min}) = \frac{1}{2}\,u_{\max}^2 + V(c/u_{\max});$$

therefore

$$\tilde{V}_{\text{eff}}(1) = \frac{h}{u_{\max}^2}.$$

For $h \to \infty$, $u_{\max} \to \infty$, $y_{\min} \to 0$ and the second term in $\tilde{V}_{\text{eff}}(y)$ is negligible. Thus

$$\frac{1}{2}\,\Delta\vartheta = \int_0^1 \frac{dy}{\sqrt{1-y^2}} = \frac{\pi}{2}.$$

Therefore, for $\alpha > 0$,

$$\Delta\vartheta = 2\,\frac{\pi}{\sqrt{2+\alpha}} = \pi,$$

and $\alpha = 2$; consequently $V(r) = kr^2$, with $k > 0$ (the case of the planar oscillator).

Let us now examine the other case: $-2 < \alpha < 0$. It is convenient to put $\beta = -\alpha$. Then

$$V(r) = kr^{-\beta} \qquad (0 < \beta < 2).$$

The limiting situation convenient in this case is that given by $h \to -0$; correspondingly $r_{\min}$ will tend to a finite value and $r_{\max}$ to $+\infty$ (therefore $u_{\min} \to 0$, $u_{\max} \to$ finite quantity). Moreover one sees that k must be negative. We shall therefore write

$$V(r) = -\mu\,r^{-\beta}, \qquad \text{with} \quad \mu > 0.$$

Reconsidering (2.44), we have in the limit

$$\frac{1}{2}\,\Delta\vartheta = \int_0^1 \frac{dy}{\sqrt{-2\,\tilde{V}_{\text{eff}}(y)}}.$$

The radical quantity is

$$-2\left[\frac{y^2}{2} - \frac{\mu}{u_{\max}^2}\left(\frac{c}{yu_{\max}}\right)^{-\beta}\right] = y^\beta\left[2\mu r_{\min}^{-\beta}\left(\frac{c}{r_{\min}}\right)^{-2}\right] - y^2.$$

But in $r_{\min}$, since $h = 0$, $c^2/2\,r_{\min}^2 + V(r_{\min}) = 0$ as well, that is

$$(c/r_{\min})^2 = -2\,V(r_{\min}) = 2\,\mu r_{\min}^{-\beta}.$$

Therefore our radical is simply reduced to $y^\beta - y^2$, and for $\Delta\vartheta$

$$\frac{1}{2}\,\Delta\vartheta = \int_0^1 \frac{dy}{\sqrt{y^\beta - y^2}} = \frac{\pi}{2-\beta}.$$

But also

$$\frac{1}{2}\,\Delta\vartheta = \frac{\pi}{\sqrt{2+\alpha}} = \frac{\pi}{\sqrt{2-\beta}},$$

and so $\sqrt{2-\beta} = 2 - \beta$ and $\beta = 1$. As a consequence $V(r) = -\mu/r$, that is, the Newtonian potential.

After having demonstrated that the only central forces giving rise to closed orbits derive from $V(r) = kr^2$ and $V(r) = -\mu/r$, Bertrand also tackled the problem of singling out the forces for which the closed orbits were ellipses. The problem was solved at the same time by Darboux and Halphen,[13] who demonstrated that the potentials $V(r) = kr^2$ and $V(r) = -\mu/r$ give elliptic closed orbits. Therefore the closed orbits can only be ellipses. The Newtonian potential, as we have seen, gives $\Delta\vartheta = 2\pi$, while for the planar oscillator we get $\Delta\vartheta = \pi$. The two different values are not in contradiction, since for the oscillator the radius vector originates from the centre of the ellipse while for the Newtonian potential the origin is one of the foci.

If we add to the Newtonian potential a "small" term (perturbation), this will cause an increase or decrease of the apsidal angle. The more frequent case in the solar system is that where the gravitational perturbation of the motion of a planet due to the action of the other planets is represented by a force of the type ε/r^4 (ε is a small quantity and such that powers higher than the first are neglected)[14]. Generally there will be a force

$$f(r) = -\frac{k}{r^2} - \frac{\varepsilon}{r^4}.$$

The angle $\Delta\vartheta$ in this case becomes

$$\Delta\vartheta = 2\pi\sqrt{\frac{V'(r_0)}{r_0 V''(r_0) + 3V'(r_0)}} = 2\pi\left[3 + r_0\frac{f'(r_0)}{f(r_0)}\right]^{-\frac{1}{2}}$$

$$= 2\pi\left[\frac{1 - \varepsilon k^{-1} r_0^{-2}}{1 + \varepsilon k^{-1} r_0^{-2}}\right]^{-\frac{1}{2}} \approx 2\pi\left(1 + \frac{\varepsilon}{kr_0^2}\right),$$

where the higher powers of ε/kr_0^2 have been neglected. Therefore, the apse advances or regresses according to whether ε is positive or negative. In the well-known case of Mercury, there is an advance of the perihelion of $574''$ in a century; calculations of the type mentioned above (obviously, with a higher accuracy) account for an advance of $531''$ in a century. There remains a discrepancy of $43''$ which, as is known, can be accounted for by Einstein's theory of general relativity.

---

[13]A demonstration, simplified with respect to the original one, can be found in P. Appel: *Traité de Mécanique Rationnelle*, 3rd edn. Vol. 1 (Gauthier-Villars, 1909), Sect. 232.

[14]A force of this kind is obtained as a first approximation of the effect exerted by a thin ring of gravitating matter (Gauss ring) in which we can consider the mass of a perturbing planet to be distributed, so to speak, along its orbit.

2.4 The Position of the Point on the Orbit

The Eccentric Anomaly

As we have seen, c and e completely determine the orbit, and, given certain initial conditions r_0 and v_0, we can obtain from them c and e and then the orbit. However, we are not able to provide an explicit expression for $r = r(t)$, that is, to give the position on the orbit at a certain instant of time (for instance t_1) of our point $r_{t=t_1} = r(t_1)$. For the solution of this problem, it is convenient to make a change of variable and to put $t = t(u)$. The new variable u, on account of the way in which it will be defined later on, is called the *eccentric anomaly*.

From the equation of energy conservation in (2.17) we obtain

$$(r\,\dot r)^2 + c^2 = 2\left(\mu\,r + h\,r^2\right).\tag{2.45a}$$

Let us now introduce the variable u such that

$$r\frac{du}{dt} = k, \qquad u = k\int_T^t \frac{d\tau}{r(\tau)},$$

where k and T are two constants. Moreover,

$$\dot r = \frac{dr}{du}\dot u = \frac{dr}{du}\frac{k}{r} = k\frac{r'}{r},$$

where the prime means derivation with respect to u. By substitution, (2.45a) becomes

$$k^2(r')^2 + c^2 = 2\left(\mu\,r + h\,r^2\right).\tag{2.45b}$$

a) $h = 0$. Let us deal first with the simplest case $h = 0$, furthermore making the choice $k^2 = \mu$. Equation (2.45b) gives

$$(r')^2 + \frac{c^2}{\mu} = 2\,r,$$

and, by taking the derivative, we get $2\,r'r'' = 2\,r'$. Since $r' \neq 0$ (otherwise r would be constant), $r'' = 1$, and then r will have the form $r = \frac{1}{2}\left(u - u_0\right)^2 + A$. By putting $u_0 = 0$ and determining A by substituting in the differential equation, one obtains

$$r = \frac{1}{2}\left(u^2 + \frac{c^2}{\mu}\right).$$

From this and from the definition of u, which, with $u = 0$ for $t = T$, implies $\int_T^t k\,dt = \int_0^u r\,du$, we get

$$\sqrt{\mu}\,(t - T) = \frac{1}{2}\int_0^u \left(u^2 + \frac{c^2}{\mu}\right)du = \frac{1}{6}u^3 + \frac{c^2}{2\mu}u.$$

To summarize,

$$r = \frac{1}{2}\left(u^2 + \frac{c^2}{\mu}\right), \qquad \sqrt{\mu}\,(t - T) = \frac{1}{6}u^3 + \frac{c^2}{2\mu}u. \tag{2.46}$$

In the second of equations (2.46), t is a strictly increasing function of u. The equation can be solved and one uniquely obtains $u = u(t)$. As to the meaning of the constant T, one must as usual distinguish the two cases $c \neq 0$ and $c = 0$.

For $c \neq 0$, $h = 0$, $e = 1$ and the orbit is given by the parabola

$$r = \frac{c^2/\mu}{1 + \cos f}. \tag{2.47}$$

Since $r_{\min} = c^2/2\mu$ (corresponding to $f = 0$) – and this is also the value of r corresponding to $u = 0$ and then to $t = T$ – we have that $t = T$ is the time of pericentre passage.

If, on the other hand, $c = 0$, we have

$$6\sqrt{\mu}\,(t - T) = u^3, \qquad r = \frac{1}{2}u^2.$$

Therefore, at $t = T$ there is a collision at the origin. If $T > 0$, the collision will occur after the instant $t = 0$ and, from that moment, the motion will change. If $T < 0$, then the point will be "emitted" from the origin and the motion will occur for $T < t < \infty$.

Returning again to the case $c \neq 0$, if we compare the first of equations (2.46) with (2.47), we get

$$u^2 = \frac{c^2}{\mu}\frac{1 - \cos f}{1 + \cos f},$$

that is,

$$u = \frac{c}{\sqrt{\mu}}\,\tan\frac{f}{2}, \tag{2.48}$$

which is the relation that enables us to pass from the *true* anomaly to the *eccentric* anomaly. Then it is possible to rewrite (2.46) by putting f in place of u by means of (2.48). There are tables which enable us to calculate f easily, if we know $(t-T)$ and $c^2/2\mu$ (Barker's tables). If instead we want to know the position by means of the eccentric anomaly u, we proceed as follows: given r_0 and v_0 (at the time $t = 0$), by differentiating the first of equations (2.46) one obtains $\dot{r} = u\dot{u} = uk/r$, so that $r\dot{r} = \boldsymbol{r} \cdot \dot{\boldsymbol{r}} = \boldsymbol{r} \cdot \boldsymbol{v} = \sqrt{\mu}\,u$, and then $\boldsymbol{r}_0 \cdot \boldsymbol{v}_0 = u_0\sqrt{\mu}$; using u_0 (at the time $t = 0$), one calculates T, and inserting it again into the second of equations (2.46) one has $u(t)$ and then, from the first, $r(t)$. At this point, if $c = 0$, the position is known since we know the line of the motion $\boldsymbol{e} = $ const. If $c \neq 0$, there are two possible values of f, since $\cos f = \cos(-f)$. Thus $f > 0$ for $t > T$, $f < 0$ for $t < T$.

b) $h \neq 0$. There are three possible kinds of motion:

1) linear, if $c = 0$,
2) hyperbolic, if $c \neq 0$, $h > 0$,
3) elliptic, if $c \neq 0$, $h < 0$.

Again consider (2.45b) and now make the choice $k^2 = \mu/a$, where the constant $a = \mu/2|h|$ has been introduced. By dividing by k^2, one gets

$$(r')^2 + \frac{a\,c^2}{\mu} = 2\,ar + \sigma\,(h)\,r^2, \qquad \text{with} \quad \sigma(h) = \frac{h}{|h|}.$$

By adding $\sigma\,(h)\,a^2$ to both sides and remembering, from (2.22), that $c^2/\mu = a\,(e^2 - 1)\,\sigma\,(h)$, one has

$$(r')^2 + a^2 e^2 \sigma\,(h) = \sigma\,(h)\,[a + \sigma(h)\,r]^2 .$$

By defining $\varrho\,(u)$ through $e\,a\,\varrho = a + \sigma\,(h)\,r$, we finally obtain

$$(\varrho')^2 - \sigma\,(h)\,\varrho^2 = -\sigma(h).$$

Apart from the solutions $\varrho = \pm 1$, which do not interest us, this equation is satisfied by

$$\varrho = \cosh\,(u + K_1), \qquad h > 0,$$
$$\varrho = \cos\,(u + K_2), \qquad h < 0.$$

Since T has not yet been fixed, we put $K_1 = K_2 = 0$. As a consequence $\varrho = \cosh u$ and $\varrho = \cos u$ in the two cases respectively. Therefore we obtain

$$r = a\,(e \cosh u - 1), \qquad h > 0, \qquad (2.49)$$
$$r = a\,(1 - e \cos u), \qquad h < 0, \qquad (2.50)$$

As $k\,dt = r\,du$, with $u = 0$ for $t = T$, by integrating we get $k(t - T) = \sqrt{\mu/a}\,(t - T) = \int_0^u r\,du$. For the two cases, respectively,

$$\frac{\sqrt{\mu}}{a^{3/2}}\,(t - T) = e \sinh u - u, \qquad \text{for} \quad h > 0$$

and

$$\frac{\sqrt{\mu}}{a^{3/2}}\,(t - T) = u - e \sin u, \qquad \text{for} \quad h < 0.$$

In the last case (elliptic motion), $\sqrt{\mu}/a^{3/2} = 2\pi/P = n$, where P is the period and n the mean motion: one defines $n\,(t - T) = l$ (*mean anomaly*).
    Also for $h \neq 0$, from equations (2.21) and (2.22), one has

$$r = \frac{a|e^2 - 1|}{1 + e \cos f}.$$

From this, $r_{\min} = a|1 - e|$ and, since $t = T$ for $u = 0$, one has that, for $c \neq 0$, T is the time of pericentre passage. On the other hand, for $c = 0$ one has $e = 1$ and then $r = 0$ for $u = 0$: in this case T is the instant of collision at the origin or of emission from the origin.

As to the relation between u and the true anomaly f, we have from (2.22) and (2.49) that

$$\frac{1 - \cosh u}{1 + \cosh u} = \frac{1 - e}{1 + e} \frac{1 - \cos f}{1 + \cos f}$$

and then, for $h > 0$,

$$\tan \frac{f}{2} = \sqrt{\frac{1 + e}{1 - e}} \tanh \frac{u}{2}. \tag{2.51}$$

For $h < 0$, from (2.22) and (2.50), we obtain

$$\tan \frac{f}{2} = \sqrt{\frac{1 + e}{1 - e}} \tan \frac{u}{2}. \tag{2.52}$$

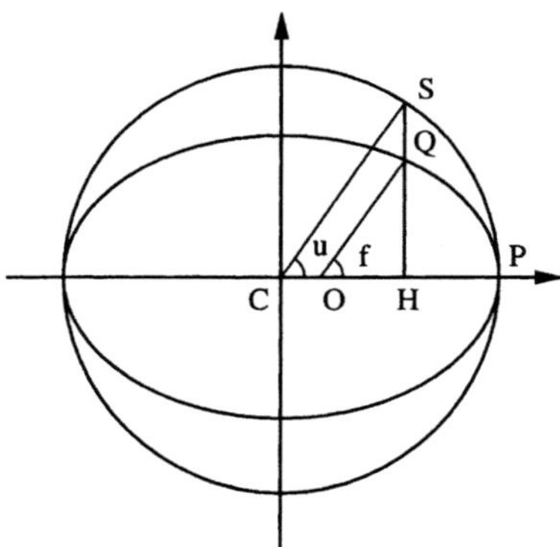

Fig. 2.5

In the last case, it is useful also to look at the geometric meaning of the eccentric anomaly.

Referring to Fig. 2.5, where the ellipse has the semi-axes a and b and the circle radius a, one has

$$\frac{\overline{QH}}{\overline{SH}} = \frac{r \sin f}{a \sin u} = \frac{b}{a},$$

from which $r \sin f = b \sin u$. Moreover $r \cos f = a(\cos u - e)$. Squaring and adding yields $r^2 = b^2 \sin^2 u + (a \cos u - ae)^2$, and as $b^2 = a^2(1 - e^2)$, $r^2 = a^2 \left(1 - 2e \cos u + e^2 \cos^2 u\right)$, that is, $r = a \left(1 - e \cos u\right)$. This proves that u is really the angle drawn in Fig. 2.5. That is, given the point Q with true anomaly f, one determines the point S by drawing from Q the line perpendicular to the semi-major axis until it intersects (on the side of increasing

f) the circle of radius a and centre C. The angle $S\widehat{C}P$ is then the eccentric anomaly u. Therefore (2.50) and

$$n(t - T) = l = u - e \sin u \qquad (h < 0,\ e < 1) \qquad (2.53)$$

can also be rewritten with the true anomaly instead of the eccentric anomaly.

For the two cases $h \neq 0$ (hyperbolic and elliptic motions) one can obtain $r = r(t)$ by proceeding along the same line of reasoning as for the case $h = 0$. Now, instead of two equations (2.46), we have (2.49) and $n\,(t - T) = e \sin h\, u - u$ for $h > 0$ and (2.50), (2.53) for $h < 0$. Unfortunately, in both cases the relation between u and t does not consist of an algebraic equation as is the case for $h = 0$, so it is not possible to obtain the solution in terms of elementary functions. We shall give a short account of the problems raised by (2.53), known as *Kepler's equation*.

Kepler's Equation

Despite its seemingly simple form, (2.53) has intrigued mathematicians for a long time and its solution by means of successive approximations is still being investigated. In the search for a solution, there are two approaches: an analytical one, which, by exploiting the properties of the sine function, allows for series expansions, and a numerical one, which, through the several methods of solution of nonlinear equations, provides approximations with different degrees of convergence and accuracy.

Before giving an outline of the problems encountered in adopting the first approach, let us establish that Kepler's equation has one and only one real solution for any value of t and l. Given (2.53), consider the function $\varphi(u) = u - e \sin u - l$. Taking into account that for $l = m\pi$ (m any integer) $u = m\pi$, consider a value of l between $m\pi$ and $(m + 1)\,\pi$. Then

$$\varphi(m\pi) = m\pi - l < 0, \qquad \varphi\,[(m + 1)\pi] = (m + 1)\pi - l > 0.$$

As a consequence, $\varphi(u) = 0$ has an odd number of real roots between $m\pi$ and $(m + 1)\pi$; but, since $\varphi'(u) = 1 - e \cos u > 0$ and so $\varphi(u)$ is ever increasing, one also infers that $\varphi(u)$ passes through the value 0 only once.

Let us turn now to the classical methods based on series expansions. In the case we have to consider, $0 < e < 1$, to obtain an expression of the function $u\,(e, l)$ we can choose between two possibilities:

1) one can expand, for any fixed positive value of the eccentricity $(e < 1)$, the deviation of the eccentric anomaly from the mean anomaly in a Fourier series in the variable l, with coefficients depending on the fixed value of e;
2) the solution $u = u\,(e, l)$ of the equation can be written, for any fixed value of the mean anomaly, as a Taylor series of powers of the variable e, with coefficients depending on the fixed value of l.

In case (1), if we consider that the function $e \sin u = u(l) - l = F(l)$ is periodic with period 2π in l, odd, tends to zero for $l \to 0$ and $l \to \pi$, and has a continuous derivative, we can expand it in a sine series

$$F(l) = u(l) - l = \sum_{i=1}^{\infty} B_i \sin i\,l,$$

with

$$B_i = \frac{1}{\pi} \int_0^{2\pi} F(l) \sin il\, dl.$$

It is fairly straightforward to show that

$$u = l + 2 \sum_{i=1}^{\infty} \frac{J_i(ie)}{i} \sin i\,l, \tag{2.54}$$

where the J_i are Bessel functions of the first kind.[15] Owing to the regularity of the function $F(l)$, series (2.54) converges uniformly for any value of l in the interval $(-\infty, +\infty)$. It is interesting to dwell for a moment upon the question of the rapidity of the convergence or, in other words, of the degree of accuracy we succeed in obtaining by means of expansion (2.54). Since in practical applications series (2.54) will obviously be truncated, the problem will consist in evaluating the error made: the more rapid the convergence, the lesser the number of terms necessary to have a good approximation. If we evaluate the first six Bessel functions[16] and substitute them in expansion (2.54), we get

$$\begin{aligned}
u = l &+ \left(e - \frac{1}{8} e^3 + \frac{1}{192} e^5 - \frac{1}{9216} e^7 \right) \sin l \\
&+ \left(\frac{1}{2} e^2 - \frac{1}{6} e^4 + \frac{1}{48} e^6 - \frac{1}{720} e^8 \right) \sin 2l \\
&+ \left(\frac{3}{8} e^3 - \frac{27}{128} e^5 + \frac{243}{5120} e^7 \right) \sin 3l + \left(\frac{1}{3} e^4 - \frac{4}{15} e^6 + \frac{4}{45} e^8 \right) \sin 4l \\
&+ \left(\frac{125}{384} e^5 - \frac{3125}{9216} e^7 \right) \sin 5l + \left(\frac{27}{80} e^6 - \frac{243}{560} e^8 \right) \sin 6l + \dots \tag{2.55}
\end{aligned}$$

If we assemble the terms having the same power of e, (2.55) can be rewritten as

$$\begin{aligned}
u = l &+ e \sin l + \frac{e^2}{2} \sin 2l + e^3 \left(\frac{3}{8} \sin 3l - \frac{1}{8} \sin l \right) \\
&+ e^4 \left(\frac{1}{3} \sin 4l - \frac{1}{6} \sin 2l \right) + e^5 \left(\frac{125}{384} \sin 5l - \frac{27}{128} \sin 3l + \frac{1}{192} \sin l \right) \\
&+ e^6 \left(\frac{27}{80} \sin 6l - \frac{4}{15} \sin 4l + \frac{1}{48} \sin 2l \right) + \dots \tag{2.56}
\end{aligned}$$

---

[15]See, for instance, W. M. Smart: *Celestial Mechanics* (Longman, Green and Co., London 1953), Chap. 3.
[16]See W. M. Smart: op. cit., Appendix.

It must be remarked that, as the eccentricity is very close to zero in most problems to do with the solar system, an expansion of the sixth order like (2.56) is a fairly good solution.

Let us now consider the second possibility, that is, to expand, for any fixed value of the mean anomaly l, the solution $u = u(e, l)$ of Kepler's equation as a Taylor series of the eccentricity around $e = 0$,

$$u(e, l) = \sum_{j=0}^{\infty} c_j(l) \, \frac{e^j}{j!}, \tag{2.57}$$

with coefficients depending on l

$$c_j(l) = \left(\frac{\partial^j u(e, l)}{\partial e^j} \right)\bigg|_{e=0}. \tag{2.58}$$

We shall see that the functions $c_j(l)$ are trigonometric polynomials in the mean anomaly. The problems presented by expansion (2.57) are:

a) to evaluate an explicit expression for the coefficients c_j;
b) to determine the radius of convergence of the power series.

Let us begin from (a) by resorting to an expansion due to Lagrange. Consider the equation

$$z = w + \alpha \, \Phi(z), \tag{2.59}$$

where the variables z, w indicate real or complex quantities and α is a parameter. If γ is a smooth closed curve such that at all its points

$$\left| \frac{\alpha \, \Phi(z)}{z - w} \right| < 1,$$

let us assume that $\Phi(z)$ is holomorphic inside γ and on the curve itself. It was shown by Lagrange that (2.59) admits one and only one root inside γ and that any function $F(z)$ holomorphic in the same domain bounded by γ, having such a z as a root, can be expanded in a power series of the parameter α, convergent in γ, as follows:

$$\begin{aligned}
F(z) &= F(w) + \alpha \Phi(w) F'(w) + \frac{\alpha^2}{2!} \frac{\partial}{\partial w} \left[\Phi^2(w) F'(w) \right] \\
&\quad + \cdots + \frac{\alpha^{n+1}}{(n+1)!} \frac{\partial^n}{\partial w^n} \left[\Phi^{n+1}(w) F'(w) \right] + \cdots,
\end{aligned} \tag{2.60}$$

that is

$$F(z) = F(w) + \sum_{n=1}^{\infty} \frac{\alpha^n}{n!} \frac{\partial^{n-1}}{\partial w^{n-1}} \left\{ F'(w) \left[\Phi(w) \right]^n \right\}. \tag{2.61}$$

If, in particular, we choose $F(z) = z$, then

$$z = w + \sum_{n=1}^{\infty} \frac{\alpha^n}{n!} \frac{d^{n-1}}{d w^{n-1}} \left[\Phi(w) \right]^n. \tag{2.62}$$

Equation (2.53), rewritten in the form

$$u = l + e \sin u,$$

is a relation of type (2.62). Therefore, putting $F(z) = z = u$, $\Phi(z) = \sin u$, $w = l$, $\alpha = e$ and demanding that the functions fulfil the required conditions, we have

$$u = l + \sum_{j=1}^{\infty} \frac{e^j}{j!} \frac{d^{j-1}}{d\,l^{j-1}} (\sin l)^j . \tag{2.63}$$

Hence the coefficients $c_j(l)$ have the form

$$c_j(l) = \frac{d^{j-1}}{d\,l^{j-1}} \sin^j l, \qquad j = 1, 2, \ldots; \qquad c_0(l) = l,$$

but, since the j-th power of $\sin l$, because of de Moivre's formula, is a linear combination of $1, \cos l, \cos 2l, \ldots, \cos jl$ or $\sin l, \sin 2l, \ldots, \sin jl$ (for even or odd j respectively), the $(j-1)$-th derivative of $\sin l$ is a linear combination of only sines; therefore we end up with

$$c_j(l) = \frac{d^{j-1}}{d\,l^{j-1}} \sin^j l = \sum_{k=0}^{j/2} (j - 2k)^{j-1} \frac{(-1)^k}{2^{j-1}} \binom{j}{k} \sin(j - 2k)\,l. \tag{2.64}$$

As a consequence, the first terms in the expansion of the solution of Kepler's equation will be

$$u = l + e \sin l + \frac{e^2}{2!} \sin 2\,l + \cdots . \tag{2.65}$$

It must be remarked, however, that the terms of higher degree are considerably more complicated to evaluate, in comparison with those written above, and the difficulties in the calculations increase rapidly. On the other hand, we usually have to deal with very small eccentricities, so that even the three terms we have written provide us with an accurate value of the eccentric anomaly. When this is not so, an option of great practical value may be that of resorting to graphical methods to solve the equation by determining the intersection of the sinusoid $y = \sin u$ with the line of the equation $y = (u-l)/e$ in the uy plane. If we now go back again to (2.65) and compare it with (2.56), we see that the two expressions practically coincide except for the fact that the coefficients of the expansion depend in one case on the fixed value of the eccentricity and, in the other, on the fixed value of the mean anomaly.

The formal identity between the two expansions was pointed out by Lagrange but his result is, however, of limited validity since it is possible to deduce that the Lagrange series converges on the whole of the real axis (that is, for any value of l on the real axis) only for small values of the eccentricity. In fact, a careful analysis of the Fourier expansion, by exploiting the properties of Bessel functions, shows that the series converges, for any real value of the mean anomaly, only if $e \leq 0.6627434\ldots$. These conclusions were the outcome of a long period of research: in fact, though Lagrange had already derived expansion (2.65), he did not come to a rigorous conclusion about the convergence. Several decades later, Laplace also tackled the problem but a rigorous proof was given only by Cauchy; finally, Rouché simplified Cauchy's demonstration remarkably. It must be emphasized that the invaluable contribution to the development of the theory of complex functions due to Cauchy originated just in the task of solving the problems of the convergence of expansion (2.65): in fact, his fundamental discovery connecting the radius of convergence with the position of the nearest singularity was made when he was studying (2.65).

It remains now to speak about the methods of solution involving successive approximations. In the last three decades, Kepler's equation has roused new interest in consequence of the requirements of space dynamics. At the same time, the increased power of computers has made people give up the analytical solutions. In fact, direct analytical solutions are rather inefficient from the computational point of view and, in particular, power series expansions are rather troublesome to calculate; the consequence is that iterative methods have practically supplanted the classical analytical ones. In spite of this, in the past a very large number of methods for the solution of Kepler's equation were proposed. In 1914 Moulton wrote:[17]

A very large number of analytical and graphical solutions have been discovered, nearly every prominent mathematician from Newton until the middle of the last century having given the subject more or less attention. A bibliography containing references to 123 papers on Kepler's equation is given in the *Bulletin Astronomique*, January 1900, and even this extended list is incomplete.

Among the various methods suggested for the solution of nonlinear equations of the type $f(u) = 0$, where f is given and one wants to find u, almost certainly the most famous is Newton's method, which has a quadratic convergence property and is also known as the Newton–Raphson method. This has been the method traditionally favoured among those making use of successive approximations to solve nonlinear equations. It is simple to apply, and, in the case of Kepler's equation, the possibilities of divergence were practically negligible, and local quadratic convergence was undoubtedly satisfactory. However, nowadays there are more efficient algorithms which ensure higher-order convergence without extra cost; the *secant method* in particular may become less expensive. Anyhow, Newton's method remains the best starting point for the systematic development of higher-order algorithms.

We cannot go into the details of the various iterative methods of solution of Kepler's equation here: it is a specialized and technical subject for which we refer the reader to the papers mentioned in the notes to this section. We confine ourselves to mentioning some general distinctive features. To solve an

---

[17]F. R. Moulton: op. cit., p. 190.

equation by means of an iterative method means to determine the limit of successive approximations constructed by making use of a uniform (iterative) procedure. One of the advantages of the use of these procedures is that they are simple and easily applicable when one uses computers. On the negative side, they present considerable difficulties for instance in the choice of the starting point of the iteration or with respect to convergence. It must indeed be said that the convergence of the procedure may become so slow that a satisfactory approximate solution cannot be obtained. It is for this reason that such methods have a limited range of applicability. Also it must not be forgotten that numerical methods are in general affected by errors (errors in the initial data, truncation errors, round-off errors). In solving Kepler's equation with numerical procedures, therefore, one will have to take into account that the solution is the result of successive approximations, and as such it will have a degree of accuracy that depends unavoidably on several factors: on the order of the method, on the number of iterations and on the choice of the starting approximation.

2.5 The Elements of the Orbit

The Motion of the Point in Space

So far we have been dealing with the determination of the orbit (its geometric shape and size), its orientation in space and the position of the point on the orbit itself: c, e and h fix the geometric shape and the size and the plane of the orbit is the one orthogonal to c; moreover through u one can determine the position of the point on the orbit. In the previous section, we argued about the determination of u as if r_0 and v_0 came out automatically from the observations. As a matter of fact, this is not a trivial problem, because r_0 and v_0 are not directly measurable but have to be obtained indirectly from the measurement of other quantities. We cannot extend our exposition by explaining the relevant procedure and so refer the reader to a textbook on theoretical astronomy.[18] In any case, this does not change the substance of our conclusions: we started with system (2.16) to determine $r = r\,(t, r_0, v_0)$ and, instead of finding ourselves with the desired result, we have determined quantities, called elements of the orbit, connected with the first integrals found, which do not exhibit any simple relation to the components of $r(t)$. On the other hand, it is clear that the elements are the quantities more immediately susceptible to measurement by means of astronomical instruments. Therefore the question is raised whether or not the Newtonian schematization of the problem (system (2.16)) is well suited to representing the features of the orbital motion.

---

[18]See also F. R. Moulton: op. cit., Chaps. V, VI.

Before answering this question, let us draw up a list of the orbital elements, which will also be useful for subsequent developments. Figure 2.6 refers to an elliptic orbit of a generic planet (different from the Earth) of the solar system. In it we have the following

S = Sun,
P = pericentre (perihelion),
A = apocentre (aphelion),
$C\Omega D\Omega'$ = orbital plane,
$E\Omega E'\Omega'$ = plane of the ecliptic,
$ECE'D$ = circle representing the celestial sphere,
$\Omega\Omega'$ = line of the nodes,
γ = vernal equinox.

The orbital elements are:

i = inclination,
Ω = longitude of the ascending node,
e = eccentricity,
a = semi-major axis,
ω = argument of the perihelion,
T = time of perihelion passage.

Alternatively, one can use $\overline{\omega} = \Omega + \omega$ (longitude of perihelion).

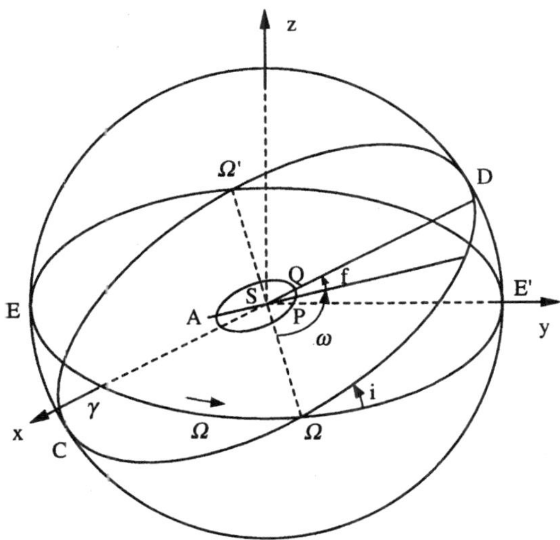

Fig. 2.6

Delaunay's Elements

We commented above on the inadequacy of the quantities of the Newtonian scheme (x, y, z, v_x, v_y, v_z), which enter at the beginning in the setting of the problem, in representing the solution of the problem itself. Furthermore, if we take into account that in any real problem (for instance, in the solar system) one must consider the variations (even if small) to which the orbit is subject in the course of time, it stands to reason that the variables of Newtonian mechanics are less suited for a precise treatment. In fact, the study of the variations of the coordinates x, y, z (quite destitute of regularity) is not the best way to understand how the orbit of a planet is changing as a consequence of the presence (not always negligible) of the other planets. Therefore we need to have, as elements of the orbit, quantities that are at the same time dynamical variables or, vice versa, to write the equations of motion directly for the elements of the orbit. This is possible, as we shall see, in the Hamilton–Jacobi theory; the fact that the equations of motion give the time derivative of the elements of the orbit directly will then allow the formulation of canonical perturbation theory.

Let us now consider, as usual, an inertial system in which x, y, z are the coordinates of our unit mass point and $p_x = \dot{x}$, $p_y = \dot{y}$, $p_z = \dot{z}$ the components of its momentum. The six quantities $(x,\ y,\ z;\ p_x,\ p_y,\ p_z)$ constitute the set of canonical variables describing our system. By means of a canonical transformation[19], we move on now to another set of canonical variables given by the spherical polar coordinates r, ϑ, φ defined by

$$x = r\ \sin\vartheta\ \cos\varphi,$$
$$y = r\ \sin\vartheta\ \sin\varphi,$$
$$z = r\ \cos\vartheta,$$

and their conjugate momenta

$$p_r = \dot{r}, \qquad p_\vartheta = r^2\,\dot{\vartheta}, \qquad p_\varphi = r^2\,\sin^2\vartheta\,\dot{\varphi}.$$

The Hamiltonian for Keplerian motion will be

$$\mathcal{H} = \frac{1}{2}\left(p_r^2 + \frac{1}{r^2}\,p_\vartheta^2 + \frac{1}{r^2\,\sin^2\vartheta}\,p_\varphi^2\right) - \frac{\mu}{r} = h \tag{2.66}$$

and the relevant Hamilton–Jacobi equation

$$\frac{1}{2}\left[\left(\frac{\partial S}{\partial r}\right)^2 + \frac{1}{r^2}\left(\frac{\partial S}{\partial \vartheta}\right)^2 + \frac{1}{r^2\,\sin^2\vartheta}\left(\frac{\partial S}{\partial \varphi}\right)^2\right] - \frac{\mu}{r} = \alpha_r = h, \tag{2.67}$$

---

[19]The treatment that follows will be repeated in more general terms, that is, for a general central potential, in Chap. 5. For these reasons, we do not dwell upon the details of the demonstrations. The reader can obtain all the missing proofs by putting $\Phi(r) = -\mu/r$ in the treatment of Sects. 5.2, 5.3.

since

$$p_r = \frac{\partial S}{\partial r}, \quad p_\vartheta = \frac{\partial S}{\partial \vartheta}, \quad p_\varphi = \frac{\partial S}{\partial \varphi}, \quad S = S\left(r, \vartheta, \varphi; J_r, J_\vartheta, J_\varphi\right). \tag{2.68}$$

If we put $S = S_r(r) + S_\vartheta(\vartheta) + S_\varphi(\varphi)$, (2.67) splits into three ordinary differential equations:

$$\frac{d\,S_\varphi}{\partial \varphi} = \alpha_\varphi,$$

$$\left(\frac{d\,S_\vartheta}{d\,\vartheta}\right)^2 + \frac{\alpha_\varphi^2}{\sin_\vartheta^2} = \alpha_\vartheta^2, \tag{2.69}$$

$$\left(\frac{d\,S_r}{d\,r}\right)^2 + \frac{\alpha_\vartheta^2}{r^2} - \frac{2\,\mu}{r} = 2\,h.$$

Immediately we see that $\alpha_\varphi = p_\varphi = r^2 \sin^2 \vartheta\, \dot\varphi$, which is in fact the component of the angular momentum relative to the polar axis ($\vartheta = 0$). Moreover, as $(-r^2\,\dot\vartheta\,\sin\varphi - r^2\,\dot\varphi\,\sin\vartheta\,\cos\vartheta\,\cos\varphi)$ and $(r^2\,\dot\vartheta\,\cos\varphi - r^2\,\dot\varphi\,\sin\vartheta\,\cos\vartheta\,\sin\varphi)$ are the other two components of the angular momentum, from (2.68) and (2.69) one obtains

$$\alpha_\vartheta = \sqrt{p_\vartheta^2 + \frac{p_\varphi^2}{\sin^2 \vartheta}} = \sqrt{(r^2\,\dot\vartheta)^2 + (r^2\,\sin\vartheta\,\dot\varphi)^2} = |\boldsymbol{r} \times \dot{\boldsymbol{r}}|,$$

that is, the modulus of the angular momentum

$$\alpha_\vartheta = c = \sqrt{\mu\,a\,(1 - e^2)}.$$

Therefore the vector $\boldsymbol{r} \times \dot{\boldsymbol{r}} = \boldsymbol{c}$ has constant length and direction, and the orbit is planar and lies in a plane orthogonal to the total angular momentum. If i is the angle between the plane of the orbit and the xy plane, then $\alpha_\varphi = \alpha_\vartheta \cos i$ (i is the inclination). Hence, the three canonical constants in our case are

$$\alpha_r = h = -\frac{\mu}{2a},$$

$$\alpha_\vartheta = c = \sqrt{\mu\,a\,(1 - e^2)}, \tag{2.70}$$

$$\alpha_\varphi = \alpha_\vartheta \cos i = \sqrt{\mu\,a\,(1 - e^2)}\,\cos i.$$

For the elliptic Keplerian orbit, φ is a cyclic coordinate, and therefore there is a rotation. As regards r and ϑ, which oscillate between a minimum and a maximum, there is libration. The limits for r are $r_0 = a\,(1 - e)$ and $r_1 = a\,(1 + e)$, and for ϑ the limits are $\pi/2 \pm i$.

To evaluate the action integrals

$$J_r = \frac{1}{2\pi} \oint p_r\,dr, \quad J_\vartheta = \frac{1}{2\pi} \oint p_\vartheta\,d\vartheta, \quad J_\varphi = \frac{1}{2\pi} \oint p_\varphi\,d\varphi, \tag{2.71}$$

we have to substitute

$$p_r = \sqrt{2h + \frac{2\mu}{r} - \frac{\alpha_\vartheta^2}{r^2}}, \quad p_\vartheta = \sqrt{\alpha_\vartheta^2 - \frac{\alpha_\varphi^2}{\sin^2\vartheta}}, \quad p_\varphi = \alpha_\varphi, \tag{2.72}$$

obtained by (2.68), (2.69). The result is (see Sects. 5.2, 5.3)

$$J_r = -\alpha_\vartheta + \sqrt{\mu a}, \qquad J_\vartheta = \alpha_\vartheta - \alpha_\varphi, \qquad J_\varphi = \alpha_\varphi \tag{2.73}$$

and, recalling that $h = -\mu/2a$, we get

$$h = -\frac{\mu^2}{2(J_r + \alpha_\vartheta)^2} = -\frac{\mu^2}{2(J_r + J_\vartheta + J_\varphi)^2}. \tag{2.74}$$

If we calculate the three frequencies ω_r, ω_ϑ, ω_φ,

$$\omega_r = \frac{\partial h}{\partial J_r}, \qquad \omega_\vartheta = \frac{\partial h}{\partial J_\vartheta}, \qquad \omega_\varphi = \frac{\partial h}{\partial J_\varphi},$$

we obtain

$$\omega_r = \omega_\vartheta = \omega_\varphi = \frac{\mu^2}{(J_r + J_\vartheta + J_\varphi)^3}. \tag{2.75}$$

The system is therefore completely degenerate (as one would expect, since the motion is simply periodic). Instead of calculating the angle variables conjugated to J_r, J_ϑ, J_φ, we first perform a canonical transformation to new action–angle variables J_1, J_2, J_3 and θ_1, θ_2, θ_3 in order to get only one frequency $\omega_1 \neq 0$ and $\omega_2 = \omega_3 = 0$ (that is, θ_2 and θ_3 will be constant).

The transformation will be generated by the function

$$W_3 = -\theta_1 (J_r + J_\vartheta + J_\varphi) - \theta_2 (J_\vartheta + J_\varphi) - \theta_3 J_\varphi \tag{2.76}$$

and will give

$$\begin{aligned}
\theta_1 &= \theta_r, & J_1 &= J_r + J_\vartheta + J_\varphi, & \text{(2.77a)} \\
\theta_2 &= \theta_\vartheta - \theta_r, & J_2 &= J_\vartheta + J_\varphi, & \text{(2.77b)} \\
\theta_3 &= \theta_\varphi - \theta_\vartheta, & J_3 &= J_\varphi, & \text{(2.77c)}
\end{aligned}$$

From (2.70), (2.72–74) and (2.77), we obtain

$$J_1 = \sqrt{\mu a}, \qquad J_2 = \alpha_\vartheta = \sqrt{\mu a (1 - e^2)}, \qquad J_3 = \alpha_\varphi = \alpha_\vartheta \cos i, \tag{2.78}$$

$$h = -\frac{1}{2}\frac{\mu^2}{J_1^2}, \qquad \omega_1 = \frac{\mu^2}{J_1^3}, \tag{2.79}$$

$$p_r = \sqrt{2h(J_1) + \frac{2\mu}{r} - \frac{J_2^2}{r^2}}, \quad p_\vartheta = \sqrt{J_2^2 - \frac{J_3^2}{\sin^2\varphi}}, \quad p_\varphi = J_3. \tag{2.80}$$

Calculating θ_1, θ_2, θ_3, we get

$$\theta_1 = \frac{\partial S}{\partial J_1} = \int \frac{\partial p_r}{\partial J_1}\, d\,r,$$

$$\theta_2 = \frac{\partial S}{\partial J_2} = \int \frac{\partial p_r}{\partial J_2}\, d\,r + \int \frac{\partial p_\vartheta}{\partial J_2}\, d\,\vartheta, \qquad (2.81)$$

$$\theta_3 = \frac{\partial S}{\partial J_3} = \int \frac{\partial p_\vartheta}{\partial J_3}\, d\,\vartheta + \int \frac{\partial p_\varphi}{\partial J_3}\, d\,\varphi.$$

For the first integral, we have from (2.80), (2.79)

$$\theta_1 = \int \frac{\omega_1\, d\,r}{\sqrt{2\,h + \frac{2\mu}{r} - \frac{J_2^2}{r^2}}}.$$

By substituting (2.50), (2.77), (2.79), we finally obtain

$$\theta_1 = \int (1 - e \cos u)\, d\,u = u - e \sin u = n\,(t - T) = l.$$

Thus the first of the angle variables is the mean anomaly: $\theta_1 = l$. For θ_2 and θ_3, the calculations are a little more complex and we refer the reader to Sect. 5.2, confining ourselves here to giving only the result:

$$\theta_2 = \widetilde{\omega} - \Omega = \omega, \qquad \theta_3 = \Omega;$$

therefore θ_3 is the longitude of the ascending node and θ_2 the argument of the pericentre. Collecting together the results obtained, we have

$$\theta_1 = n\,(t - T) = l, \qquad J_1 = \sqrt{\mu\,a} = L,$$

$$\theta_2 = \widetilde{\omega} - \Omega = g, \qquad J_2 = \sqrt{\mu\,a\,(1 - e^2)} = G, \qquad (2.82)$$

$$\theta_3 = \Omega = h, \qquad J_3 = \sqrt{\mu\,a\,(1 - e^2)} \cos i = H.$$

The six symbols l, g, h, L, G, H were introduced by Delaunay and are named after him. According to what we have said in Sect. 1.17, the Delaunay elements (2.82) are determined uniquely up to a linear transformation with integer coefficients; then any linear combination with integer coefficients of these elements again gives a set of action–angle variables. Five of the six elements (2.82) are obviously constant and characterize the orbit; let us now write the canonical equation for the non-constant element.

The Hamiltonian will be

$$\mathcal{H} = -\frac{1}{2}\frac{\mu^2}{J_1^2} = -\frac{1}{2}\frac{\mu^2}{L^2} = -\frac{1}{2}\frac{\mu}{a} \qquad (2.83)$$

and the equation

$$\frac{dl}{dt} = \frac{\partial \mathcal{H}}{\partial L} = \frac{\mu^2}{L^3} = \frac{\mu^2}{(\mu\,a)^{3/2}}, \qquad (2.84a)$$

that is, $n = \sqrt{\mu/a^3}$; squaring this yields

$$\mu = n^2\, a^3. \tag{2.84b}$$

Therefore, the canonical equation (2.84a) is Kepler's third law (2.84b), and so the close connection between the Delaunay elements and the dynamical problem clearly appears.

2.6 The Problem of Regularization

In the last hundred years the regularization (of the equations of motion) has been studied and obtained with motivations that have changed as time has gone on. The initial demand for demonstrating the existence of solutions for the equations of motion, represented by convergent series expansions, has altered little by little in the quest for formulae to be used in the best way in modern computers. The fundamental research which focused the problem goes back to the studies of Levi-Cìvita (1903 and subsequents) and Sundman (1907, 1912) on the regularization of the three-body problem.[20]

In recent years a method has been introduced (K–S transformation) by Kustaanheimo and Stiefel (1964, 1965), which extends in a four-dimensional parametric space the Levi-Cìvita transformation for the planar case. The method of the K–S matrix enables one to obtain great accuracy in those numerical calculations concerned mainly with the motion of satellites. In this section we shall deal exclusively with the two-body problem; therefore the aim is that of attaining a more suitable form (for numerical calculations) of the equations of motion. If we rewrite the relevant system as follows:

$$\dot{\boldsymbol{r}} = \boldsymbol{v}, \quad \dot{\boldsymbol{v}} = -\mu\,\frac{\boldsymbol{r}}{r^3}, \tag{2.85}$$

we see that it has a singularity at the origin $\boldsymbol{r} = 0$; in other words, at this point the Newtonian force becomes infinitely great. Even if, in practice, it never occurs that one reaches $\boldsymbol{r} = 0$ since the real bodies (both natural and artificial ones) have finite size, it is nonetheless necessary to eliminate the singularities so as to make the numerical integrations feasible without difficulty. This is particularly important for the motion of artificial satellites when they are "very close" to the centre of attraction, whether on leaving or on arriving. The same can occur, in the case of highly eccentric orbits, for the pericentre passage. The procedure for eliminating the singularities from the equations of motion, or more generally from a system of differential equations, is called *regularization*. If we have, for instance, a differential equation

$$\frac{dy}{dx} = f\,(x,y),$$

---

[20]See T. Levi-Cìvita: *Questioni di Meccanica Classica e Relativista* (Zanichelli, Bologna, 1924).

we say that it is "regular" in (x_0, y_0) if in an open domain enclosing (x_0, y_0) both $f(x, y)$ and $\partial f / \partial y$ are continuous and bounded. In this case, the theorem of existence and uniqueness guarantees that a solution exists and is the only one satisfying a given initial condition. In the opposite case, the equation is singular in (x_0, y_0) and, in general, one cannot guarantee the existence of the solution. The argument can be immediately extended to a system of differential equations. It is quite natural to ask if it is possible to eliminate the singularities from the equations in which one is interested by means of a change of variables, which enables one to transform the equation, or the system of equations, into a regular equation (or system). Sundman (1912) gave a positive answer in the case of the three-body problem. It must also be stressed that the "regularization" we are speaking of concerns the equations, not their solutions: in fact singular equations may even occur which have nonsingular solutions.[21]

As a first step in tackling the regularization of the two-body problem, we will show that the introduction of the eccentric anomaly is actually a procedure of regularization. If, for system (2.85), we introduce polar coordinates r, ϑ in the plane of motion, for the radial motion, the integral of areas and the integral of energy, we have respectively

$$\ddot{r} - r\dot{\vartheta}^2 = -\frac{\mu}{r^2}, \tag{2.86}$$

$$r^2 \dot{\vartheta} = c, \tag{2.87}$$

$$\frac{1}{2}\dot{r}^2 - \frac{\mu}{r} + \frac{1}{2}\frac{c^2}{r^2} = h. \tag{2.88}$$

By substituting (2.87) into (2.86), we get

$$\ddot{r} - \frac{c^2}{r^3} = -\frac{\mu}{r^2}. \tag{2.89}$$

If we define u (eccentric anomaly), as in Sect. 2.4,

$$r \frac{du}{dt} = \sqrt{2|h|}, \tag{2.90}$$

and substitute

$$\frac{d}{dt} = \frac{\sqrt{2|h|}}{r} \frac{d}{du}$$

in (2.89) and (2.88), we obtain

$$2|h| \left(-\frac{r'^2}{r^3} + \frac{r''}{r^2} \right) - \frac{c^2}{r^3} = -\frac{\mu}{r^2}, \qquad \frac{1}{2} \left(\frac{\sqrt{2|h|}}{r} r' \right)^2 - \frac{\mu}{r} + \frac{1}{2}\frac{c^2}{r^2} = h,$$

where the prime means derivation with respect to u. Substituting the second equation into the first, yields

---

[21]See, for instance, E. L. Stiefel, G. Scheifele: *Linear and Regular Celestial Mechanics* (Springer, 1971), Sect. 4.

$$r'' + r = \frac{\mu}{2|h|} = a, \tag{2.91}$$

where we have taken into account that we are considering the case $h < 0$. Equation (2.91) is the regularized equation of motion (in comparison with the "singular" one (2.89)). It is easy to verify that, if we impose the initial conditions $r(0) = a(1-e)$, (i.e. for $u = 0$, r must be the pericentric distance) and $\dot{r}(0) = 0$ (in the apses the radial velocity vanishes), then from the general solution of (2.91)

$$r = A \cos u + B \sin u + a$$

we have immediately that $r = a(1 - e \cos u)$, which is (2.50). Therefore the eccentric anomaly u is a regularizing variable for Keplerian motion.

The Levi-Cìvita Transformation

At this point, we can even aim at more general results, by considering system (2.85) expressed through rectangular coordinates in the plane of motion. From now, therefore, we shall consider that $\boldsymbol{r} \equiv (x_1, x_2)$, and $\boldsymbol{v} = \dot{\boldsymbol{r}} \equiv (\dot{x}_1, \dot{x}_2)$. If we also put $dt = r\,ds$ and denote by a prime derivation with respect to the new independent variable s (the "fictitious time"), the equation of motion and energy integral will be

$$\boldsymbol{r}'' - \frac{r'}{r}\,\boldsymbol{r}' + \mu\frac{\boldsymbol{r}}{r} = 0, \tag{2.92}$$

$$\frac{1}{2}\frac{|\boldsymbol{r}'|^2}{r^2} - \frac{\mu}{r} = h. \tag{2.93}$$

It is immediately seen that neither is (2.92) regularized nor can it be regularized by substituting (2.93), since now (2.92) represents two scalar equations and the energy integral by itself is no longer equivalent to the equation of motion as in the case of the radial motion. Thus a further transformation is required. Levi-Cìvita obtained the regularization of (2.92) by means of the transformation

$$x_1 = u_1^2 - u_2^2, \qquad x_2 = 2u_1 u_2 \tag{2.94a}$$

from the vector $\boldsymbol{r} \equiv (x_1, x_2)$ to the vector $\boldsymbol{u} \equiv (u_1, u_2)$; the $x_1 x_2$ plane is called the physical plane and the plane $u_1 u_2$ the parametric plane.[22] Equations (2.94a) become particularly meaningful if we consider the relevant complex planes. In place of (2.94a), we have the (conformal) transformation

$$z = x_1 + ix_2 = (u_1 + iu_2)^2 = w^2. \tag{2.94b}$$

---

[22]Equations (2.94a) are the same relations as those we introduced in (2.30) to define parabolic coordinates. However, in this section, unlike what done in Sect. 2.2, we shall look at the curves in the $u_1 u_2$ plane corresponding to $x_1 = \text{const.}$, $x_2 = \text{const.}$, and then they will be hyperbolas in the $u_1 u_2$ plane.

We can immediately check that (2.94b) is a conformal[23] transformation by ascertaining that the Cauchy–Riemann conditions

$$\frac{\partial x_1}{\partial u_1} = \frac{\partial x_2}{\partial u_2}, \qquad \frac{\partial x_1}{\partial u_2} = -\frac{\partial x_2}{\partial u_1} \tag{2.95}$$

are satisfied owing to (2.94a).

In the $u_1 u_2$ plane, the curves corresponding to $x_1 = \text{const.}$, $x_2 = \text{const.}$ are two families of mutually orthogonal hyperbolas; moreover, if we consider the portion of the $u_1 u_2$ plane bounded by two hyperbolas of a family, it is easy to see that it is transformed in an infinite strip parallel to one of the axes in the $x_1 x_2$ plane. From (2.94a) and (2.94b), $|r| = |u|^2$, that is, the transformation of the $u_1 u_2$ plane into $x_1 x_2$ "squares" the distances from the origin; as regards the polar angles, they are doubled.[24] In fact, if ϑ is the polar angle in the $x_1 x_2$ plane and ψ the corresponding angle in the $u_1 u_2$ plane, then

$$\tan \vartheta = \frac{x_2}{x_1} = \frac{2 u_1 u_2}{u_1^2 - u_2^2} = \frac{2\, u_2/u_1}{1 - (u_2/u_1)^2} = \frac{2\, \tan \psi}{1 - \tan^2 \psi} = \tan 2\psi.$$

Then, to obtain the image of a point, one has to square its distance from the origin and to double its polar angle. For a closed curve described by a point moving in the parametric plane, there will correspondingly be a curve described twice in the physical plane. Inversely, to a complete cycle in the physical plane there will correspond a half-cycle in the parametric plane. On the other hand (2.94b) is obviously not one to one. To a point w_0 one and only one point $z_0 = u_0^2$ corresponds, while to a point z_0 two values of w correspond: $w = \sqrt{z_0}$, $w = -\sqrt{z_0}$.

It is convenient now, also in view of future extensions, to rewrite the Levi-Civita transformation in matrix form. By representing r and u by means of column vectors, (2.94a) becomes

$$\begin{pmatrix} x_1 \\ x_2 \end{pmatrix} = \begin{pmatrix} u_1 & -u_2 \\ u_2 & u_1 \end{pmatrix} \begin{pmatrix} u_1 \\ u_2 \end{pmatrix}, \tag{2.96a}$$

which can be written as

$$r = L(u)\, u. \tag{2.96b}$$

One can also immediately verify that

$$r' = 2\, L(u)\, u'. \tag{2.97}$$

Moreover the matrix $L(u)$ has the following properties:[25]

a) $\qquad L^T(u)\, L(u) = |u|^2 1, \qquad L^{-1}(u) = \frac{1}{|u|^2}\, L^T(u)$

---

[23]Obviously, only at the points at which $dz/dw \neq 0$.
[24]At the origin $dz/dw = 0$; therefore there the transformation is not conformal.
[25]See Stiefel-Scheifele: op. cit.

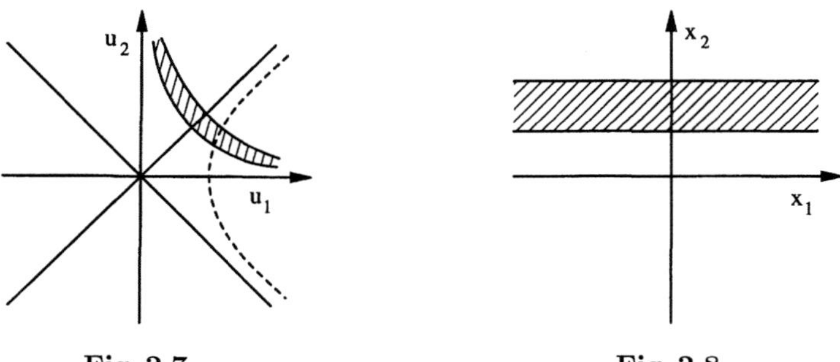

Fig. 2.7   **Fig. 2.8**

(where **1** is the identity matrix), and then it is an orthogonal matrix. Therefore one can solve (2.97) and get

$$u' = \frac{1}{2\,|u|^2}\,L^T(u)\,r'.$$

b) The elements of $L(u)$ are linear and homogeneous functions of the u_i, and then

$$L(u)' = L(u').$$

c) The first column $L(u)$ is given by the vector u itself.

Putting off discussion of how the L matrix can be extended to three-dimensional space, let us again turn to our problem, that is, the regularization of (2.92). By differentiating (2.97) and substituting into (2.92), one finally obtains

$$u'' + \frac{\frac{\mu}{2} - (u' \cdot u')}{(u \cdot u)}\,u = 0. \tag{2.98a}$$

By squaring (2.97) and substituting into (2.93), one also has

$$\frac{\frac{\mu}{2} - |u'|^2}{|u|^2} = -\frac{1}{2}\,h.$$

Consequently (2.98a) becomes

$$u'' - \frac{1}{2}\,h\,u = 0. \tag{2.98b}$$

For Keplerian elliptic motion, $h < 0$, the components will be

$$u_1'' + \frac{1}{2}\,|h|\,u_1 = 0, \qquad u_2'' + \frac{1}{2}\,|h|\,u_2 = 0, \tag{2.99}$$

that is, the equations of motion of a planar oscillator in the parametric $u_1 u_2$ plane. To study the peculiarities of the Levi-Cìvita transformation, we shall confine ourselves to considering the case where (2.99) corresponds to an elliptic orbit, that is, the two oscillations have a phase difference of $\pi/2$:

$$u_1 = \alpha\,\cos\omega s, \qquad u_2 = \beta\,\sin\omega s, \tag{2.100a}$$

with $\omega = \sqrt{|h|/2}$. From the definition of the eccentric anomaly,[26] we have $\omega s = u/2$, and (2.100a) become

$$u_1 = \alpha \cos \frac{u}{2}, \qquad u_2 = \beta \sin \frac{u}{2}. \tag{2.100b}$$

Let us assume that ellipse (2.100a) has its minor axis on the u_1 axis ($\beta > \alpha$). From (2.94a,b)

$$x_1 = \alpha^2 \cos^2 \frac{u}{2} - \beta^2 \sin^2 \frac{u}{2} = \frac{\alpha^2 - \beta^2}{2} + \frac{\alpha^2 + \beta^2}{2} \cos u, \tag{2.101}$$

$$x_2 = \alpha \beta \sin u.$$

Moreover

$$r = \sqrt{x_1^2 + x_2^2} = u_1^2 + u_2^2 = \frac{\alpha^2 - \beta^2}{2} \cos u + \frac{\alpha^2 + \beta^2}{2}$$

Equations (2.101) are the parametric equations of an ellipse in the $x_1 x_2$ plane. It has its centre at the point

$$x_1 = \frac{\alpha^2 - \beta^2}{2}, \qquad x_2 = 0,$$

and the semiaxes are respectively $a = (\alpha^2 + \beta^2)/2$, $b = \alpha \beta$ (see Figs. 2.9 and 2.10, which are not, obviously, at the same scale).

Also $2ae = \beta^2 - \alpha^2$ and then the expression for r becomes

$$r = \frac{\alpha^2 + \beta^2}{2} + \frac{\alpha^2 - \beta^2}{2} \cos u = a \left(1 - e \cos u\right),$$

which is again (2.50).

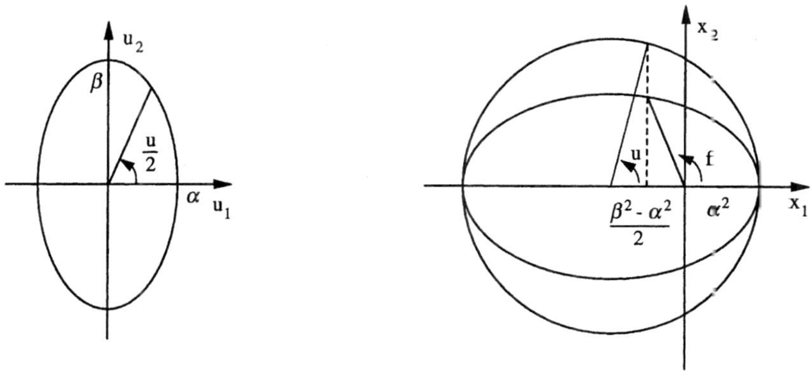

Fig. 2.9   Fig. 2.10

---

[26]The eccentric anomaly u should not be mistaken for the modulus of the vector $\boldsymbol{u}$.

Regularization by Means of Conformal Transformations in the Lagrangian Formalism

Besides (2.94), a certain number of conformal transformations have been set up to regularize the equations of motion. The search for these transformations has been particularly sedulous in the case of the restricted three-body problem, which will be discussed in Chap. 4. However, we now want to show how, in the Lagrangian formalism, it is possible to state the problem in a completely general manner for systems with two degrees of freedom. We shall confine ourselves to natural systems, deferring to Chap. 4 the case where the kinetic energy contains terms linear in the velocities. Therefore, consider a conservative system described by means of the Lagrangian coordinates $q_1 = x_1$, $q_2 = x_2$, subjected to forces deriving from a potential $U = U(x_1, x_2)$. Thus we shall have the Lagrange equations

$$\frac{d}{dt}\left(\frac{\partial \mathcal{L}}{\partial \dot{x}_1}\right) - \frac{\partial \mathcal{L}}{\partial x_1} = 0, \qquad \frac{d}{dt}\left(\frac{\partial \mathcal{L}}{\partial \dot{x}_2}\right) - \frac{\partial \mathcal{L}}{\partial x_2} = 0, \qquad (2.102)$$

with $\mathcal{L} = \mathcal{T} + U$ and $h = \mathcal{T} - U = $ constant. Now, let us consider the complex variable $z = x_1 + i\,x_2$ and the transformation $z = f(w)$, with $w = u_1 + i\,u_2$. If the Cauchy–Riemann relations (2.95) are valid, then the transformation $z = f(w)$ is a conformal transformation at all the points where $f'(w) \neq 0$. If we assume u_1, u_2 to be new Lagrangian coordinates, owing to the invariance of the Lagrange equations under a transformation of variables, we can write the new system as

$$\frac{d}{dt}\left(\frac{\partial \mathcal{L}}{\partial \dot{u}_1}\right) - \frac{\partial \mathcal{L}}{\partial u_1} = 0, \qquad \frac{d}{dt}\left(\frac{\partial \mathcal{L}}{\partial \dot{u}_2}\right) - \frac{\partial \mathcal{L}}{\partial u_2} = 0, \qquad (2.103)$$

where

$$\mathcal{L} = \frac{1}{2}\left\{\left[\frac{dx_1(u_1, u_2)}{dt}\right]^2 + \left[\frac{dx_2(u_1, u_2)}{dt}\right]^2\right\} + U[x_1(u_1, u_2), x_2(u_1, u_2)].$$

If we call J the Jacobian determinant of the transformation

$$J = \left|\frac{dz}{dw}\right|^2 = \left(\frac{\partial x_1}{\partial u_1}\right)^2 + \left(\frac{\partial x_2}{\partial u_1}\right)^2 = \left(\frac{\partial x_1}{\partial u_2}\right)^2 + \left(\frac{\partial x_2}{\partial u_2}\right)^2, \qquad (2.104)$$

then

$$\mathcal{T} = \frac{1}{2}\left\{\left(\frac{\partial x_1}{\partial u_1}\dot{u}_1 + \frac{\partial x_1}{\partial u_2}\dot{u}_2\right)^2 + \left(\frac{\partial x_2}{\partial u_1}\dot{u}_1 + \frac{\partial x_2}{\partial u_2}\dot{u}_2\right)^2\right\} = \frac{1}{2}\,J\left(\dot{u}_1^2 + \dot{u}_2^2\right),$$

because of (2.95). Therefore,

$$\mathcal{L} = \frac{1}{2}\,J\left(\dot{u}_1^2 + \dot{u}_2^2\right) + U. \qquad (2.105)$$

By substituting (2.105) into (2.103), we obtain

$$\frac{d}{dt}\left[J\,\dot{u}_1\right] - \frac{\partial U}{\partial u_1} - \frac{1}{2}\frac{\partial J}{\partial u_1}\left(\dot{u}_1^2 + \dot{u}_2^2\right) = 0,$$

$$\frac{d}{dt}\left[J\,\dot{u}_2\right] - \frac{\partial U}{\partial u_2} - \frac{1}{2}\frac{\partial J}{\partial u_2}\left(\dot{u}_1^2 + \dot{u}_2^2\right) = 0.$$

If, finally, we transform the independent variable by putting

$$dt = J\,d\tau, \tag{2.106}$$

we have

$$\frac{d^2 u_1}{d\tau^2} = J\,\frac{\partial(U+h)}{\partial u_1} + \frac{\partial J}{\partial u_1}\,(U+h),$$

$$\frac{d^2 u_2}{d\tau^2} = J\,\frac{\partial(U+h)}{\partial u_2} + \frac{\partial J}{\partial u_2}\,(U+h),$$

taking into account that h is a constant and $\mathcal{T} = U + h$. Hence

$$\frac{d^2 u_1}{d\tau^2} = \frac{\partial}{\partial u_1}[J\,(U+h)], \qquad \frac{d^2 u_2}{d\tau^2} = \frac{\partial}{\partial u_2}[J\,(U+h)]. \tag{2.107a}$$

Clearly, if the transformation we are considering is such that $JU = \text{const.}$, the equations of motion become

$$\frac{d^2 u_1}{d\tau^2} = \frac{\partial}{\partial u_1}(J\,h) = h\,\frac{\partial J}{\partial u_1}, \qquad \frac{d^2 u_2}{d\tau^2} = \frac{\partial}{\partial u_2}(J\,h) = h\,\frac{\partial J}{\partial u_2}. \tag{2.107b}$$

In addition, whether (2.107b) are regularized or not will depend on the form of J (as a function of u_1, u_2). In the case of Keplerian motion, as in (2.94a), (2.94b), we have

$$J = 4\left(u_1^2 + u_2^2\right), \qquad U = \frac{\mu}{r} = \frac{\mu}{(x_1^2 + x_2^2)^{1/2}} = \frac{\mu}{u_1^2 + u_2^2}.$$

Therefore JU is constant and (2.107) give

$$\frac{d^2 u_1}{d\tau^2} = h\,\frac{\partial}{\partial u_1}\left[4\left(u_1^2 + u_2^2\right)\right] = 8\,h\,u_1,$$

$$\frac{d^2 u_2}{d\tau^2} = h\,\frac{\partial}{\partial u_2}\left[4\left(u_1^2 + u_2^2\right)\right] = 8\,h\,u_2.$$

Denoting with the prime derivation with respect to τ, for $h < 0$, these become

$$u_1'' + 8\,|h|\,u_1 = 0, \qquad u_2'' + 8\,|h|\,u_2 = 0, \tag{2.108}$$

that is, the equations of motion of the planar oscillator with frequency $\omega = \sqrt{8\,|h|}$. The difference, by a factor of 16, with respect to (2.98) is due to the fact that the two fictitious times s and τ differ by a factor of 4.

It is interesting now to look at the transformations undergone by the first integrals. Let us begin with the energy integral, in the transformation from the variables x_1, x_2, t to the variables u_1, u_2, τ. Then

$$\frac{1}{2} J \left(\dot{u}_1^2 + \dot{u}_2^2\right) - U = h,$$

and, multiplying this by J, we get

$$\frac{1}{2} J^2 \left(\dot{u}_1^2 + \dot{u}_2^2\right) = \frac{1}{2}\left[\left(\frac{d\,u_1}{d\tau}\right)^2 + \left(\frac{d\,u_2}{d\tau}\right)^2\right] = J\left(U + h\right).$$

When the transformation is such that $J U = \text{const.} = K$, then $\widetilde{\mathcal{T}} - J h = K$, where

$$\widetilde{\mathcal{T}} = \frac{1}{2}\left[(u_1')^2 + (u_2')^2\right] \tag{2.109}$$

is the "new" kinetic energy and $-h\,J = \widetilde{V}$ the "new" potential energy. In the Kepler case seen above

$$\frac{1}{2}\left[\left(\frac{d\,u_1}{d\tau}\right)^2 + \left(\frac{d\,u_2}{d\tau}\right)^2\right] + 4|h|(u_1^2 + u_2^2) = 4\mu;$$

therefore the total energy of the oscillator is given by 4μ.

In the case of central potentials $U(x_1, x_2) = U(\sqrt{x_1^2 + x_2^2}) = U(r)$, one also has conservation of angular momentum, equation (2.87), where ϑ is the polar angle measured from the positive x_1 axis. Recall that under a conformal transformation $z = f(w)$, every linear size is subject to a magnification by a factor $|dz/dw|^{-1}$. The angles obviously remain unchanged (the magnitude as well as the sense of rotation are preserved), except for the points where $dz/dw = 0$. Let us suppose that at the point a one has $(dz/dw)|_{w=a} = 0$ and the derivative has a "zero" of order α (when a function $f(w)$ has the expansion

$$f(w) = \sum_{n=0}^{\infty} f_n (w - a)^n$$

about any point $w = a$, one says that $f(w)$ has a zero of order α in a when $f_0 = f_1 = \ldots f_{\alpha-1} = 0, f_\alpha \neq 0$); in this case the angle ϑ is transformed into the angle $\varphi = \alpha\vartheta$. If we now consider the planes $x_1 x_2$ and $u_1 u_2$ and the angles $d\vartheta$ and $d\varphi$ with the vertices at their respective origins, the relevant elementary areas dS and $d\widetilde{S}$ will be given by $2dS = r^2 d\vartheta$, $2d\widetilde{S} = R^2 d\varphi$, where $R^2 = u_1^2 + u_2^2$. Because of what was said above,

$$r^2 d\vartheta = \left|\frac{dz}{dw}\right|^2 R^2 \frac{d\varphi}{\alpha} = J R^2 \frac{d\varphi}{\alpha}. \tag{2.110}$$

If now $c = |\boldsymbol{c}| = r^2\dot{\vartheta}$ and $C = |\boldsymbol{C}| = R^2\varphi'$ are the angular momenta in the variables x_1, x_2, t and u_1, u_2, τ respectively (recall that $dt = Jd\tau$), then

$$c\,dt = r^2 d\vartheta = J R^2 \frac{d\varphi}{d\tau}\frac{1}{\alpha}d\tau = \frac{J}{\alpha}C d\tau,$$

and thus

$$c = \frac{C}{\alpha}. \tag{2.111}$$

Obviously, if at the origin $dz/dw \neq 0$, then $\alpha = 1$ and $c = C$. In the case of the Kepler problem, $z = w^2$, and then $\alpha = 2$, and consequently $C = 2c$.

2.7 Topology of the Two-Body Problem

We have seen in Sect. 2.5 that the Delaunay elements for Keplerian motion represent a set of action–angle variables. This entails, as we have seen in Sect. 1.17, that the phase curves wind round a two-dimensional torus; moreover, as the motion is a periodic motion, these curves are closed: projection on the plane of motion yields for each of them an ellipse (we are dealing only with the case $h < 0$). It is therefore evident that the topological structure of the manifold round which the phase curves of the Kepler problem are wound is that of a two-dimensional torus. In fact, the topological properties of a manifold are those properties which are invariant under *homeomorphisms*, that is, under transformations which are one to one and bicontinuous: this is the case here. However, we want to reobtain this result in a more direct way, without passing through the action–angle variables, but looking, from the beginning, at the geometric aspect of the problem.

Let the $x_1 x_2$ plane be the plane of motion and consider a unit mass; moreover, for the sake of simplicity, we choose a system of units in which $\mu = 1$. If we put $\dot{x}_1 = x_3$, $\dot{x}_2 = x_4$, the system of equations will be:

$$\dot{x}_1 = x_3, \quad \dot{x}_2 = x_4, \quad \dot{x}_3 = -\frac{x_1}{r^3}, \quad \dot{x}_4 = -\frac{x_2}{r^3}, \quad r = \sqrt{x_1^2 + x_2^2}. \tag{2.112}$$

The phase space consists of the four-dimensional space $x_1 x_2 x_3 x_4$: in this space, deprived of the plane $x_1 = 0$, $x_2 = 0$, (2.112) define a nonsingular family of phase curves. For system (2.112), the first integrals of energy and areas are

$$x_3^2 + x_4^2 = \frac{2}{r} + 2h, \tag{2.113a}$$

$$x_1 x_4 - x_2 x_3 = c. \tag{2.114a}$$

Equations (2.113a) and (2.114a) each define a three-dimensional manifold: the integral manifold of the energy (which will be called E) and the integral manifold of the areas (which will be called A). Since (2.113a) and (2.114a) are independent first integrals of system (2.112), if we simultaneously fix the constants h and c, we come to define a new manifold (intersection of E and A), which we shall call EA, having dimension two, on which the phase curves of the system will develop in full.

Equations (2.112), as we know, in addition have another first integral, corresponding to the existence of the Laplace–Runge–Lenz vector, which we shall write in the most general form:

$$\arccos \frac{x_1}{r} - \arccos \left[\frac{-r + (x_1 x_4 - x_2 x_3)^2}{r\sqrt{(x_1 x_4 - x_2 x_3)^2 \left(x_3^2 + x_4^2 - \frac{2}{r} \right) + 1}} \right] = \omega. \quad (2.115)$$

By introducing the polar coordinates r and ϑ in the $x_1 x_2$ plane and fixing the constants h and c, (2.115) becomes

$$\vartheta - \omega = \arccos \left[\frac{-r + c^2}{r\sqrt{2\,h\,c^2 + 1}} \right],$$

from which

$$r = \frac{c^2}{1 + e \, \cos (\vartheta - \omega)}, \quad (2.116a)$$

where $e^2 = 1 + 2\,h\,c^2$. For $h < 0$, to have real solutions, it must be $c^2 < -1/2h$. In the case $c = 0$, from (2.115), we have $\vartheta = \omega + \pi$, that is, rectilinear motion. For $c \neq 0$, (2.116a) guarantees that the projections of the phase curves in the $x_1 x_2$ plane are conic sections and, in particular, for $h < 0$, ellipses.

Let us now see what is the topological structure of the manifold E. We shall prove that E is *homeomorphic* with the (Euclidean) three-dimensional space minus a straight line. If we consider the phase space of our system, which is the four-dimensional space $x_1 \, x_2 \, x_3 \, x_4$ minus the points of the $x_3 x_4$ plane, and define in it the transformation

$$r' = \frac{1}{r}, \qquad \vartheta' = \vartheta, \qquad x_3' = x_3, \qquad x_4' = x_4, \quad (2.117)$$

this transformation is a homeomorphism of the four-space (minus the $x_3 \, x_4$ plane) on itself. In fact the first two of equations (2.117) define an inversion with respect to the unit circle in the $x_1 x_2$ plane, while the remaining two leave the coordinates $x_3 x_4$ unchanged. By substituting (2.117) in (2.113a), we get

$$x_3'^{\,2} + x_4'^{\,2} = 2r' + 2\,h. \quad (2.113b)$$

Since ϑ' does not appear in (2.113b), the hypersurface having (2.113b) as its equation is a hypersurface of revolution. It is obtained by rotating, around the $x_3' \, x_4'$ plane, the paraboloid whose equation in the three-space $x_3' \, x_4' \, r'$ is given by (2.113b) itself. It must be stressed that now we are considering r' as an orthogonal Cartesian coordinate, with the condition $r' \geq 0$, and the paraboloid does not intersect the line $r' = 0$, since $h < 0$. As it is known that a paraboloid is homeomorphic with the interior of a half-plane, it follows that the hypersurface of revolution is homeomorphic with that obtained by rotating a half-plane around its boundary line. Therefore one obtains a three-dimensional space minus a straight line.

Now we shall deduce, as a corollary, that A, in the case $c \neq 0$, is also a manifold which is homeomorphic with the three-space minus a straight line. First of all, note that, for $c \neq 0$, according to (2.114a) A does not contain any one of the points of $x_1 = 0$, $x_2 = 0$ we have deleted. Furthermore, let us suppose $c < 0$; the same method can be applied for $c > 0$. If we now perform the rotations, in the planes $x_1 x_4$ and $x_2 x_3$, defined by

$$\delta\, x_1 = x_1' - x_1 = x_4'', \qquad \delta\, x_4 = x_4'' - x_4 = -x_1''$$

(which imply that $x_4 = x_4'' + x_1''$ and $x_1 = x_1'' - x_4''$, from which $x_1 x_4 = x_1''^2 - x_4''^2$) and by

$$\delta\, x_2 = x_2'' - x_2 = x_3'', \qquad \delta\, x_3 = x_3'' - x_3 = -x_2''$$

(which imply that $x_2 = x_2'' - x_3''$ and $x_3 = x_3'' + x_2''$, from which $x_2 x_3 = x_2''^2 - x_3''^2$), we get

$$c = x_1 x_4 - x_2 x_3 = x_1''^2 - x_4''^2 - x_2''^2 + x_3''^2. \tag{2.118}$$

By writing ϱ and $\varPhi$ for the polar coordinates in the $x_2'' x_4''$ plane, we can write (2.118) as

$$x_1''^2 + x_3''^2 = \varrho^2 + c. \tag{2.119a}$$

By means of the transformation $x_1''' = x_1''$, $x_3''' = x_3''$, $\varrho''' = \varrho^2$, $\varPhi''' = \varPhi$, which is a homeomorphism of the four-space, (2.119a) becomes

$$x_1'''^2 + x_3'''^2 = \varrho''' + c. \tag{2.119b}$$

As $c < 0$, (2.119b) is analogous to (2.113b), and then one can draw the same conclusions: A is also homeomorphic with the three-space minus a straight line. For $c > 0$, the same method applies; moreover one can show that for $c = 0$ the manifold A (minus the points $x_1 = 0$, $x_2 = 0$) continues to have the same topological structure.

Finally, let us move on to study the manifold EA. Suppose we have fixed the two constants c and h, with $h < 0$, and consider the intersection EA of the two manifolds E and A. We shall prove that

1) If $0 < c^2 < -\frac{1}{2h}$, the manifold EA is homeomorphic with the surface of a torus.
2) If $c = 0$, EA is homeomorphic with the interior of an annulus.
3) If $c^2 = -\frac{1}{2h}$, EA is homeomorphic with a circle.

The equations for EA are given by system (2.113a), (2.114a); in the coordinates $r', \vartheta', x_3', x_4'$, they become the already seen (2.113b) and

$$x_4' \cos \vartheta' - x_3' \sin \vartheta' = c\, r'. \tag{2.114b}$$

The latter can be written

$$x_4' \frac{\cos \vartheta'}{\sqrt{1 + c^2}} - x_3' \frac{\sin \vartheta'}{\sqrt{1 + c^2}} = \frac{c}{\sqrt{1 + c^2}}\, r', \tag{2.114c}$$

which, for any fixed ϑ', represents in normal form the equation of a plane passing through the origin of the system of coordinates $x_3'\, x_4'\, r'$ and forming an angle given by

$$\arcsin \frac{c}{\sqrt{1+c^2}}$$

with the r' axis. This plane intersects the $x_3'\, x_4'$ plane along a line which makes an angle ϑ' with the x_3' axis. If we now make the angle ϑ' vary from 0 to 2π, the plane starts from the position given by $x_4' = c\,r'$ in the sense from the positive x_3' axis to the positive x_4' axis. Since the paraboloid given by (2.113b) is, in the space $x_3'\, x_4'\, r'$, a paraboloid of revolution, the type of intersection with the plane will not depend on ϑ'. For $\vartheta' = 0$, the intersection curve reduces to equation (2.113b) and to $x_4' = c\,r'$, whose projection on the plane $x_3'\, x_4'$ is given by

$$x_3'^{\,2} + \left(x_4' - \frac{1}{c}\right)^2 = \frac{1+2\,c^2\,h}{c^2}. \tag{2.120}$$

For $1 + 2\,c^2\,h > 0$, the intersection is represented by an ellipse. As ϑ' varies from 0 to 2π, the ellipse rotates around the r' axis and at last again reaches the starting position. The surface generated by the rotation of the ellipse is that of a torus. On this torus we can make use, as coordinates, of ϑ' (longitude) and of the angle ψ (latitude) that the vector which specifies a point on the ellipse makes with the x_4' axis (which encloses the centre of the ellipse).

In the case $c = 0$, (2.113b) and (2.114b) become (for $\vartheta' = 0$)

$$x_3'^{\,2} = 2\,r' + 2\,h, \qquad x_4' = 0. \tag{2.121}$$

Instead of an ellipse, we have a parabola. As ϑ' varies, the generated surface will be the topological product of a parabola and a circle. Since the parabola is homeomorphic with an open interval, the generated surface will be homeomorphic with the interior of an annulus.

Finally, if $c = \pm\sqrt{-1/2h}$, the ellipse degenerates in the point with coordinates

$$x_3' = 0, \qquad x_4' = \frac{1}{c}, \qquad r' = \frac{x_4'}{c} = \frac{1}{c^2}.$$

As ϑ' varies from 0 to 2π, this point will describe a circle. Therefore, points (1), (2) and (3) are demonstrated.

Once the nature of the manifold EA corresponding to a given value of c has been clarified, it remains to see how, as c varies, the manifolds EA decompose the manifold E corresponding to the value of h fixed at the beginning. To do this, refer again to the coordinates defined in (2.117). By keeping ϑ' fixed, consider the three-dimensional space $r'\, x_3'\, x_4'$ and the angular momentum c that varies in the interval $[-\sqrt{-1/2h},\, +\sqrt{-1/2h}]$. There will be two families of paraboloids corresponding, in the two half-spaces of $r'\, x_3'\, x_4'$, to the negative and positive values of c respectively. In Fig. 2.11 we consider $\vartheta' = 0$ and the paraboloids viewed from along the r' axis. If we make ϑ' vary, the whole figure rotates around the origin, and then, as ϑ' varies from 0 to 2π, the figure is topologically multiplied by a circle. The situation may be more clearly understood if first the $x_3'\,x_4'$ plane of Fig. 2.11 is mapped into the half-plane $\xi\,\eta$, $\xi > 0$ of Fig. 2.12. By rotating the families of curves in Fig. 2.12, the degenerate cases included, around the η axis, one obtains the families of tori, and relevant degenerate cases, which fill the three-dimensional space minus a straight line. The phase curves are such as to give, for any fixed triple h, c, ω (with $h < 0$, $c \neq 0$), a projection on the $x_1\,x_2$ plane consisting of the ellipse given by (2.116a). If we perform the transformation (2.117), (2.116a) becomes

$$r' = \frac{1 + \sqrt{1 + 2\,h\,c^2}\,\cos\left(\vartheta' - \omega\right)}{c^2}, \qquad (2.116b)$$

which is the equation of a limaçon. From (2.116b), we see that r' is a one-valued function of ϑ', and then, as the latter varies, it varies between two extreme values (a minimum and a maximum). Equation (2.116b) represents the projection of a curve which is on the torus EA given by (2.113b) and (2.114b). The intersection of EA with the plane $\vartheta' = \mathrm{const.}$ is an ellipse (see Fig. 2.11); to every value of ϑ', (2.116b) makes a value of r' correspond: then there is a point $P_{\vartheta'}$ of the trajectory on every ellipse $\vartheta' = \mathrm{const.}$ As ϑ' varies, the point varies on the torus; therefore we shall have a situation of the type $\Phi = f(\vartheta')$, with $f(\vartheta' + 2\pi) = f(\vartheta')$, mod 2π. The same will be true for all the trajectories, except for a translation of ϑ (see (2.116b)). From (2.116b) we also see that f is a monotonic function, either strictly increasing or strictly decreasing, and therefore one can invert it to obtain $\vartheta' = g(\Phi)$. In the cases $c = \pm\sqrt{-1/2h}$, EA reduces to a closed curve whose projection is a circle.

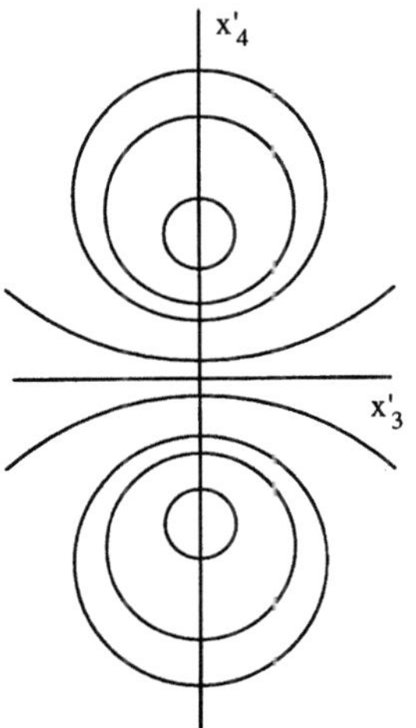

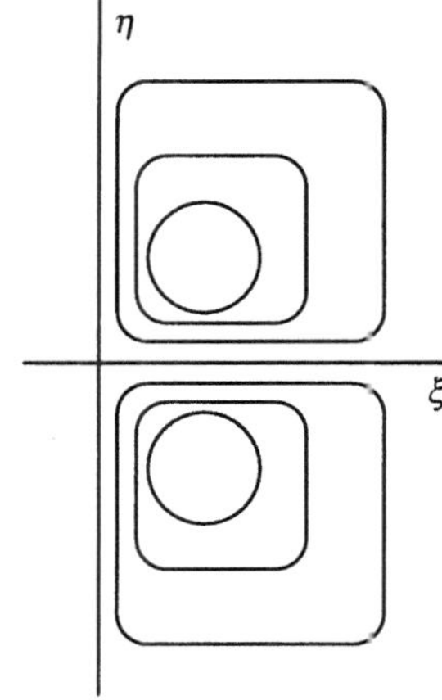

Fig. 2.11   Fig. 2.12

Chapter 3

The N-Body Problem

The order of the chapters, and thus the framework of our treatment, reflects the fact that we have followed the traditional route; that is, to examine first the two-body problem and then the N-body problem. In abstract terms, it might appear more sensible to proceed backwards, from arbitrary N to the particular case $N = 2$. The fact is that the two problems, $N = 2$ and $N > 2$ are *qualitatively* and not just *quantitatively* different, and therefore the traditional route, beside not running against the ideal logical course, also appears more promising from the didactical point of view. The subject of the present chapter is thus the problem of the motion of N pointlike masses, interacting among themselves through no other forces than their mutual gravitational attraction according to Newton's law.

3.1 Equations of Motion and the Existence Theorem

When we refer to a system of N bodies interacting through their mutual gravitational attraction and confined to a delimited region of space, we are actually speaking of a mathematical model we think appropriate to represent some aspects of real systems like the solar system and stellar systems (open and globular clusters, galaxies, clusters of galaxies, etc.). In the first instance, with N quite small ($N \sim 10$), we are in the framework of the discipline traditionally called celestial mechanics; in the second instance we have $N > 10^4$ and we can refer to the approach proper of stellar dynamics. Obviously, in both cases the problem is to understand if and how the mathematical model of N material points (that is, of bodies endowed with mass, but with no extension) is suitable for describing the motion of N bodies with finite and, in general, different sizes.

In his mighty treatise on celestial mechanics, F. Tisserand, dividing the problem into two distinct sub-problems, set the question in the following way:

... Considérons l'un des corps de notre système; nous pouvons décomposer son mouvement en deux autres: le mouvement de son centre de gravité et le mouvement du corps autour de son centre de gravité. De là les deux principaux problèmes de la Mécanique Céleste:

1) Déterminer les mouvements des centres de gravité des corps célestes;
2) Déterminer les mouvements des corps célestes autour de leurs centres de gravité.[1]

It remains now to see if the decomposition proposed by Tisserand is justified, that is, if the "external" and the "internal problem"[2] are really separable. Analysis of the influence of the internal structure on the external problem[3] shows that it is of the order of $\varepsilon\,\alpha^2$ with respect to the Newtonian attraction of two pointlike masses, $G\,m^2/R^2$, where $\alpha = L/R$, with L the characteristic linear extension of the bodies, R the scale of their separation and ε a number between 0 and 1. Since in planetary and stellar systems it is almost always the case that $R \gg L$ and, in general, $\varepsilon \ll 1$ as well, we get $\varepsilon\,\alpha^2 \ll 1$ and we say that there is an *effacement* [4] of the internal structure. On these foundations

---

[1] Let us consider one of the bodies of our system; we can decompose its motion in two components: the motion of its centre of mass and the motion of the body about its centre of mass. From here, the two main problems of Celestial Mechanics follow: 1) To find the motion of the centres of mass of celestial bodies; 2) To find the motion of celestial bodies about their centres of mass. (F. Tisserand: *Traité de Mécanique Céleste*, Vol. I (Paris, Gauthier-Villars, 1889), pp. 51–52, English translation is ours).

[2] This is how Fock described Tisserand's points (1) and (2). See, V. Fock: *The Theory of Space-time and Gravitation* (Pergamon Press, 1959).

[3] See the clear and exhaustive treatment given by T. Damour: The problem of motion in Newtonian and Einsteinian gravity, in *Three Hundred Years of Gravitation* S. Hawking, W. Israel eds. (Cambridge University Press, 1987).

[4] This term is due to Brillouin and Levi-Cìvita (see the paper by T. Damour cited above).

we will apply our mathematical model with the well-grounded feeling that it is suitable for the real cases mentioned above.

We are thus given a system of $N (\geq 2)$ material points $P_1, P_2, \ldots, P_N$ in motion under the action of the Newtonian gravitational force. This implies that we specify:

a) the positive constants $G; m_1, m_2, \ldots, m_N$;
b) a Cartesian coordinate system, suitably chosen in three-dimensional Euclidean space;
c) an independent variable t.

The equations of the motion are then

$$m_k \ddot{\boldsymbol{r}}_k = \sum_{j=1}^{N} \frac{G m_j m_k}{r_{jk}{}^3} (\boldsymbol{r}_j - \boldsymbol{r}_k), \qquad (j \neq k), \quad r_{jk} = |\boldsymbol{r}_j - \boldsymbol{r}_k|, \qquad (3.1\text{a})$$

where the vector $\boldsymbol{r}_k$ gives the position of the point $P_k = P_k(x_k, y_k, z_k)$ and the right-hand side of the k-th equation of (3.1a) represents the total force acting on the k-th point by the other $N-1$ material points. The coordinate system $Oxyz$ is an inertial system, which we will call, for brevity, the *fixed system*, the variable t is the *absolute time*[5] of Newtonian mechanics, and the positive constants $m_1, m_2, \ldots, m_N$ are the masses of the material points. All these notions are inseparable ingredients of the Newtonian mechanics. Let us also assume that it is always possible to define the inertial reference frame uniquely. Defining the potential function

$$U = \sum_{1 \leq j < k \leq N} G \frac{m_j m_k}{r_{jk}} \qquad (3.2)$$

and the velocities $\boldsymbol{v}_k = \dot{\boldsymbol{r}}_k$, we can rewrite (3.1a) in the form

$$\dot{\boldsymbol{r}}_k = \boldsymbol{v}_k, \quad \dot{\boldsymbol{v}}_k = \frac{1}{m_k} \frac{\partial U}{\partial \boldsymbol{r}_k} \qquad (k = 1, 2, \ldots, N). \qquad (3.1\text{b})$$

Set (3.1b) is a conservative, reversible, dynamical system, with $3N$ degrees of freedom. From the point of view of the theory of systems of differential equations, it is a system of $6N$ first-order equations. It is of the same type as (1.A.1b), with the vector $\boldsymbol{X}$ independent of time, and therefore it is an autonomous system:

$$\dot{x}_i = X_i(x_1, x_2, \ldots, x_m) \qquad (i = 1, 2, \ldots, m = 6N). \qquad (3.3)$$

If the m functions X_i are Lipschitzian in a real neighborhood of $\boldsymbol{x} = \boldsymbol{\xi}$, then, for given initial data $\boldsymbol{x}(t_0) = \boldsymbol{\xi}$, set (3.3) admits one and only one

______

[5] From the first Scholium of the *Principia*, op. cit. p. 6: "Absolute, true, and mathematical time, of itself and from its own nature, flows equably without relation to anything external".

solution. Moreover, if the functions X_i are regular analytic in the variables $x_1, x_2, \ldots, x_m$ and are bounded by a positive constant M in the domain $|x_i - \xi_i| < r$, $i = 1, 2, \ldots, m$, then the solution $\boldsymbol{x}(t)$ of set (3.3), which satisfies the initial condition $\boldsymbol{x}(t_0) = \boldsymbol{\xi}$, is a regular analytic function of t in the complex neighbourhood

$$|t - t_0| < \frac{r}{(m+1)\,M}$$

and satisfies

$$|x_i - \xi_i| < r, \qquad i = 1, 2, \ldots, m.$$

This statement of the existence and uniqueness (Cauchy) theorem comes from Siegel and Moser,[6] to which we refer the reader for the proof.

We will also follow Siegel and Moser for the application of the Cauchy theorem to set (3.1b) and will choose units so that $G = 1$ in (3.2). The initial time, $\boldsymbol{r}_k(t_0)$ and $\boldsymbol{v}_k(t_0)$ are real but, to find the positive constants r and M, we will assume that the values of these variables are complex. Suppose that at the starting time t_0 we have $r_{jk}(t_0) > 0$ $(j \neq k)$ and let A be an upper bound of U at the time $t = t_0$:

$$U_{t_0} \leq A. \tag{3.4}$$

Let us make the definitions

$$\varrho = \min_{j \neq k} r_{jk}(t_0), \quad \mu = \min_k m_k. \tag{3.5}$$

From (3.2), we have directly that

$$\frac{\mu^2}{\varrho} \leq U_{t_0} \leq A,$$

from which

$$\varrho \geq \frac{\mu^2}{A}. \tag{3.6}$$

We now want to find an upper bound for

$$\left| \frac{\partial U}{\partial x_k} \right|, \quad \left| \frac{\partial U}{\partial y_k} \right|, \quad \left| \frac{\partial U}{\partial z_k} \right|,$$

with $k = 1, 2, \ldots, N$ and $\boldsymbol{r}_k \equiv (x_k, y_k, z_k)$. To this end, we look for a lower bound of r_{jk}, for complex $\boldsymbol{r}_k$ near to $\boldsymbol{r}_k(t_0)$. Letting

$$\begin{aligned}
X &= [x_k - x_k(t_0)] - [x_j - x_j(t_0)], \\
Y &= [y_k - y_k(t_0)] - [y_j - y_j(t_0)], \\
Z &= [z_k - z_k(t_0)] - [z_j - z_j(t_0)],
\end{aligned}$$

---

[6] C. L. Siegel, J. K. Moser: *Lectures on Celestial Mechanics* (Springer, 1971) Sects. 4, 5.

we also have

$$x_k - x_j = X + [x_k(t_0) - x_j(t_0)],$$
$$y_k - y_j = Y + [y_k(t_0) - y_j(t_0)],$$
$$z_k - z_j = Z + [z_k(t_0) - z_j(t_0)].$$

For r_{jk}^2 we have

$$r_{jk}^2 = (x_k - x_j)^2 + (y_k - y_j)^2 + (z_k - z_j)^2$$
$$= r_{jk}^2(t_0) + (X^2 - Y^2 + Z^2)$$
$$+ 2X[x_k(t_0) - x_j(t_0)] + 2Y[y_k(t_0) - y_j(t_0)] + 2Z[z_k(t_0) - z_j(t_0)].$$

Using the Schwarz inequality to majorize the sum of the last three terms and substituting the second term with the sum of the squared moduli, we finally get

$$r_{jk}^2 \geq r_{jk}^2(t_0) - 2r_{jk}(t_0)\sqrt{|X|^2 + |Y|^2 + |Z|^2} - \left(|X|^2 + |Y|^2 + |Z|^2\right). \quad (3.7)$$

Then, if we impose the condition that

$$|x_i - x_i(t_0)|, \ |y_i - y_i(t_0)|, \ |z_i - z_i(t_0)| < \frac{\varrho}{14}, \qquad \forall i = 1, 2, \ldots, N, \quad (3.8)$$

so that

$$|X|, \ |Y|, \ |Z| < \frac{\varrho}{7},$$

$$|X|^2 + |Y|^2 + |Z|^2 < \frac{3}{49}\varrho^2 < \frac{\varrho^2}{16},$$

inequality (3.7) becomes

$$r_{jk}^2 \geq r_{jk}^2(t_0) - 2r_{jk}(t_0)\frac{\varrho}{4} - \frac{\varrho^2}{16} \geq \frac{7}{16}r_{jk}^2(t_0) > \frac{1}{4}r_{jk}^2(t_0), \qquad (3.9)$$

as a result of the first of equations (3.5). Finally

$$r_{jk} > \frac{1}{2}r_{jk}(t_0). \qquad (3.10)$$

To satisfy inequality (3.8), taking into account (3.6), we require that

$$|x_i - x_i(t_0)|, \ |y_i - y_i(t_0)|, \ |z_i - z_i(t_0)| < \frac{\mu^2}{14\,A}, \qquad \forall i = 1, 2, \ldots, N. \quad (3.11)$$

From (3.8) we also have

$$|x_j - x_k| = |[x_j - x_j(t_0)] + x_j(t_0) - [x_k - x_k(t_0)] - x_k(t_0)|$$
$$\leq |x_j - x_j(t_0)| + |x_k - x_k(t_0)| + |x_j(t_0) - x_k(t_0)|$$
$$\leq \frac{1}{7}\varrho + r_{jk}(t_0) \leq \frac{8}{7}r_{jk}(t_0)$$

and analogously

$$|y_j - y_k|, \ |z_j - z_k| < \frac{8}{7}r_{jk}(t_0), \qquad \forall j, k = 1, 2, \ldots, N.$$

From these and (3.6) and (3.10), it follows that

$$\frac{|x_j - x_k|}{r_{jk}{}^3} \leq \left(\frac{2}{r_{jk}(t_0)}\right)^3 \frac{8}{7}\, r_{jk}(t_0) = \frac{64}{7} \frac{1}{[r_{jk}(t_0)]^2} \leq \frac{64\,A^2}{7\,\mu^4}$$

and analogously

$$\frac{|y_j - y_k|}{r_{jk}{}^3}, \quad \frac{|z_j - z_k|}{r_{jk}{}^3} \leq \frac{64\,A^2}{7\,\mu^4}.$$

As a consequence

$$\left| m_k^{-1} \frac{\partial U}{\partial x_k} \right| = \left| \sum_{j \neq k} \frac{m_j}{r_{jk}{}^3} (x_j - x_k) \right| < c_1 A^2$$

and

$$\left| \sum_{j \neq k} \frac{m_j}{r_{jk}{}^3} (y_j - y_k) \right|, \quad \left| \sum_{j \neq k} \frac{m_j}{r_{jk}{}^3} (z_j - z_k) \right| < c_1 A^2,$$

where the constant c_1 depends only on the masses. Denoting by v one of the $3N$ components of the velocity $\dot{x}_1, \ldots, \dot{x}_N, \ldots, \dot{y}_1, \ldots, \dot{z}_N$, we find that the energy integral (the system is conservative) gives

$$\frac{1}{2}\, m\, v^2(t_0) \leq \mathcal{T}_{t_0} = U_{t_0} + h \leq A + h,$$

where $\mathcal{T}$ is the total kinetic energy, h the total energy constant and m the mass of the body having one of the components of its velocity equal to v. Therefore

$$|v(t_0)| \leq c_2 \sqrt{A + h},$$

with c_2 another constant depending only on the masses. Now writing

$$\frac{\mu^2}{14\,A} = r \tag{3.12}$$

and requiring, together with (3.11), that the inequality $|v - v(t_0)| < r$ is satisfied, we finally obtain

$$|v| \leq |v - v(t_0)| + |v(t_0)| < r + c_2 \sqrt{A + h}.$$

We can conclude, therefore, that, if the constants r and M in the statement of Cauchy's theorem are given by (3.12) and by

$$M = c_1\,A^2 + \frac{\mu^2}{14\,A} + c_2 \sqrt{A + h}, \tag{3.13}$$

in such a way that the functions on the right hand side of the set (3.1b) are regular and bounded in absolute value in the domains

$$|x_i - x_i(t_0)|,\ |y_i - y_i(t_0)|,\ |z_i - z_i(t_0)| < r,$$
$$|\dot{x}_i - \dot{x}_i(t_0)|,\ |\dot{y}_i - \dot{y}_i(t_0)|,\ |\dot{z}_i - \dot{z}_i(t_0)| < r,$$

then the solutions $r_k(t)$ and $v_k(t)$, corresponding to the initial conditions $r_k(t_0)$ and $v_k(t_0)$, are regular analytical functions of t in the complex neighbourhood:

$$|t - t_0| < \frac{r}{(6\,N + 1)\,M} = \delta.$$

In particular this will be true in the interval

$$t_0 \leq t \leq t_0 + \delta,$$

where δ depends only on A, on h and on the masses.

It is possible to prove[7] that, in the case in which in place of system (3.1b) we have the Hamiltonian equations

$$\dot{q}_i = \frac{\partial \mathcal{H}}{\partial p_i}, \quad \dot{p}_i = -\frac{\partial \mathcal{H}}{\partial q_i} \qquad (i = 1, 2, \ldots, 3N), \tag{3.14}$$

with $\mathcal{H} = \mathcal{T} - U$, with the bounding conditions

$$|\mathcal{H}(\boldsymbol{p}, \boldsymbol{q})| \leq M,$$
$$|q_i - q_i(t_0)| < 2\,\varrho,$$
$$|p_i - p_i(t_0)| < 2\,\varrho,$$

we find that the solution is regular for

$$|t - t_0| < \frac{\varrho^2}{(6\,N + 1)\,M}$$

and that it satisfies

$$|q_i(t) - q_i(t_0)| < \varrho,$$
$$|p_i(t) - p_i(t_0)| < \varrho.$$

Returning to the solution of set (3.1b), we see that, up to $t - t_0 < \delta$, the solution $r_k(t)$ remains regular as a function of t and, in particular, there can be no collisions. In fact it is obvious that, in the case of a collision between the k-th and the i-th point, it should be the case that $r_{ki} = 0$ with a corresponding singularity in the potential and, with h remaining constant, at least one of the components of the velocities would become infinite and this clearly would be in contrast with the asserted regularity of the solution. If we analytically prolong the solution for $t > t_0$, it can happen that all the N vectors r_k stay regular for every finite $t > t_0$ or that a $t_1 > t_0$ is passed such that there at least one $r_k(t)$ becomes singular: in such a case, for $t > t_1$ the potential U tends to infinity. If this were not the case, in fact, we should have

$$U_{t_1} \leq A,$$

and, again applying the existence theorem with t_1 in place of t_C, we would find that the solution is regular, contrary to the hypothesis. This shows that

---

[7] C. L. Siegel, J. K. Moser: op. cit., Sect. 4.

$$\lim_{t \to t_1} U = \infty,$$

that is to say, for the definition of U, the least of the $n\,(n-1)/2$ distances $r_{jk}\ (j \neq k)$ must tend to zero when $t \to t_0$.

3.2 The Integrals of the Motion

From what we have said in the preceding section, it appears that the solution of the N-body problem depends on:

1) the condition of applicability of Cauchy's theorem (only if these conditions are fulfilled does the theorem insure the existence of the solution);
2) the actual construction of the solution, i.e., the integration of (3.1a) or (3.1.b).

Historically, the second point, was the first faced by eighteenth-century and later mathematicians.

The first milestone was established by Lagrange in a celebrated paper on the three-body problem.[8] We should first remark that the number of first integrals whose existence can be demonstrated does not depend on N but on the nature of forces; therefore the results obtained for $N = 3$ are valid for any N. That work concentrated on the three-body problem was due both to the fact that it was the easiest approach to the problem (for $N > 2$) and to the possibility of concrete application to the solar system. The system of differential equations obtained by Lagrange was of order seven and did not explicitly contain the time: it could already have been concluded that, by applying the procedure of eliminating time (see Sect. 1.11), the final system was of order six. In other words, the order of the initial system can be lowered by 12. This result, although it does not appear explicitly in Lagrange's paper as far as the latter operation is concerned, was never improved upon in the next two centuries. By this we mean that the order of the system to be integrated has remained $6\,N - 12$: over two hundred years later, the best efforts of the most eminent mathematicians have not modified this result. As we shall see, at the end of the last century, Bruns demonstrated, in fact, that this is the maximum number of integrals which can be obtained.

In this section, we shall therefore try to demonstrate how the first integrals of the N-body problem may be obtained and we shall also discuss their nature and meaning from different points of view, while in the next section we shall deal with the singularities, that is, when the Cauchy theorem cannot be applied and consequently the existence of the solution cannot be assured. We

---

[8] J. L. Lagrange: Essai sur le problème des trois corps [Recueil des pièces qui ont remporté le prix de l'Acadèmie Royale des Sciences de Paris, tome IX, 1772], reprinted in J. L. Lagrange: *Oeuvres*, Vol. VI (Gauthier-Villars, Paris, 1873), pp. 229–324.

shall start by considering the problem in the Newtonian formalism; we will then look for the integrals in a direct way (see the remark at the beginning of Sect. 2.2).

Let us begin with system (3.1a); summing the right-hand sides over k, we obtain

$$\sum_k \left[\sum_{j \neq k} \frac{G \, m_j \, m_k}{r_{jk}^3} \, (\boldsymbol{r}_j - \boldsymbol{r}_k) \right],$$

which turns out to be equal to zero, because in the sum to each term $(\boldsymbol{r}_i - \boldsymbol{r}_\ell)$ there will correspond a term $(\boldsymbol{r}_\ell - \boldsymbol{r}_i)$ that is equal and of opposite sign. That is,

$$\sum_k m_k \, \ddot{\boldsymbol{r}}_k = 0. \tag{3.15}$$

This relation can be interpreted in the following way. Define the vector

$$\boldsymbol{r}_{\text{c.m.}} = \frac{1}{M} \sum_k m_k \, \boldsymbol{r}_k, \tag{3.16}$$

with $M = \sum_k m_k$, which identifies the centre of mass of the system; then (3.15) is equivalent to

$$\ddot{\boldsymbol{r}}_{\text{c.m.}} = 0. \tag{3.17}$$

This can be immediately integrated giving

$$\boldsymbol{r}_{\text{c.m.}} = \boldsymbol{a} \, t + \boldsymbol{b}, \tag{3.18}$$

where $\boldsymbol{a}$ and $\boldsymbol{b}$ are constant vectors that can be expressed as a function of the initial data: relation (3.18) then provides six first integrals. From a physical point of view, (3.18) expresses the conservation of the total momentum of the system. The centre of mass of the system moves with uniform rectilinear motion. The centre of mass can therefore be taken as the origin of a new inertial reference system with axes parallel with those of the previous fixed system, and thus the motion of the N bodies can be referred to their own centre of mass.

The equations of motion do not change, since $\ddot{\boldsymbol{r}}_{\text{c.m.}} = 0$, and therefore we shall keep using the same notations, bearing in mind, however, that between $\boldsymbol{r}_k$ and $\boldsymbol{v}_k$

$$\sum_k m_k \, \boldsymbol{r}_k = 0 \tag{3.19}$$

(the origin of the reference system is at the centre of mass) and

$$\sum_k m_k \, \boldsymbol{v}_k = 0 \tag{3.20}$$

(the total momentum referred to the centre of mass is equal to zero). These correspond to six scalar relations and as a consequence the order of the system is reduced to $6N - 6$.

Let us now consider the system in the form (3.1b) and multiply scalarly for $\dot{\boldsymbol{r}}_k$; by summing over k one gets

$$\sum_k m_k\, \dot{\boldsymbol{r}}_k \cdot \ddot{\boldsymbol{r}}_k = \sum_k \frac{\partial U}{\partial \boldsymbol{r}_k} \cdot \frac{d\boldsymbol{r}_k}{dt},$$

from which

$$\frac{d}{dt}\left[\frac{1}{2}\sum_k m_k\, v_k{}^2\right] = \frac{dU}{dt},$$

and then, since the quantity in square brackets is the total kinetic energy of the system $\mathcal{T}$,

$$\mathcal{T} = U + h, \tag{3.21}$$

where h is constant and equal to the total energy. By applying the energy integral, the order is reduced to $6N - 7$.

Starting from (3.21), we can obtain an important relation known as the *Lagrange–Jacoby identity.*[9] Defining the moment of inertia of the system as

$$I = \sum_k m_k\, r_k{}^2 = \sum_k m_k\, \boldsymbol{r}_k \cdot \boldsymbol{r}_k, \tag{3.22}$$

and differentiating twice, we have

$$\ddot{I} = 2\sum_k m_k\,(\boldsymbol{v}_k \cdot \boldsymbol{v}_k) + 2\sum_k \boldsymbol{r}_k \cdot m_k\, \ddot{\boldsymbol{r}}_k = 4\,\mathcal{T} + 2\sum_k \frac{\partial U}{\partial \boldsymbol{r}_k} \cdot \boldsymbol{r}_k.$$

As U is a homogeneous function of degree -1 in the variables $(x_k, y_k, z_k) \equiv \boldsymbol{r}_k$, Euler's theorem enables us to write

$$\sum_k \frac{\partial U}{\partial \boldsymbol{r}_k} \cdot \boldsymbol{r}_k = -U.$$

Therefore $\ddot{I} = 4\,\mathcal{T} - 2\,U = 2\,U + 4\,h$, so that

$$\frac{1}{2}\ddot{I} = U + 2\,h \tag{3.23}$$

(Lagrange–Jacobi identity).

We will now obtain a further reduction of the order of the system by 3, taking into account angular momentum conservation. Let us vectorially multiply system (3.1a) by $\boldsymbol{r}_k$ and sum over k:

$$\sum_k m_k\,(\boldsymbol{r}_k \times \ddot{\boldsymbol{r}}_k) = \sum_k \left[\sum_{j \neq k} \frac{G\, m_j\, m_k}{r_{jk}{}^3}\,(\boldsymbol{r}_j \times \boldsymbol{r}_k)\right].$$

---

[9] This equation was obtained by Lagrange for the three-body problem in the paper mentioned in Footnote 8 and generalized by Jacobi (see the fourth lecture of Jacobi: op. cit. p. 22)

The sum over k to the right gives a null result, because of the antisymmetry of the vector product and, for the same reason, we have for the left-hand side

$$\frac{d}{dt}\left[\sum_k m_k \ (\boldsymbol{r}_k \times \boldsymbol{v}_k)\right] = 0.$$

That is,

$$\sum_k m_k \ (\boldsymbol{r}_k \times \boldsymbol{v}_k) = \boldsymbol{c}, \tag{3.24a}$$

where $\boldsymbol{c}$ is a constant vector that represents the total angular momentum of the system. Equation (3.24a) corresponds to the three scalar relations

$$\sum_k m_k \ (y_k \, \dot{z}_k - z_k \, \dot{y}_k) = c_1,$$

$$\sum_k m_k \ (z_k \, \dot{x}_k - x_k \, \dot{z}_k) = c_2, \tag{3.24b}$$

$$\sum_k m_k \ (x_k \, \dot{y}_k - y_k \, \dot{x}_k) = c_3,$$

and $\boldsymbol{c} \equiv (c_1, c_2, c_3)$. Hence the order of the system is reduced to $6N-10$. The ten constants of the motion $a_1, a_2, a_3, \ b_1, b_2, b_3, \ c_1, c_2, c_3, \ h$ are algebraic functions of the coordinates and velocities of the N mass points and of the time and one can demonstrate that they are independent of each other.

As we have already said, in addition Bruns[10] demonstrated that further independent algebraic integrals do not exist. Later on, Painlevé[11] was able to extend the demonstration to the case where the algebraic dependence is required only for the dependence on the velocities, while the dependence on the coordinates can be anything. Furthermore, it has been demonstrated that this result also applies when integrals are substituted for the algebraic functions, that is Abelian functions can also be included. Bruns's theorem, although it ended the search for integrals of the type mentioned above, did not stop the question being asked. As well as the tendency to interpret Bruns's result as a proof of the nonexistence, tout court, of integrals in addition to the known ones, we also mention the position of those who considered it irrelevant from the point of view of dynamical systems. As an example, here is Wintner's conclusion:[12]

---

[10]H. Bruns: Über die Integrale des Vielkörper-Problems, *Acta Math.*, **11**, 25–96 (1887). The demonstration of Bruns's theorem can also be found, for instance, in Whittaker: op cit. pp. 358–379.

[11]P. Painlevé: Mémoire sur les intégrales premières du problème des N corps, *Bull. Astr.*, **15**, 81–113 (1898).

[12]A. Wintner: *The Analytical Foundations of Celestial Mechanics* (Princeton University Press, 1947), p. 97 (the notations in the text have been adapted according to our Sect. 3.1).

"... it must, however, be said that the elegant negative results of this arithmetical type do not have any dynamical significance. For all that is of dynamical interest is an enumeration of all those independent integrals $F(x)$ which are isolating. Now, even if $X(x)$ is algebraic, the algebraic character of an integral $F(x)$ of (3.3), though sufficient, is by no means necessary for an $F(x)$ which is an isolating integral."

The integrals of dynamical interest, Wintner maintains, are those which have the property of being *isolating*, i.e., those which enable one to make predictions concerning the possible future (or past) positions of the solution path which goes, for instance, at $t = t_0$, through $x = x_0$.

A more careful characterization of the nature of isolating integrals will be given in Sect. 5.1, to which we refer the reader. Taking up the thread of our argument again, we remark that it is possible to obtain a further lowering of two in the order $6N$ and to reach the order $6N - 12$, by applying two procedures, the first being the already mentioned (Sect. 1.11) elimination of time, and the second, due to Jacobi,[13] the *elimination of nodes*. The procedure of eliminating nodes was introduced by Jacobi for the three-body problem and later on extended by other people to the case of any N; it has been reformulated in various ways, and a satisfactory explanation would take up too much space. We limit ourselves to taking it up as follows, using the words out of Whittaker's book (p. 341): when one of the coordinates which define the position of the system is taken to be the azimuth ϕ of one of the bodies with respect to some fixed axis (say the axis of z), and the other coordinates define the position of the system relative to the plane having this azimuth, the coordinate ϕ is an ignorable coordinate, and consequently the corresponding integral (which is one of the integrals of angular momentum) can be used to depress the order of the system by *two* units (with a net profit of one). We shall again find this result later on when determining what are the integrals in involution admitted by the N-body problem.

Let us now tackle the problem from a Lagrangian point of view, by applying Noether's theorem,[14] as we have already done in the case of the two-body problem in Sect. 2.2. In that case, having simply considered the system in the plane of motion, the transformations involved were dependent on only one parameter. Now, on the other hand, since we are obliged to consider the N-body system in three-dimensional space, this will no longer be possible; it is enough, in fact, to think about the example of rotations: in three-dimensional space, a rotation will, in general, depend on three parameters. So it will be better to write (1.B.35) and (1.B.37) in a more general form, making use of a parameter $\varepsilon \equiv (\varepsilon_1, \varepsilon_2, \ldots, \varepsilon_r)$, with $r \leq n$. Thus, we will replace (1.B.37a)

---

[13]K. G. J. Jacobi: Sur l'elimination des noeuds dans le problème des trois corps, *Crelle's Journal f.r. und ang. Mathematik Bd.*, **XXVI**, 115–131 (1843).

[14]The first application of Noether's theorem to the N-body problem goes back to Bessel-Hagen. See: E. Bessel-Hagen: Über die Erhaltungssätze der Electrodynamik, *Math. Ann.*, **84**, 258–276 (1921).

with

$$\bar{t} = t + \varepsilon_s\,\tau_s\,(t, \boldsymbol{q}, \dot{\boldsymbol{q}}) + o\,(\varepsilon), \qquad \bar{q}_k = q_k + \varepsilon_s\,\xi_s^k\,(t, \boldsymbol{q}, \dot{\boldsymbol{q}}) + o\,(\varepsilon), \qquad (3.25)$$

where $\varepsilon = |\boldsymbol{\varepsilon}|$, $s = 1, 2, \ldots, r \le n$, $k = 1, 2, \ldots, n$. It is easy to check that the Killing equations will now govern the generators τ_s and ξ_s^k and that (1.B.48), which defines the first integrals, will become

$$I_s = \mathcal{L}\,\tau_s + \frac{\partial \mathcal{L}}{\partial \dot{q}_i}\,(\xi_s^i - \dot{q}_i\,\tau_s) - f_s, \qquad s = 1, 2, \ldots, r \quad (r \le n). \qquad (3.26)$$

As the Lagrangian of the system, we define

$$\mathcal{L} = \mathcal{T} + U = \frac{1}{2} \sum_{k=1}^{N} m_k\,\left(\dot{x}_k{}^2 + \dot{y}_k{}^2 + \dot{z}_k{}^2\right) + U, \qquad (3.27)$$

where U is given by (3.2) and the Cartesian coordinates will be taken as Lagrangian coordinates according to the rule

$$x_1 = q_1,\, y_1 = q_2,\, z_1 = q_3;\, x_2 = q_4,\, \ldots,\, y_N = q_{n-1},\, z_N = q_n, \qquad n = 3N.$$

In the same way, the conjugate momenta $p_i = \partial \mathcal{L}/\partial \dot{q}_i$ are

$$p_1 = m_1\,\dot{x}_1,\, p_2 = m_1\,\dot{y}_1,\, \ldots,\, p_{n-1} = m_N\,\dot{y}_N,\, p_n = m_N\,\dot{z}_N.$$

Let us now proceed to verify with what types of symmetry and quasi-symmetry our system is endowed. We begin with *time translations*. If $\tau_1 = -1$, $f_1 \equiv 0$, $\xi_1^i \equiv 0$, $\forall i$, then the integral

$$I_1 = \frac{\partial \mathcal{L}}{\partial \dot{q}_i}\,\dot{q}_i - \mathcal{L} = \mathcal{T} - U = h \qquad (3.28)$$

exists and represents the total energy. Then define the *space translations*:

$$\bar{t} = t, \qquad \bar{x}_k = x_k + \varepsilon_2, \qquad \bar{y}_k = y_k + \varepsilon_3, \qquad \bar{z}_k = z_k + \varepsilon_4,$$

with $\tau_s \equiv 0$, $\xi_s^i = 1$, $s = 2, 3, 4$, $\forall i = 1, 2, \ldots, n$. The corresponding Killing equations are satisfied by $f_2, f_3, f_4, \equiv 0$. For the sake of brevity, we omit, here and in what follows, the details of the calculations. Equations (3.26) give

$$I_2 = \sum_k m_k\,\dot{x}_k = \text{const.},$$

$$I_3 = \sum_k m_k\,\dot{y}_k = \text{const.}, \qquad (3.29)$$

$$I_4 = \sum_k m_k\,\dot{z}_k = \text{const.},$$

that is, the conservation of the total momentum. Integrals of the type (3.29) could not be present in the case of the two-body problem, since we studied it in the field of a fixed centre and so precluded a priori invariance under translations.

Let us now study the behaviour of the system under *spatial rotations*. By considering a spatial rotation as the result of three rotations in the planes xy, xz, yz, and denoting by ε_5, ε_6, ε_7 respectively the corresponding parameters, we have

$$\begin{aligned}
\overline{x}_k &= x_k - \varepsilon_5\, y_k + \varepsilon_6\, z_k, \\
\overline{y}_k &= y_k + \varepsilon_5\, x_k - \varepsilon_7\, z_k, \\
\overline{z}_k &= z_k - \varepsilon_6\, x_k + \varepsilon_7\, y_k.
\end{aligned} \tag{3.30}$$

The generators will then be given by $\tau_s \equiv 0$, for $s = 5, 6, 7$, and

$$\xi_s^{xk} = \begin{cases} -y_k, \\ z_k, \\ 0, \end{cases} \qquad
\xi_s^{yk} = \begin{cases} x_k, \\ 0, \\ -z_k \end{cases} \qquad
\xi_s^{zk} = \begin{cases} 0, & s = 5, \\ -x_k, & s = 6, \\ y_k, & s = 7, \end{cases} \tag{3.31}$$

where we have labelled with suffixes $\boldsymbol{x}$, $\boldsymbol{y}$, $\boldsymbol{z}$ the three groups of generators corresponding to the transformations of the coordinates x_k, y_k, z_k (with $k = 1, 2, \ldots, N$) respectively. Also in this case, it is possible to check that the Killing equations are satisfied by $f_s \equiv 0$ ($s = 5, 6, 7$). Equations (3.26) give

$$\frac{\partial \mathcal{L}}{\partial \dot{x}_k}\,\xi_s^{xk} + \frac{\partial \mathcal{L}}{\partial \dot{y}_k}\,\xi_s^{yk} + \frac{\partial \mathcal{L}}{\partial \dot{z}_k}\,\xi_s^{zk} = I_s = \text{const.} \qquad (s = 5, 6, 7).$$

By inserting (3.31), one obtains

$$\begin{aligned}
I_5 &= \sum_k m_k\,(x_k\,\dot{y}_k - y_k\dot{x}_k) = c_3, \\
I_6 &= \sum_k m_k\,(z_k\,\dot{x}_k - x_k\dot{z}_k) = c_2, \\
I_7 &= \sum_k m_k\,(y_k\,\dot{z}_k - z_k\dot{y}_k) = c_1,
\end{aligned} \tag{3.32}$$

that is, the components of the angular momentum.

Lastly, let us consider a final type of transformation. Since the motion of our system has been referred to an inertial reference system, one expects invariant behaviour under a transformation that is equivalent to passing to another inertial system. The transformation will be of the type

$$\overline{t} = t, \quad \overline{x}_k = x_k + \varepsilon_8\, t, \quad \overline{y}_k = y_k + \varepsilon_9\, t, \quad \overline{z}_k = z_k + \varepsilon_{10}\, t, \tag{3.33}$$

which represents a translatory motion at constant velocity of components ε_8, ε_9, ε_{10}. The corresponding generators will be $\tau_s \equiv 0$, for $s = 8, 9, 10$ and

$$\xi_s^{xk} = \begin{cases} t, & s = 8, \\ 0, & s = 9,10, \end{cases} \qquad
\xi_s^{yk} = \begin{cases} t, & s = 9, \\ 0, & s = 8,10, \end{cases} \qquad
\xi_s^{zk} = \begin{cases} t, & s = 10 \\ 0, & s = 8,9. \end{cases} \tag{3.34}$$

Unlike the transformations considered up to now, in (3.34) we have generators depending on time, and so the Killing equations will also contain partial time derivatives, both of generators and of the functions f_s. They turn out to be satisfied by (3.34) if

$$f_8 = \sum_k m_k\, x_k, \quad f_9 = \sum_k m_k\, y_k, \quad f_{10} = \sum_k m_k\, z_k; \tag{3.35}$$

hence the case in hand is a case of quasi-symmetry. The corresponding first integrals are

$$I_8 = t \sum_k m_k\, \dot{x}_k - \sum_k m_k\, x_k,$$

$$I_9 = t \sum_k m_k\, \dot{y}_k - \sum_k m_k\, y_k, \tag{3.36}$$

$$I_{10} = t \sum_k m_k\, \dot{z}_k - \sum_k m_k\, z_k.$$

As for their physical meaning, if we compare (3.36) with (3.16), (3.18), we immediately get

$$M\, \mathrm{a}_x = \sum_k m_k\, \dot{x}_k, \quad M\, \mathrm{a}_y = \sum_k m_k\, \dot{y}_k, \quad M\, \mathrm{a}_z = \sum_k m_k\, \dot{z}_k$$

and

$$M\, \mathrm{a}_x\, t - \sum_k m_k\, x_k = -M\mathrm{b}_x = I_8,$$

$$M\, \mathrm{a}_y\, t - \sum_k m_k\, x_k = -M\mathrm{b}_y = I_9, \tag{3.37}$$

$$M\, \mathrm{a}_z\, t - \sum_k m_k\, x_k = -M\mathrm{b}_z = I_{10};$$

therefore I_8, I_9, I_{10} are the first integrals connected with the uniformity of the centre of mass motion. Furthermore, it is possible to demonstrate that the set of transformations we have considered constitutes a group, a 10-parameter group, usually named the *Galilei group*.

If we now consider the Hamiltonian of the system, $\mathcal{H} = \mathcal{T} - U = I_1$, and evaluate the Poisson bracket,

$$\begin{aligned}
(I_8, \mathcal{H}) = \sum_k &\left[\frac{\partial I_8}{\partial x_k} \frac{\partial I_1}{\partial (m_k\, \dot{x}_k)} - \frac{\partial I_8}{\partial (m_k\, \dot{x}_k)} \frac{\partial I_1}{\partial x_k} \right] \\
+ \sum_k &\left[\frac{\partial I_8}{\partial y_k} \frac{\partial I_1}{\partial (m_k\, \dot{y}_k)} - \frac{\partial I_8}{\partial (m_k\, \dot{y}_k)} \frac{\partial I_1}{\partial y_k} \right] \\
+ \sum_k &\left[\frac{\partial I_8}{\partial z_k} \frac{\partial I_1}{\partial (m_k\, \dot{z}_k)} - \frac{\partial I_8}{\partial (m_k\, \dot{z}_k)} \frac{\partial I_1}{\partial z_k} \right],
\end{aligned}$$

and analogously (I_9, I_1) and (I_{10}, I_1), it is easy to verify that they are different from zero. Therefore the integrals I_8, I_9, I_{10} are not in involution with the Hamiltonian. In the same way, one can easily verify that I_8, I_9, I_{10} are in involution with each other. Passing to the components of the angular momentum, we find that each of them is in involution with $c^2 = c_1^2 + c_2^2 + c_3^2$; if we pick out one of them, we get, together with c^2, a pair of independent integrals in involution. If now, at the end, we evaluate (I_8, c^2), (I_9, c^2), (I_{10}, c^2), we get a result different from zero. Nevertheless the three brackets will vanish if we fix the origin of the coordinate system at the centre of mass (in this case $I_8 = I_9 = I_{10} = 0$). Therefore, by putting the origin at the centre of mass, our system gets five integrals in involution. This is equivalent, as we know from Lie's theorem (see Sect. 1.18), to ten first integrals. If we add to these ten integrals the energy integral and eliminate the time, again we reduce the order of the system to $6N - 12$. To sum up: the possibility, implicitly contained in Lagrange's paper, of reducing the order of the system to $6N - 12$ has been checked by different methods by a number of mathematicians over more than two centuries. In particular, the most recent reductions (Poincaré, Whittaker, Levi-Cìvita) leave the canonical form of the equations of motion invariant.[15]

3.3 The Singularities

We saw in Sect. 3.1 that, if the initial conditions are such that ϱ, as defined in (3.5), is positive, then, for Cauchy's existence theorem, there exists a real analytical solution for system (3.1b) in an open time interval $|t - t_0| < \delta$. This *local* result can be extended, analytically prolonging the solution for $t > t_0 + \delta$. If this analytical continuation of the solution is no longer possible at a certain time $t = t_1$, we say that the problem has a *singularity* at $t = t_1$.

At this point we need to clarify the nature of this singularity, since one might wonder if it could be eliminated in some way (changing variables or by some other trick). We ask ourselves if the singularity at time t_1 is due to a physical collision between two or more material points or to an unlimited motion in a finite time or, finally, to uncontrolled unlimited oscillations of some points. Here and in the following, by a "collision" at the time $t = t_1$ we mean that, for $t \to t_1$, the position vectors of the N material points tend to well-defined limits of which at least two are coincident.

The systematic analysis of the singularities for the N-body problem was begun by P. Painlevé, who in 1895 gave at the University of Stockholm a series of lectures on differential equations that at the time were exceptionally well received.[16] Painlevé was the first to raise the question if all singularities are due to collisions. Before illustrating his result, let us introduce some useful relations. First we calculate the quantities

---

[15]See, for instance, E. T. Whittaker: op. cit., Chap. XIII.

[16]The lectures were published in 1897 – P. Painlevé: *Leçons sur la Théorie Analitique des Equations Differentielles Professées à Stockholm* (A. Hermann, Paris, 1897). The N-body problem constitutes the final part of the treatise (pp. 582–589).

$$\sum_j m_j \left(\boldsymbol{r}_j - \boldsymbol{r}_k\right)^2 = \sum_j m_j\, r_j^2 + \sum_j m_j\, r_k^2 - 2\boldsymbol{r}_k\cdot\sum_j m_j\,\boldsymbol{r}_j, \quad j,k = 1,2,\ldots,N.$$

If the coordinates are referred to the centre of mass, the last term on the right is zero, so that

$$\sum_j m_j \left(\boldsymbol{r}_j - \boldsymbol{r}_k\right)^2 = I + M\, r_k^2.$$

Multiplying by m_k and summing over k we find that $\sum_k \sum_j m_k\, m_j\, r_{jk}^2 = M I + M I = 2 I M$, which can be written as

$$\sum_{1\le j<k\le N} m_j\, m_k\, r_{jk}^2 = I M. \tag{3.38}$$

Let us define the following

$$r\left(t\right) = \min_{j\ne k} r_{jk}(t), \qquad R\left(t\right) = \max_{j\ne k} r_{jk}(t). \tag{3.39}$$

Remembering definition (3.2) of the potential U, we soon obtain (in units such that $G = 1$)

$$U \le \sum_{1\le j<k\le N} \frac{m_j\, m_k}{r},$$

and since $\sum_{1\le j<k\le N} m_j\, m_k \le M^2/2$, we also have $U \le M^2/2\,r$. Moreover,

$$U \ge \frac{m_j\, m_k}{r_{jk}} \ge \frac{\mu^2}{r_{jk}}, \qquad \forall\ 1 \le j,k \le N,$$

where μ was defined in (3.5), and, since at every instant at least one r_{jk} is equal to r, it will be also $U \ge \mu^2/r$. There therefore exist two positive constants A and B ($A = \mu^2$, $B = M^2/2$) such that $A\, r^{-1} \le U \le B\, r^{-1}$, or

$$A\, U^{-1} \le r \le B\, U^{-1}, \tag{3.40}$$

so that U^{-1} represents an estimate of the minimum distance among the points in the system.

Let us see now the role of the moment of inertia I. From (3.38) and (3.39), we have the following chain of inequalities:

$$\frac{\mu^2}{M}R^2 \le \frac{\mu^2}{M} \sum_{j<k} r_{jk}^2 \le I = \frac{1}{M} \sum_{j<k} m_j\, m_k\, r_{jk}^2 \le \frac{1}{M} \sum_{j<k} m_j\, m_k\, R^2 \le \frac{1}{2}MR^2$$

so that

$$\frac{\mu^2}{M} R^2 \le I \le \frac{1}{2}MR^2.$$

Finally,

$$\tilde{A}\, I^{1/2} \le R \le \tilde{B}\, I^{1/2}, \qquad \text{with}\quad \tilde{A} = \sqrt{\frac{2}{M}}, \quad \tilde{B} = \frac{\sqrt{M}}{\mu}, \tag{3.41}$$

and thus $\sqrt{I}$ is an estimate of the greatest separation between the points.

We now arrive at the results obtained by Painlevé, quoting his statement of the theorem on the existence of the singularities in the N-body problem:

Soit $r(t)$ la plus petite des distances r_{ij} à l'instant t. Si le mouvement est régulier pour $t < t_1$ mais point au déla, $r(t)$ tend vers zéro quand t tend vers t_1: c'est à dire que pour t suffisamment voisin de t_1, $r(t)$ *reste* inferieur à toute quantité donnée à l'avance ε.[17]

That is, a necessary and sufficient condition for there to be a singularity at the instant $t = t_1$ is that $r \to 0$ for $t \to t_1$; therefore, from (3.40), we also have $U \to \infty$ for $t \to t_1$. That the condition is sufficient follows from the definition of singularity; that it is necessary follows instead from the considerations at the end of Sect. 3.1. Incidentally, Painlevé noted that the condition $r \to 0$ for $t \to t_1$ does not necessarily imply that the singularity was a collision. In fact, it may happen that some of the distances r_{ij} oscillate in such a way that

$$\lim \inf r_{ij} = 0, \qquad \lim \sup r_{ij} > 0,$$

whereas $r \to 0$; that is, r may tend to zero, and so there can be a singularity without a collision. Following a suggestion by Poincaré, Painlevé spoke in this case of *pseudocollisions*.[18] In any case, he succeeded in proving a theorem relative to the three-body case:

For $N = 3$, all the singularities are collisions.

The solution to the question posed by Painlevé, for $N \geq 4$, was provided about a century later, as we will see at the end of this section. But, to follow strict chronological order, note that an important step forward was taken in 1908 by Von Zeipel[19] with the following theorem:

If some of the material points do not tend to a finite limiting position, for $t \to t_1$, then necessarily $R \longrightarrow \infty$, in the limit $t \to t_1$, being R the greatest distance among the points.

This means that, if the singularity is due to a collision, the system stays bounded and the only way to have a pseudocollision is for the system to "blow up" to infinity in a finite time. The conclusions by Painlevé and Von Zeipel were taken up again lately and, after more than half a century, new results were obtained that further clarified the conditions in which a singularity occurs. The subject is further considered in a series of theorems, some of which will be stated here without proof, which would in general be quite

---

[17]Painlevé: op. cit., p. 583.

[18]Painlevé: op. cit., p. 588.

[19]H. Von Zeipel: Sur les singularités du problème des n corps, *Arkiv för Matematik, Astronomi och Fysik*, **4**, 32, 1–4 (1908). The theorem was redemonstrated in "modern" language by McGehee – R. McGehee: Von Zeipel's theorem on singularities in celestial mechanics, *Expo. Math.*, **4**, 335–345, (1986).

involved and "technical". Since we are treating an autonomous system, and therefore can "translate" the solutions by changing the origin of the time axis, it is convenient to take $t_1 = 0$ and so always consider the limit for $t \to 0$. Let us moreover assume the following definitions:

a) A function $f(t)$ is said to *slowly* vary for $t \to 0$ if $f(\beta t)/f(t) \to 1$ for $t \to 0$, with β an arbitrary positive constant.
b) A function $g(t)$ is said to *regularly* vary for $t \to 0$, if $g(t) = t^\alpha f(t)$, where f is a slowly varying function for $t \to 0$ and α is bounded from below by a suitable constant A.

Let us now move on to the statement of the first theorem:[20]

> *If there is a singularity at the time $t = 0$ and if I is a slowly varying function for $t \to 0$, then the singularity is due to a collision.*

As a corollary of this theorem we have Von Zeipel's theorem mentioned above:

> *If $I = O(1)$ for $t \to 0$, then the singularity at the time $t = 0$ is a collision.*

The proof of the corollary is simple: if there is a singularity at $t = 0$, then, in agreement with Painlevé's result, $U \to \infty$ for $t \to 0$. From the Lagrange–Jacobi identity (3.23), it follows that $\ddot{I} \to \infty$ for $t \to 0$; therefore the function $I(t)$ has an upward concavity and may tend either to a positive limit $(I \geq 0)$ or to infinity, for $t \to 0$. The infinite limit contradicts the hypothesis $I = O(1)$; only $I \geq 0$ remains finite. If $I \to 0$ for $t \to 0$, then $r_k \to 0, \forall k$ as well, and therefore there is a collision. If I stays greater than zero for $t \to 0$, then it is a slowly varying function and the theorem by Saari can be applied, so that again there is a collision. The corollary is then proved. Let us state the second theorem:[21]

> *If the N material points are placed on a straight line, then all the singularities are due to collisions.*

We may assume all the points on the x axis of the barycentric frame; as $\sum_{i=1}^{N} m_i x_i = 0$, if we denote by x_1 the coordinate of the first point (on the left of the origin) and by x_N that of the last one (on the right of the origin), then $x_1 \leq 0$ and $x_N \geq 0$. Let us prove the theorem by finding a contradiction supposing that it is untrue. If it is true, from Von Zeipel's theorem, the existence of a pseudocollision at $t = 0$ implies that $I \to \infty$ for $t \to 0$. From (3.41), as a consequence, $R \to \infty$ too, for $t \to t_0$. Since the N points must always lie on the x axis and cannot change their relative order, $R = |x_N - x_1|$ and therefore also $|x_N - x_1| \to \infty$. From the condition $\sum_i m_i x_i = 0$, we see that x_N does not have a limit for $t \to 0$ (actually

---

[20]D. G. Saari: Singularities and collisions of Newtonian gravitational systems, *Arch. Mech. Math.*, **49**, 311–320, (1973).
[21]Saari: loc. cit., p. 319.

$x_N \to \infty$ for $t \to 0$). In fact, supposing that this is not so and then that x_N is bounded, from $|x_N - x_1| \to \infty$ we will have $x_1 \to -\infty$. Since the centre of mass is fixed, this implies that

$$\sum_{i=2}^{N} m_i\, x_i = -m_1\, x_1 \to \infty.$$

But this means that there is some point x_i such that $\limsup x_i = \infty$. Since x_N is to the right of all other particles, then $\limsup x_N = \infty$ too, in contradiction with what we assumed. Then, if the singularity is a pseudocollision, x_N does not have a limit. Let us now consider the question from another point of view. The equation of motion for m_N is

$$m_N\, \ddot{x}_N = \sum_{j \neq N} \frac{m_N\, m_j\, (x_j - x_N)}{|x_j - x_N|^3}.$$

Any point x_i, with $i = 1, 2, \ldots, N-1$, will be to the left of x_N, so that it will be always the case that $\ddot{x}_N < 0$: the function $x_N(t)$ will be concave downwards, and, as a consequence, x_N will be limited for $t \to 0$, but this is in contradiction with what was just established, and so the theorem is proved. Let us now consider the third theorem:[22]

A singularity at the time $t = 0$ is due to collision if and only if $U \approx \alpha\, t^{-2/3}$ for $t \to 0$, for some positive constant α.

The notation $f \approx g$ used is intended in the sense that, if f and g are positive functions, after some time t there exist positive constants A and B such that $A\,g(t) \leq f(t) \leq B\,g(t)$. Referring to the cited works for a proof of the necessity of this condition, let us see briefly how the stated condition is sufficient. From the energy integral $\mathcal{T} - U = h$ and from the hypothesis of the theorem, it follows that $\mathcal{T} = \mathrm{O}\left(t^{-2/3}\right)$ for $t \to 0$, from which $v_k^2 = \mathrm{O}\left(t^{-2/3}\right)$, that is, $\boldsymbol{v}_k = \dot{\boldsymbol{r}}_k = \mathrm{O}\left(t^{-1/3}\right)$ and, integrating between t_1 and t_2 $(0 < t_1 < t_2)$, we have $|\boldsymbol{r}_k(t_1) - \boldsymbol{r}_k(t_2)| \leq A|t_1^{2/3} - t_2^{2/3}|$ for some constant A. For t_1 and $t_2 \to 0$, the right-hand side of this inequality tends to zero and the left-hand side does too; Cauchy's criterion for the existence of the limit gives that every $\boldsymbol{r}_k$ tends to a limit for $t \to 0$. Therefore the singularity can be due only to a collision. Finally, we quote a result due to Saari[23] as the fourth theorem:

The set of initial conditions leading to a collision in a finite time has zero (Lebesgue) measure.

In other words, in an N-body self-gravitating system, *the collisions have a low probability of occuring*. In summary, collisions, however improbable, are

[22] A. Wintner: op. cit., pp. 255–257. H. Pollard, D. G. Saari: Singularities of the N-body problem. I, *Arch. Rational Mech. Anal.*, **30**, 263–269, (1968).

[23] D. G. Saari: Improbability of collisions in Newtonian gravitational systems, *Trans. Amer. Math. Soc.*, **162**, 267–271, (1971); **181**, 351–368, (1973).

a considerable hindrance to the treatment of the N-body problem, since, in general, we do not know how to exclude the initial conditions leading to one or more collisions. The only situation in which this is possible is in the case in which all the points collide at the centre of mass: this is the *global collapse* we will examine in the next section.

Let us return as we said we would at the start of the section, to the question of the singularities that are not due to collisions. In 1974, McGehee[24] announced the construction of a non-collisional singularity in the rectilinear five-body problem; later on, Mather and McGehee[25] proved that there can be a noncollisional singularity in the rectilinear four-body problem. This result, however, must be intended (see Saari's theorem, proved above) in the sense that this singularity occurs after an infinite number of (regularized) collisions. Finally in 1988, Xia[26] demonstrated the existence of a non-collisional singularity, not preceded by an infinite number of collisions as in the result by Mather and McGehee, in the spatial five-body problem. Xia considers two pairs, each constituted by two equal masses, plus a small fifth body. The two binary systems are on opposite sides with respect the xy plane and rotate in opposite directions. The fifth particle moves along the z axis (so that the total angular momentum is zero) and oscillates between the two binaries, determining an unlimited motion in a finite time. The system is represented in Fig. 3.1. The example given by Xia can be extended to cases with the same type of symmetry with any $N > 5$. Another case of noncollisional singularity has been found by Gerver,[27] who considers $3n$ particles in the plane. $2n$ of them (all of the same mass) constitute n pairs covering circular orbits whose centres are at the vertices of a regular n-agon. The other n have very small equal masses and move from one pair to the other (see Fig. 3.2). When each of them moves near a binary, it gets kinetic energy from the pair and releases linear momentum, so that the binary becomes more bounded and its centre moves away from the centre of the n-agon. If one iterates the process, for suitable values of the initial conditions, the extension of the configuration will grow at each close encounter of a particle with one of the binaries and the system will become unlimited in a finite time.

---

[24]R. McGehee: Triple collisions in the collinear three body problem, *Invent. Math.*, **27**, 191–227 (1974).

[25]J. Mather, R. McGehee: Solutions of the Collinear Four Body Problem which Become Unbounded in Finite Time, Lecture Notes in Physics **38**, ed. by J. Moser (Springer, 1975), pp. 573–597.

[26]Z. Xia: The existence of non collision singularities in the Newtonian systems, *Ann. Math.*, **135**, 411–468 (1992).

[27]J. L. Gerver: The existence of pseudocollisions in the plane, *J. Diff. Eq.*, **89**, 1–68 (1991).

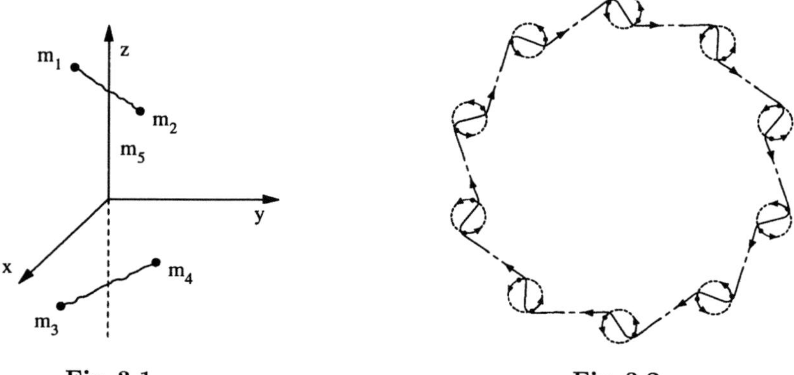

Fig. 3.1   Fig. 3.2

3.4 Sundman's Theorem

As we have already said, for any N (the case $N = 3$ will be examined separately), it is in general unknown how to exclude binary collisions. However, there is a celebrated theorem due to Sundman,[28] whose content, incidentally, was already familiar to Weierstrass,[29] that allows us to exclude the global collapse of the system. This theorem states that:

> The global collapse of the system may occur only in the case in which the total angular momentum vanishes.

Let us, as a preliminary, establish an important inequality. From (3.24a),

$$c = |\boldsymbol{c}| = \sum_k m_k\, r_k\, v_k\, \sin\vartheta_k,$$

θ_k being the angles between the vectors $\boldsymbol{r}_k$ and $\boldsymbol{v}_k$; from this we obtain

$$c \le \sum_k m_k\, r_k\, v_k\, |\sin\vartheta_k|,$$

and, from Schwarz's inequality,

$$c^2 \le \sum_k m_k\, r_k^2 \sum_k m_k\, v_k^2 \sin^2\vartheta_k. \tag{3.42}$$

From (3.22) we then have, by differentiating with respect to t,

$$\dot{I} = 2\sum_k m_k\, (\boldsymbol{r}_k \cdot \boldsymbol{v}_k) = 2\sum_k m_k\, r_k\, v_k\, \cos\vartheta_k,$$

---

[28]K. F. Sundman: Mémoire sur le probléme de trois corps, *Acta Mathematica*, **36**, 105–179 (1912). The results reported in this review had already been published in the *Acta Societatis Scientiarum Fennicae*, **34, 35**.

[29]The letter by Weierstrass to Mittag-Leffler in which he speaks of this result (without proof) is published in *Acta Mathematica*, **35**, 55–58 (1912).

from which

$$|\dot{I}| \le 2 \sum_k m_k\, r_k\, v_k\, |\cos \vartheta_k|$$

and, again applying Schwarz's inequality, we get

$$\dot{I}^2 \le 4 \sum_k m_k\, r_k^2 \sum_k m_k\, v_k^2\, \cos^2 \vartheta_k. \tag{3.43}$$

Summing (3.42) and (3.43), we finally obtain

$$\frac{1}{4}\dot{I}^2 + c^2 \le \sum_k m_k\, r_k^2 \sum_k m_k\, v_k^2,$$

that is,

$$\frac{1}{4}\dot{I}^2 + c^2 \le 2I\mathcal{T}, \tag{3.44}$$

which we call *Sundman's inequality*.

Let us now prove that, if the collapse occurs, it must occur in a finite time: that is, it cannot be the case that $I \to 0$ for $t \to \infty$. In fact, referring to the expression of the potential U, if every distance tends to zero ($r_{jk} \to 0$) for $t \to \infty$, then $U \to \infty$ as $t \to \infty$. On the other hand, the Lagrange–Jacobi identity implies that $\ddot{I} \to \infty$, since h (the constant energy) is finite. From a certain instant onwards, it must be the case that $\ddot{I} \ge 1$. Integrating twice yields

$$I \ge \frac{1}{2}t^2 + At + B,$$

where A and B are integration constants. This means that $I \to \infty$ for $t \to \infty$. But this is the opposite of $I \to 0$ for $t \to \infty$: the collapse must therefore occur in a finite time t_1.

We now give the proof of Sundman's theorem. Assume that $t_1 > 0$ (the proof can obviously also be worked out in the case $t_1 < 0$). Then

$$U \to \infty, \quad \ddot{I} \to \infty, \quad I \to 0, \quad \text{for} \quad t \to t_1,$$

in the case of collapse. Also $\ddot{I} > 0$, at least for a time interval $t_2 \le t < t_1$; moreover, since $I > 0$ and $I(t_1) = 0$, in such an interval we also have $\dot{I} < 0$. Therefore, in the case of collapse, $I(t)$ is a decreasing function. Sundman's inequality (3.44) furthermore yields

$$c^2 \le 2I\mathcal{T}.$$

By substituting the Lagrange–Jacobi identity in this, we have

$$c^2 \le 2I\left(\frac{1}{2}\ddot{I} - h\right).$$

Multiplying by the (positive) quantity $-\dot{I}/I$, we obtain

$$-\frac{1}{2}c^2\frac{\dot{I}}{I} \le h\dot{I} - \frac{1}{2}\dot{I}\ddot{I}.$$

Integrating for $t \in [t_2, t_1)$ gives

$$\frac{1}{2} c^2 \ln\left(I^{-1}\right) \leq h I - \frac{1}{4} \dot{I}^2 + k,$$

where k is an integration constant. Furthermore,

$$c^2 \leq \frac{2 I h + 2 k}{\ln\left(I^{-1}\right)}. \tag{3.45}$$

Then, if $I \to 0$ for $t \to t_1$, c also must tend to zero; c being a constant of the motion, $c = 0$ always. With this the theorem is proved.

Referring our mathematical model of N material points to the solar system, we find that the total angular momentum can be calculated with sufficient precision and can therefore be determined an invariant plane perpendicular to the angular momentum vector. Such a plane can be taken as a reference: for example, we can set the origin at the centre of mass and the z axis along the angular momentum vector. A reference frame of this kind, already suggested by Laplace, did not find great favour among astronomers. It might be thought that, since the planets are not actually material points but extended bodies rotating around their own axes and therefore endowed with intrinsic angular momentum, the aforementioned plane, passing through the centre of mass, could assume various orientations, induced by subsequent changes of the rotation axes of the planets (precession, tidal friction, etc.). But, according to present estimates, about 98% of the angular momentum of the solar system is due to the orbital motion of the major planets (Jupiter, Saturn, Uranus and Neptune), and so this is not the case, and therefore the invariant plane remains effectively fixed with respect to the centre of mass of the system.

3.5 The Evolution of the System for $t \to \infty$

Up to now, we have approached the N-body problem with the disposition of one who wants to "solve" set (3.1b). We stated the conditions under which one can be sure that a solution does exist (*existence theorem*), we investigated the cases in which the theorem can(not?) be applied (*occurrence of a singularity*), and we reviewed all the possible first integrals or at least the eventuality that the order of the system could be reduced. At this point it is clear that, even in the most favourable case ($N = 3$) we are very far from believing it possible to integrate (3.1b) in the usual meaning of the term. Waiting for the study of particular solutions characterized by symmetry properties such as to allow their detailed analysis and the possible generalization to any N of some result we will find for $N = 3$, let us now see what can be said about the evolution of the system for $t \to \infty$.

The problem is to try to assess the long time behaviour of the mutual distances of the material points and whether their number stays constant or not, assuming, at the same time, that *the solution exists in the interval* $[0, \infty)$. Such studies have been undertaken only recently, if we excluce the $N = 3$ case, studied in depth by Chazy[30] in the 1920s. In essence, apart some hints in the book by Wintner, the paper that, as it were, founded this field of research is that by Pollard in 1967,[31] followed by works by Saari and others. The bulk of the results depends on the application of the so called *Tauberian Theorem*,[32] whose importance for celestial mechanics Wintner had already suggested.[33] The work by Pollard provided a classification of the final motions of a system of N material points on the basis of the sign of the total energy and it was shown that the behaviour of N-body systems at non-negative energy was essentially a generalization of the two- and three-body problems.

In 1970 Saari,[34] again taking Pollard results, gave a classification of the final motions that did not depend on the sign of the energy. Moreover, he showed that, in the absence of so-called oscillatory or pulsating motions, the system evolves in a regular fashion: the N bodies divide into groups, inside which the mutual distances among the components are limited for $t \to \infty$. These groups form subsystems characterized by a separation going as $t^{2/3}$; in turn, the centres of mass of the subsystems asymptotically separate from each other as ct, where c is a constant vector. In the following, we give the results summarized above: the treatment is extremely sketchy, for most theorems we simply state them, referring to the original works for the proofs.

The Behaviour of r and R for $t \to \infty$

For the minimum distance r the following is relevant

Theorem 1: *If h is negative, r stays limited. If $h = 0$, then $r = \mathrm{O}\left(t^{2/3}\right)$ for $t \to \infty$. If $h > 0$, $r = \mathrm{C}\left(t\right)$ for $t \to \infty$.*

---

[30] J. Chazy: Sur l'allure du mouvement dans le problème des trois corps quand le temps croît indefinitement, *Ann. Sci. École Norm. Sup.*, **39**, 29–130 (1922).

[31] H. Pollard: The behavior of gravitational systems, *J. Math. Mech.*, **17**, 601–611 (1967).

[32] The theorem by Tauber (1897) states that: if the series $a_0 + a_1 x + a_2 x^2 + \ldots$ converges for $|x| < 1$ and its sum converges to ℓ for $x \to 1$, then $a_0 + a_1 + a_2 + \ldots$ converges to ℓ if $n a_n$ is infinitesimal for $n \to \infty$. In general we say that a theorem is "Abelian" if it asserts something about an average of a sequence from a hypothesis on its ordinary limit; it is "Tauberian" if conversely the implication goes from average to limit. By extension theorems concerning integrals are also called Tauberian, being the continuous analogs of the series. See, D. V. Widder: *An Introduction to Transform Theory* (Academic Press, 1971), Chap. 8.

[33] A. Wintner: op. cit., pp. 428–429.

[34] D. G. Saari: Expanding gravitational systems, *Trans. Amer. Math. Soc.*, **156**, 219–240 (1971).

If $h < 0$, the energy integral will be $T = U - |h|$, and since $T \geq 0$, $U \geq |h|$ as well so that, by (3.40), the minimum distance r remains limited.

Let us look at the cases $h = 0$ and $h > 0$. From the definitions of U and r, differentiating and using (3.40) we get

$$|\dot{U}| \leq A U^2 \sum_{1 \leq j < k \leq N} (m_j m_k)^{1/2} |\dot{r}_{jk}|,$$

so that, using Schwarz's inequality,

$$|\dot{U}|^2 \leq A U^4 \sum_{1 \leq j < k \leq N} m_j m_k \dot{r}_{jk}{}^2.$$

Since the centre of mass is at the origin,

$$T = \frac{1}{2M} \sum_{1 \leq j < k \leq N} m_j m_k \dot{r}_{jk}{}^2,$$

and

$$|\dot{U}|^2 \leq A U^4 T,$$

and therefore

$$|\dot{U}| \leq A U^2 T^{1/2} \tag{3.46}$$

(in the preceding lines A indicates a positive constant, differing, in general, from the original one). Setting $\varrho = U^{-1}$, substituting in (3.46) and using the energy integral, we obtain

$$|\dot{\varrho}| \leq A \left(\frac{1}{\varrho} + h \right)^{1/2}. \tag{3.47}$$

If $h = 0$, then $\varrho^{1/2} |\dot{\varrho}| \leq A$; integrating this yields $\varrho^{3/2} \leq A t$, or $U^{-3/2} \leq A t$ so that, by (3.40), we obtain $r = \mathrm{O}\,(t^{2/3})$, as was to be proved.

Now consider the case $h > 0$. We introduce the definition $p = \varrho$ when $\varrho \geq 1$ and, $p = 1$ when $\varrho < 1$. For $\varrho \geq 1$, (3.47) immediately implies that $\dot{p}$ is limited; in each interval in which $\varrho < 1$, on the other hand, $\dot{p} = 0$, p being constant. Therefore $\dot{p}$ is limited everywhere, so that $p = \mathrm{O}\,(t)$. As $\varrho \leq p$, it follows that $\varrho = \mathrm{O}\,(t)$, or $U^{-1} = \mathrm{O}\,(t)$ and then $r = \mathrm{O}\,(t)$. This theorem, therefore, provides upper bounds on the increase of r, whereas nothing is known about the lower bounds. Equally, it is possible to obtain information about the lower bounds of R, but not on its upper bounds.

Theorem 2: *If* $h > 0$, $\quad R \geq A t$; $\quad$ *if* $h = 0$, $\quad R \geq A t^{2/3}$.

If $h > 0$, from the Lagrange–Jacobi identity, since U is always positive, we have $\ddot{I} \geq 2h$. A twofold integration gives $I \geq A t^2$ so that, from (3.41), $R \geq A t$ (A always indicates a positive constant). If $h = 0$, theorem 1 implies that $U \geq A t^{-2/3}$, so that also $\ddot{I} \geq A t^{-2/3}$. A twofold integration gives $I \geq A t^{4/3}$. Inequalities (3.41) readily give $R \geq A t^{2/3}$. This procedure cannot be applied in the case $h < 0$.

Relations Between r and R

We will now try to establish some connection between the behaviour of r and R. The following holds.

Theorem 3: *If $R = O(t)$ for $t \to \infty$, then r is limited in the mean from below; that is:*

$$\frac{1}{t} \int_0^t r(u)\, du \geq A > 0, \qquad \forall\, t > 0.$$

From Taylor's expansion

$$I(2t) = I(t) + t\,\dot{I}(t) + \frac{t^2}{2}\,\ddot{I}(\xi), \qquad t < \xi < 2t,$$

and from the Lagrange–Jacobi identity (from which $\ddot{I} \geq 2\,h$) it follows that

$$I(2t) \geq t\,\dot{I} + h\,t^2 \tag{3.48}$$

always. As by hypothesis, $R = O(t)$, from (3.41), we have $I = O(t^2)$. From this and (3.48), we deduce that $t\,\dot{I} \leq A\,t^2$, or $\dot{I} \leq A\,t$. Integrating the Lagrange–Jacobi identity gives

$$\frac{1}{2}\,\dot{I} = \int_0^t U(u)\, du + 2\,h\,t + \text{const.}$$

As $\dot{I} \leq A\,t$, this yields

$$\int_0^t U(u)\, du \leq A\,t$$

and, from (3.40),

$$\int_0^t \frac{du}{r(u)} \leq A\,t.$$

This last equation and Schwarz's inequality imply that

$$t^2 = \left\{ \int_0^t [r(u)]^{1/2}\, \frac{du}{[r(u)]^{1/2}} \right\}^2 \leq \int_0^t r(u)\, du \int_0^t \frac{du}{r(u)} \leq A\,t \int_0^t r(u)\, du.$$

With this the theorem is proved.

Theorem 4: *There exists a time t_0, depending on initial conditions, such that:*

$$\sqrt{\frac{1}{2}I} < t\left(\sqrt{U} + \sqrt{T}\right), \qquad t \geq t_0. \tag{3.49}$$

We observe that, if I_0 is the initial value of I for $t = 0$, the relation

$$\frac{1}{2}I - \frac{1}{2}t\,\dot{I} = \frac{1}{2}I_0 - \int_0^t u\,U(u)\, du - h\,t^2 \tag{3.50}$$

is valid, since it is obviously true at $t = 0$ and can be verified for $t > 0$ by differentiating. In fact (3.23) is obtained. Since $U^{-1} \leq A\,r$, as seen in theorem 1, U is always limited from below for $t \to \infty$. This implies that the integral in (3.50) diverges for $t \to \infty$; therefore, there will exist a t_0 such that, for $t > t_0$,

$$\frac{1}{2}I - \frac{1}{2}t\dot{I} < -h\,t^2. \tag{3.51}$$

But, from Sundman's inequality (3.44), we also get

$$\frac{1}{2}\dot{I} \leq \sqrt{2IT},$$

so that

$$\frac{1}{2}I - t\sqrt{2\,IT} < -h\,t^2, \qquad \forall\, t \geq t_0.$$

Adding $t^2\,T$ to both sides of this last inequality and remembering that $T - U = h$, we obtain

$$\left(\sqrt{\frac{1}{2}I} - t\sqrt{T}\right)^2 < U\,t^2, \qquad \forall\, t \geq t_0,$$

which implies (3.49). As a consequence, we can prove the following theorem.

Theorem 5: *For $t \to \infty$, $\frac{R}{t} \to \infty$ if and only if $r \to 0$.*

Assume, from the start, that $r \to 0$. Then, from (3.40), $U \to \infty$ and, from (3.23), $\ddot{I} \to \infty$. This implies that, $\forall\, L > 0$, there exists a t_1 such that $\ddot{I} \geq L$, $\forall\, t \geq t_1$. Integrating twice and dividing by t^2, we have

$$\liminf \frac{I}{t^2} \geq \frac{1}{2}L,$$

and, since L can be chosen arbitrarily large, $I/t^2 \to \infty$ so that, from (3.41), $R/t \to \infty$.

Suppose now, instead, that $R/t \to \infty$; by (3.41) therefore $I^{1/2}/t \to \infty$, and, from (3.49), theorem 4, it follows that

$$\sqrt{U} + \sqrt{T} \to \infty.$$

As $T = U + h$, then $U \to \infty$ also and $r \to 0$.

Observe, at this point, that the catastrophic event $R/t \to \infty$ cannot be due to the fast "explosion" of all the distances r_{jk}, it not being possible, that is, that $r_{jk}/t \to \infty$ for every mutual distance r_{jk}. This would imply, in fact, that $t/r_{jk} \to 0$ for every pair (j, k), so that $t\,U \to 0$, and, moreover, $r/t \to \infty$. But this would be in contradiction with what was proved in theorem 1. On the other hand, we cannot exclude that $\limsup (r_{jk}/t) = \infty$ for every pair of points.

Coming back to the hypothesis of theorem 3, namely $R = O(t)$, and therefore also $R^3 = O(t^3)$, it is obvious that, with greater reason, (since $r < R$),

$$R^2 r = O(t^3), \qquad t \to \infty. \tag{3.52}$$

This can be proved with the following theorem.

Theorem 6: *Estimate (3.52) is always valid; moreover, except when $h > 0$ and r is unlimited,*

$$R^2 r = O(t^2), \qquad t \to \infty. \tag{3.53}$$

We omit the proof, for the sake of brevity.

Null Energy and Positive Energy Systems

For $h = 0$, there are exact results only in the cases $N = 2$ and $N = 3$. In the first case the motion is parabolic and (2.46) hold. From the first of these we have that, for large t, r goes as u^2; from the second we then have that u^3 goes as t, so that $r = A t^{2/3}$ with $A > 0$, for large t. For $N = 3$, Chazy[35] showed that there are two possibilities:

$$r \sim A t^{2/3}, \qquad R \sim A t^{2/3}$$

or

$$r = O(1), \qquad R \sim A t.$$

For any N the following can be stated.

Theorem 7: *If $h = 0$, then either*

$$U \sim \alpha\, t^{-2/3} \qquad I \sim \frac{9}{4}\,\alpha\, t^{4/3}$$

for some positive constant α, or $r = O(t^{2/3})$ and $I\, t^{-4/3} \to \infty$.

For the proof, exploiting a nonlinear Tauberian theorem, we refer to the already cited paper by Pollard.

In the case $h > 0$, fundamental results were obtained by Chazy.[36] He proved that the limit

$$\lim_{t \to \infty} \frac{r}{R} = L$$

always exists and is finite. Moreover, if $L = 0$, it holds that

$$\frac{r}{R} \leq At^{-1/3}. \tag{3.54}$$

In the case $L \neq 0$, a complete and exact theory about the behaviour of the systems can be obtained, and, in particular

---

[35] J. Chazy: loc. cit.
[36] J. Chazy: loc. cit., Footnote 30

$$U \sim \frac{A}{t}, \qquad I \sim h\,t^2.$$

In the case $L = 0$, however, it can be inferred from (3.54) only for $N = 3$ that

$$r = \mathrm{O}(t^{2/3}), \qquad t \to \infty, \tag{3.55}$$

holds.

Up to now we have considered results by Chazy. Pollard proved that (3.55) always holds when $L = 0$, even if $N \neq 3$. In addition, he obtained some new results, starting from the following lemma, which we only state.

Lemma: *Whatever the sign of the energy, the limit*

$$\lim_{t \to \infty} \frac{1}{t\,U} = \ell. \tag{3.56}$$

always exists. If $\ell = 0$, then (3.55) holds.

This lemma allows us to prove the first part of the following theorem.

Theorem 8: *If $h > 0$, then either $U \sim \alpha/t$ for some positive constant α, or $r = \mathrm{O}(t^{2/3})$. In the first circumstance, $I = h\,t^2 + \alpha\,t\,\ln t + \mathrm{O}(\ln t)$, for $t \to \infty$; in the second $I > h\,t^2 + A\,t^{4/3}$.*

Obviously the second part follows from the first; integrating twice and the use of the Lagrange–Jacobi identity is sufficient.

Escape from the System

In the case for which $h \geq 0$, integrating the Lagrange–Jacobi identity (3.23) twice, gives $I \to \infty$ for $t \to \infty$. As a consequence, $\limsup r_k = \infty$, for some k. However, this does not necessarily imply the escape of a particle from the system; this occurrence is warranted only if $r_k \to \infty$, for some k, so that further conditions must be added. In this respect there is the following.

Theorem 9: *If $h > 0$ and $\int_0^\infty U^2(u)\,du < \infty$, then at least one particle leaves the system.*

From the equations of motion (3.1a), we have

$$|\ddot{\boldsymbol{r}}_k| \leq \sum_{j \neq k} \frac{A}{r_{jk}^2} \leq \frac{A'}{r^2} \leq A'' U^2.$$

Integrating this gives

$$|\boldsymbol{v}_k(t_1) - \boldsymbol{v}_k(t_2)| \leq A'' \int_{t_1}^{t_2} U^2(u)\,du.$$

For t_1 and $t_2 \to \infty$, the right-hand side of this inequality vanishes; the same must happen for the left-hand side. For every k then, the finite limit

$$\lim_{t\to\infty} \boldsymbol{v}_k = \boldsymbol{\ell}_k$$

exists. As $\boldsymbol{v}_k = \dot{\boldsymbol{r}}_k$, we deduce that $\boldsymbol{r}_k/t \to \boldsymbol{\ell}_k$, for each k, and similarly for the absolute values $r_k/t \to \ell_k$. It remains to show that the ℓ_k are not all zero, that is, that at least one of them does not vanish. Actually, if they should all vanish, then $\boldsymbol{v}_k \to 0$, $\forall k$, and then $\mathcal{T} \to 0$ and thus $U \to -h$, which is impossible, U having always to be positive.

Theorem 10: *If $h > 0$, in the case for which $U \sim \alpha/t$, with $\alpha > 0$, at least $N - 1$ particles leave the system.*

The first part of the proof (that is, the existence of

$$\lim_{t\to\infty} \boldsymbol{v}_k = \boldsymbol{\ell}_k$$

for each k) proceeds exactly as in the previous case. It remains to prove that, now, at most only one of the ℓ_k can be zero. If two were vanishing, e.g. $\ell_i = 0$ and $\ell_j = 0$, we would have $r_i/t \to 0$ and $r_j/t \to 0$, so that $r_{ij}/t \to 0$ and then $(t\,U)^{-1} \to 0$, which contrasts with the hypothesis $U \sim \alpha/t$

Theorem 11: *If $I \to \infty$, for $t \to \infty$, and the condition*

$$\int_0^\infty \frac{\sqrt{4I\mathcal{T} - \frac{1}{4}\dot{I}^2}}{I}\, d\tau < \infty$$

is satisfied, then at least two particles leave the system.

We again emphasize that we speak of the escape of a particle from the system only when its distance from the centre of mass tends to infinity, when $t \to \infty$. We say that the system disintegrates or is dispersive when all the mutual distances become infinite. It has never been proved that a particle must escape a disintegrating system, since it is not true that a system from which all particles escape, disintegrates. A classical counterexample is that of the three-body problem (the hyperbolic–elliptic case) where a pair of particles "runs" away in one direction and the third particle in the opposite direction.

Final Motions

Generally speaking, we can classify the final motions, that is, those for $t \to \infty$, in three broad groups: limited motions, oscillatory motions and unlimited motions. By limited motions we mean those for which:

a) $r_{ij}(t) \neq 0$ for every $i \neq j$ and any t;
b) $|r_{ij}(t)| < A$, with A a positive constant, for any t.

In this case, therefore, the system remains confined to a limited region of space; the motion is also said to be "stable". For this to happen, the total energy (h) must be negative (Jacobi stability criterion). The proof is straightforward. We have already seen that, for $h \geq 0$, integrating the Lagrange–Jacobi identity yields $I \to \infty$, and therefore, from (3.38), it cannot always be the case that $|r_{ij}(t)| < A$, $\forall i \neq j$. Obviously the condition is only necessary; therefore $h < 0$ is not sufficient to warrant the "stability".

Let us now introduce the definition of oscillatory motion in the case of any N as a generalization of the case $N = 3$. In the three-body problem there can be initial conditions that lead to motions with the following properties:[37] for $t \to \infty$

$$\limsup r_{23} = \infty,$$

$$\limsup \left(\frac{r_{12}}{r_{23}} \right) > 0,$$

$$\liminf \left(\frac{r_{12}}{r_{23}} \right) = 0.$$

In the case of any N whatsoever, the masses m_k, m_j, m_i undergo oscillatory motion if, for $t \to \infty$,

$$\limsup r_{ij} = \infty,$$

$$\limsup \left(\frac{r_{ik}}{r_{ij}} \right) > 0,$$

$$\liminf \left(\frac{r_{ik}}{r_{ij}} \right) = 0.$$

Finally, there will be unlimited oscillatory motion if $R \to \infty$ and $r \to 0$; if this happens in a limited interval of time, we speak of unlimited motion in a finite time; otherwise we speak of *super-hyperbolic expansion.*

The table aside[38] lists the various possible types of final motion. It must be noted, however, that such considerations cannot be taken as definitive for all cases.

The systems with a hyperbolic expansion separate into two or more subsystems whose mutual distances grow as t, whereas the distances between the components of the subsystems go, at most, as $t^{2/3}$. For every subsystem, the energy and the angular momentum tend to a limit; for the energy, the limit will be negative or zero; in this last eventuality the subsystem will tend to a central configuration (see Sect. 3.8). The central configuration will be the limit in other cases too: motions of the N-parabolic type or triple or multiple collisions.

---

[37]The existence of oscillatory motions in the three-body problem was conjectured by Chazy (see Footnote 30) to account for the fact that, for $h < 0$, I can at the same time be limited and unlimited for $t \to \infty$. The proof was given in 1960, by Sitnikov – K. Sitnikov: The existence of oscillatory motion in the three-body problem, *Dokl. Akad. Nauk. SSR*, **133**, 303–306 (1960).

[38]The table, due to C. Marchal, is in C. Marchal: Qualitative analysis in the few body problem, *Proceedings of the 96th Colloquium of the IAU*, ed. by M. J. Valtonen (Kluwer, 1988), pp. 5–25.

Final motions in the N-body problem	
Singular types	Triple or multiple collisions $R \to R_f$; $r \to 0$ Unlimited expansion in a finite time $R \to \infty$; $r \to 0$ Super-hyperbolic expansion $R/t \to \infty$; $r \to 0$
Hyperbolic expansion	$\boldsymbol{r}_j = \boldsymbol{A}_j t + \mathrm{O}(t^{2/3})$; from 2 to N subsystems, $R \sim t$, $\forall j$
Parabolic and sub-parabolic types	N-parabolic types $h = 0$; $R, r \sim t^{2/3}$ Parabolic types, $h < 0$, $R = \mathrm{O}(t^{2/3})$, r limited subsystems, $h < 0$, $R = \mathrm{O}(t^{2/3})$, r limited limited motions, $h < 0$, $R = \mathrm{O}(t^{2/3})$, r limited oscillatory motions, $h < 0$, $R = \mathrm{O}(t^{2/3})$, r limited

3.6 The Virial Theorem

Sometimes it happens that theorems, formulated in a given context with the aim of providing a theoretical assessment of some specified ideas, find extensions and applications in fields apparently very different from the original one. This is the case for the so-called *virial theorem*. With this theorem, published in 1870, Clausius[39] intended to arrive at an almost definitive statement about a long time of research on the mechanical origin of heat. Very soon, however, its importance was appreciated in applications to the N-body problem, particularly with regard to stellars dynamics.[40] It should be emphasized, however, that the theorem differs, in some features, from typical mechanical theorems, because Clausius introduces a concept with a *statistical* nature: averaging over time. This fundamental feature is not always taken into account, and, as a consequence, it may happen that the virial theorem as described by some authors is mistaken for a particular case of the Lagrange–Jacobi identity (3.23).

To demonstrate the theorem, we shall repeat the procedure already followed to demonstrate (3.23); we start from system (3.1b), which we again write as follows

$$m_k \ddot{\boldsymbol{r}}_k = \frac{\partial U}{\partial \boldsymbol{r}_k}. \tag{3.57}$$

[39]R. Clausius: On a mechanical theorem applicable to heat, *Phil. Mag. Ser. 4*, **XL**, 122–127 (1870).

[40]A. Eddington: The kinetic energy of a star cluster, *Mon. Not. Roy. Astron. Soc.* **76**, 525–528 (1916); J. Jeans: *Problems of Cosmogony and Stellar Dynamics*, (Cambridge University Press, 1919), pp. 188 ff.; H. Poincaré: *Leçons sur les Hypotèses Cosmogoniques*, (Hermann, Paris, 1913), p. 94.

Taking into account the identity

$$\frac{1}{2}\frac{d^2}{dt^2}\left(\boldsymbol{r}_k \cdot \boldsymbol{r}_k\right) = |\dot{\boldsymbol{r}}_k|^2 + \boldsymbol{r}_k \cdot \ddot{\boldsymbol{r}}_k$$

and multiplying (3.57) by $\boldsymbol{r}_k$, we obtain

$$\frac{1}{4}\frac{d^2}{dt^2}\left(m_k\, r_k^2\right) = \frac{1}{2}\,m_k\,|\dot{\boldsymbol{r}}_k|^2 + \frac{1}{2}\,\boldsymbol{r}_k \cdot \frac{\partial U}{\partial \boldsymbol{r}_k}.$$

Summing over k from 1 to N, we get

$$\frac{1}{4}\frac{d^2}{dt^2}\left(\sum_{k=1}^{N} m_k\, r_k^2\right) = \sum_{k=1}^{N}\frac{1}{2}\,m_k\,|\dot{\boldsymbol{r}}_k|^2 + \frac{1}{2}\sum_{k=1}^{N}\boldsymbol{r}_k \cdot \frac{\partial U}{\partial \boldsymbol{r}_k},$$

i.e.

$$\frac{1}{4}\frac{d^2 I}{dt^2} = \mathcal{T} + \frac{1}{2}\sum_{k=1}^{N}\boldsymbol{r}_k \cdot \frac{\partial U}{\partial \boldsymbol{r}_k}. \tag{3.58}$$

We now introduce the idea of time averaging. Given a quantity $A(t)$, its average over the time interval $(0,t)$ is given by

$$\langle A \rangle_t = \frac{1}{t}\int_0^t A(\tau)\,d\tau. \tag{3.59}$$

For t tending to ∞, we also have

$$\langle A \rangle_\infty = \lim_{t\to\infty}\frac{1}{t}\int_0^t A(\tau)\,d\tau. \tag{3.60}$$

In typical applications, average (3.60) is replaced by that defined by (3.59), taking into consideration "very large" times. As we shall see, in some cases, matters are very tricky and erroneous extrapolations may be generated.

If now, in (3.58), we average over a time t, we get

$$\frac{1}{4}\frac{\dot{I}(t) - \dot{I}(0)}{t} = \langle \mathcal{T} \rangle_t + \frac{1}{2}\left\langle \sum_{k=1}^{N}\boldsymbol{r}_k\frac{\partial U}{\partial \boldsymbol{r}_k}\right\rangle_t. \tag{3.61}$$

In the case for which our N-body system is subjected to a periodic motion of period $\bar{t}$, it is clear that, for $t = \bar{t}$ or multiples,

$$\langle \mathcal{T} \rangle_{\bar{t}} = -\frac{1}{2}\left\langle \sum_{k=1}^{N}\boldsymbol{r}_k \cdot \frac{\partial U}{\partial \boldsymbol{r}_k}\right\rangle_{\bar{t}}.$$

The same result may be obtained, however, for any motion (not periodic), when $\dot{I}(t) - \dot{I}(0)$ is negligible compared with a very large time (at the limit, infinite) in the denominator. The result will be exact when

$$\lim_{t\to\infty}\frac{\dot{I}(t) - \dot{I}(0)}{t} = 0. \tag{3.63a}$$

Then

$$\langle \mathcal{T} \rangle_\infty = -\frac{1}{2} \left\langle \sum_{k=1}^{N} \boldsymbol{r}_k \cdot \frac{\partial U}{\partial \boldsymbol{r}_k} \right\rangle_\infty \tag{3.62}$$

holds if and only if (3.63a) is valid. Equation (3.63a) can be rewritten as $\dot{I}(t) = \mathrm{o}(t)$, $t \to \infty$, or $I(t) = \mathrm{o}(t^2)$, $t \to \infty$; again, for (3.41), (3.63a) is equivalent to

$$R(t) = \mathrm{o}(t), \qquad t \to \infty. \tag{3.63b}$$

The quantity on the right in (3.62) was called by Clausius the *virial*. The content of the theorem is then as follows:

The average total kinetic energy (over an infinite time) equals the virial of the system if and only if (3.63a, b) *is valid.*

As one can see, condition (3.63a,b) is far weaker than the condition that the system be bounded, which, as a rule, is given as necessary for the validity of (3.62); for instance, $R \sim t^{2/3}$ satisfies (3.63a,b) and thus (3.62) is valid.

If now, as in the case of the Lagrange–Jacobi identity, we apply Euler's theorem, taking into account that U is the Newtonian potential, we have

$$\langle \mathcal{T} \rangle_\infty = \frac{1}{2} \langle U \rangle_\infty, \tag{3.64}$$

which is the expression of the virial theorem in the case of the Newtonian N-body system. Applying energy conservation (3.21) and taking into account that $\langle h \rangle_\infty \equiv h$, h being constant, we also get

$$\langle \mathcal{T} \rangle_\infty = -h, \tag{3.65}$$
$$\langle U \rangle_\infty = -2h, \tag{3.66}$$

which are equivalent to (3.64).

It was shown by Milne[41] that (3.64) maintains its validity even if dissipative forces are present, if these are proportional to the velocity of the mass points involved. If we add to the right of (3.57) a term $-c_k \dot{\boldsymbol{r}}_k$ (where c_k is a positive constant), we have, applying the same procedure as above, instead of (3.58)

$$\frac{1}{4} \frac{d^2 I}{dt^2} + \frac{1}{4} \frac{d}{dt} \left(\sum_{k=1}^{N} c_k r_k^2 \right) = \mathcal{T} + \frac{1}{2} \sum_{k=1}^{N} \boldsymbol{r}_k \cdot \frac{\partial U}{\partial \boldsymbol{r}_k}.$$

For

$$\lim_{t \to \infty} \frac{\sum_{k=1}^{N} c_k \left[r_k^2(t) - r_k^2(0) \right]}{t} = 0, \tag{3.67}$$

one again obtains (3.64) for averages over an infinite time. It is clear, however, that (3.67) is a stronger condition than (3.63). In fact, now $R^2(t) = \mathrm{o}(t)$

---

[41]E. A. Milne: An extension of the theorem of the virial, *Phil. Mag. Ser. 6*, **50**, 409–414 (1925).

must be true. Dissipative forces proportional to the velocity can then modify the system, by decreasing the value of h, but not the relation between the averages of $\mathcal{T}$ and U.

Before discussing the use of the virial theorem in the case of large N (dynamics of stellar systems), we shall first take a look at how to apply it in the two body-problem. It appears from (3.65) that the case $h > 0$ does not fulfil the conditions of the theorem; we shall therefore consider only the elliptic and parabolic motions. Since the first case refers to periodic motion, we can confine our consideration to the period $\bar{t} = 2\pi/n$, where n is the mean motion. The equations to be taken into consideration are (2.50) and Kepler's equation (2.53). From the definition of the eccentric anomaly

$$\sqrt{\frac{\mu}{a}}\, dt = r\, du,$$

we obtain for the potential

$$U(t)\, dt = \frac{\mu}{r}\, dt = \sqrt{\mu a}\, du,$$

and

$$\langle U\rangle_t = \frac{1}{t}\int_0^t U(\tau)\, d\tau = \frac{1}{t}\int_0^{u(t)} \sqrt{\mu a}\, du = \sqrt{\mu a}\,\frac{u(t)}{t}.$$

Taking as a reference the apse line $(T = 0)$, Kepler's equation becomes

$$n t = u - e \sin u,$$

and, by substituting it in $\langle U\rangle_t$, we shall have:

$$\langle U\rangle_t = \frac{\mu}{a}\,\frac{u}{u - e \sin u}. \tag{3.68}$$

In the same way, from $\langle \mathcal{T}\rangle_t = h + \langle U\rangle_t$, we get

$$\langle \mathcal{T}\rangle_t = \frac{\mu}{2a}\,\frac{(u + e \sin u)}{u - e \sin u}. \tag{3.69}$$

For $t = \bar{t} = 2\pi/n$ or multiples thereof, $\sin u = 0$, and then

$$\langle U\rangle_{\bar{t}} = \frac{\mu}{a}, \qquad \langle \mathcal{T}\rangle_{\bar{t}} = \frac{\mu}{2a},$$

which, since $h = -\mu/2a$, coincide respectively with (3.66) and (3.65), with $\bar{t}$ or multiples of $\bar{t}$ instead of ∞.

The case $h = 0$ (parabolic motion) must be handled with greater caution. First of all, for any finite t it will be always the case that $\langle \mathcal{T}\rangle_t = \langle U\rangle_t$, since $h \equiv 0$. This could lead to the conclusion that the virial theorem is not valid, since for t however large (3.64) could never be approximated. It is not like that, however. Let us, then, calculate $\langle U\rangle_t$ and $\langle \mathcal{T}\rangle_t$ explicitly. From (2.46), referring also in this case to the apse line, we have

$$\sqrt{\mu}\, t = \frac{1}{6} u^3 + \frac{c^2}{2\mu}\, u,$$

and, recalling that $\sqrt{\mu}\, dt = r\, du$, we get $U(t)\, dt = \sqrt{\mu}\, du$, so that

$$\langle U \rangle_t = \sqrt{\mu}\, \frac{u(t)}{t} = \frac{\mu}{\frac{1}{6} u^2 + \frac{c^2}{2\mu}} \sim t^{-2/3}, \qquad t \to \infty.$$

Therefore

$$\lim_{t \to \infty} \langle U \rangle_t = 0, \qquad (3.70)$$

and also

$$\lim_{t \to \infty} \langle \mathcal{T} \rangle_t = 0. \qquad (3.71)$$

Both (3.70) and (3.71) satisfy (3.65), (3.66), since $h \equiv 0$. Thus in the parabolic case as well the virial theorem is valid. On the other hand, since asymptotically $\langle U \rangle_t \sim t^{-2/3}$, also $R \sim t^{2/3}$, and consequently condition (3.63b) is satisfied. The seeming contradiction lies in the fact that whenever one looks for an approximate verification of the theorem, considering the values of $\langle \mathcal{T} \rangle_t$ and $\langle U \rangle_t$ for finite t, one has, for the averages, values equal and different from zero in disagreement with (3.64). The passage to the limit, stated in the theorem, on the other hand, allows us to verify that (3.65) and (3.66) are equivalent to (3.64). If the case $N \geq 3$ is considered, instead of $N = 2$, from what we have said in the preceeding section, the virial theorem is satisfied for motions characterized by $h \leq 0$ (parabolic and sub-parabolic motions).

As we have said, the most typical application of the virial theorem refers to the case of very large N. We shall now seek to outline this case and thereafter pinpoint what the risks of careless application can be. It is clear that, for any application, averages over a finite time, for instance the real age t_0 of the system, should replace $\langle \mathcal{T} \rangle_\infty$ and $\langle U \rangle_\infty$; this, however, is not yet sufficient, since we cannot verify the soundness of the hypothesis

$$\langle \mathcal{T} \rangle_{t_0} \sim \langle \mathcal{T} \rangle_\infty, \qquad \langle U \rangle_{t_0} \sim \langle U \rangle_\infty$$

in the cases which one can actually meet in nature. What we can verify, albeit with a very rough approximation, is wether the system is in a steady state, i.e. if it is confined to a given region of space, and the positions and velocities of its mass points have a stationary distribution. Let us try to clarify this concept. We assume that our system may be represented in a 6-dimensional Euclidean space $\mathbb{R}^6$, by introducing a correspondence for every sextuple $(x_k, y_k, z_k, \dot{x}_k, \dot{y}_k, \dot{z}_k)$, which gives the position and velocity of a point mass m_k, with a point of the space so defined. If the system is kept bounded in ordinary space, the velocities will also remain bounded, and consequently the representative region of the system in $\mathbb{R}^6$ will be bounded. Let us split this region into a suitable number of small 6-dimensional cells; if, as time passes, the representative points in each cell remain constant in number, we may conclude that the distribution of the positions and velocities has not changed and that therefore the system is in a steady state.

It is immediately apparent that a system can be considered steady or not depending on the dimension chosen for the side of the cells. The cells must be large enough to contain a sufficient number of representative points to provide meaning for a statistical approach, and at the same time, however, there must be a sufficient number of cells to allow evidence to be given for possible evolution of the system. It is clear, therefore, that the steadiness of the system is a concept that is strictly dependent on the "graininess" assigned to it. This being the case, if we judge that a system has reached a steady state, then its total kinetic energy and total potential will stay practically constant in time (with the chosen "grain"). The adverb "practically" means that, in fact, both $\mathcal{T}$ and U will fluctuate, with fluctuations of the order of $N^{-1/2}$, about fixed values $\overline{\mathcal{T}}$ and $\overline{U}$. We shall use these values $\overline{\mathcal{T}}$ and $\overline{U}$ in the form $\langle \mathcal{T} \rangle_\infty$ and $\langle U \rangle_\infty$ and we then say that the system is *virialized*.

Let us now define, at a certain instant of time,

$$\overline{r} = \frac{\left(\sum_{k=1}^{N} m_k\right)^2}{\sum_{j \neq k} \frac{m_j m_k}{|r_j - r_k|}}, \qquad \overline{v^2} = \frac{\sum_{k=1}^{N} m_k |\dot{\boldsymbol{r}}_k|^2}{\sum_{k=1}^{N} m_k};$$

then let us take

$$\overline{U} = \frac{1}{2} \frac{G M^2}{\overline{r}},$$

with, as usual, $M = \sum_{k=1}^{N} m_k$. In the same way,

$$\overline{\mathcal{T}} = \frac{1}{2} \sum_{k=1}^{N} m_k |\dot{\boldsymbol{r}}_k|^2 = \frac{1}{2} M \overline{v^2}.$$

We adopt for the virial theorem the form

$$2\overline{\mathcal{T}} = \overline{U}. \tag{3.72}$$

Then

$$M \overline{v^2} = \frac{1}{2} G \frac{M^2}{\overline{r}}$$

and

$$M = \frac{2 \overline{r} \, \overline{v^2}}{G}. \tag{3.73}$$

Relation (3.73) as a rule is applied to evaluate the total mass of a stellar system (globular cluster, galaxy, etc.) considered, within a given approximation, stationary and consequently virialized.[42] Equation (3.72), by which we "practically" express the virial theorem, hides, if not interpreted with due care, a trap. Let us see what that trap is.

---

[42]For the technical details on the evaluation of $\overline{r}$ and $\overline{v^2}$ and for a comparative discussion with the results obtained by the mass–luminosity ratio, we refer the reader to C. W. Saslaw: *Gravitational Physics of Stellar and Galactic Systems* (Cambridge University Press, 1985), Sect. 41.

Suppose that, rather than considering $\mathcal{T}$ and U fluctuating around the values $\overline{\mathcal{T}}$ and $\overline{U}$, we think that "really" $2\mathcal{T}(t) = U(t)$ for any non-degenerate interval of time. In this case the following theorem is valid:

If $2\mathcal{T} = U$ for any finite time interval, then $\mathcal{T} = -h$, $U = -2h$, $\dot{U} = 0$, $\dot{I} = 0$, $I(t) = I(0)$, $\forall t$.

From energy conservation, it follows immediately that $\mathcal{T} = -h$, $U = -2h$, and then $\dot{U} = 0$. Note, by the way, that in this case h *must* be negative. By inserting the results in the Lagrange–Jacobi identity, we get $\ddot{I}(t) = 0$. Integrating this yields

$$I(t) = \dot{I}(0)\,t + I(0), \qquad \forall t.$$

If $\dot{I}(0) \neq 0$, then, either for $t \to \infty$ or for $t \to -\infty$ (depending on the sign of $\dot{I}(0)$), $I(t) \to -\infty$ as well. This contradicts the fact that $I \geq 0$ always. It follows, then, that $\dot{I}(0) = 0$, and therefore $I(t) = I(0)$, $\forall t$. Furthermore, we can demonstrate the following corollary:

If $2\mathcal{T} = U$ over a finite time interval, then, $\forall t$,

$$\frac{G\,m_i\,m_j}{2|h|} \leq r_{ij}(t) \leq \sqrt{\frac{I(0)}{m_i}} + \sqrt{\frac{I(0)}{m_j}}. \tag{3.74}$$

The result is that the components of the system can neither get closer than a given limit nor leave the system itself. Let us see a demonstration of this. First

$$2|h| = U \geq \frac{G\,m_i\,m_j}{r_{ij}(t)},$$

from which

$$r_{ij}(t) \geq \frac{G\,m_i\,m_j}{2|h|}.$$

In the same way

$$I(0) = I(t) \geq m_i\,r_i^2,$$

from which

$$r_i(t) \leq \sqrt{\frac{I(0)}{m_i}}, \qquad r_j(t) \leq \sqrt{\frac{I(0)}{m_j}},$$

and, owing to the triangle inequality,

$$r_{ij} \leq \sqrt{\frac{I(0)}{m_i}} + \sqrt{\frac{I(0)}{m_j}}.$$

A typical motion which satisfies (3.74) is provided by a relative equilibrium solution (see Sect. 3.7). One can further introduce the hypothesis[43] (not yet

---

[43]The above considerations and the theorem are due to D. Saari. See D. Saari: On bounded solutions of the n-body problem, in *Periodic Orbits, Stability and Resonances*, ed. by G. E. O. Giacaglia, (Reidel, 1970), pp. 76–81.

demonstrated) that the motions which satisfy (3.74) come without exception from relative equilibrium solutions. In any case, it appears, from what we have shown that a "naive" interpretation of (3.72) leads us to approximate the system under study with a model that:

1) has negative energy,
2) satisfies (3.74);

i.e. with a model characterized by very peculiar features.

Let us add that, when the system is studied in the Hamiltonian formalism, the virial theorem is formulated as

$$\left\langle \sum_i p_i \frac{\partial \mathcal{H}}{\partial p_i} \right\rangle_\infty = \left\langle \sum_i q_i \frac{\partial \mathcal{H}}{\partial q_i} \right\rangle_\infty , \qquad (3.75)$$

the virial being given by

$$\frac{1}{2} \left\langle \sum_i q_i \frac{\partial \mathcal{H}}{\partial q_i} \right\rangle_\infty .$$

3.7 Particular Solutions of the N-Body Problem

Since a general solution of system (3.1a,b) cannot be given, great importance has been attached from the very beginning to the search for particular solutions where the N mass points fulfilled certain initial conditions. Historically, mainly with Euler and Lagrange, the work has centred on the three-body problem and thus on the exact solutions that could be obtained with particular symmetry properties. In the celebrated paper we have already mentioned,[44] Lagrange demonstrated that, in the case of three bodies of finite mass undergoing mutual Newtonian attraction, as many as five different configurations exist such that, under suitable initial conditions, the ratios of the mutual distances stay constant. In all these configurations the bodies describe similar conic sections around the common centre of mass: in three configurations the bodies always lie along the same straight line, while in the other two they are at the corners of an equilateral triangle. These configurations are called, respectively, the *collinear solution* and *equilateral solution*; the first one had been previously found by Euler.[45] For $N > 3$, there are then solutions which extend to any N the Euler and Lagrange solutions: these are *homographic solutions*. This is the topic which we will deal with in this section, postponing to the section devoted to the three-body problem a detailed study of the Euler and Lagrange solutions. First of all, following Wintner,[46] we list some definitions and considerations.

---

[44]See Footnote 8 in Sect. 3.2.
[45]L. Euler: De motu rectilineo trium corporum se mutuo attrahentium, *Novi Comm. Acad. Sci. Imp. Petrop.* **11**, 144–151 (1767).
[46]A. Wintner: op. cit., Chap. V.

Planar Solutions

We call a solution of the problem of N-body motion *planar*, if a plane Π^* containing all the bodies at any time exists and furthermore has a time-independent fixed position with respect to the inertial barycentric system. One can immediately see that, in the case of $c \neq 0$, i.e. when the invariant plane exists, then Π^* coincides with the invariant plane. In fact, if Π^* exists, it contains the centre of mass, and we can take it as the xy plane of the barycentric inertial system; since the motion occurs entirely in this plane, all the z_k and the velocities $\dot{z}_k$ will be zero as well as the components c_x and c_y of the total angular momentum. Therefore $|\boldsymbol{c}| = c_z$ at any time: the plane Π^* is at any time perpendicular to $\boldsymbol{c}$ and coincides with the invariable plane. Obviously, planar solutions may even exist in the case $\boldsymbol{c} = 0$.

Flat Solutions

We call a solution of the N-body problem *flat* if at any time a plane $\Pi = \Pi(t)$ exists which contains all the bodies at that time. Of course, every flat solution is not necessarily planar: for $N = 3$ this is trivial, since every solution is flat. For examples of non-planar flat solutions for $N \geq 4$, see Wintner.[47]

For the case where the invariable plane does not exist ($\boldsymbol{c} = 0$), the following proposition is true:

> If the invariable plane does not exist, any flat solution is necessarily planar.

It must be stressed that the condition $\boldsymbol{c} = 0$ is a sufficient condition for the planar motion, but *not* necessary. From the proposition one obtains as a corollary that, in the case $N = 3$, the condition $\boldsymbol{c} = 0$ implies that the motion occurs in a fixed plane. Referring to Wintner[48] for the demonstration for any N, we give below the demonstration for $N = 3$.

Let us consider the plane that contains the three bodies at an instant $t = t_0$. Since the centre of mass also obviously belongs to this plane, we can take it as the fixed xy plane of our inertial barycentric system. If the three-body plane does not remain fixed, it will rotate around the centre of mass, and then at the time $t = t_0$, although $z_k = 0$ $(k = 1, 2, 3)$, in general $\dot{z}_k \neq 0$ $(k = 1, 2, 3)$. Therefore, the conservation of angular momentum (equal to zero) $\sum_{k=1}^{3} m_k \boldsymbol{r}_k \times \dot{\boldsymbol{r}}_k = 0$, gives us, at $t = t_0$

$$\sum_{k=1}^{3} m_k \, y_k \, \dot{z}_k = 0, \qquad \sum_{k=1}^{3} m_k \, x_k \, \dot{z}_k = 0$$

---

[47] A. Wintner: op. cit., note to Sect. 325.
[48] A. Wintner: op. cit., Sect. 326.

and furthermore the centre of mass being at rest, we get

$$\sum_{k=1}^{3} m_k \dot{z}_k = 0.$$

We then have a homogeneous system in the three unknowns $m_1 \dot{z}_1, m_2 \dot{z}_2, m_3 \dot{z}_3$

$$y_1 \left(m_1 \dot{z}_1 \right) + y_2 \left(m_2 \dot{z}_2 \right) + y_3 \left(m_3 \dot{z}_3 \right) = 0,$$
$$x_1 \left(m_1 \dot{z}_1 \right) + x_2 \left(m_2 \dot{z}_2 \right) + x_3 \left(m_3 \dot{z}_3 \right) = 0,$$
$$m_1 \dot{z}_1 + m_2 \dot{z}_2 + m_3 \dot{z}_3 = 0.$$

As is known, we can have:

a) the trivial solution

$$m_1 \dot{z}_1 = m_2 \dot{z}_2 = m_3 \dot{z}_3 = 0,$$

or

b) infinite solutions $\neq 0$ if

$$\begin{vmatrix} y_1 & y_2 & y_3 \\ x_1 & x_2 & x_3 \\ 1 & 1 & 1 \end{vmatrix} = 0.$$

As the latter is the condition that implies that the three points lie on a straight line at the time $t = t_0$, we are allowed to rotate the system around this line (which obviously contains the centre of mass) in order to have $\dot{z}_3 = 0$. Since the case $\dot{z}_1 = 0$, $\dot{z}_2 = 0$, $\dot{z}_3 = 0$ is already included in solution (a), all that is left is (from the remaining equations) $y_1 = y_2$, $x_1 = x_2$. However, this case, the point P_1 coinciding with P_2, contradicts the assumption made at the beginning. It is therefore demonstrated that the only acceptable solution is $\dot{z}_1 = \dot{z}_2 = \dot{z}_3 = 0$. The existence and uniqueness theorem then warrants that our plane stays fixed.

Collinear Solutions

We say that the N mass points are in *syzygy*, or are collinear, at a certain instant $t = t_0$, if at that instant they all lie on the same straight line. Obviously, this does not ensure that they will remain there for $t \neq t_0$. However, one can demonstrate that, if the invariable plane exists, then at the time $t = t_0$ that line lies in the invariable plane. In fact, since at $t = t_0$ every pair of points is collinear with the origin, at that instant $\boldsymbol{r}_i \times \boldsymbol{r}_k = 0$, $\forall\ i, k = 1, 2, \ldots, N$. Scalarly multiplying by $\boldsymbol{v}_i = \dot{\boldsymbol{r}}_i$, also $\boldsymbol{r}_i \times \boldsymbol{r}_k \cdot \dot{\boldsymbol{r}}_i = 0$, and then $\boldsymbol{r}_i \times \dot{\boldsymbol{r}}_k \cdot \boldsymbol{r}_i = 0$. Finally, summing over i we get

$$\left(\sum_{i=1}^{N} \boldsymbol{r}_i \times \dot{\boldsymbol{r}}_i \right) \cdot \boldsymbol{r}_k = \boldsymbol{c} \cdot \boldsymbol{r}_k = 0.$$

Since $c \cdot r_k = 0$, $\forall k$, is, for $c \neq 0$, the equation of the invariable plane, our thesis is thus proved.

If all the N points always remains on the same straight line $\Lambda = \Lambda(t)$, which, however, may vary with time, we say that we have a *collinear solution*. A collinear solution is obviously flat; according to what we have seen above, at any instant, $\Lambda(t)$ (which, for instance, may rotate around the barycentre) must lie in the invariable plane, if $c \neq 0$, and therefore it is planar. On the other hand, if $c = 0$, already we know that a flat solution is planar. Therefore, in all cases a collinear solution is planar.

Rectilinear Solutions

A solution will be called *rectilinear* when the straight line Λ^* on which all the points lie (see the preceding case) stays fixed with time with respect to the inertial barycentric system. In such a case, however, the solution cannot exist for all times $-\infty < t < +\infty$, without there being a collision at a finite instant $t = t_1$ between at least two of the N bodies. In fact, if we choose the x axis of the reference system to coincide with Λ^* and consider the masses $m_1, m_2, \ldots, m_N$ arranged from the left to the right, the N-th mass will have a coordinate $x_N > 0$, the barycentre being at the origin. On the other hand, as m_N is attracted towards the left by the remaining $N - 1$ masses, $\ddot{x}_N < 0$ always. If we now consider the function $x_N(t)$, it is clear that neither $x_N > 0$ nor $\ddot{x}_N < 0$ for $-\infty < t < +\infty$ can hold; therefore either the solution exists only for a limited interval of time or at least one of the denominators of the potential vanishes at a finite time $t = t_1$, and then the equations of motion are no longer defined.

Homographic Solutions

A *homographic solution* $r_k = r_k(t)$ of system (3.1a,b) is characterized by imposing the condition that the configuration formed by the N bodies at the instant t with respect to the inertial barycentric system remains similar to itself as t varies. This implies the existence of a scalar function $\lambda = \lambda(t)$ and of an orthogonal 3×3 matrix $\boldsymbol{\Omega} = \boldsymbol{\Omega}(t)$ such that, for any value of k and t,

$$r_k(t) = \lambda(t)\,\boldsymbol{\Omega}(t)\,r_k^\circ + \boldsymbol{\tau}. \qquad (3.76)$$

Here $\lambda(t)$ represents a *dilatation* and $\boldsymbol{\Omega}(t)$ a *rotation*; the vector $\boldsymbol{\tau}$ which represents a *translation* must be considered identically zero because of the condition

$$\sum_{k=1}^{N} m_k\,r_k = 0$$

for the coordinates of the barycentre. Finally r_k° represents the value of r_k $(k = 1, 2, \ldots, N)$ at an instant $t = t_0$ conventionally assumed as the initial instant. Therefore it must also be the case that $\lambda^\circ = \lambda(t_0) = 1$ and $\Omega^\circ = \Omega(t_0) = 1$.

Clearly, the homographic solutions represent a rather restricted class of solutions of the N-body problem, since system (3.1a,b), which is of order $6\,N$, must be satisfied by $1+3$ scalar functions and by the $3\,N$ integration constants represented by $\lambda(t)$, $\Omega(t)$, r_k° respectively. In homographic motion, the relative positions of the bodies are independent of the time; that is, their radii vectors have, at any instant, a constant ratio. If a homographic motion is such that $\Omega(t) = 1$, $\forall\, t > 0$, and then $r_k(t) = \lambda(t)\, r_k^\circ$, that is, if the motion consists of a pure dilatation or contraction, the solution is said to be *homothetic*; if, on the other hand, the motion is such that $\lambda(t) = 1$, $\forall\, t > 0$, then it consists of a pure rotation and the solution is said to be a *relative equilibrium* solution: the points are at rest with respect to a reference system rotating around the barycentre.

Let us now determine the fundamental equations of homographic motion; to do this, let us introduce, in addition to the inertial barycentric system whose axes have been called (x, y, z), a second system (ξ, η, ζ), whose axes can expand or shrink, rotating around the origin of the former. The vector pertaining to the k-th particle at the generic instant t is given by $\varrho_k \equiv (\xi_k, \eta_k, \zeta_k)$; at the instant $t = t_0$, we have

$$r_k(t_0) = r_k^\circ = \varrho_k(t_0) = \varrho_k^\circ.$$

For a generic t, instead we have

$$\varrho_k(t) = \lambda(t)\, r_k^\circ, \qquad k = 1, 2, \ldots, N. \tag{3.77}$$

If we call $\omega \equiv (\omega_1, \omega_2, \omega_3)$ the angular velocity of the system (ξ, η, ζ) with respect to the inertial barycentric one (x, y, z), we have that:

a) for $\lambda = $ const. > 0 and $\omega = $ const. $\neq 0$, the motion is of relative equilibrium;

b) for $\lambda = \lambda(t) > 0$ and $\omega = 0$, the motion is homothetic.

For the moment of inertia,

$$I^\circ = \sum_{k=1}^{N} m_k \, |r_k^\circ|^2 = \sum_{k=1}^{N} m_k \left[(x_k^\circ)^2 + (y_k^\circ)^2 + (z_k^\circ)^2 \right]$$

at the initial instant t_0. Instead, in the system (ξ, η, ζ),

$$I = \sum_{k=1}^{N} m_k \, |\varrho_k|^2 = \sum_{k=1}^{N} m_k \left(\xi_k^2 + \eta_k^2 + \zeta_k^2 \right),$$

the moment of inertia at the generic instant t. From (3.77), one sees immediately that

$$I = \lambda^2 \, I^\circ. \tag{3.78}$$

The components of the velocity of the k-th mass point, in the system (ξ, η, ζ), will be[49]

$$(v_{\xi k},\, v_{\eta k},\, v_{\zeta k}) = \frac{d}{dt}\,[(\xi_k,\, \eta_k,\, \zeta_k)] - [(\xi_k,\, \eta_k,\, \zeta_k) \times (\omega_1,\, \omega_2,\, \omega_3)], \qquad (3.79a)$$

that is,

$$
\begin{aligned}
v_{\xi k} &= \frac{d\,\xi_k}{dt} - (\eta_k\,\omega_3 - \zeta_k\,\omega_2) = x_k^\circ\,\frac{d\lambda}{dt} - \lambda\,(y_k^\circ\,\omega_3 - z_k^\circ\,\omega_2)\,, \\[2mm]
v_{\eta k} &= \frac{d\,\eta_k}{dt} - (\zeta_k\,\omega_1 - \xi_k\,\omega_3) = y_k^\circ\,\frac{d\lambda}{dt} - \lambda\,(z_k^\circ\,\omega_1 - x_k^\circ\,\omega_3)\,, \qquad (3.79b) \\[2mm]
v_{\zeta k} &= \frac{d\,\zeta_k}{dt} - (\xi_k\,\omega_2 - \eta_k\,\omega_1) = z_k^\circ\,\frac{d\lambda}{dt} - \lambda\,(x_k^\circ\,\omega_2 - y_k^\circ\,\omega_1)\,,
\end{aligned}
$$

and the kinetic energy is

$$T = \frac{1}{2} \sum_{k=1}^{N} m_k \left(v_{\xi k}^2 + v_{\eta k}^2 + v_{\zeta k}^2 \right).$$

If we call ϱ_{jk} the distance between any two masses m_j and m_k, then

$$\varrho_{jk}^2 = (\xi_j - \xi_k)^2 + (\eta_j - \eta_k)^2 + (\zeta_j - \zeta_k)^2\,, \qquad (j, k = 1, 2, \ldots, N)$$

and, from (3.77), $\varrho_{jk} = \lambda\, r_{jk}^\circ$. Therefore, for the potentials,

$$U = \frac{1}{\lambda}\,U^\circ, \qquad (3.80)$$

since

$$U = \sum_{1 \le j < k \le N} \frac{m_j\, m_k}{\varrho_{jk}}, \qquad U^\circ = \sum_{1 \le j < k \le N} \frac{m_j\, m_k}{r_{jk}^\circ},$$

with $G = 1$. Let us now consider Lagrange's equations:

$$\frac{d}{dt}\left(\frac{\partial \mathcal{L}}{\partial \dot\varrho_k} \right) - \frac{\partial \mathcal{L}}{\partial \varrho_k} = 0,$$

where $\mathcal{L} = T + U$. By substituting and performing the calculations, we obtain

$$
\begin{aligned}
x_j^\circ &\left[\lambda\,\frac{d^2\lambda}{dt^2} - \lambda^2\,(\omega_2^2 + \omega_3^2) \right] + y_j^\circ \left[\lambda^2\,\omega_1\,\omega_2 - \frac{d}{dt}\,(\lambda^2\,\omega_3) \right] \\[2mm]
&+ z_j^\circ \left[\lambda^2\,\omega_1\,\omega_3 + \frac{d}{dt}\,(\lambda^2\,\omega_2) \right] = \frac{1}{\lambda} \sum_{k \ne j} m_k\,\frac{x_k^\circ - x_j^\circ}{(r_{jk}^\circ)^3}, \qquad (3.81a)
\end{aligned}
$$

---

[49]E. T. Whittaker: op. cit., p. 17

$$x_j^\circ \left[\lambda^2 \omega_1 \omega_2 + \frac{d}{dt}(\lambda^2 \omega_3) \right] + y_j^\circ \left[\lambda \frac{d^2 \lambda}{dt^2} - \lambda^2 (\omega_3^2 + \omega_1^2) \right]$$

$$+ z_j^\circ \left[\lambda^2 \omega_2 \omega_3 - \frac{d}{dt}(\lambda^2 \omega_1) \right] = \frac{1}{\lambda} \sum_{k \neq j} m_k \frac{y_k^\circ - y_j^\circ}{(r_{jk}^\circ)^3}, \qquad (3.81b)$$

$$x_j^\circ \left[\lambda^2 \omega_1 \omega_3 - \frac{d}{dt}(\lambda^2 \omega_2) \right] + y_j^\circ \left[\lambda^2 \omega_2 \omega_3 + \frac{d}{dt}(\lambda^2 \omega_1) \right]$$

$$+ z_j^\circ \left[\lambda \frac{d^2 \lambda}{dt^2} - \lambda^2 (\omega_1^2 + \omega_2^2) \right] = \frac{1}{\lambda} \sum_{k \neq j} m_k \frac{z_k^\circ - z_j^\circ}{(r_{jk}^\circ)^3}. \qquad (3.81c)$$

Equations (3.81) are the equations of the homographic motion for the j-th mass point. Now we shall study them, trying to draw out all possible information, in the three cases of:

a) collinear homographic motions,
b) flat homographic motions,
c) spatial homographic motions, for which $N \geq 4$.

a) Collinear Homographic Motions

If all the N points are on a straight line, which we take as the ξ axis, then $y_k^\circ = z_k^\circ = 0$, $\forall k$, at the initial instant $t = t_0$, and also $\eta_k = \zeta_k = 0$, $\forall k$, at any subsequent instant of time. In this case, it is possible to choose the η axis in such a way that $\omega_2 = 0$. Then (3.81) become

$$\frac{d^2 \lambda}{dt^2} - \lambda \omega_3^2 = \frac{1}{\lambda^2} \frac{1}{x_j^\circ} \sum_{k \neq j} m_k \frac{x_k^\circ - x_j^\circ}{(r_{jk}^\circ)^3},$$

$$\frac{d}{dt}(\lambda^2 \omega_3) = 0, \qquad \lambda^2 \omega_1 \omega_3 = 0. \qquad (3.82)$$

From the last two equations one has that either $\omega_3 = 0$ and $\omega_1 \neq 0$, or $\omega_3 \neq 0$ and $\omega_1 = 0$. In the first case, since $\omega_2 = \omega_3 = 0$, the ξ axis is fixed and the motion then occurs on a straight line fixed in space; in fact, the total angular momentum of the system vanishes, as can be seen since $\eta_k = \zeta_k = 0$ and $v_{\eta k} = v_{\zeta k} = 0$, and the motion is homothetic. In the second case $\lambda^2 \omega_3 = \text{const.} = \alpha$, and the ξ axis will no longer be fixed in space but rotate around the ζ axis, which is now fixed. The motion will take place in the plane $\zeta = 0$ and therefore it will be a planar motion, the $\xi\eta$ plane being fixed in space. Since the first of equations (3.82) does not contain ω_1, we can deal with both cases by substituting for ω_3 the expression α/λ^2, meaning that the first case corresponds to $\alpha = 0$. Then

$$\frac{d^2 \lambda}{dt^2} - \frac{\alpha^2}{\lambda^3} = \frac{1}{\lambda^2} \frac{1}{x_j^\circ} \sum_{k \neq j} m_k \frac{x_k^\circ - x_j^\circ}{(r_{jk}^\circ)^3}, \qquad (3.83)$$

which, since $\lambda \neq 0$, we can rewrite in the form

$$\lambda^2 \frac{d^2 \lambda}{dt^2} - \frac{\alpha^2}{\lambda} = \frac{1}{x_j^\circ} \sum_{k \neq j} m_k \frac{x_k^\circ - x_j^\circ}{(r_{jk}^\circ)^3},$$

where it is evident that, the right-hand side being constant, the left-hand side will be as well. Let us call this constant $-\beta^2$, the negative sign signifying that the force is attractive. We can then write, in place of (3.82),

$$\frac{d^2 \lambda}{dt^2} = -\frac{\beta^2}{\lambda^2} + \frac{\alpha^2}{\lambda^3}, \qquad \lambda^2 \omega_3 = \alpha. \tag{3.84}$$

By multiplying both sides of the first equation by $d\lambda/dt$ and integrating, we obtain

$$\frac{1}{2}\left(\frac{d\lambda}{dt}\right)^2 - \frac{\beta^2}{\lambda} + \frac{\alpha^2}{2\lambda^2} = \gamma,$$

with γ an integration constant. By substituting the second equation here, we finally get

$$\frac{1}{2}\left(\frac{d\lambda}{dt}\right)^2 + \frac{1}{2}\lambda^2 \omega_3^2 - \frac{\beta^2}{\lambda} = \gamma,$$

which, if we take into account (3.77) and compare with (2.9b), looks like the energy integral for Keplerian motion. Hence (3.84) represent the equations of Keplerian motion. For constant λ, ω_3 will also be constant and the Keplerian orbits will be circular; then there will be a collinear solution of relative equilibrium: the N points stay fixed on a straight line which rotates with angular velocity ω_3 around the barycentre. For $\alpha = 0$ ($\omega_3 = 0$), the line will stay fixed (rectilinear solution) and the motion will be homothetic; we have already seen that in the case of a rectilinear solution the motion ends in a collision.

Going back to (3.83) and setting the right-hand side equal to $-\beta^2/\lambda^2$, we obtain

$$X_j = \sum_{k \neq j} m_k \frac{x_k^\circ - x_j^\circ}{(r_{jk}^\circ)^3} + \beta^2 x_j^\circ = 0, \qquad j, k = 1, 2, \ldots, N. \tag{3.85}$$

The quantities X_j in (3.85) depend only on the masses and on the initial coordinates, and so (3.85) represent N conditions which they must satisfy for the motion to be homographic. Since $\sum_j m_j X_j = 0$ also holds for the N quantities X_j, there are actually only $N - 1$ independent relations.

For $N = 2$, from (3.85) we have

$$m_2 \frac{x_2^\circ - x_1^\circ}{(r_{12}^\circ)^3} + \beta^2 x_1^\circ = 0, \qquad m_1 \frac{x_1^\circ - x_2^\circ}{(r_{12}^\circ)^3} + \beta^2 x_2^\circ = 0,$$

which, reexpressed in the variables ξ, give the circular Keplerian motion around the barycentre of the two bodies with angular velocity $\omega_3 = \beta\,\lambda^{-3/2}$.

For $N = 3$, from (3.85), assuming, for instance, $x_1^\circ < x_2^\circ < x_3^\circ$, we can obtain an equation of fifth degree whose unknown is the ratio $\delta = r_{12}^\circ/r_{23}^\circ$ between two of the three mutual distances of the bodies. The solution turns out to be unique for $0 < \delta < 1$, and this solution is just the collinear solution found by Euler in 1767. We have not carried out the calculations explicitly, since we shall obtain the fifth-degree equation in another way in Sect. 4.1. As there are three possible ways in which to arrange the three aligned bodies, there are just three possible collinear solutions in the case of three bodies.

We can summarize what we have learnt about collinear homographic motions by referring to (3.85). The relative positions of the N bodies at the initial instant are the same for both $\alpha = 0$ and $\alpha \neq 0$. For $\alpha = 0$, the total angular momentum vanishes and the motion will be homothetic on the $\xi = x$ axis fixed in space. In the second case, $\alpha \neq 0$, the purely homothetic motion is not possible: one has planar motion and the straight line on which the N bodies are placed rotates around the barycentre; in the case of constant λ, the N bodies will rotate all with the same constant angular velocity, and therefore there will be a solution of relative equilibrium. We have seen that for $N = 3$ the possible solutions are $3!/2 = 3$; in the case $N = 2$, there is only one possibility, that is, $2!/2$. Moulton[50] demonstrated that, if n_p is the number of the possible solutions for $N = p$, then $(p + 1)\, n_p$ is the number of solutions for $N = p + 1$; by induction, the number of solutions for any N is then given by $N!/2$ (Moulton's theorem).

b) Flat Homographic Motions

Let us assume as the initial plane of the motion the $\zeta = 0$ plane; then $z_k^\circ = 0$, $\forall\, k$ and also $\zeta_k = 0$, $\forall\, k$, at any subsequent instant. Equations (3.81) become

$$x_j^\circ \left[\lambda\, \frac{d^2\lambda}{dt^2} - \lambda^2 \left(\omega_2^2 + \omega_3^2\right) \right] + y_j^\circ \left[\lambda^2\, \omega_1\omega_2 - \frac{d}{dt}(\lambda^2\omega_3) \right] = \frac{1}{\lambda} \sum_{j\neq k} m_k\, \frac{x_k^\circ - x_j^\circ}{(r_{jk}^\circ)^3},$$

$$x_j^\circ \left[\lambda^2\omega_1\omega_2 - \frac{d}{dt}(\lambda^2\omega_3) \right] + y_j^\circ \left[\lambda\, \frac{d^2\lambda}{dt^2} - \lambda^2 \left(\omega_3^2 + \omega_1^2\right) \right] = \frac{1}{\lambda} \sum_{k\neq j} m_k\, \frac{y_k^\circ - y_j^\circ}{(r_{jk}^\circ)^3},$$

$$x_j^\circ \left[\lambda^2\omega_1\omega_3 - \frac{d}{dt}(\lambda^2\omega_2) \right] + y_j^\circ \left[\lambda^2\omega_2\omega_3 + \frac{d}{dt}(\lambda^2\,\omega_1) \right] = 0. \tag{3.86}$$

From the third of these

$$-\frac{\lambda^2\,\omega_1\,\omega_3 - \frac{d}{dt}\left(\lambda^2\,\omega_2\right)}{\lambda^2\,\omega_2\,\omega_3 + \frac{d}{dt}\left(\lambda^2\,\omega_1\right)} = \frac{x_1^\circ}{y_1^\circ} = \frac{x_2^\circ}{y_2^\circ} = \cdots = \frac{x_N^\circ}{y_N^\circ},$$

---

[50]F. R. Moulton: The straight line solutions of the problem of N-bodies, *Ann. Math. II Ser.* **12**, 1–17 (1910).

from which one concludes that the N bodies lie on the same straight line, and the motion is collinear homographic. We exclude this case, having already dealt with it; consequently, it must be the case that

$$\lambda^2 \omega_1 \omega_3 - \frac{d}{dt}(\lambda^2 \omega_2) = 0, \qquad \lambda^2 \omega_2 \omega_3 + \frac{d}{dt}(\lambda^2 \omega_1) = 0. \tag{3.87}$$

By multiplying the first of equations (3.86) by $\lambda^2 \omega_2$ and the second by $\lambda^2 \omega_1$ and subtracting them, we get

$$\lambda^2 \omega_2 \frac{d}{dt}(\lambda^2 \omega_2) + \lambda^2 \omega_1 \frac{d}{dt}(\lambda^2 \omega_1) = 0,$$

which, integrated, gives

$$\lambda^4 (\omega_1^2 + \omega_2^2) = \text{const.} \tag{3.88}$$

Furthermore, inspection of the first two equations shows that the quantities

$$\lambda \frac{d^2 \lambda}{dt^2} - \lambda^2 (\omega_2^2 + \omega_3^2), \quad \lambda \frac{d^2 \lambda}{dt^2} - \lambda^2 (\omega_1^2 + \omega_3^2), \quad \lambda^2 \omega_1 \omega_2, \quad \frac{d}{dt}(\lambda^2 \omega_3)$$

are all of the form $\text{const.} \times \lambda^{-1}$. By subtracting, we also find that $\lambda^2 (\omega_2^2 - \omega_1^2)$ and $\lambda^2 (\omega_1^2 + \omega_2^2)$ are of the same form. Therefore, by multiplying the last quantity by λ, we get

$$\lambda^3 (\omega_1^2 + \omega_2^2) = \text{const.} \tag{3.89}$$

Since (3.88) and (3.89) must hold simultaneously, there are two possibilities: $\omega_1 = \omega_2 = 0$, or $\lambda = $ constant. In the latter case, also $\omega_1^2 + \omega_2^2$, $\omega_2^2 + \omega_3^2$, $\omega_1^2 + \omega_3^2$ are all constant and, from (3.87), $\omega_1 \omega_3 = \omega_2 \omega_3 = 0$ as well. As a consequence, either $\omega_1 = \omega_2 = 0$ (which comes into the first case, not yet examined), or $\omega_3 = 0$. If $\omega_3 = 0$, the motion is a rotation about a line of the $\zeta = 0$ plane; we can assume this line as the ξ axis, and then $\omega_2 = 0$ also. Equation (3.86) then reduce to

$$0 = \sum_{k \neq j} m_k \frac{x_k^\circ - x_j^\circ}{(r_{jk}^\circ)^3}, \qquad -\omega_1^2 \, y_j^\circ = \frac{1}{\lambda^3} \sum_{k \neq j} m_k \frac{y_k^\circ - y_j^\circ}{(r_{jk}^\circ)^3}.$$

From the first of these,

$$\sum_j m_j x_j^\circ \sum_{k \neq j} m_k \frac{x_k^\circ - x_j^\circ}{(r_{jk}^\circ)^3} = \frac{1}{2} \sum_{1 \leq j < i \leq N} \frac{m_j m_i (x_j^\circ - x_i^\circ)^2}{(r_{ij}^\circ)^3} = C,$$

but this is absurd, since it is a sum of terms which are all positive. Therefore, it must be that $\omega_3 \neq 0$, and, at the same time, $\omega_1 = \omega_2 = 0$; the total angular momentum will be different from zero and the motion will not be homothetic. The $\zeta = 0$ plane will rotate around the ζ axis remaining fixed in space, and therefore the motion will be planar. In the case of constant ω_3 ($\lambda = \text{const.}$) the solution will be of relative equilibrium.

Summarizing, we conclude that a flat homographic motion is also planar and cannot be homothetic. Once we have ascertained that the motion takes place in a fixed plane, to study the solutions it is convenient to take this plane as a complex one and to represent the point masses m_k $(k = 1, 2, \ldots, N)$ in the complex plane. To signify that the geometric figures formed by these points must remain similar as time passes, we shall represent our points by means of

$$x_k + i\, y_k = z_k = a_k\, q(t) \qquad (k = 1, 2, \ldots, N), \tag{3.90}$$

where a_k are complex constants, in general different, and $q(t)$ a complex function of the real variable t (the symbol z_k has a meaning different from before). Now

$$r_{kl} = |z_k - z_l| = |a_k - a_l|\,|q|.$$

The equations of motion (3.1a) become, in the planar case and in complex form,

$$\ddot{z}_k = G \sum_{l \neq k} m_l\, \frac{z_l - z_k}{r_{kl}^3} \qquad (k, l = 1, 2, \ldots, N). \tag{3.91}$$

By substituting, for $q \neq 0$, one has

$$a_k\, \ddot{q} = q\, |q|^{-3}\, G \sum_{l \neq k} m_l\, \frac{a_l - a_k}{|a_l - a_k|^3},$$

and, since not all the constants a_k vanish, one also obtains $\ddot{q}\, q^{1/2}\, \overline{q}^{3/2} = b$ (a constant not depending on t), because $|q| = (q\,\overline{q})^{1/2}$. In fact,

$$a_k\, \ddot{q} = q\, |q|^{-3}\, G \sum_{l \neq k} m_l\, \frac{a_l - a_k}{|a_l - a_k|^3} = q\, q^{-3/2}\, \overline{q}^{-3/2}\, G \sum_{l \neq k} m_l\, \frac{a_l - a_k}{|a_l - a_k|^3},$$

and then, multiplying this by $q^{1/2}\overline{q}^{3/2}$, one obtains on the right-hand side a quantity that does not depend on t, as we have said. Therefore, the problem is turned back into that of the solution of the differential equation

$$\ddot{q}\, q^{1/2}\, \overline{q}^{3/2} = b \tag{3.92}$$

(which is independent of N) coupled with the algebraic system

$$a_k\, b = G \sum_{l \neq k} m_l\, \frac{a_l - a_k}{|a_l - a_k|^3} \qquad (k = 1, 2, \ldots, N). \tag{3.93}$$

Since (3.92) is the same for any N, it is clear that we can consider its solution known, since the solution of the two-body problem is known. We shall make use of this to get information about the constant b.

For $N = 2$, $m_1\, a_1 + m_2\, a_2 = 0$, which can be reexpressed by putting $a_1 = m_2 a$ and $a_2 = -m_1 a$ (where a is a not vanishing complex number). From system (3.93) we immediately obtain $b = -(m_1 + m_2)^{-2}\, |a|^{-3}$. Therefore b is a real and negative quantity. Then, if we consider the function $q = q(t)$, the solution of the differential equation (3.92), to be a known function with $b < 0$ and $a \neq 0$, $x + iy = z = a\,q(t)$ will be the parametric representation on the xy plane of a conic having one of the foci at the barycentre.

If we come back to the N-body problem and suppose that $a_1, a_2, \ldots, a_N$ be an N-tuple of complex numbers satisfying the algebraic system with $b < 0$, then any solution of (3.92) coupled to it will provide a solution of the equation of the motion such that every point will follow a conic and in the course of time the polygons formed by the N material points will remain similar to themselves. In the particular case in which the conic is an ellipse, the function $q(t)$ will be a periodic function so that there will be a periodic planar homographic solution. It is easy to check that the case of circular orbits corresponds to $q = e^{i\omega t}$, where ω is the angular velocity. In such a case we obtain

$$\ddot{q}\,q^{1/2}\,\overline{q}^{3/2} = -\omega^2 = b.$$

We will again use this argument when we study in detail the equilateral solution for $N = 3$.

c) Homographic Spatial Motions

Let us now assume that the N bodies do not all lie, at a given instant, in the same plane. Taking the ζ axis as the instantaneous rotation axis, we then get $\omega_1 = \omega_2 = 0$ and $\omega_3 \neq 0$. The equations of motion (3.81) become

$$x_j^{\circ}\left[\lambda\,\frac{d^2\lambda}{dt^2} - \lambda^2\,\omega_3^2\right] - y_j^{\circ}\,\frac{d}{dt}(\lambda^2\,\omega_3) = \frac{1}{\lambda}\sum_{k \neq j} m_k\,\frac{x_k^{\circ} - x_j^{\circ}}{(r_{jk}^{\circ})^3},$$

$$x_j^{\circ}\,\frac{d}{dt}(\lambda^2\,\omega_3) + y_j^{\circ}\left[\lambda\,\frac{d^2\lambda}{dt^2} - \lambda^2\,\omega_3^2\right] = \frac{1}{\lambda}\sum_{k \neq j} m_k\,\frac{y_k^{\circ} - y_j^{\circ}}{(r_{jk}^{\circ})^3}, \qquad (3.94)$$

$$z_j^{\circ}\,\lambda\,\frac{d^2\lambda}{dt^2} = \frac{1}{\lambda}\sum_{k \neq j} m_k\,\frac{z_k^{\circ} - z_j^{\circ}}{(r_{jk}^{\circ})^3}.$$

Multiplying the first of (3.94) by $-m_j\,y_j^{\circ}$ and the second by $m_j\,x_j^{\circ}$ and summing them, we get

$$m_j\left[(x_j^{\circ})^2 + (y_j^{\circ})^2\right]\frac{d}{dt}(\lambda^2\,\omega_3) = \frac{1}{\lambda}\sum_{k \neq j}\frac{m_j\,m_k}{(r_{jk}^{\circ})^3}\left[x_j^{\circ}(y_k^{\circ} - y_j^{\circ}) - y_j^{\circ}(x_k^{\circ} - x_j^{\circ})\right],$$

and, summing over j,

$$\sum_{j=1}^{N} m_j\left[(x_j^{\circ})^2 + (y_j^{\circ})^2\right]\frac{d}{dt}(\lambda^2\,\omega_3) = \frac{1}{\lambda}\sum_{j=1}^{N}\sum_{k \neq j}\frac{m_j\,m_k}{(r_{jk}^{\circ})^3}\left(x_j^{\circ}\,y_k^{\circ} - y_j^{\circ}\,k_k^{\circ}\right) = 0,$$

from which

$$\lambda^2\,\omega_3 = \text{const.} = \alpha. \qquad (3.95)$$

Substituting in the first two of (3.94), we obtain

$$\lambda \frac{d^2\lambda}{dt^2} - \lambda^2\omega_3^2 = \frac{1}{\lambda}\frac{1}{x_j^0}\sum_{k\neq j} m_k \frac{x_k^\circ - x_j^\circ}{(r_{jk}^\circ)^3} = \frac{1}{\lambda}\frac{1}{y_j^0}\sum_{k\neq j} m_k \frac{y_k^\circ - y_j^\circ}{(r_{jk}^\circ)^3} = \frac{\text{const.}}{\lambda}.$$

From the third of (3.94), we also have

$$\lambda \frac{d^2\lambda}{dt^2} = \frac{1}{\lambda}\frac{1}{z_j^0}\sum_{k\neq j} m_k \frac{z_k^\circ - z_j^\circ}{(r_{jk}^\circ)^3}.$$

Then, finally, $\lambda^2\omega_3^2 = \text{const.} \times \lambda^{-1}$; that is,

$$\lambda^3\omega_3^2 = \text{const.} \tag{3.96}$$

Comparing (3.95) with (3.96), we see that there are two possibilities: either $\lambda = \text{const.}$ with $\omega_3 \neq 0$ or λ arbitrary with $\omega_3 = 0$. In the first circumstance, from the third of (3.94), we have

$$\sum_{k\neq j} m_k \frac{z_k^\circ - z_j^\circ}{(r_{jk}^\circ)^3} = 0,$$

which, multiplied by $m_j z_j^0$ and summed over j, gives

$$\sum_{j=1}^{N}\sum_{k\neq j} m_j z_j^0 m_k \frac{z_k^\circ - z_j^\circ}{(r_{jk}^\circ)^3} = \frac{1}{2}\sum_{1\leq j<k\leq N} \frac{m_j m_k \left(z_j^\circ - z_k^\circ\right)^2}{(r_{jk}^\circ)^3} = 0,$$

which is absurd, being a sum of positive terms. As a consequence λ cannot be constant with $\omega_3 \neq 0$. It therefore remains the case that $\omega_3 = 0$, for which, from (3.94), we have

$$\lambda^2 \frac{d^2\lambda}{dt^2} = \text{const.} = -\beta^2, \tag{3.97}$$

where, as usual, we insert the minus sign to signify that the force is attractive. From (3.97) we deduce that every material point moves on a straight line, fixed in space and passing through the barycentre, and follows a Keplerian motion. As $\omega_1 = \omega_2 = \omega_3 = 0$, the total angular momentum of the system vanishes and the motion is homothetic.

Conclusions

From what we have seen in (a), (b), (c) we can conclude that every homographic motion is flat (in particular it can be collinear), if the angular momentum c is different from zero; or homothetic, if the angular momentum vanishes; or, finally, both flat and homothetic (as happens in the case of the collinear homographic solution with a fixed straight line). A homographic motion with vanishing angular momentum is a homothetic motion in which every body moves on a straight line passing through the barycentre of the system and stands on one of the vertices of a polyhedron which always remains similar to itself in the course of time and always takes an analogous position with respect to the barycentre.

3.8 Homographic Motions and Central Configurations

What we have got up to now is the classification of all the possible homographic solutions; however, the question of the existence of such solutions remains open. From what we have seen, if a homographic solution exists, it is determined on one side by the function $\lambda(t)$ and the matrix $\Omega(t)$, and on the other side by the N vectors r_k^0 representing the initial positions. Our concern here is to find the necessary and sufficient condition for the existence of a homographic solution.

The sufficient condition was formulated in rigorous terms by Laplace[51] and is known as *Laplace's theorem*:

> If the initial configuration of the N bodies is a central configuration, then the motion is homographic.

As we will see, this condition is necessary too. Let us first specify what we mean by *central configuration*: at a given instant $t = t_0$, the N-body system gives rise to a central configuration if the gravitational force acting on every mass point m_k is proportional to m_k itself and to its position vector r_k referred to the centre of mass. This means that

$$\frac{\partial U}{\partial r_k} = \sigma\, m_k\, r_k, \qquad \forall\, k = 1, 2, \ldots, N, \tag{3.98}$$

with the scalar σ not depending on k. It is clear that, if σ remains constant in a solution, the motion is that of an oscillator. The quantity σ can be found in the following way. Multiplying (3.98) by r_k and summing over k we get

$$\sum_k r_k \cdot \frac{\partial U}{\partial r_k} = \sigma \sum_k m_k\, |r_k|^2;$$

that is, $-U = \sigma\, I$, from which

$$\frac{\partial U}{\partial r_k} = -\frac{U}{I}\, m_k\, r_k, \tag{3.99}$$

where we have made use of Euler's theorem and the definition of the moment of inertia. Moreover, since $\partial I/\partial r_k = 2\, m_k\, r_k$, (3.99) can be written, if we multiply by $I\,U$, in the form

---

[51]P. S. de Laplace: *Traité de Mécanique Céleste*, Tome quatrième, Livre X (Paris, 1805), pp. 307–313.

$$2\,I\,U\,\frac{\partial U}{\partial r_k} + U^2\,\frac{\partial I}{\partial r_k} = 0,$$

or even

$$\frac{\partial}{\partial r_k}\left(I\,U^2\right) = 0. \tag{3.100}$$

It follows that the central configurations are specified by the critical points of the function $I\,U^2$.

Before proving Laplace's theorem, we must establish some relations that will be useful for the proof. If we indicate by

$$r_k \equiv \begin{pmatrix} x_k \\ y_k \\ z_k \end{pmatrix}$$

the position vector of the k-th material point in the inertial barycentric frame, we have

$$\varrho_k \equiv \begin{pmatrix} \xi_k \\ \eta_k \\ \zeta_k \end{pmatrix} = \Omega^{-1}\,r_k = \Omega^{-1}\begin{pmatrix} x_k \\ y_k \\ z_k \end{pmatrix} \tag{3.101}$$

for the position vector of the same point in the rotating frame. The matrix $\Omega = \Omega\,(t)$ is the orthogonal matrix (with determinant $+1$) introduced in the previous section; we assume that both Ω and the vectors r_k and ϱ_k are C^2 functions of t. As Ω is an orthogonal matrix,

$$\Omega^T = \Omega^{-1};$$

that is,

$$\Omega\,\Omega^T = \Omega^T\,\Omega = 1. \tag{3.102}$$

Differentiating, we get

$$\frac{d}{dt}\left(\Omega^T\,\Omega\right) = 0,$$

from which

$$\dot{\Omega}^T\,\Omega + \Omega^T\,\dot{\Omega} = 0,$$

$$\dot{\Omega}^T\left(\Omega^T\right)^{-1} + \Omega^{-1}\,\dot{\Omega} = 0,$$

$$\Omega^{-1}\,\dot{\Omega} = -\left(\Omega^{-1}\,\dot{\Omega}\right)^T.$$

Therefore $\Omega^{-1}\,\dot{\Omega}$ is an antisymmetric matrix, which we shall denote by Σ, and, introducing a vector[52] ω, we can write

$$\omega \equiv \begin{pmatrix} \omega_1 \\ \omega_2 \\ \omega_3 \end{pmatrix}, \quad \Sigma \equiv \Omega^{-1}\,\dot{\Omega} = \begin{pmatrix} 0 & -\omega_3 & \omega_2 \\ \omega_3 & 0 & -\omega_1 \\ -\omega_2 & \omega_1 & 0 \end{pmatrix} = -\Sigma^T. \tag{3.103}$$

---

[52]See A. Wintner: op. cit., p. 50.

Differentiating, we get

$$\dot{\Sigma} = \frac{d}{dt}\left(\Omega^{-1}\,\dot{\Omega}\right) = \frac{d}{dt}\left(\Omega^{T}\,\dot{\Omega}\right) = \Omega^{T}\,\ddot{\Omega} + \dot{\Omega}^{T}\,\dot{\Omega}$$
$$= \Omega^{-1}\,\ddot{\Omega} + (\Omega\,\Sigma)^{T}\,\dot{\Omega} = \Omega^{-1}\,\ddot{\Omega} - \Sigma\,\Omega^{T}\,\dot{\Omega} = \Omega^{-1}\,\ddot{\Omega} - \Sigma^{2},$$

where $\Sigma^{2} = \Sigma\,\Sigma$ is the matrix of elements $(\omega_i\,\omega_k - \omega^2\,\delta_{ik})$ and $\omega^2 = \omega_1^2 + \omega_2^2 + \omega_3^2$. Then

$$\Omega^{-1}\,\ddot{\Omega} = \dot{\Sigma} + \Sigma^{2}. \tag{3.104}$$

It can be also verified that

$$\Sigma\,\boldsymbol{\varrho}_k = \boldsymbol{\omega} \times \boldsymbol{\varrho}_k \tag{3.105}$$

$$\dot{\Sigma}\,\boldsymbol{\varrho}_k = \dot{\boldsymbol{\omega}} \times \boldsymbol{\varrho}_k \tag{3.106}$$

$$\Sigma^{2}\,\boldsymbol{\varrho}_k = (\boldsymbol{\omega} \cdot \boldsymbol{\varrho}_k)\,\boldsymbol{\omega} - \omega^2\,\boldsymbol{\varrho}_k. \tag{3.107}$$

Multipying (3.101) by Ω and differentiating, we have $\dot{\boldsymbol{r}}_k = \dot{\Omega}\,\boldsymbol{\varrho}_k + \Omega\,\dot{\boldsymbol{\varrho}}_k$, from which

$$\Omega^{-1}\,\dot{\boldsymbol{r}}_k = \Omega^{-1}\,\dot{\Omega}\,\boldsymbol{\varrho}_k + \dot{\boldsymbol{\varrho}}_k = \dot{\boldsymbol{\varrho}}_k + \boldsymbol{\omega} \times \boldsymbol{\varrho}_k, \tag{3.108}$$

where we have used (3.103) with (3.105). Comparing (3.108) with (3.79a), we promptly see that the vector $\boldsymbol{\omega}$ is nothing but the angular velocity of the rotating frame with respect to the barycentric inertial frame. The left-hand side of (3.108) represents, in fact, the velocity of the k-th point referred to the rotating frame, obtained by operating with the matrix Ω^{-1} on the vector $\dot{\boldsymbol{r}}_k$ (velocity referred to the inertial frame). A further differentiation of $\dot{\boldsymbol{r}}_k$, with the use of (3.105–107), provides the acceleration

$$\Omega^{-1}\,\ddot{\boldsymbol{r}}_k = \ddot{\boldsymbol{\varrho}}_k + 2\,\boldsymbol{\omega} \times \dot{\boldsymbol{\varrho}}_k + \dot{\boldsymbol{\omega}} \times \boldsymbol{\varrho}_k + (\boldsymbol{\omega} \cdot \boldsymbol{\varrho}_k)\,\boldsymbol{\omega} - \omega^2\,\boldsymbol{\varrho}_k. \tag{3.109}$$

In the case of a planar solution, once the plane $\zeta = 0$ has been chosen as the plane of the solution, it can be made to coincide with the $z = 0$ plane of the inertial frame, and therefore the matrix $\Omega = \Omega(t)$ can be expressed in the form

$$\Omega = \begin{pmatrix} \cos\phi & -\sin\phi & 0 \\ \sin\phi & \cos\phi & 0 \\ 0 & 0 & 1 \end{pmatrix}, \tag{3.110}$$

ϕ being the rotation angle. Using (3.103), we obtain for the matrix Σ

$$\Sigma = \begin{pmatrix} 0 & -\dot{\phi} & 0 \\ \dot{\phi} & 0 & 0 \\ 0 & 0 & 0 \end{pmatrix} = \begin{pmatrix} 0 & -\omega_3 & 0 \\ \omega_3 & 0 & 0 \\ 0 & 0 & 0 \end{pmatrix}, \tag{3.111}$$

since, in this case,

$$\boldsymbol{\omega} = \begin{pmatrix} 0 \\ 0 \\ \dot{\phi} \end{pmatrix}. \tag{3.112}$$

From (3.111) and (3.112), we obtain in this case the expressions for the velocity (3.108) and acceleration (3.109):

$$\boldsymbol{\Omega}^{-1}\dot{\boldsymbol{r}}_k = \begin{pmatrix} \dot{\xi}_k \\ \dot{\eta}_k \\ 0 \end{pmatrix} + \begin{pmatrix} 0 \\ 0 \\ \dot{\phi} \end{pmatrix} \times \begin{pmatrix} \xi_k \\ \eta_k \\ 0 \end{pmatrix} = \begin{pmatrix} \dot{\xi}_k - \eta_k\dot{\phi} \\ \dot{\eta}_k + \xi_k\dot{\phi} \\ 0 \end{pmatrix}, \tag{3.113}$$

$$\boldsymbol{\Omega}^{-1}\ddot{\boldsymbol{r}}_k = \begin{pmatrix} \ddot{\xi}_k \\ \ddot{\eta}_k \\ 0 \end{pmatrix} + 2 \begin{pmatrix} 0 \\ 0 \\ \dot{\phi} \end{pmatrix} \times \begin{pmatrix} \dot{\xi}_k \\ \dot{\eta}_k \\ 0 \end{pmatrix} + \begin{pmatrix} 0 \\ 0 \\ \ddot{\phi} \end{pmatrix} \times \begin{pmatrix} \xi_k \\ \eta_k \\ 0 \end{pmatrix}$$

$$+ \left[\begin{pmatrix} 0 \\ 0 \\ \dot{\phi} \end{pmatrix} \cdot \begin{pmatrix} \xi_k \\ \eta_k \\ 0 \end{pmatrix} \right] \begin{pmatrix} 0 \\ 0 \\ \dot{\phi} \end{pmatrix} - \left[\begin{pmatrix} 0 \\ 0 \\ \dot{\phi} \end{pmatrix} \cdot \begin{pmatrix} 0 \\ 0 \\ \dot{\phi} \end{pmatrix} \right] \begin{pmatrix} \xi_k \\ \eta_k \\ 0 \end{pmatrix}$$

$$= \begin{pmatrix} \ddot{\xi}_k - 2\dot{\eta}_k\dot{\phi} - \xi_k\dot{\phi}^2 - \eta_k\ddot{\phi} \\ \ddot{\eta}_k + 2\dot{\xi}_k\dot{\phi} - \eta_k\dot{\phi}^2 + \xi_k\ddot{\phi} \\ 0 \end{pmatrix}. \tag{3.114}$$

Returning to the general case, we can write (3.77) in the form $\boldsymbol{\varrho}_k = \lambda\, \boldsymbol{r}_k^\circ = \lambda\mathbf{1}\, \boldsymbol{r}_k^\circ$, where, as usual, $\mathbf{1}$ is the unit matrix. Differentiating, yields $\dot{\boldsymbol{\varrho}}_k = \dot{\lambda}\mathbf{1}\, \boldsymbol{r}_k^\circ$ and $\ddot{\boldsymbol{\varrho}}_k = \ddot{\lambda}\mathbf{1}\, \boldsymbol{r}_k^\circ$. In place of (3.108) and (3.109), we have

$$\boldsymbol{\Omega}^{-1}\dot{\boldsymbol{r}}_k = \left(\dot{\lambda}\mathbf{1} + \lambda\,\boldsymbol{\Sigma} \right) \boldsymbol{r}_k^\circ, \tag{3.115}$$

$$\boldsymbol{\Omega}^{-1}\ddot{\boldsymbol{r}}_k = \left[\ddot{\lambda}\mathbf{1} + 2\dot{\lambda}\,\boldsymbol{\Sigma} + \lambda(\dot{\boldsymbol{\Sigma}} + \boldsymbol{\Sigma}^2) \right] \boldsymbol{r}_k^\circ. \tag{3.116}$$

Moreover, from the definition of potential,

$$\boldsymbol{\Omega}^{-1}\frac{\partial U}{\partial \boldsymbol{r}_k} = \frac{1}{\lambda^2}\frac{\partial U^\circ}{\partial \boldsymbol{r}_k} = \frac{1}{\lambda^2}m_k\,\boldsymbol{a}_k^\circ, \tag{3.117}$$

where by $\boldsymbol{a}_k^\circ$, we refer to the value of the acceleration at the time t_0, so that it is a constant vector. Therefore, if $\boldsymbol{r}_k(t)$ is a homographic solution, we have

$$m_k\,\ddot{\boldsymbol{r}}_k = \frac{\partial U}{\partial \boldsymbol{r}_k} = m_k\,\boldsymbol{\Omega}\left[\ddot{\lambda}\mathbf{1} + 2\dot{\lambda}\,\boldsymbol{\Sigma} + \lambda(\dot{\boldsymbol{\Sigma}} + \boldsymbol{\Sigma}^2) \right] \boldsymbol{r}_k^\circ$$

$$= \boldsymbol{\Omega}\,\frac{1}{\lambda^2}\frac{\partial U^\circ}{\partial \boldsymbol{r}_k} = \boldsymbol{\Omega}\,\frac{1}{\lambda^2}m_k\,\boldsymbol{a}_k^\circ,$$

from which,

$$\boldsymbol{a}_k^\circ = \boldsymbol{K}(t)\,\boldsymbol{r}_k^\circ, \tag{3.118}$$

with

$$\boldsymbol{K}(t) = \lambda^2\left[\ddot{\lambda}\mathbf{1} + 2\dot{\lambda}\,\boldsymbol{\Sigma} + \lambda(\dot{\boldsymbol{\Sigma}} + \boldsymbol{\Sigma}^2) \right]. \tag{3.119}$$

In the case of a planar solution, with $\omega_1 = \omega_2 = 0$ and $\omega_3 = \dot{\phi}$, the kinetic energy and the angular momentum will be

$$T = \frac{1}{2}\left(\dot{\lambda}^2 + \lambda^2\,\omega_3^2\right)I^\circ = \frac{1}{2}\left(\dot{\lambda}^2 + \lambda^2\,\dot{\phi}^2\right)I^\circ,$$
$$c = \lambda^2\,\omega_3\,I^\circ = \lambda^2\,\dot{\phi}\,I^\circ. \tag{3.120}$$

From the first of (3.120) and from (3.80), the energy integral will be

$$T - U = \frac{1}{2}\left(\dot{\lambda}^2 + \lambda^2\,\dot{\phi}^2\right)I^\circ - \frac{U^\circ}{\lambda} = h. \tag{3.121}$$

The matrix $K(t)$, in turn, will be:

$$K(t) = \begin{pmatrix} \lambda^2(\ddot{\lambda} - \lambda\dot{\phi}^2) & -\lambda^2(\lambda\ddot{\phi} + 2\dot{\lambda}\dot{\phi}) & 0 \\ \lambda^2(\lambda\ddot{\phi} + 2\dot{\lambda}\dot{\phi}) & \lambda^2(\ddot{\lambda} - \lambda\dot{\phi}^2) & 0 \\ 0 & 0 & \lambda^2\ddot{\lambda} \end{pmatrix}. \tag{3.122}$$

In the general, not planar, case of homographic motions, we saw in Sect. 3.7 that the only possible motions are those corresponding to $\omega_1 = \omega_2 = \omega_3 = 0$ (and $c = 0$) so that equations (3.120–122) are still valid if we assume $\omega_3 = 0$, $c = 0$. The Lagrange–Jacobi identity will also continue to hold in the form

$$\ddot{I} = 2U + 4h = \frac{d^2}{dt^2}(\lambda^2\,I^\circ) = \frac{2\,U^\circ}{\lambda} + 4h,$$

from which

$$I^\circ\left(\dot{\lambda}^2 + \lambda\,\ddot{\lambda}\right) - \frac{U^\circ}{\lambda} = 2h. \tag{3.123}$$

We can now prove that a necessary condition for the existence of homographic motions is that the configuration be central, namely that, if homographic solutions exist for the N-body problem, then the configuration is central. Let us introduce the following constants:

$$m_\circ = \frac{U^\circ}{I^\circ}, \tag{3.124}$$

$$h_\circ = \frac{h}{I^\circ}, \tag{3.125}$$

$$c_\circ = \frac{c}{I^\circ}. \tag{3.126}$$

By substituting (3.124) and (3.125) into (3.121), we obtain

$$\left(\dot{\lambda}^2 + \lambda^2\,\dot{\phi}^2\right) - \frac{2\,m_\circ}{\lambda} = 2\,h_\circ. \tag{3.127}$$

From (3.126) and the second of equations (3.120),

$$c_\circ = \lambda^2\,\dot{\phi}. \tag{3.128}$$

Finally, putting (3.124) and (3.125) into (3.123) yields

$$\left(\dot{\lambda}^2 + \lambda\,\ddot{\lambda}\right) - \frac{m_\circ}{\lambda} = 2\,h_\circ. \tag{3.129}$$

Subsequently, from (3.127) and (3.129), we get

$$\ddot{\lambda} - \lambda\,\dot{\phi}^2 + \frac{m_\circ}{\lambda^2} = 0,$$

and, differentiating (3.128), $2\,\lambda\,\dot{\lambda}\,\dot{\phi} + \lambda^2\,\ddot{\phi} = 0$. As a consequence, substituting into $\boldsymbol{K}(t)$, we shall obtain

$$\boldsymbol{K}(t) = \begin{pmatrix} -m_\circ & 0 & 0 \\ 0 & -m_\circ & 0 \\ 0 & 0 & -m_\circ + \lambda^3\,\dot{\phi}^2 \end{pmatrix}.$$

Since, by (3.118), we have

$$\boldsymbol{a}_k^\circ = \boldsymbol{K}(t)\,\boldsymbol{r}_k^\circ = m_k^{-1}\,\frac{\partial U^\circ}{\partial \boldsymbol{r}_k},$$

it follows that, both in the case of planar homographic motion, with $z_k^\circ = 0$ and $\partial U^\circ/\partial z_k = 0$, and in the general $(\dot{\phi} = 0)$ case, condition (3.98) is satisfied for $\sigma = -m_\circ = -U^\circ/I^\circ$ at the initial instant $t = t_0$. As the choice of t_0 is arbitrary, the necessity of the condition is proved.

We can finally prove Laplace's theorem, that is, that condition (3.98) is sufficient, besides being necessary for the existence of homographic motions. We have already observed that the definition of homographic motion can be written, both in the planar and in the general case, in the form

$$\boldsymbol{r}_k = \lambda(t)\,\boldsymbol{\Omega}(t)\,\boldsymbol{r}_k^\circ,$$

with the matrix $\boldsymbol{\Omega}(t)$ given by (3.110) and setting $\dot{\phi} = 0$ in the general case. Therefore every homographic solution turns out to be determined by the N initial position vectors $\boldsymbol{r}_k^\circ$ and by the two functions $\lambda(t)$ and $\phi(t)$, which we will assume to be C^2 and subject to the constraints

$$\lambda_0 = \lambda(t_0) = 1, \qquad \phi_0 = \phi(t_0) = 0, \qquad \dot{\phi}_0 = \dot{\phi}(t_0) \geq 0.$$

To prove Laplace's theorem it is therefore sufficient to prove that every solution $\boldsymbol{r}_k = \lambda\,\boldsymbol{\Omega}\,\boldsymbol{r}_k^\circ$, for which the consequences of (3.127–129) can be assumed to be valid, satisfying the condition (3.98) is also a solution of the N-body problem. In fact, it is clear that, if the vectors $\boldsymbol{r}_k$ are of the form $\lambda\,\boldsymbol{\Omega}\,\boldsymbol{r}_k^\circ$, they either represent homographic solutions or are not solutions. By hypothesis,

$$\frac{\partial U^\circ}{\partial \boldsymbol{r}_k} = -m_\circ\,m_k\,\boldsymbol{r}_k^\circ.$$

From (3.117), the equations of the homographic motion can be written in the form

$$m_k\,\ddot{\boldsymbol{r}}_k = \frac{\partial U}{\partial \boldsymbol{r}_k} = \boldsymbol{\Omega}\,\frac{1}{\lambda^2}\,\frac{\partial U^\circ}{\partial \boldsymbol{r}_k}.$$

Inserting into this the relations implicit in the hypothesis, we have $m_k\,\ddot{\boldsymbol{r}}_k = -\lambda^{-2}\,\boldsymbol{\Omega}\,m_\circ\,m_k\,\boldsymbol{r}_k^\circ$, so that:

$$\lambda^2\,\boldsymbol{\Omega}^{-1}\,\ddot{\boldsymbol{r}}_k = -m_\circ\,\boldsymbol{r}_k^\circ. \tag{3.130}$$

For our thesis be proved, we need (3.130) to be an identity in t, while at the same time (3.127), (3.128) and (3.129) hold. Whit this goal in mind, let $\lambda(t)$ and $\phi(t)$ be two given functions of class C^2, $\boldsymbol{\Omega}(t)$ a matrix function of ϕ according to (3.110), and finally $\boldsymbol{K}(t)$ a matrix given by (3.122). Differentiating and using the multiplication rules among matrices, we then find that

$$\lambda^2 \, \boldsymbol{\Omega}^{-1} \, \frac{d^2}{dt^2}(\lambda \, \boldsymbol{\Omega}) \equiv \boldsymbol{K}. \tag{3.131}$$

On the other hand, differentiating the homographic solution twice gives

$$\ddot{\boldsymbol{r}}_k = \frac{d^2}{dt^2}(\lambda \, \boldsymbol{\Omega}) \, \boldsymbol{r}_k^\circ,$$

which, by virtue of (3.131), provides the identity (in t) $\lambda^2 \, \boldsymbol{\Omega}^{-1} \, \ddot{\boldsymbol{r}}_k = \boldsymbol{K} \, \boldsymbol{r}_k^\circ$. Comparing this with (3.130), we have only to prove that

$$\boldsymbol{K} \, \boldsymbol{r}_k^\circ = -m_\circ \, \boldsymbol{r}_k^\circ \tag{3.132}$$

is an identity in t. But this has already been proved: in fact, we found that, both in the planar and in the general case, $\boldsymbol{K} \, \boldsymbol{r}_k^\circ = -m_k^{-1} \, (\partial U^\circ / \partial \boldsymbol{r}_k)$ holds, so that, since now by hypothesis $\partial U^\circ / \partial \boldsymbol{r}_k = -m_\circ \, m_k \, \boldsymbol{r}_k^\circ$, (3.132) follows straightforwardly.

Chapter 4

The Three-Body Problem

Historically, the three-body problem is the most important problem of celestial mechanics and for about two centuries has also been the most extensively studied problem of the whole mathematical physics. In this chapter, we consider only the "classical" subjects, while the more "modern" ones (KAM theory, chaotic solutions, etc.) find their place in the Volume 2. Of course, we have tried to fashion the treatment of the subject in such a way that it seems quite natural to insert subsequent developments. It is very hard, indeed impossible, in so little room, to succeed in introducing all the main problems; therefore, we have made a choice, consistent with the general intention of the book. For the rest, we have endeavoured to provide the reader with the tools and the information necessary for continuing to study the subject in specialized books or original papers.

4.1 The General Three-Body Problem

As Wintner already warned,[1] "... every new generation is usually compelled to reinterpret what the "problem" of three bodies actually is". As a matter of fact, the three-body problem, for more than two centuries the most famous (unsolved) problem of mathematical physics, has spurred on the study and the progress of several branches of mathematics and, as a consequence, it has been examined from different, yet complementary viewpoints. Wintner's statement remains valid, as we shall now try to summarize.

The system of differential equations for the three-body problem (see below: (4.1)) cannot be reduced to quadratures, and hence the problem *is not solved*; on the other hand, one can find a convergent series expansion along the whole of the time axis for the solution and consequently state that the problem is "solved". The first assertion (negative) has to be considered inadequate and the second one (positive) meaningless. The appropriate way, nowadays, to face the problem is to study the motion of the representative point on the manifold defined by the integrals of energy and angular momentum (the time having not been eliminated, this manifold is seven dimensional). In this section we shall give an account about the information which can be obtained from the set of equations, and also evaluate its homographic solutions. In the next section we shall present the main ideas of Sundman's method to obtain the above-mentioned series expansion (regularization) and we shall briefly explain the subsequent treatment by Levi-Cìvita of the same problem. The study of the flow on the manifold of energy and angular momentum and the treatment of periodic orbits will be dealt with in Volume 2.

The System

For $N = 3$, the set of equations (3.1a) becomes

$$m_1 \ddot{\boldsymbol{r}}_1 = \frac{G\, m_2\, m_1}{(r_{21})^3}\, (\boldsymbol{r}_2 - \boldsymbol{r}_1) + \frac{G\, m_3\, m_1}{(r_{31})^3}\, (\boldsymbol{r}_3 - \boldsymbol{r}_1)\,,$$

$$m_2 \ddot{\boldsymbol{r}}_2 = \frac{G\, m_3\, m_2}{(r_{32})^3}\, (\boldsymbol{r}_3 - \boldsymbol{r}_2) + \frac{G\, m_1\, m_2}{(r_{12})^3}\, (\boldsymbol{r}_1 - \boldsymbol{r}_2)\,, \qquad (4.1)$$

$$m_3 \ddot{\boldsymbol{r}}_3 = \frac{G\, m_1\, m_3}{(r_{13})^3}\, (\boldsymbol{r}_1 - \boldsymbol{r}_3) + \frac{G\, m_2\, m_3}{(r_{23})^3}\, (\boldsymbol{r}_2 - \boldsymbol{r}_3)\,.$$

It is a system of order 18 and, as we have already said, may be reduced, by means of the 10 first integrals and the two subsequent reductions, by 12 orders. In general, there remains, therefore, a system of order 6: the solution of such a system is what has constituted, historically, the *three-body problem*. The actual reduction of the system to the sixth order has very little interest, since we do not know how to solve the reduced system. It is interesting,

---

[1] A. Wintner, op. cit., pp. 345–346.

on the other hand, to study the particular solutions which we mentioned in Sects. 3.7 and 3.8, that is, the *homographic solutions*. In our case, they can be either *collinear* or *planar*, i.e. the Lagrangian solutions recalled above.

The Lagrangian Solutions

Let us start from the *collinear* case. The conditions that the initial coordinates should satisfy, for instance at the time $t = 0$, to have a homographic motion are given by (3.85), which become, for $N = 3$,

$$m_2 \frac{(x_2^\circ - x_1^\circ)}{(r_{12}^\circ)^3} + m_3 \frac{(x_3^\circ - x_1^\circ)}{(r_{13}^\circ)^3} + \beta^2 x_1^\circ = 0,$$

$$m_3 \frac{(x_3^\circ - x_2^\circ)}{(r_{23}^\circ)^3} + m_1 \frac{(x_1^\circ - x_2^\circ)}{(r_{21}^\circ)^3} + \beta^2 x_2^\circ = 0, \tag{4.2}$$

$$m_1 \frac{(x_1^\circ - x_3^\circ)}{(r_{31}^\circ)^3} + m_2 \frac{(x_2^\circ - x_3^\circ)}{(r_{32}^\circ)^3} + \beta^2 x_3^\circ = 0.$$

Of the three equations (4.2), only two will be independent, since the coordinates x_i° refer to the inertial barycentric frame. Without loss of generality, we shall put $x_1^\circ < x_2^\circ < x_3^\circ$ so that $r_{12}^\circ = r_{21}^\circ = x_2^\circ - x_1^\circ$, $\quad r_{13}^\circ = r_{31}^\circ = x_3^\circ - x_1^\circ$ and $\quad r_{23}^\circ = r_{32}^\circ = x_3^\circ - x_2^\circ$. By subtracting now the second of equations (4.2) from the first one, we now obtain

$$\beta^2 \, r_{12}^\circ = \frac{(m_1 + m_2)}{(r_{12}^\circ)^2} + m_3 \left[\frac{1}{(r_{13}^\circ)^2} - \frac{1}{(r_{23}^\circ)^2} \right]$$

and, by subtracting the third equation from the second,

$$\beta^2 \, r_{23}^\circ = \frac{(m_2 + m_3)}{(r_{23}^\circ)^2} + m_1 \left[\frac{1}{(r_{13}^\circ)^2} - \frac{1}{(r_{12}^\circ)^2} \right].$$

Furthermore, by defining

$$\delta = \frac{x_2^\circ - x_1^\circ}{x_3^\circ - x_2^\circ} > 0,$$

so that $x_3^\circ - x_1^\circ = (1 + 1/\delta)\,(x_2^\circ - x_1^\circ)$ and $x_3^\circ - x_2^\circ = (x_2^\circ - x_1^\circ)/\delta$, we get

$$\beta^2 \, r_{12}^\circ = \frac{(m_1 + m_2)}{(r_{12}^\circ)^2} + m_3 \left[\frac{\delta^2}{(1 + \delta)^2 \, (r_{12}^\circ)^2} - \frac{\delta^2}{(r_{12}^\circ)^2} \right],$$

$$\beta^2 \frac{r_{12}^\circ}{\delta} = \frac{(m_1 + m_3)\,\delta^2}{(r_{12}^\circ)^2} + m_1 \left[\frac{\delta^2}{(1 + \delta)^2 \, (r_{12}^\circ)^2} - \frac{1}{(r_{12}^\circ)^2} \right]. \tag{4.3}$$

By eliminating r_{12}° from (4.3), we obtain an equation which is of fifth degree in δ:

$$(m_2 + m_3)\,\delta^5 + (2\,m_2 + 3\,m_3)\,\delta^4 + (m_2 + 3\,m_3)\,\delta^3$$
$$- (3\,m_1 + m_2)\,\delta^2 - (3\,m_1 + 2\,m_2)\,\delta + (m_1 + m_2) = 0. \tag{4.4}$$

If we write $f(\delta)$ for the left-hand side, then

$$\lim_{\delta \to 0} f(\delta) = -(m_1 + m_2), \qquad \lim_{\delta \to \infty} f(\delta) = +\infty.$$

Hence, (4.4) will have at least one positive real root. Since the coefficients of the powers of δ in (4.4) show only one change of sign, according to Descartes's rule, we shall have one and only one positive root. Since there are three ways to arrange three points on a straight line, there are three possible collinear solutions. In the case where λ is constant (see Sect. 3.7), the three bodies will all rotate at the same constant angular velocity, and then there will be a relative equilibrium solution (see Fig. 4.1). When λ is not constant, on the other hand, the three bodies will describe conics having one focus in common. Let us move on to the *planar* case.

We start from system (3.93), where the unknowns are the complex constants a_k $(k = 1, 2, 3)$, which, together with the function $q(t)$, characterize the positions of m_1, m_2, m_3 in the complex plane. If at the time $t = 0$ we normalize the function $q(t)$, i.e. we make the real part equal to 1 and the imaginary part equal to zero, then $q(0) = 1$, and in addition

$$a_1 = x_1^\circ + i\, y_1^\circ, \qquad a_2 = x_2^\circ + i\, y_2^\circ, \qquad a_3 = x_3^\circ + i\, y_3^\circ,$$

where $x_k^\circ = x_k(0)$. System (3.93), supplemented by the condition that the coordinates are barycentric, will therefore give

$$
\begin{aligned}
x_1^\circ\, b &= G \left[m_2 \frac{(x_2^\circ - x_1^\circ)}{(r_{12}^\circ)^3} + m_3 \frac{(x_3^\circ - x_1^\circ)}{(r_{13}^\circ)^3} \right], \\[4pt]
x_2^\circ\, b &= G \left[m_3 \frac{(x_3^\circ - x_2^\circ)}{(r_{23}^\circ)^3} + m_1 \frac{(x_1^\circ - x_2^\circ)}{(r_{21}^\circ)^3} \right], \\[4pt]
x_3^\circ\, b &= G \left[m_1 \frac{(x_1^\circ - x_3^\circ)}{(r_{31}^\circ)^3} + m_2 \frac{(x_2^\circ - x_3^\circ)}{(r_{32}^\circ)^3} \right], \\[4pt]
0 &= m_1\, x_1^\circ + m_2\, x_2^\circ + m_3\, x_3^\circ \\[4pt]
y_1^\circ\, b &= G \left[m_2 \frac{(y_2^\circ - y_1^\circ)}{(r_{12}^\circ)^3} + m_3 \frac{(y_3^\circ - y_1^\circ)}{(r_{13}^\circ)^3} \right], \\[4pt]
y_2^\circ\, b &= G \left[m_3 \frac{(y_3^\circ - y_2^\circ)}{(r_{23}^\circ)^3} + m_1 \frac{(y_1^\circ - y_2^\circ)}{(r_{21}^\circ)^3} \right], \\[4pt]
y_3^\circ\, b &= G \left[m_1 \frac{(y_1^\circ - y_3^\circ)}{(r_{31}^\circ)^3} + m_2 \frac{(y_2^\circ - y_3^\circ)}{(r_{32}^\circ)^3} \right], \\[4pt]
0 &= m_1\, y_1^\circ + m_2\, y_2^\circ + m_3\, y_3^\circ.
\end{aligned}
\tag{4.5}
$$

In (4.5), there are obviously only six independent equations, since the equations for the coordinates of the barycentre can be obtained from the other six by multiplying them for the corresponding masses and by summing them.

Let us now demonstrate that the only admissible configuration, besides the collinear one, is the configuration where the three masses are at the corners of an equilateral triangle. In fact, we start by supposing that it will not be like that: for instance, $r_{13}^\circ \neq r_{23}^\circ$. Since the orientation of the axes is completely arbitrary, we can fix it by letting the x axis pass through the mass point m_3. Then $y_3^\circ = 0$, and from the equation of the barycentre $m_1\, y_1^\circ = -m_2\, y_2^\circ$, which is consistent with

$$\frac{m_1\, y_1^\circ}{(r_{31}^\circ)^3} + \frac{m_2\, y_2^\circ}{(r_{32}^\circ)^3} = 0$$

only for $y_1^\circ = y_2^\circ = 0$, if $r_{31}^\circ \neq r_{32}^\circ$. Repeating the procedure, by making the x axis pass through one of the other two points, one immediately finds that the only solutions of the system are those corresponding to:

1) $r_{12}^\circ = r_{23}^\circ = r_{31}^\circ$,
2) the three points on the same straight line.

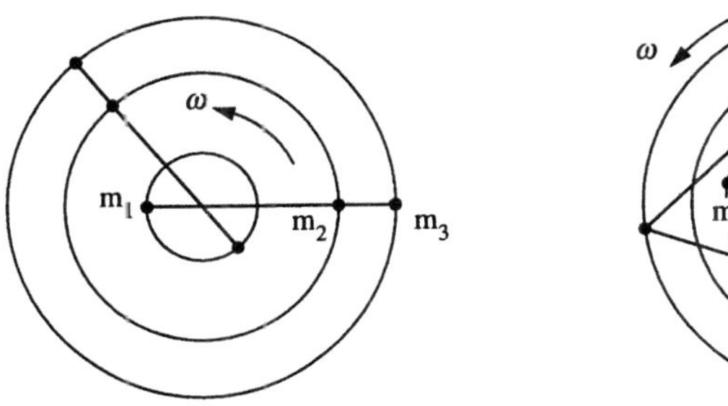

Fig. 4.1   Fig. 4.2

In conclusion, the only homographic solutions of the three-body problem are given by the Lagrangian solutions: the collinear solution and the equilateral solution. Writing r_0 for the side of the equilateral triangle at the time $t = 0$, from (4.5) we get

$$x_1^\circ b = G\,\frac{-m_2\, x_1^\circ - m_3\, x_1^\circ + (m_2\, x_2^\circ + m_3\, x_3^\circ)}{r_0^3} = -\frac{G(m_1 + m_2 + m_3)\, x_1^\circ}{r_0^3},$$

$$x_2^\circ b = -\frac{G(m_1 + m_2 + m_3)\, x_2^\circ}{r_0^3}, \quad \text{etc.}$$

If we put $m_1 + m_2 + m_3 = M$, then

$$-b = \frac{G M}{r_0^3}, \tag{4.6}$$

it not being possible to have $x_k = y_k = 0,\ \forall\, k$.

We have already said in Sect. 3.7 that in the case of circular orbits one has $-b = \omega^2$, where ω is the angular velocity common to the three bodies (rotating in the inertial barycentric system around the barycentre, see Fig. 4.2). Hence

$$\omega = \pm \sqrt{\frac{G M}{r_0^3}}, \qquad q = e^{i\omega t}. \tag{4.7}$$

In the general case (λ not constant) the orbits will be given by conics; in the case of elliptic orbits (see Fig. 4.3) $q(t)$ will be a periodic function.

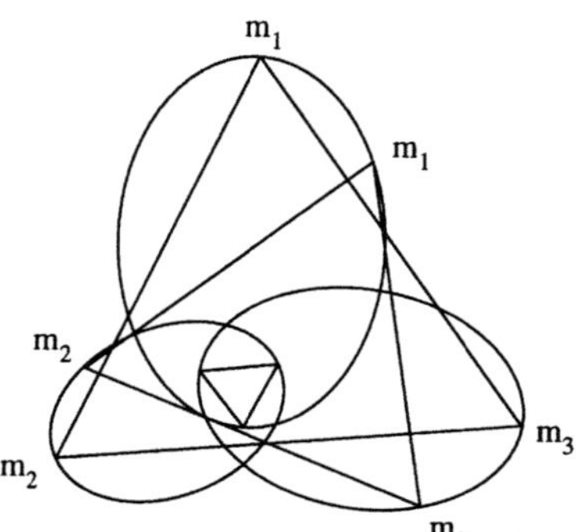

Fig. 4.3

The Jacobi Coordinate System

In the above-mentioned paper referring to the elimination of nodes, Jacobi introduced a particular system of coordinates so as to reduce the study of the N-body problem to that of an equivalent problem of $N - 1$ bodies of a suitable mass. Furthermore, the final system of equations is still a canonical system. Let us examine it for the case $N = 3$. First we refer the motion of m_2 to the body m_1, by setting $\boldsymbol{r} = \boldsymbol{r}_2 - \boldsymbol{r}_1$, and the motion of m_3 to the barycentre of m_1 and m_2. If we define $\mu = m_1 + m_2$, the coordinates of this barycentre are given by $(m_1 \boldsymbol{r}_1 + m_2 \boldsymbol{r}_2)/\mu$ or $-m_3 \boldsymbol{r}_3/\mu$, if we choose barycentric coordinates, and thus $m_1 \boldsymbol{r}_1 + m_2 \boldsymbol{r}_2 + m_3 \boldsymbol{r}_3 = 0$. If we denote by $\boldsymbol{\varrho}$ the vector distance of m_3 from this barycentre, then

$$\boldsymbol{\varrho} = \boldsymbol{r}_3 + \frac{m_3 \boldsymbol{r}_3}{m_1 + m_2} = \frac{(m_1 + m_2 + m_3) \boldsymbol{r}_3}{m_1 + m_2} = \frac{M}{\mu} \boldsymbol{r}_3.$$

In addition, one can easily verify that

$$\boldsymbol{r}_3 - \boldsymbol{r}_1 = \boldsymbol{\varrho} + \frac{m_2}{\mu} \boldsymbol{r}, \qquad \boldsymbol{r}_3 - \boldsymbol{r}_2 = \boldsymbol{\varrho} - \frac{m_1}{\mu} \boldsymbol{r}.$$

By dividing the first of equations (4.1) by m_1, the second by m_2, and by substituting one into the other, one gets

$$\ddot{\boldsymbol{r}} = -\frac{G \mu}{r^3} \boldsymbol{r} + G m_3 \left[\frac{\boldsymbol{\varrho} - (m_1/\mu) \boldsymbol{r}}{(r_{23})^3} - \frac{\boldsymbol{\varrho} + (m_2/\mu) \boldsymbol{r}}{(r_{13})^3} \right]. \tag{4.8}$$

By next multiplying the third of the equations of motion by $M\,(\mu\,m_3)^{-1}$, one eventually obtains

$$\ddot{\boldsymbol{\varrho}} = -\frac{G\,M\,m_1}{\mu\,(r_{13})^3}\,(\boldsymbol{\varrho}+(m_2/\mu)\,\boldsymbol{r}) - \frac{G\,M\,m_2}{\mu\,(r_{23})^3}\,(\boldsymbol{\varrho}-(m_1/\mu)\,\boldsymbol{r})\,. \tag{4.9}$$

If we define the relative velocities $\boldsymbol{v}=\dot{\boldsymbol{r}}$, $\boldsymbol{V}=\dot{\boldsymbol{\varrho}}$ and the masses

$$M_2 = \frac{m_1\,m_2}{m_1+m_2}\,, \qquad M_3 = \frac{m_3\,(m_1+m_2)}{m_1+m_2+m_3} = \frac{m_3\,\mu}{M}\,,$$

we obtain in the new coordinates

$$\boldsymbol{c} = M_2\,(\boldsymbol{r}\times\boldsymbol{v})+M_3\,(\boldsymbol{\varrho}\times\boldsymbol{V})\,, \quad I = M_2 r^2 + M_3\varrho^2\,, \quad \mathcal{T} = \frac{1}{2}M_2 v^2 + \frac{1}{2}M_3 V^2\,.$$

If furthermore $\boldsymbol{p}=M_2\,\boldsymbol{v}$, $\boldsymbol{P}=M_3\,\boldsymbol{V}$, and

$$\mathcal{H} = \mathcal{T}-U = \frac{1}{2}\frac{p^2}{M_2} + \frac{1}{2}\frac{P^2}{M_3} - \frac{G\,m_1 m_2}{r} - \frac{G\,m_2 m_3}{r_{23}} - \frac{G\,m_3 m_1}{r_{31}}\,,$$

we can obtain the Hamiltonian form of the previous system:

$$\dot{\boldsymbol{r}} = \frac{\partial\mathcal{H}}{\partial\boldsymbol{p}}\,, \qquad \dot{\boldsymbol{p}} = -\frac{\partial\mathcal{H}}{\partial\boldsymbol{r}}\,, \qquad \dot{\boldsymbol{\varrho}} = \frac{\partial\mathcal{H}}{\partial\boldsymbol{P}}\,, \qquad \dot{\boldsymbol{P}} = -\frac{\partial\mathcal{H}}{\partial\boldsymbol{\varrho}}\,. \tag{4.10}$$

System (4.10), of order 12, is then a Hamiltonian system relative to two bodies with fictitious masses M_2 and M_3. By exploiting in the usual way the integrals of energy and angular momentum and eliminating the time and the nodes one of course gets to order 6. The adoption of the Jacobi coordinates, then, does not imply any variation in the number of the reductions. However, as we shall see, it is the most straightforward way to go from the general problem to the restricted one. As we stated at the beginning, Jacobi's method applies in general to the N-body problem and consists mainly in the choice of two vectors: one given by the position vector of the barycentre, which will be assumed equal to zero by fixing the origin at the barycentre itself, the other given by the distance between two of the three bodies.

Let us see how the method is applied in practice, finding system (4.10) once more. Since this is a canonical system, we may obviously take it as obtained by means of a canonical transformation performed on an initial system, equally canonical. Consider system (4.1), which can certainly be written in Hamiltonian form, assuming the components of $\boldsymbol{r}_1$, $\boldsymbol{r}_2$, $\boldsymbol{r}_3$ as coordinates and the components of $\boldsymbol{p}_1 = m_1\dot{\boldsymbol{r}}_1$, $\boldsymbol{p}_2 = m_2\dot{\boldsymbol{r}}_2$, $\boldsymbol{p}_3 = m_3\dot{\boldsymbol{r}}_3$ as conjugate momenta. Let us now look for a canonical transformation which gives as new coordinate vectors

$$\boldsymbol{r} = \boldsymbol{r}_2-\boldsymbol{r}_1\,, \quad \boldsymbol{\varrho} = \boldsymbol{r}_3 - \frac{m_1\,\boldsymbol{r}_1+m_2\,\boldsymbol{r}_2}{\mu}\,, \quad \boldsymbol{r}_{\text{c.m.}} = \frac{m_1\,\boldsymbol{r}_1+m_2\,\boldsymbol{r}_2+m_3\,\boldsymbol{r}_3}{M}\,,$$

where μ and M have their previously defined meanings. In so doing, we have not yet chosen the position of the origin, by not fixing it at the barycentre.

Assume that $W_2 = W_2\left(\boldsymbol{r},\, \boldsymbol{\varrho},\, \boldsymbol{r}_{\text{c.m.}};\, \boldsymbol{p}_1,\, \boldsymbol{p}_2,\, \boldsymbol{p}_3\right)$ is the generating function of the desired transformation. Hence

$$\boldsymbol{r}_1 = \frac{\partial W_2}{\partial \boldsymbol{p}_1}, \qquad \boldsymbol{r}_2 = \frac{\partial W_2}{\partial \boldsymbol{p}_2}, \qquad \boldsymbol{r}_3 = \frac{\partial W_2}{\partial \boldsymbol{p}_3}. \tag{4.11}$$

On the other hand, by means of a bit of algebra, one can also easily obtain

$$\boldsymbol{r}_1 = \boldsymbol{r}_{\text{c.m.}} - \frac{m_3}{M}\,\boldsymbol{\varrho} - \frac{m_2}{\mu}\,\boldsymbol{r}, \quad \boldsymbol{r}_2 = \boldsymbol{r}_{\text{c.m.}} - \frac{m_3}{M}\,\boldsymbol{\varrho} + \frac{m_1}{\mu}\,\boldsymbol{r}, \quad \boldsymbol{r}_3 = \boldsymbol{r}_{\text{c.m.}} + \frac{\mu}{M}\,\boldsymbol{\varrho}. \tag{4.12}$$

From (4.11) and (4.12), one then has that W_2 is given by

$$W_2 = \left(\boldsymbol{r}_{\text{c.m.}} - \frac{m_3}{M}\,\boldsymbol{\varrho} - \frac{m_2}{\mu}\,\boldsymbol{r}\right)\boldsymbol{p}_1 + \left(\boldsymbol{r}_{\text{c.m.}} - \frac{m_3}{M}\,\boldsymbol{\varrho} + \frac{m_1}{\mu}\,\boldsymbol{r}\right)\boldsymbol{p}_2 + \left(\boldsymbol{r}_{\text{c.m.}} + \frac{\mu}{M}\,\boldsymbol{\varrho}\right)\boldsymbol{p}_3. \tag{4.13}$$

The new conjugate momenta will be obtained by differentiating (4.13) with respect to $\boldsymbol{r}$, $\boldsymbol{\varrho}$, $\boldsymbol{r}_{\text{c.m.}}$. The result is

$$\frac{\partial W_2}{\partial \boldsymbol{r}} = -\frac{m_2}{\mu}\,\boldsymbol{p}_1 + \frac{m_1}{\mu}\,\boldsymbol{p}_2 = \frac{m_1\, m_2}{m_1 + m_2}\left(\dot{\boldsymbol{r}}_2 - \dot{\boldsymbol{r}}_1\right) = M_2\left(\dot{\boldsymbol{r}}_2 - \dot{\boldsymbol{r}}_1\right) = \boldsymbol{p};$$

$$\frac{\partial W_2}{\partial \boldsymbol{\varrho}} = \frac{m_3}{M}\left(\mu\dot{\boldsymbol{r}}_3 - m_2\dot{\boldsymbol{r}}_2 - m_1\dot{\boldsymbol{r}}_1\right) = M_3\,\boldsymbol{V} = \boldsymbol{P};$$

$$\left.\frac{\partial W_2}{\partial \boldsymbol{r}}\right|_{\text{c.m.}} = \boldsymbol{p}_1 + \boldsymbol{p}_2 + \boldsymbol{p}_3 = \boldsymbol{p}_{\text{c.m.}}.$$

If now, lastly, we put the origin of the coordinate system at the barycentre, i.e. by setting $\boldsymbol{r}_{\text{c.m.}} = 0$ and $\boldsymbol{p}_{\text{c.m.}} = 0$, we get the system described by the canonical vectors $\boldsymbol{r}$, $\boldsymbol{\varrho}$, $\boldsymbol{p}$, $\boldsymbol{P}$ and consequently have system (4.10) as the Hamiltonian equations of the motion.

4.2 Existence of the Solution –
Sundman and Levi-Cìvita Regularization

As we have said several times, in the case where the conditions for the applicability of Cauchy's theorem are satisfied, there will also be a solution which can be represented by a convergent series expansion. Since the radius of convergence of a series is determined by the position of the nearest singularity, it appears that the first difficulty that one faces, when trying to build a solution given by a convergent series expansion, is given by the singularities due to binary collisions. In the case of the three-body problem this difficulty will also be the only one, provided that the case of zero angular momentum is not taken into consideration. In Sect. 3.3, we recalled the theorem which states that collisions of whatever type are not probable in Newtonian gravitational systems; however, they do exist and, when they occur, they cause a decrease of the convergence radius of any series solution.

Needless to say, we are trying to study the collisions in the *mathematical model* of three mass points subjected to their mutual Newtonian attraction. In physical reality, it would be meaningless to deal with the analytical continuation of the solution after the collision, since the collision itself, doubtless inelastic, could substantially modify the problem. The problem is then exclusively the problem of guaranteeing the existence of the solution of system (4.1), for any value of t.

The other feature of the problem of regularization, that is, that of the numerical calculations already mentioned in Sect. 2.6, does not have great importance in the case of the general three-body problem, since actual applications in celestial mechanics deal almost entirely with the restricted problem, which we shall analyse in the following sections. The complete regularization of the three-body problem was obtained for the first time by Sundman in 1907,[2] by applying rather cumbersome procedures.

Sundman's approach is not direct; it requires the introduction of a rather great number of auxiliary variables, and clumsy calculations, in order to obtain a regularized system which can no longer be included in the framework of the dynamics; this certainly is a serious inconvenience, because it is no longer permissible (at least without preliminary discussion) to apply to such a system either theoretical results, or the methods of calculation of analytical mechanics.[3]

Levi-Cìvita's criticism is addressed mainly to the consideration that in Sundman's theory it is not possible in the end to "read" the equations as dynamical equations, that is, equations which still keep their initial features, for instance, those of a Hamiltonian system. This property, as we shall see, will, however, be maintained in Levi-Cìvita regularization.

We should add, to give the whole truth of the matter, that the route initiated by Sundman is still followed when looking for global solutions of the N-body problem.[4] Taking into account the complexity of Sundman's method, we shall confine ourselves to dealing only with the one-dimensional problem, which, however, demonstrates of the method fully. We shall follow step by step the nice presentation given by Saari.[5]

---

[2] See Footnote 28, Sect. 3.4.

[3] T. Levi-Cìvita: op. cit. in Footnote 20, Sect. 2.6, p. 12.

[4] See, for instance, W. Qiu-Dong: The global solution of the N-body problem, *Celestial Mech. and Dyn. Astron.* **50**, 73–78, (1991).

[5] D. G. Saari: A visit to the Newtonian N-body problem via elementary complex variables, *Amer. Math. Monthly* **97**, 105–119, (1990).

Sundman's Method

Consider a Newtonian force field in which the particles collide collinearly. The motion occurs along the positive x semiaxis and, leaving aside the constants, the equations of motion are $\ddot{x} = -x^{-2}$; as the right-hand side is always negative, the solution $x(t)$ has a downward concavity. This forces a collision to occur in the system, going backwards in time; furthermore, if $\dot{x}(t) \leq 0$, then $\dot{x}$ will also remain as such in subsequent instants of time. Multiply both sides by $\dot{x}$ and integrate to obtain the energy integral:

$$\dot{x}^2 = \frac{2}{x} + 2\,h. \tag{4.14}$$

If a collision occurs at a particular instant, which for the sake of simplicity we shall assume to be $t_1 = 0$, then the energy integral becomes $x\,(\dot{x})^2 = 2+2\,h\,x \sim 2$, that is, $x^{1/2}\,\dot{x} \sim -2^{1/2}$. By integrating this, one gets $x^{3/2}(t) \sim -3t/\sqrt{2}$, that is:

$$x(t) = \left(\frac{9}{2}\right)^{1/3} t^{2/3}, \qquad \text{for} \quad t \to 0. \tag{4.15}$$

We saw in Sect. 3.3 that in the general problem of N bodies, a necessary and sufficient condition for there to be a collision at the time $t = 0$ is that the potential goes as $t^{-2/3}$. The behaviour of (4.15) then corresponds to the general case. What the nature of the singularity is remains to be discovered. It is possible to demonstrate (here we shall do it only for particles colliding collinearly, but the result is true in general) that a binary collision always corresponds to an algebraic branching point. Let us put $x(t) = X(t)\,t^{2/3}$ and substitute in $\ddot{x} = -x^{-2}$; this yields

$$t^2\,\ddot{X} + \frac{4}{3}\,t\,\dot{X} - \frac{2}{9}\,X = -\frac{1}{X^2}. \tag{4.16a}$$

To transform this equation into an equation which admits analytical solutions, we make a change of the independent variable: $s = t^{1/3}$. Then

$$s^2\,X'' + 2\,s\,X' - 2\,X = -\frac{9}{X^2}, \tag{4.16b}$$

where the prime indicates differentiation with respect to s. Using standard methods (series solutions and the method of majorizing to demonstrate the convergence) one can show that (4.16b) has a solution which is analytical in s. Furthermore, in a neighbourhood of the instant of collision the solution is

$$x(t) = \sum_k a_k\, t^{\frac{2k}{3}}. \tag{4.17}$$

From (4.17) it follows that a binary collision is an algebraic branching point; locally $s = t^{1/3}$ is suitable for regularizing the equation of motion. Equation (4.17) also suggests, at least from a heuristic point of view, an explanation of how the change of independent variable operated by Sundman works. Unlike what we have supposed (binary collision for $t \to t_1$), we do not know a priori when the collision will occur. It is known, however, that it occurs if and only if r tends to zero. Therefore, the growth properties of r enable us to establish the instant at which the collision will occur. Consequently, it is necessary to choose a suitable exponent α for r, just as r^α can effectively replace the term $(t_1 - t)^{2/3}$ in the change of independent variable. The asymptotic relation (4.15) shows that it must be $\alpha = 1$, so that a natural choice for the change of the independent variable is $ds = dt/r(t)$. This is the very change operated by Sundman. In this way, the real singularities are eliminated; however, we must take into account that complex singularities may exist, a further cause of the decrease in the radius of convergence of the series. Indeed, imaginary collisions among particles for imaginary values of the time exist. To understand how these imaginary collisions are connected with the actual behaviour of the system, we shall consider the elliptical solutions of Kepler's problem:

$$\ddot{\boldsymbol{r}}(t) = -\frac{\boldsymbol{r}}{r^3}. \tag{4.18}$$

The solution $r(t)$ is implicitly defined by (2.50):

$$r = a\,(1 - e\,\cos u),$$

with u a function of the independent variable t through Kepler's equation (2.53). The elliptical solutions are well behaved and there is no possibility that collisions occur; one might then think that the series expansion converges for every value of the time t. This, on the contrary, does not happen, because there are complex singularities. Putting $u = u_1 + i\,u_2$ and $t = t_1 + i\,t_2$, not only has it been calculated that an imaginary collision occurs when $r = 0$ (for complex values of u and t), but it has also been found that the real part of t is determined exactly where $r(t)$ reaches its minimum value on the real axis. Furthermore, the smaller the value of this minimum of $r(t)$ (that is the greater is the value of the eccentricity e), the closer the complex singularity is to the real axis. This suggests that, if somehow the three bodies are forced to remain far from a triple collision, then perhaps the complex singularity remains far from the real axis.

Sundman demonstrated that, if $\boldsymbol{c} \neq 0$, then for each value of t

$$\max_{i \neq j}\left\{|\boldsymbol{r}_i\,(t) - \boldsymbol{r}_j\,(t)|\right\} > D\,(\boldsymbol{c}) > 0.$$

This means that, not only do triple collisions not occur, but the system keeps itself far from such collisions. By applying the above argument and Cauchy's existence theorem, Sundman showed that the system does not admit complex singularities in a strip of the complex plane (dependent on the value of $\boldsymbol{c}$) containing the real axis. This strip is the analytical domain of the existence interval (that is, the real axis) of the solution which, as we have seen, is well behaved, since both the real collisions and the complex ones have been eliminated. Now, it is sufficient to find the conformal transformation able to bring the strip obtained in this way into a unit disk. For instance, if the strip is defined by $|\mathrm{Im}(s)| \leq \beta$, the change of independent variable is

$$\sigma = \left(e^{\pi s/2\beta} - 1\right)\left(e^{\pi s/2\beta} + 1\right)^{-1}.$$

In this new system, the motion equations do not have singularities in the unit disk, and so the series expansion converges. Along the real axis in the unit disk, σ corresponds in a one to one fashion to all the real values of time. In this way, Sundman demonstrated the existence, for the three-body problem, of a convergent series expansion for every real value of time. However, the solution thus obtained is not of any practical use, because it converges too slowly and hence too many terms have to be taken into account to achieve an acceptable accuracy. Remember that two changes of independent variable have been performed, the first $(s = t^{1/3})$ having been used to eliminate the binary collisions. This results in a "slowing down" of the dynamics: the numerical value of s tends to be greater than the corresponding numerical value of t. By the second change, the greatest values of s are exponentially transferred into the unit disk of the complex plane σ. Then small values of t may be identified with values of σ close to the boundary of the unit disk. When this happens, the speed of convergence of the series is slowed down.

Levi-Cìvita Regularization

To introduce Levi-Cìvita regularization, it is convenient to consider first the case of planar motion. Let us assume then that the three points P_1, P_2, P_3 always move in a plane, which we shall assume is the $x\,y$ plane (see Fig. 4.4). Moreover, we shall set the origin of the coordinate system $(x\,y)$ at the point P_1; our coordinate system will have a moving origin (with respect to the inertial barycentric system); however, we shall assume fixed directions for the coordinate axes. As coordinates of P_2 and P_3, we shall take the Cartesian coordinates x_2, y_2, x_3, y_3 respectively; as conjugate momenta we shall take instead the components of the momenta with respect to the barycentre B of the three bodies. We shall call $\boldsymbol{p}_2$ and $\boldsymbol{p}_3$ the respective vectors; the analogous momentum of P_1 will consequently be given by $-(\boldsymbol{p}_2 + \boldsymbol{p}_3)$. The potential function will be given, as always, by

$$U = G\left[\frac{m_1\,m_2}{r} + \frac{m_2\,m_3}{r_{23}} + \frac{m_3\,m_1}{r_{31}}\right], \tag{4.19}$$

with the symbols having their usual meanings. If we denote by $\mathcal{T}_1, \mathcal{T}_2, \mathcal{T}_3$ the kinetic energies of the three bodies, then

$$\mathcal{T}_2 = \frac{1}{2\,m_2}\left(p_{2x}^2 + p_{2y}^2\right), \qquad \mathcal{T}_3 = \frac{1}{2\,m_3}\left(p_{3x}^2 + p_{3y}^2\right),$$

$$\mathcal{T}_1 = \frac{1}{2\,m_1}\left[(p_{2x} + p_{3x})^2 + (p_{2y} + p_{3y})^2\right],$$

and $\mathcal{T} = \mathcal{T}_1 + \mathcal{T}_2 + \mathcal{T}_3$ is the total kinetic energy.

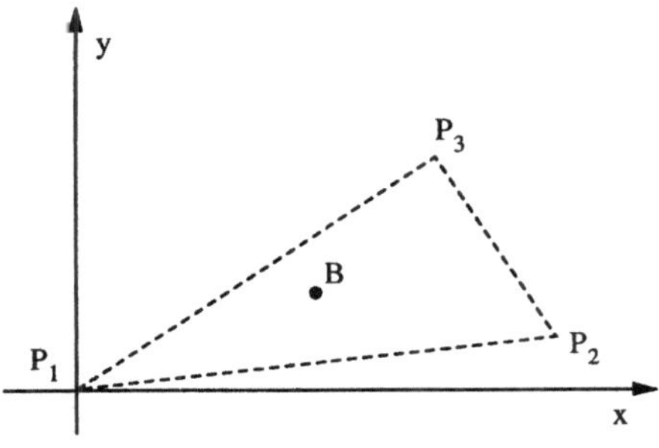

Fig. 4.4

We shall define as the Hamiltonian the function $\mathcal{H} = \mathcal{T} - U$, and the Hamiltonian equations of motion of P_2 and P_3 with respect to P_1 will be given by

$$\frac{d\,x_i}{dt} = \frac{\partial\mathcal{H}}{\partial p_{ix}}, \qquad \frac{d\,p_{ix}}{dt} = -\frac{\partial\mathcal{H}}{\partial x_i},$$

$$\frac{d\,y_i}{dt} = \frac{\partial\mathcal{H}}{\partial p_{iy}}, \qquad \frac{d\,p_{iy}}{dt} = -\frac{\partial\mathcal{H}}{\partial y_i} \qquad (i = 2, 3), \tag{4.20a}$$

with the energy integral

$$\mathcal{H} = \mathcal{T} - U = h = \text{const.} \tag{4.21}$$

From what we have learned from Sundman's and Painlevé's theorems, we know that for $c \neq 0$ and $N = 3$ the singularities that we may meet are exclusively due to binary collisions. Let us concentrate on the collision of P_1 and P_2, meaning that whatever conclusion is reached it can be repeated exactly for the other two kinds of possible collisions. From previous discussions about the nature of the singularities and also from the study of the regularization of the two-body problem, we obtain the elements with which we can schematize the question as follows. With t tending to the instant t_1:

a) $\displaystyle\lim_{t\to t_1} r = 0$;

b) position and (barycentric) velocity of the third body P_3, i.e. x_3, y_3 and p_{3x}, p_{3y}, tend to finite and well-defined limits when t tends to t_1; in particular

$$\lim_{t\to t_1} r_{23} = \lim_{t\to t_1} r_{31} > 0;$$

c) the fraction $1/r$ becomes infinite for $t \to t_1$, because of (a), but remains integrable, and, by putting $ds = dt/r$, one defines (up to an irrelevant additive constant) a parameter s always increasing with t and tending to a finite value s_1 when t tends to t_1;

d) from the energy integral, by multiplying by r and passing to the limit, one obtains

$$\lim_{t \to t_1} r\,\mathcal{T} = \lim_{t \to t_1} \frac{r}{2} \left(\frac{1}{m_1} + \frac{1}{m_2} \right) \left(p_{2x}^2 + p_{2y}^2 \right) = G\, m_1\, m_2$$

(the velocity of P_2 becomes infinite because of (4.21)).

That having been stated, we pass from system (4.20a) to another system more suitable for subsequent considerations. Since system (4.20a) does not change if a constant is added to $\mathcal{H}$, we can substitute the quantity $\mathcal{H} - h$ for $\mathcal{H}$. By doing this, one confines oneself to that kind of solution which corresponds to a well-defined, however entirely arbitrary, value of h; the advantage then is that $\mathcal{H} - h$ vanishes along each one of the solutions under consideration. If now we put

$$\mathcal{H}^* = r\,(\mathcal{H} - h) = r\,\mathcal{T}_3 + r\,(\mathcal{T}_1 + \mathcal{T}_2) - r\,U - r\,h \qquad (4.22)$$

and replace the independent variable t by s, as defined in (c), system (4.20a) becomes

$$\frac{d\,x_i}{ds} = \frac{\partial \mathcal{H}^*}{\partial p_{ix}}, \qquad \frac{d\,p_{ix}}{ds} = -\frac{\partial \mathcal{H}^*}{\partial x_i},$$

$$\frac{d\,y_i}{ds} = \frac{\partial \mathcal{H}^*}{\partial p_{iy}}, \qquad \frac{d\,p_{iy}}{ds} = -\frac{\partial \mathcal{H}^*}{\partial y_i} \qquad (i = 2, 3), \qquad (4.20b)$$

with the first integral $\mathcal{H}^* = \text{const.}$

We are concerned, however, only with the particular kind of solution where $\mathcal{H}^* = 0$, which corresponds to that of system (4.20a), where $\mathcal{H} = h$. Let us now see what improvement has been made by the introduction of the function $\mathcal{H}^*$. If we look at the analytical structure of $\mathcal{H}^*$ for $t \to t_1$, we see that something has been obtained: the infinities actually disappeared, both in U and in $\mathcal{T}_1 + \mathcal{T}_2$ (since they have been multiplied by r); however, we are not yet dealing with a function that is regular with respect to all the variables. In fact, since $\mathcal{H}^*$ contains r, the function presents a critical point with respect to x_2 and y_2, when these vanish; furthermore, $p_{2x}^2 + p_{2y}^2$, because of (d), grows indefinitely, and then, also with regard to the variables p_{2x} and p_{2y}, the collision of P_1 and P_2 is not included in the domain of regularity. We must therefore resort to a further transformation which involves the variables x_2, y_2 and p_{2x}, p_{2y}. The required transformation is the one we have already dealt widely with in Sect. 2.6 (the *Levi-Civita transformation*). We shall therefore write

$$z_2 = x_2 + i\,y_2 = \left(\xi_2 + i\,\eta_2 \right)^2 = \zeta_2^2. \qquad (4.23)$$

This is merely a point transformation; we shall correspondingly choose conjugate momenta $\Pi_{2\xi}, \Pi_{2\eta}$ so as to ensure that the transformation is canonical. We shall impose the condition that

$$p_{2x}\,dx_2 + p_{2y}\,dy_2 = \Pi_{2\xi}\,d\xi_2 + \Pi_{2\eta}\,d\eta_2 \qquad (4.24)$$

becomes an identity by substituting (4.23) into it. Explicitly,

$$p_{2x} + i\,p_{2y} = \frac{\Pi_{2\xi} + i\,\Pi_{2\eta}}{2\,(\xi_2 - i\,\eta_2)}. \tag{4.25}$$

Equations (4.24) can actually be obtained by differentiating (4.23), changing i to $-i$ in (4.25) and then multiplying the resulting equations and taking the real part of the product. As $r = |z_2|$, $r = |\zeta_2|^2$ and, from (4.25), by equating the squared moduli,

$$p_{2x}^2 + p_{2y}^2 = \frac{\Pi_{2\xi}^2 + \Pi_{2\eta}^2}{4\,|\zeta_2|^2},$$

that we can also write

$$\frac{r}{2}\left(p_{2x}^2 + p_{2y}^2\right) = \frac{\Pi_{2\xi}^2 + \Pi_{2\eta}^2}{8}. \tag{4.26}$$

From (4.26) and from (d), we finally get

$$\frac{r}{2}\left(p_{2x}^2 + p_{2y}^2\right) = \frac{\Pi_{2\xi}^2 + \Pi_{2\eta}^2}{8} = G\,\frac{m_1^2\,m_2^2}{m_1 + m_2},$$

which guarantees that $\Pi_{2\xi}^2 + \Pi_{2\eta}^2$ remains finite when the two bodies P_1 and P_2 tend to collide. This, together with the fact that one can see that the direction $P_1\,P_2$ tends to a well-defined limit, ensures that the same occurs for $\Pi_{2\xi}$ and $\Pi_{2\eta}$ separately. Multiplying both sides of (4.25) by $r = (\xi_2 + i\,\eta_2)\,(\xi_2 - i\,\eta_2)$, one obtains

$$r\,(p_{2x} + i\,p_{2y}) = \frac{1}{2}\,(\Pi_{2\xi} + i\,\Pi_{2\eta})\,(\xi_2 + i\,\eta_2). \tag{4.27}$$

From (4.23), (4.26) and (4.27), the result is therefore that

$$x_2,\; y_2,\; r,\; r\left(p_{2x}^2 + p_{2y}^2\right),\; rp_{2x},\; rp_{2y}$$

are all second-degree functions, in the new variables ξ_2, η_2, $\Pi_{2\xi}$ and $\Pi_{2\eta}$, and are regular in the neighbourhood of $\xi_2 = 0$, $\eta_2 = 0$ and of the finite values of $\Pi_{\xi 2}$ and $\Pi_{2\eta}$ corresponding to a collision. As far as $1/r_{23}$ and $1/r_{31}$ are concerned, which already behaved regularly in the neighbourhood of a collision as functions of x_2, y_2 and x_3, y_3, it is clear that such a condition is also maintained by the substitution of ξ_2, η_2 for x_2, y_2.

In conclusion, therefore, the complete regularization of $\mathcal{H}^*$ has been achieved, and, together with it, also the regularization of the system of the equations of motion, which is still a canonical system, since a canonical transformation has been performed. The new system with the Hamiltonian[6] $\mathcal{H}^*$ expressed in the new variables will be

$$\frac{d\xi_2}{ds} = \frac{\partial \mathcal{H}^*}{\partial \Pi_{2\xi}}, \quad \frac{d\Pi_{2\xi}}{ds} = -\frac{\partial \mathcal{H}^*}{\partial \xi_2}; \quad \frac{d x_3}{ds} = \frac{\partial \mathcal{H}^*}{\partial p_{3x}}, \quad \frac{d p_{3x}}{ds} = -\frac{\partial \mathcal{H}^*}{\partial x_3},$$

$$\frac{d\eta_2}{ds} = \frac{\partial \mathcal{H}^*}{\partial \Pi_{2\eta}}, \quad \frac{d\Pi_{2\eta}}{ds} = -\frac{\partial \mathcal{H}^*}{\partial \eta_2}; \quad \frac{d y_3}{ds} = \frac{\partial \mathcal{H}^*}{\partial p_{3y}}, \quad \frac{d p_{3y}}{ds} = -\frac{\partial \mathcal{H}^*}{\partial y_3}. \tag{4.28}$$

---

[6] Remember that for us $\mathcal{H}^* = 0$. See also the discussion in Sects. 1.14 and 2.6.

If we pass from the planar problem to the spatial one, clearly the number of degrees of freedom increases by two, since now three coordinates are needed to characterize each of the two points P_2 and P_3. We shall write now x_2, y_2, z_2 and x_3, y_3, z_3 for the coordinates of the two points and p_{2x}, p_{2y}, p_{2z} and p_{3x}, p_{3y}, p_{3z} for the conjugate momenta, still given by the components of the momenta with respect to the barycentre. Moreover,

$$
\mathcal{T}_1 = \frac{1}{2\,m_1} \left[(p_{2x} + p_{3x})^2 + (p_{2y} + p_{3y})^2 + (p_{2z} + p_{3z})^2 \right],
$$

$$
\mathcal{T}_2 = \frac{1}{2\,m_2} \left[p_{2x}^2 + p_{2y}^2 + p_{2z}^2 \right], \qquad \mathcal{T}_3 = \frac{1}{2\,m_3} \left[p_{3x}^2 + p_{3y}^2 + p_{3z}^2 \right], \qquad (4.29)
$$

$$
\mathcal{T} = \mathcal{T}_1 + \mathcal{T}_2 + \mathcal{T}_3, \qquad \mathcal{H} = \mathcal{T} - U.
$$

Since the transformation given by (4.23), owing to its intrinsic nature, is not extensible to three dimensions, we ought now to find a canonical transformation, and not just a point transformation, and hence a transformation more general than (4.23), which enables us to regularize the new Hamiltonian system built up starting from (4.29). As we know, after unsuccessful attempts, Levi-Cìvita[7] attained the desired transformation about ten years after the publication of the results obtained by Sundman. Levi-Cìvita's method consists in the introduction of a fictitious intermediate motion of parabolic type.

Let us now see what this means. If we consider the body P_2 at an instant t, with momentum $\boldsymbol{p}_2$ with respect to the barycentre, we can also imagine the fictitious motion of a point mass which occupies at the instant t the same position as P_2 but for which the momentum $\boldsymbol{p}_2$ is the momentum with respect to the point P_1, where we have located the origin of the coordinate axes. We can furthermore consider it subjected only to the Newtonian attraction exerted by P_1, that is, subjected to a potential given by $G\,m_1\,m_2/r$. We then have the motion of a body subjected only to the Newtonian attraction of a fixed centre. If h is the total energy of the three-body system, we can consider the fictitious motion having the same total energy, for example

$$
\frac{1}{2} \frac{|\boldsymbol{p}_2|^2}{m_2} - \frac{G\,m_1\,m_2}{r} = h. \tag{4.30}
$$

Even more generally, one can consider the motion of a body of unit mass subjected to a potential whose constant is defined by means of the equation, analogous to (4.30),

$$
\frac{1}{2} |\boldsymbol{p}_2|^2 - \frac{k}{r} = h. \tag{4.31}
$$

If $h < 0$, we have an intermediate elliptic motion with the same energy as the true motion of the system. If $h = 0$, we have instead a parabolic motion as the fictitious motion, which, as we shall see, regularizes the true motion.

---

[7] T. Levi-Cìvita: Sur la régularisation du problème des trois corps, *Acta Math.* **42**, 99–144 (1918).

Let us therefore consider the fictitious motion described by the Hamiltonian

$$\mathcal{H} = \frac{1}{2}\left(p_{2x}^2 + p_{2y}^2 + p_{2z}^2\right) - \frac{k}{r} = 0. \tag{4.32}$$

By defining $dt = r\,ds$ as before, the solutions for the parabolic motion will satisfy the system

$$\frac{d\,p_{2x}}{ds} = -\frac{\partial\,(r\,\mathcal{H})}{\partial x_2}, \qquad \frac{d\,x_2}{ds} = \frac{\partial\,(r\,\mathcal{H})}{\partial p_{2x}},$$

$$\frac{d\,p_{2y}}{ds} = -\frac{\partial\,(r\,\mathcal{H})}{\partial y_2}, \qquad \frac{d\,y_2}{ds} = \frac{\partial\,(r\,\mathcal{H})}{\partial p_{2y}}, \tag{4.33}$$

$$\frac{d\,p_{2z}}{ds} = -\frac{\partial\,(r\,\mathcal{H})}{\partial z_2}, \qquad \frac{d\,z_2}{ds} = \frac{\partial\,(r\,\mathcal{H})}{\partial p_{2z}}.$$

As we know (Sect. 1.15), all the solutions of system (4.33), taking into account (4.32), are represented by a complete integral of

$$\frac{1}{2}\,r\left(p_{2x}^2 + p_{2y}^2 + p_{2z}^2\right) = F\left(\xi_2,\,\eta_2,\,\zeta_2\right) = \text{const.},$$

where $\xi_2,\,\eta_2,\,\zeta_2$ are three constants which must satisfy $F\left(\xi_2,\,\eta_2,\,\zeta_2\right) = k$, since $\mathcal{H} = 0$. If we introduce polar coordinates r, ϑ, φ in the usual way, the corresponding Hamilton–Jacobi equation will be

$$\frac{r}{2}\left\{\left(\frac{\partial W_1}{\partial r}\right)^2 + \frac{1}{r^2}\left(\frac{\partial W_1}{\partial\vartheta}\right)^2 + \frac{1}{r^2\sin^2\vartheta}\left(\frac{\partial W_1}{\partial\varphi}\right)^2\right\} = \text{const.} = k. \tag{4.34}$$

One can see that (4.34) has a complete integral independent of φ given by

$$W_1 = W_1\left(x_2, y_2, z_2;\, \xi_2, \eta_2, \zeta_2\right) = \sqrt{2\,\varrho\,r}\,\sqrt{1 - \cos\vartheta} = \sqrt{2}\,\sqrt{\varrho\,r - \boldsymbol{\varrho}\cdot\boldsymbol{r}}, \tag{4.35}$$

where

$$\varrho = \sqrt{\xi_2^2 + \eta_2^2 + \zeta_2^2}, \qquad \cos\vartheta = \frac{\xi_2\,x_2 + \eta_2\,y_2 + \zeta_2\,z_2}{\varrho\,r} = \frac{\boldsymbol{\varrho}\cdot\boldsymbol{r}}{\varrho\,r}.$$

By substituting this in the Hamilton–Jacobi equation (4.34), one obtains $\varrho/2$ on the left-hand side and so

$$\varrho = 2\,k. \tag{4.36}$$

The equations which define the motion will then be (see Sect. 1.15)

$$-\frac{\partial W_1}{\partial\xi_2} = \Pi_{2\xi} = \frac{r}{W_1}\left(\frac{x_2}{r} - \frac{\xi_2}{\varrho}\right), \qquad \frac{\partial W_1}{\partial x_2} = p_{2x} = \frac{\varrho}{W_1}\left(\frac{x_2}{r} - \frac{\xi_2}{\varrho}\right),$$

$$-\frac{\partial W_1}{\partial\eta_2} = \Pi_{2\eta} = \frac{r}{W_1}\left(\frac{y_2}{r} - \frac{\eta_2}{\varrho}\right), \qquad \frac{\partial W_1}{\partial y_2} = p_{2y} = \frac{\varrho}{W_1}\left(\frac{y_2}{r} - \frac{\eta_2}{\varrho}\right), \tag{4.37a}$$

$$-\frac{\partial W_1}{\partial\zeta_2} = \Pi_{2\zeta} = \frac{r}{W_1}\left(\frac{z_2}{r} - \frac{\zeta_2}{\varrho}\right), \qquad \frac{\partial W_1}{\partial z_2} = p_{2z} = \frac{\varrho}{W_1}\left(\frac{z_2}{r} - \frac{\zeta_2}{\varrho}\right);$$

that is

$$\boldsymbol{\Pi}_2 = \frac{r}{W_1}\left(\frac{\boldsymbol{r}}{r} - \frac{\boldsymbol{\varrho}}{\varrho}\right), \qquad \boldsymbol{p}_2 = \frac{\varrho}{W_1}\left(\frac{\boldsymbol{r}}{r} - \frac{\boldsymbol{\varrho}}{\varrho}\right), \qquad (4.37\text{b})$$

and then

$$r\,\boldsymbol{p}_2 = \varrho\,\boldsymbol{\Pi}_2. \qquad (4.38)$$

On the other hand, since (4.34) is of type (1.C.97),

$$\dot{\Pi}_{2\xi} = -\frac{\partial F}{\partial \xi_2} = -\frac{1}{2}\frac{\xi_2}{\varrho}, \quad \dot{\Pi}_{2\eta} = -\frac{\partial F}{\partial \eta_2} = -\frac{1}{2}\frac{\eta_2}{\varrho}, \quad \dot{\Pi}_{2\zeta} = -\frac{\partial F}{\partial \zeta_2} = -\frac{1}{2}\frac{\zeta_2}{\varrho}$$

and, in vector form,

$$\dot{\boldsymbol{\Pi}}_2 = -\frac{1}{2}\frac{\boldsymbol{\varrho}}{\varrho}. \qquad (4.39)$$

The integration of (4.39) immediately gives

$$\boldsymbol{\Pi}_2 = -\frac{1}{2}\frac{\boldsymbol{\varrho}}{\varrho}t + \boldsymbol{a} \qquad (4.40)$$

where $\boldsymbol{\varrho}$ and $\boldsymbol{a}$ are constant vectors. If we vectorially multiply (4.40) by $\boldsymbol{\varrho}$, we get

$$\boldsymbol{\varrho} \times \boldsymbol{\Pi}_2 = \boldsymbol{\varrho} \times \boldsymbol{a} = \boldsymbol{c}_2, \qquad (4.41)$$

where the vector integration constant has been called $\boldsymbol{c}_2$ in view of its meaning, which will now be clarified. In fact, by vectorially multiplying the first of equations (4.37b) by $\boldsymbol{\varrho}$ and the second by $\boldsymbol{r}$ we also get

$$\boldsymbol{\varrho} \times \boldsymbol{\Pi}_2 = \boldsymbol{c}_2 = \frac{1}{W_1}\left(\boldsymbol{\varrho} \times \boldsymbol{r}\right), \qquad \boldsymbol{r} \times \boldsymbol{p}_2 = \frac{1}{W_1}\left(-\boldsymbol{r} \times \boldsymbol{\varrho}\right) = \frac{1}{W_1}\left(\boldsymbol{\varrho} \times \boldsymbol{r}\right),$$

that is, $\boldsymbol{r} \times \boldsymbol{p}_2 = \boldsymbol{\varrho} \times \boldsymbol{\Pi}_2 = \boldsymbol{c}_2$, which expresses the constancy of the angular momentum $\boldsymbol{r} \times \boldsymbol{p}_2$, simply given by the vector $\boldsymbol{c}_2$.

The equation of the plane of motion will be obviously provided by $\boldsymbol{c}_2 \cdot \boldsymbol{r} = 0$. Equation (4.41) also ensures that the vectors $\boldsymbol{\varrho}$ and $\boldsymbol{\Pi}_2$ belong to the plane of motion. From the expression (4.35) for W_1, we have $W_1^2 = 2\left(\varrho r - \boldsymbol{\varrho} \cdot \boldsymbol{r}\right)$, and from the first of (4.37b),

$$|\boldsymbol{\Pi}_2|^2 = \frac{2\left(r^2 - \frac{r}{\varrho}\boldsymbol{\varrho} \cdot \boldsymbol{r}\right)}{W_1^2} = \frac{r}{\varrho}.$$

Also, because of (4.37b), since $\boldsymbol{\varrho} \cdot \boldsymbol{\Pi}_2 = \left(\boldsymbol{\varrho} \cdot \boldsymbol{r} - \varrho r\right)/W_1 = -W_1/2$, we finally obtain

$$\varrho^2\,|\boldsymbol{\Pi}_2|^2 - \left(\boldsymbol{\varrho} \cdot \boldsymbol{\Pi}_2\right)^2 = \varrho^2\,|\boldsymbol{\Pi}_2|^2 - \frac{1}{4}W_1^2 = r\,\varrho - \frac{1}{2}\left(\varrho r - \boldsymbol{\varrho} \cdot \boldsymbol{r}\right)$$

$$= \frac{1}{2}\left(\varrho r + \boldsymbol{\varrho} \cdot \boldsymbol{r}\right) = \frac{1}{2}\varrho r\left(1 + \cos\vartheta\right).$$

But because of (4.41), $\varrho^2\,|\boldsymbol{\Pi}_2|^2 - \left(\boldsymbol{\varrho} \cdot \boldsymbol{\Pi}_2\right)^2 = |\boldsymbol{c}_2|^2$. The last relation can be written as $|\boldsymbol{c}_2|^2 = \frac{1}{2}\varrho r\left(1 + \cos\vartheta\right)$, that is, defining $c = |\boldsymbol{c}_2|$,

$$r = \frac{2\,c^2/\varrho}{1 + \cos\vartheta}, \qquad (4.42\text{a})$$

which is evidently the polar equation of the trajectory, and which, because of (4.36), will become

$$r = \frac{c^2/k}{1 + \cos \vartheta}. \tag{4.42b}$$

The constant vector ϱ, whose modulus is $2k$, has the direction of the axis of the parabola and points from the focus towards the vertex (epicentre); the parameter of the parabola is given by $2\,c^2/\varrho$.

Let us now go back to our canonical transformation generated by the function

$$W_1 = \sqrt{2} \sqrt{\sqrt{(x_2^2 + y_2^2 + z_2^2)\,(\xi_2^2 + \eta_2^2 + \zeta_2^2)} - x_2\,\xi_2 - y_2\,\eta_2 - z_2\,\zeta_2},$$

which satisfies

$$p_{2x}\,d\,x_2 + p_{2y}\,d\,y_2 + p_{2z}\,d\,z_2 - \Pi_{2\xi}\,d\,\xi_2 - \Pi_{2\eta}\,d\,\eta_2 - \Pi_{2\zeta}\,d\,\zeta_2 = d\,W_1,$$

as one can see from (4.37a). By inverting the first three of equation (4.37a) and by replacing r/ϱ by $|\boldsymbol{\Pi}_2|^2$, we obtain

$$x_2 = |\boldsymbol{\Pi}_2|^2\,\xi_2 + W_1\,\Pi_{2\xi}, \quad y_2 = |\boldsymbol{\Pi}_2|^2\,\eta_2 + W_1\,\Pi_{2\eta}, \quad z_2 = |\boldsymbol{\Pi}_2|^2\,\zeta_2 + W_1\,\Pi_{2\zeta},$$

$$\tag{4.43}$$

which, together with the last three of (4.37a), are expressions for the "old" variables as functions of the "new" ones. In the new variables, we can then write a Hamiltonian system, which will be the three-dimensional equivalent of system (4.28). In this system, $\mathcal{H}^* = r\,(\mathcal{H} - h)$, where $\mathcal{H}$ is given by (4.29) but expressed in the new variables, will be the Hamiltonian. This system appears to be completely regularized. In fact, when P_2 gets close to P_1, the modulus of p_2 tends to infinity, but the product $r\,|\boldsymbol{p}_2|^2$ remains finite and different from zero. Similarly ξ_2, η_2, ζ_2, tend to finite and well determined limits, because ϱ has been introduced as a vector whose length is $2k$; $\Pi_{2\xi}, \Pi_{2\eta}, \Pi_{2\zeta}$, tend instead to zero, since the modulus $|\boldsymbol{\Pi}_2| = (r/\varrho)\,|\boldsymbol{p}_2|$ tends to zero. If, instead of the form (of Poincaré) chosen for the Hamiltonian system, one considers the form expressed in the Jacobi coordinates as defined in the last section, Levi-Civita's method is applicable as well.[3] Of the three possible collisions (P_1, P_2), (P_2, P_3), (P_3, P_1), we have dealt with (P_1, P_2). It is clear, as we have already said, that, on repeating our reasoning (changing the canonical transformations) for (P_2, P_3), (P_3, P_1), one reaches an analogous result. Obviously, to do that, we have also to perform in advance another change of the independent variable. The last change could be avoided by choosing, once and for all, a parameter which is symmetrical with respect to the three kinds of collision. For the parameter s, we had $r\,ds = dt$; if we now take into account that

---

[3] T. Levi-Civita: loc. cit., pp. 139–140.

$$r\,U = G\,m_1\,m_2 + \frac{G\,m_1\,m_3}{r_{13}}\,r + G\,\frac{m_2\,m_3}{r_{23}}\,r,$$

considered as a function of the transformed variables $(\xi_2, \eta_2, \zeta_2, \Pi_{2\xi}, \Pi_{2\eta}, \Pi_{2\zeta})$ and also of x_3, y_3, z_3 behaves regularly in a neighbourhood of a collision (P_1, P_2), collision point included, we can even decide to replace s by the parameter τ, defined by:

$$d\tau = r\,U\,ds = U\,dt. \tag{4.44}$$

The parameter τ can play the same role as s for collision (P_1, P_2), and furthermore, given the symmetrical structure of U, it can play a similar role for collisions (P_2, P_3) and (P_1, P_3). Therefore we shall write a new canonical system, by replacing $\mathcal{H}^*$ by

$$\mathcal{K} = \frac{1}{r\,U}\,\mathcal{H}^*.$$

It must be always kept in mind that the above considerations apply exclusively to the solutions corresponding to $\mathcal{H}^* = 0$ and hence $\mathcal{K} = 0$. In that case, the Hamiltonian system will be

$$\frac{d\boldsymbol{p}_2}{d\tau} = -\frac{\partial \mathcal{K}}{\partial \boldsymbol{x}_2}, \qquad \frac{d\boldsymbol{x}_2}{d\tau} = \frac{\partial \mathcal{K}}{\partial \boldsymbol{p}_2},$$
$$\frac{d\boldsymbol{p}_3}{d\tau} = -\frac{\partial \mathcal{K}}{\partial \boldsymbol{x}_3}, \qquad \frac{d\boldsymbol{x}_3}{d\tau} = \frac{\partial \mathcal{K}}{\partial \boldsymbol{p}_3}. \tag{4.45}$$

In (4.45), $\mathcal{K}$ is a regular function of the variables $\boldsymbol{x}_2, \boldsymbol{p}_2, \boldsymbol{x}_3, \boldsymbol{p}_3$ while the positions of the three bodies remain distinct. In the neighbourhood of a collision, (P_1, P_2), (P_2, P_3), (P_3, P_1), a suitable canonical transformation, of type (4.37), restores the regularity. System (4.45) is thus to be considered, however unregularized, to be always amenable to being regularized without difficulty. It should then also be regarded as "solved", in the sense that a solution exists, given by a convergent series expansion for any value of τ, and then of t.

4.3 The Restricted Three-Body Problem

As we have already stressed by writing the equations of motion for the three-body system in Jacobi form, by working out all the possible reductions one can reach the minimum order six for the system of the equations of motion. If one confine oneself to taking into consideration only planar motions, then the order of the system goes down to four: however, it is still a problem of the utmost complexity. It may be made simpler, albeit not completely solvable, if one introduces the approximation of considering one of the three bodies uninfluential on the motion of the other two (called primaries): clearly this is equivalent to considering m_3 negligible in comparison with m_1 and m_2 and the point P_3 moving "sufficiently far" from P_1 and P_2. We shall lastly consider the case where the primaries move along circular orbits with constant angular velocity ω around their barycentre (*planar circular restricted problem*).[9] To become acquainted with the simplification which occurs in this case, let us again consider the equations in Jacobi form (4.8) and (4.9)

By putting $m_3 = 0$, the first becomes

$$\ddot{\boldsymbol{r}} = -\frac{G\mu}{r^3}\boldsymbol{r}, \tag{4.46}$$

and the second, if for the sake of simplicity we write $\boldsymbol{\varrho} + (m_2/\mu)\,\boldsymbol{r} = \boldsymbol{\varrho}_1$ and $\boldsymbol{\varrho} - (m_1/\mu)\,\boldsymbol{r} = \boldsymbol{\varrho}_2$, becomes

$$\ddot{\boldsymbol{\varrho}} = -\frac{G\,m_1}{\varrho_1^3}\,\boldsymbol{\varrho}_1 - \frac{G\,m_2}{\varrho_2^3}\,\boldsymbol{\varrho}_2. \tag{4.47}$$

Then, of the two equations, the first describes the relative motion of the primaries and can be considered solved already, since we know the general solution of the two-body problem; the second equation, since $\boldsymbol{r}$ is known from the first one, merely describes the motion of the third body $(m_3 \sim 0)$ with respect to the barycentre of the first two. The problem is therefore brought down to the solution of this second equation. Note that, since the approximation $m_3 \sim 0$ has been made, we can no longer resort to the usual first integrals to lower the order. We may, however, resort to other kinds of simplification. First of all, since both the primaries rotate with constant angular velocity around their barycentre, we can take as the reference frame in which to study the motion of m_3 the plane (rotating at constant angular velocity ω with respect to the inertial barycentric plane) where the primaries are at rest and the x axis coincides with the straight line containing m_1 and m_2 (*synodic system*). If, to spare symbols, we call x_1 and x_2 the abscissae of P_1 and P_2 and x and y the coordinates of P_3, (4.47) will have as components in the rotating system

$$\ddot{x} - 2\omega\,\dot{y} - \omega^2\,x = G\,\frac{m_1}{\varrho_1^3}\,(x_1 - x) + G\,\frac{m_2}{\varrho_2^3}\,(x_2 - x),$$

$$\ddot{y} + 2\omega\,\dot{x} - \omega^2\,y = -G\,\frac{m_1}{\varrho_1^3}\,y - G\,\frac{m_2}{\varrho_2^3}\,y. \tag{4.48a}$$

A further formal simplification of these equations can be obtained by a suitable choice of measurement units:

- choose the unit of mass such that $m_1 + m_2 = 1$;
- choose the unit of length such that $r = 1$;
- choose the unit of time such that $G = 1$.

---

[9] The term "restricted" was introduced by Poincaré (1892), but the schematization of the problem goes back to Jacobi (1836) and Euler (1772).

Let us call the smallest of the two masses (for instance, m_2) $\overline{\mu}$ (which is dimensionless) and locate it to the right of the origin. Obviously, the other mass (m_1) will be given by $1 - \overline{\mu}$. Since, in the old measurement units, $m_1 x_1 + m_2 x_2 = 0$, and since now $x_2 - x_1 = r = 1$, it will also be the case that $x_1 = -\overline{\mu}$, $x_2 = 1 - \overline{\mu}$. In these units, we also have $\omega = 1$; in fact, for the two primaries,

$$\frac{G\,m_1\,m_2}{r^2} = m_2\,|x_2|\,\omega^2 = m_1\,|x_1|\,\omega^2$$

(*uniform circular motion*); and then also

$$\frac{G\,(m_1 + m_2)}{r^2} = (|x_1| + |x_2|)\,\omega^2 = r\,\omega^2,$$

from which $G\,(m_1 + m_2) = \omega^2\,r^3$ (*Kepler's third law*) and therefore, in the chosen dimensionless system, $\omega = 1$.

If we now call the potential in these units U, then

$$U = \frac{1 - \overline{\mu}}{\varrho_1} + \frac{\overline{\mu}}{\varrho_2},$$

where

$$\varrho_1 = \sqrt{(x + \overline{\mu})^2 + y^2}, \qquad \varrho_2 = \sqrt{(x + \overline{\mu} - 1)^2 + y^2}$$

and system (4.48a) will become

$$\ddot{x} - 2\dot{y} - x = \frac{\partial U}{\partial x}, \qquad \ddot{y} + 2\dot{x} - y = \frac{\partial U}{\partial y}. \tag{4.48b}$$

If, lastly, we define a new potential function modified as follows:

$$\Phi(x, y) = \frac{1}{2}\,(x^2 + y^2) + U + \frac{1}{2}\overline{\mu}\,(1 - \overline{\mu}), \tag{4.49}$$

the equations become

$$\ddot{x} - 2\dot{y} = \frac{\partial \Phi}{\partial x}, \qquad \ddot{y} + 2\dot{x} = \frac{\partial \Phi}{\partial y}. \tag{4.50}$$

Although in this case one cannot have, as we have already warned, the first integrals corresponding to the usual conservation laws, one can demonstrate immediately that (4.50) admit the first integral known as *Jacobi integral*.[10]

By multiplying the first equation of (4.50) by $\dot{x}$ and the second by $\dot{y}$ and adding we obtain $\dot{x}\,\ddot{x} + \dot{y}\,\ddot{y} = d\Phi/dt$, from which

$$\frac{d}{dt}\,[2\,\Phi - \dot{x}^2 - \dot{y}^2] = 0,$$

and then

$$2\,\Phi - \dot{x}^2 - \dot{y}^2 = C = \text{const.} \tag{4.51}$$

We shall call the constant C the *Jacobi constant*.

---

[10]See Sect. 1.9.

The Jacobi integral is therefore a relation which connects, in the *dimensionless synodic system*, the square of the velocity with the coordinates of the point P_3 (whose mass has been taken equal to zero); in this system the equations of motion contain only one parameter: the dimensionless mass $\overline{\mu}$. It appears that one can reduce the fourth-order system above, and in various ways, to a second order one by means of the Jacobi integral and the elimination of the time. Unfortunately this has only formal interest, because then it is not possible to solve[11] the second-order system obtained.

Tisserand's Criterion

Let us recall now, as an application of the Jacobi integral, an (approximate) criterion of identification of comets due to Tisserand.[12] The problem is as follows: if it happens that a comet passes very close to a planet (for instance, Jupiter), then the elements of its orbit will undergo a significant change, and later on it may not be easy to recognize the comet itself, unless one resorts to properties of the elements which remain unchanged even after the perturbation of the planet. Tisserand suggested regarding the system Sun–planet–comet as a restricted three-body system, where the comet represented the third body of negligible mass. If we take into consideration the system Sun–Jupiter–comet, in fact, we are allowed to consider the origin of the axes of the inertial system located in the Sun and consequently the orbit of the comet in the inertial system will be its heliocentric orbit. Furthermore, since the eccentricity of Jupiter is very small (0.048435), we consider its orbit to be circular. Finally we take the inclination of the comet to be negligible. At this point one can apply to our system the Jacobi integral, expressed, however, in the inertial system. If we call the coordinates of P_3 in this system $\overline{x}$ and $\overline{y}$, we have

$$x = \overline{x}\cos t + \overline{y}\sin t, \qquad y = \overline{y}\cos t - \overline{x}\sin t, \tag{4.52}$$

and then we can write the Jacobi integral (4.51) in the inertial (heliocentric) system:

$$\dot{\overline{x}}^2 + \dot{\overline{y}}^2 = 2\left(\overline{x}\,\dot{\overline{y}} - \overline{y}\,\dot{\overline{x}}\right) + \frac{2\left(1-\overline{\mu}\right)}{\varrho_1} + \frac{2\,\overline{\mu}}{\varrho_2} - C. \tag{4.53}$$

Putting $\overline{\mu} = 0$ (Jupiter's mass negligible), the system Sun–comet will have the energy integral

$$\frac{1}{2}\left(\dot{\overline{x}}^2 + \dot{\overline{y}}^2\right) - \frac{1}{\varrho} = h = -\frac{1}{2\,a}, \tag{4.54}$$

ϱ being the Sun–comet distance and a the semimajor axis of the orbit of the comet.

---

[11]See, for instance, V. Szebehely: *Theory of Orbits. The Restricted Problem of Three Bodies* (Academic Press, 1967), Sects. 2.2 and 2.3.
[12]See F. Tisserand: op. cit., Vol. 4, p. 203.

With $\overline{\mu} = 0$, one will also have the integral of "areas" and then

$$c = \overline{x}\,\dot{\overline{y}} - \overline{y}\,\dot{\overline{x}} = \sqrt{a\,(1-e^2)}, \tag{4.55}$$

where e is the eccentricity of the orbit of the comet. By substituting (4.54) and (4.55) into (4.53), one has

$$\frac{2}{\varrho} - \frac{1}{a} = 2\,\sqrt{a\,(1-e^2)} + \frac{2\,(1-\overline{\mu})}{\varrho_1} + \frac{2\,\overline{\mu}}{\varrho_2} - C.$$

By making the approximations $\varrho = \varrho_1$, $\overline{\mu}\,({\varrho_2}^{-1} - {\varrho_1}^{-1}) = 0$, one finally obtains

$$\frac{1}{a} + 2\,\sqrt{a\,(1-e^2)} = C. \tag{4.56}$$

Equation (4.56) is the fundamental relation of Tisserand's criterion. As C is constant, at the times t_0 and t_1

$$\frac{1}{a_0} + 2\,\sqrt{a_0\,(1-e_0^2)} = \frac{1}{a_1} + 2\,\sqrt{a_1\,(1-e_1^2)},$$

with the symbols having obvious meanings. In the case where one must take account of the inclination of the orbit of the comet,

$$\frac{1}{a_0} + 2\,\sqrt{a_0\,(1-e_0^2)}\,\cos i_0 = \frac{1}{a_1} + 2\,\sqrt{a_1\,(1-e_1^2)}\,\cos i_1.$$

As we have said, this is an approximate criterion; notwithstanding, if we insert the elements of "two" comets observed at different times into (4.56), and notice a sizeable difference, we can be sure that they are not the same comet.

The Equilibrium Solutions

Let us now go back to equations (4.50), with first integral (4.51). Since we have seen that the general problem of three bodies admits planar solutions where the three bodies rotate with constant angular velocity around their barycentre, we may expect correspondingly in the restricted problem $(m_3 \sim 0)$ the same configurations as equilibrium solutions (with respect to the synodic system). Equation (4.51), once the Jacobi constant has been fixed by means of the initial conditions, will give us the velocity of the point P_3 as a function of the coordinates; or, alternatively, for every value of the velocity they provide the corresponding position where the point may be located. In particular, at the zero value of the velocity there will be a corresponding family of curves (zero velocity curves or *Hill curves*) given by

$$\Phi\,(x,y) = \frac{C}{2}. \tag{4.57}$$

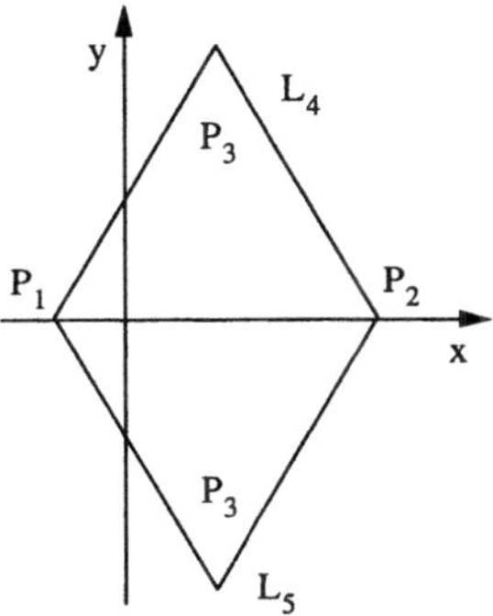

Fig. 4.5

If we rewrite (4.49) by substituting $x^2 + y^2 = (1 - \overline{\mu})\,\varrho_1^2 + \overline{\mu}\,\varrho_2^2 - \overline{\mu}\,(1 - \overline{\mu})$, we obtain

$$
\begin{aligned}
\Phi = \Phi\,(\varrho_1,\,\varrho_2) &= (1 - \overline{\mu})\left(\frac{1}{2}\,\varrho_1^2 + \frac{1}{\varrho_1}\right) + \overline{\mu}\left(\frac{1}{2}\,\varrho_2^2 + \frac{1}{\varrho_2}\right) \\
&= \frac{3}{2} + (1 - \overline{\mu})\left(\frac{1}{2} + \frac{1}{\varrho_1}\right)(\varrho_1 - 1)^2 + \overline{\mu}\left(\frac{1}{2} + \frac{1}{\varrho_2}\right)(\varrho_2 - 1)^2.
\end{aligned}
\tag{4.58}
$$

One can easily see that Φ is always positive, and, from the last expression above, that it has a minimum $\Phi_{\min} = 3/2$ for $\varrho_1 = \varrho_2 = 1$. That is, Φ has a minimum when P_3 is at a vertex of an equilateral triangle with P_1 and P_2 at the other two vertices. There are then two possible positions (see Fig. 4.5).

This case corresponds to $C = 3$. Therefore, for $C = 3$ the Hill curve degenerates into the two points, which we call L_4 and L_5 and which represent the equilateral solution of the restricted problem. From system (4.50), it can be seen that, if one puts $\dot{x} = 0$, $\dot{y} = 0$ (velocity equal to zero), nonetheless $\ddot{x} = \partial\Phi/\partial x$, $\ddot{y} = \partial\Phi/\partial y$; therefore, if one also wants the acceleration to be zero (equilibrium solutions), one must have $\partial\Phi/\partial x = \partial\Phi/\partial y = 0$. These are also the conditions which have to be satisfied by the coordinates of the stationary points of $\Phi\,(x, y)$; the minimum points already found then belong to this category. Let us now try to look for such points and then for the equilibrium solutions, by following the standard procedure. Hence, let us set

$$
\frac{\partial\Phi}{\partial x} = \frac{\partial\Phi}{\partial\varrho_1}\frac{\partial\varrho_1}{\partial x} + \frac{\partial\Phi}{\partial\varrho_2}\frac{\partial\varrho_2}{\partial x} = 0, \qquad \frac{\partial\Phi}{\partial y} = \frac{\partial\Phi}{\partial\varrho_1}\frac{\partial\varrho_1}{\partial y} + \frac{\partial\Phi}{\partial\varrho_2}\frac{\partial\varrho_2}{\partial y} = 0.
$$

By using the first of equations (4.58), we have

$$
\begin{aligned}
(1 - \overline{\mu})\left(\varrho_1 - \frac{1}{\varrho_1^2}\right)\frac{x + \overline{\mu}}{\varrho_1} + \overline{\mu}\left(\varrho_2 - \frac{1}{\varrho_2^2}\right)\frac{x + \overline{\mu} - 1}{\varrho_2} &= 0, \\
(1 - \overline{\mu})\left(\varrho_1 - \frac{1}{\varrho_1^2}\right)\frac{y}{\varrho_1} + \overline{\mu}\left(\varrho_2 - \frac{1}{\varrho_2^2}\right)\frac{y}{\varrho_2} &= 0.
\end{aligned}
\tag{4.59}
$$

Consider first the case $y \neq 0$; the second equation will give us

$$(1 - \overline{\mu}) \left(\varrho_1 - \frac{1}{\varrho_1^2} \right) \frac{1}{\varrho_1} + \overline{\mu} \left(\varrho_2 - \frac{1}{\varrho_2^2} \right) \frac{1}{\varrho_2} = 0,$$

which, when inserted in the first one, gives $\varrho_2^3 = 1$ and then also $\varrho_1^3 = 1$. The only solution of these equations (and of the system) is thus $\varrho_1 = \varrho_2 = 1$. Then, for the case $y \neq 0$, there are only two possible solutions for the position of P_3: the two vertices of the equilateral triangles whose basis is the segment $P_1 P_2$. By solving

$$\sqrt{(x + \overline{\mu})^2 + y^2} = 1, \qquad \sqrt{(x + \overline{\mu} - 1)^2 + y^2} = 1,$$

we obtain

$$x = \frac{1}{2} - \overline{\mu}, \qquad y = \pm \frac{\sqrt{3}}{2}$$

for the coordinates of L_4 and L_5 respectively.

Consider now the case $y = 0$. We are left with only the first equation

$$(1 - \overline{\mu}) \left(\varrho_1 - \frac{1}{\varrho_1^2} \right) \frac{x + \overline{\mu}}{\varrho_1} + \overline{\mu} \left(\varrho_2 - \frac{1}{\varrho_2^2} \right) \frac{x + \overline{\mu} - 1}{\varrho_2} = 0,$$

with $\varrho_1 = |x + \overline{\mu}|$, $\varrho_2 = |x + \overline{\mu} - 1|$. There are three possible cases:

a) P_3 to the left of P_1: $x < -\overline{\mu}$,
b) P_3 between P_1 and P_2: $-\overline{\mu} < x < 1 - \overline{\mu}$,
c) P_3 to the right of P_2: $x > 1 - \overline{\mu}$.

For these,

$$\begin{aligned}
&\text{a)} \quad \varrho_1 = -x - \overline{\mu}, &&\varrho_2 = 1 - x - \overline{\mu} = 1 + \varrho_1, \\
&\text{b)} \quad \varrho_1 = x + \overline{\mu}, &&\varrho_2 = 1 - x - \overline{\mu} = 1 - \varrho_1, \\
&\text{c)} \quad \varrho_1 = x + \overline{\mu}, &&\varrho_2 = x + \overline{\mu} - 1 = \varrho_1 - 1.
\end{aligned}$$

Let us substitute these expressions in the equation to be solved. We see that, if we put

a) $\varrho_1 = \varrho$, $\varrho_2 = 1 + \varrho$, one has

$$(1 - \overline{\mu}) \left[\varrho - \frac{1}{\varrho^2} \right] + \overline{\mu} \left[\varrho + 1 - \frac{1}{(\varrho + 1)^2} \right] = 0;$$

b) $\varrho_1 = \varrho$, $\varrho_2 = 1 - \varrho$, one has

$$(1 - \overline{\mu}) \left(\varrho^2 - \frac{1}{\varrho^2} \right) - \overline{\mu} \left[1 - \varrho - \frac{1}{(1 - \varrho)^2} \right] = 0;$$

c) $\varrho_2 = \varrho$, $\varrho_1 = 1 + \varrho$, one has

$$\overline{\mu} \left[1 + \varrho - \frac{1}{(1 + \varrho)^2} \right] + (1 - \overline{\mu}) \left[\varrho - \frac{1}{\varrho^2} \right] = 0.$$

In all three cases the corresponding equation has only one (positive) solution for ϱ. Let us look at cases (a) and (c) represented by the same equation, which can be written in the form

$$F(\varrho) = \frac{\varrho - \frac{1}{\varrho^2}}{1 + \varrho - \frac{1}{(1+\varrho)^2}} = -\frac{\overline{\mu}}{1 - \overline{\mu}} < 0.$$

One can verify that

$$F'(\varrho) > 0, \qquad F(0_+) = -\infty, \qquad F(1) = 0.$$

Then $F = -\overline{\mu}/(1 - \overline{\mu}) < 0$ for only one value of ϱ between 0 and 1. Let us call the two points corresponding to (a) and (c) L_1 and L_2.

In case (b) the equation becomes

$$G(\varrho) = \frac{1 - \varrho - \frac{1}{(1-\varrho)^2}}{\varrho - \frac{1}{\varrho^2}} = \frac{1 - \overline{\mu}}{\overline{\mu}} \geq 1$$

($\overline{\mu} \leq 1/2$ by hypothesis); the function $G(\varrho)$ is increasing in the interval $1/2 \leq \varrho < 1$; moreover,

$$G(1/2) = 1, \qquad G(1_-) = \infty.$$

Then G takes the value $(1 - \overline{\mu})/\overline{\mu}$ only once in the interval $1/2 \leq \varrho < 1$. We call the corresponding position L_3 (which is then closer to m_1 than to m_2). The points L_1, L_2, L_3, L_4, L_5 which we have determined can be given an interpretation that is particularly meaningful from the geometrical point of view. If, in fact, we consider the four-dimensional manifold which constitutes the phase space of our system with two degrees of freedom we see that the Jacobi integral reduces the dimensions of the manifold available for the motion of the system to three. Any one 4-tuple of numbers x, y, $\dot{x}$, $\dot{y}$ which satisfies the Jacobi equation $\dot{x}^2 + \dot{y}^2 = 2\,\Phi(x, y) - C$ represents a possible motion for a given C. The Jacobi equation is of the type $F(x, y, \dot{x}, \dot{y}, C) = 0$. It is then a three-dimensional manifold whose singular points will be given by $\partial F/\partial x = \partial F/\partial y = \partial F/\partial \dot{x} = \partial F/\partial \dot{y} = 0$ (we exclude from these considerations possible singularities of $\Phi(x, y)$ which we shall deal with when we study the regularization of our problem). In our case, the written equations become $\partial \Phi/\partial x = \partial \Phi/\partial y = \dot{x} = \dot{y} = 0$. By substituting these in the equations of motion, one additionally obtains $\ddot{x} = \ddot{y} = 0$. Then the singular points of the three-dimensional manifold of the states of motion are the stationary points, or equilibrium points, which we have already determined.

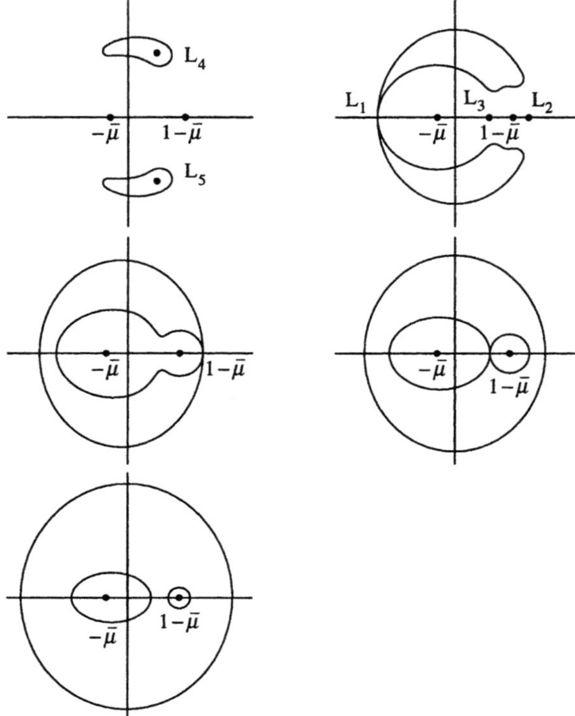

Fig. 4.6

The Hill Curves

We have already defined the zero velocity curves, or Hill curves,[13] as being
the curves belonging to the family $\Phi(x,y) = C/2$, with $C \geq 3$. Taking
into account that Φ is somehow a potential function, the Hill curves are the
equipotential lines. We have already seen that for $C = 3$ the corresponding
curve degenerates in the points L_4 and L_5. Moreover, the symmetry prop-
erty $\Phi(x,y) = \Phi(x,-y)$ is clear, since in ϱ_1 and ϱ_2 y is squared; then the
curves $\Phi = $ const. are symmetrical with respect to the x axis. Let us now
study the behaviour of the curves as the Jacobi constant varies (Fig. 4.6). Increasing C a little above the value 3, one has two closed curves which enclose
the points L_4 and L_5. The part of the plane inside these curves corresponds
to $v^2 = 2\Phi - C < 0$, the part outside to $v^2 > 0$. Therefore these curves
separate the regions in which the motion is possible (real v) from those in
which the motion is not possible (imaginary v). If C increases further on,
the two curves will join at L_1. After a further increase which will cause the

---

[13]G. W. Hill: Researches in the lunar theory, *Am. J. Math.* **1**, 5–26, 129–147, 245–
261, (1878).

forming of a single closed curve containing L_4, L_5, L_1, there will also be a joining at L_2 and finally at L_3, encircling the primaries. A further increase of C makes the joining at L_3 disappear, leaving the two primaries encircled. Figure 4.7[14] (which must be completed symmetrically) together with Fig. 4.6 gives an (approximate) idea of what we have said; the last phase is not represented in the figure. The curves refer to $\overline{\mu}/(1 - \overline{\mu}) = 0.1$.

One can apply what has been explained above to the motion of the Moon. If we neglect the eccentricity of the orbit of the Earth and the inclination of the Moon, one can schematize the Earth–Sun–Moon system as the circular restricted problem we have so far dealt with. Therefore the question is that of calculating the Jacobi constant for the Moon: $C = 2\Phi - v^2$, v being the velocity of the Moon with respect to the rotating coordinate system. One obtains a value $C_{\text{Moon}} > C_3$ for the Moon, C_3 being the value of the Jacobi constant for the curve passing through L_3. Therefore the Moon remains inside the zero velocity oval, which in turn is inside the eight-shaped limiting figure: hence it cannot leave the Earth. For a thorough analysis of the Hill curves we refer the reader to the already mentioned textbook by Szebehely.

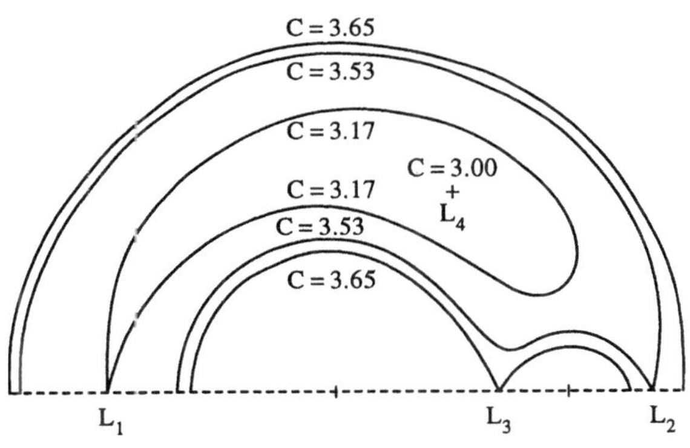

Fig. 4.7

4.4 The Stability of the Equilibrium Solutions

System (4.50) with Jacobi integral (4.51) admits, as we have seen, the equilibrium solutions made up of the five Lagrangian points grouped into the collinear solution and into the equilateral solution. In view of their applications, studying the stability of such solutions is of the utmost importance. From a physical point of view, to say that one of the Lagrangian points is a position of stable equilibrium simply means that a particle which occupies

---

[14]Taken from G. H. Darwin: Periodic Orbits, *Acta math.* **21**, 99–242 (1897)

that position tends to go back to it whenever it is taken away from it by a perturbation. Given the approximations made to schematize the restricted problem, with the consequent impossibility of resorting to the usual conservation laws, the classical Lagrange–Dirichlet theorem, which states that a position of stable equilibrium is an isolated minimum for the potential energy, does not help at all. We shall study the stability of our solutions by applying the definition given in Sect. 1.2 (L-stability). As we shall see, it is not a quick matter; in fact, in the case of the equilateral solution it was only in recent years (the 1960s), two centuries after the formulation of the problem, that an exhaustive and definitive answer has been obtained.

It is convenient, especially for the study of the last case, to make use of the Hamiltonian formalism. If we define

$$q_1 = x, \qquad p_1 = \dot{x} - y,$$
$$q_2 = y, \qquad p_2 = \dot{y} + x, \qquad (4.60)$$

and the Hamiltonian function

$$\mathcal{H} = \frac{1}{2}\left(p_1^2 + p_2^2\right) - \left(q_1\, p_2 - q_2\, p_1\right) + \frac{1}{2}\left(q_1^2 + q_2^2\right) - \varPhi\left(q_1, q_2\right), \qquad (4.61)$$

then system (4.50) is equivalent to the Hamiltonian system:

$$\dot{q}_1 = \frac{\partial \mathcal{H}}{\partial p_1} = p_1 + q_2, \qquad \dot{p}_1 = -\frac{\partial \mathcal{H}}{\partial q_1} = p_2 - q_1 + \frac{\partial \varPhi}{\partial q_1},$$
$$\dot{q}_2 = \frac{\partial \mathcal{H}}{\partial p_2} = p_2 - q_1, \qquad \dot{p}_2 = -\frac{\partial \mathcal{H}}{\partial q_2} = -p_1 - q_2 + \frac{\partial \varPhi}{\partial q_2}. \qquad (4.62)$$

The equilibrium solutions correspond to the critical points of the function $\varPhi\left(q_1, q_2\right) = \varPhi\left(x, y\right)$. If we generically call the coordinates of these points $x = a$, $y = b$, then $q_1 = a$, $q_2 = b$, $p_1 = -b$, $p_2 = a$ will correspond to them. If we perform the translation $q_1 - a = \xi_1$, $q_2 - b = \xi_2$ and, as a consequence, $p_1 + b = \eta_1$, $p_2 - a = \eta_2$, we find that this is a canonical transformation, and that we can then write a new canonical system in the variables ξ_1, ξ_2, η_1, η_2, for which the equilibrium solutions will correspond to $\xi_1 = \xi_2 = \eta_1 = \eta_2 = 0$. This system is

$$\dot{\xi}_1 = \frac{\partial \mathcal{H}}{\partial \eta_1} = \eta_1 + \xi_2, \qquad \dot{\eta}_1 = -\frac{\partial \mathcal{H}}{\partial \xi_1} = \eta_2 - \xi_1 + \frac{\partial \varPhi}{\partial \xi_1},$$
$$\dot{\xi}_2 = \frac{\partial \mathcal{H}}{\partial \eta_2} = \eta_2 - \xi_1, \qquad \dot{\eta}_2 = -\frac{\partial \mathcal{H}}{\partial \xi_2} = -\eta_1 - \xi_2 + \frac{\partial \varPhi}{\partial \xi_2}, \qquad (4.63)$$

with

$$\mathcal{H} = \frac{1}{2}\left(\eta_1^2 + \eta_2^2\right) - \left(\xi_1\, \eta_2 - \xi_2\, \eta_1\right) + \frac{1}{2}\left(\xi_1^2 + \xi_2^2\right) - \varPhi\left(\xi_1, \xi_2\right). \qquad (4.64)$$

To spare symbols, we have maintained both $\mathcal{H}$ and $\varPhi$ for the functions in the new variables.

To study the stability of the solution $\xi_1 = \xi_2 = \eta_1 = \eta_2 = 0$, we must linearize system (4.63); by expanding it in a Taylor series in the neighbourhood of the origin, we get

$$\dot{\xi}_1 = \xi_2 + \eta_1, \quad \dot{\xi}_2 = -\xi_1 + \eta_2,$$

$$\dot{\eta}_1 = -\xi_1 + \left(\frac{\partial^2 \Phi}{\partial \xi_1^2}\right)\bigg|_0 \xi_1 + \left(\frac{\partial^2 \Phi}{\partial \xi_1 \partial \xi_2}\right)\bigg|_0 \xi_2 + \eta_2 + O_2,$$

$$\dot{\eta}_2 = -\xi_2 + \left(\frac{\partial^2 \Phi}{\partial \xi_1 \partial \xi_2}\right)\bigg|_0 \xi_1 + \left(\frac{\partial^2 \Phi}{\partial \xi_2^2}\right)\bigg|_0 \xi_2 - \eta_1 + \overline{O}_2,$$

(4.65)

where O_2 and $\overline{O}_2$ will not contain terms of the first degree in ξ and η. For the Hamiltonian, one has the expansion

$$\mathcal{H} = \frac{1}{2}\left(\eta_1^2 + \eta_2^2\right) - \left(\xi_1\,\eta_2 - \xi_2\,\eta_1\right) + \frac{1}{2}\left(\xi_1^2 + \xi_2^2\right)$$
$$- \frac{1}{2}\left[\left(\frac{\partial \Phi}{\partial \xi_1^2}\right)\bigg|_0 \xi_1^2 + 2\left(\frac{\partial^2 \Phi}{\partial \xi_1 \partial \xi_2}\right)\bigg|_0 \xi_1\xi_2 + \left(\frac{\partial^2 \Phi}{\partial \xi_2^2}\right)\bigg|_0 \xi_2^2\right] + \mathcal{H}_3 + \mathcal{H}_4 - \cdots,$$

(4.66)

where $\mathcal{H}_l$ represents a homogeneous polynomial of degree l in the canonical variables, and the constant term has been omitted, because it will be uninfluential in the equations which we shall use. The linear part of system (4.65) is of type (1.A.19), with a matrix of coefficients

$$\begin{pmatrix} 0 & 1 & 1 & 0 \\ -1 & 0 & 0 & 1 \\ -1 + \Phi_{\xi_1\xi_1} & \Phi_{\xi_1\xi_2} & 0 & 1 \\ \Phi_{\xi_1\xi_2} & -1 + \Phi_{\xi_2\xi_2} & -1 & 0 \end{pmatrix}$$

(4.67)

and the corresponding characteristic equation

$$\lambda^4 + \left(4 - \Phi_{\xi_1\xi_1} - \Phi_{\xi_2\xi_2}\right)\lambda^2 + \left(\Phi_{\xi_1\xi_1}\Phi_{\xi_2\xi_2} - \Phi_{\xi_1\xi_2}^2\right) = 0. \tag{4.68}$$

We shall now calculate the three second derivatives of $\Phi\left(\xi_1, \xi_2\right)$ at the five Lagrangian points. Because

$$\left(\Phi_{\xi_1\xi_1}\right)\big|_0 = \left(\Phi_{xx}\right)\big|_{\substack{x=a \\ y=b}},$$

$$\left(\Phi_{\xi_1\xi_2}\right)\big|_0 = \left(\Phi_{xy}\right)\big|_{\substack{x=a \\ y=b}},$$

$$\left(\Phi_{\xi_2\xi_2}\right)\big|_0 = \left(\Phi_{yy}\right)\big|_{\substack{x=a \\ y=b}},$$

we shall start from the expressions of Φ_x and Φ_y in (4.59) to calculate the second derivatives. Then

$$\Phi_{xx} = 1 - \frac{(1-\overline{\mu})\left[\varrho_1^2 - 3\,(x+\overline{\mu})^2\right]}{\varrho_1^5} - \frac{\overline{\mu}\left[\varrho_2^2 - 3\,(x+\overline{\mu}-1)^2\right]}{\varrho_2^5},$$

$$\Phi_{xy} = \frac{3\,(x+\overline{\mu})\,(1-\overline{\mu})}{\varrho_1^5}\,y + \frac{3\,(x+\overline{\mu}-1)\,\overline{\mu}\cdot y}{\varrho_2^5}, \qquad\qquad (4.69)$$

$$\Phi_{yy} = 1 - \frac{(1-\overline{\mu})\,(\varrho_1^2 - 3\,y^2)}{\varrho_1^5} - \frac{\overline{\mu}\,(\varrho_2^2 - 3\,y^2)}{\varrho_2^5}.$$

Let us first examine the collinear solution. Immediately we see that $\Phi_{xy} = 0$ and

$$\begin{aligned}
\Phi_{xx} &= 1 - \frac{(1-\overline{\mu})\,(\varrho_1^2 - 3\,\varrho_1^2)}{\varrho_1^5} - \frac{\overline{\mu}\,(\varrho_2^2 - 3\,\varrho_2^2)}{\varrho_2^5} \\
&= 1 + \frac{2\,(1-\overline{\mu})}{\varrho_1^3} + \frac{2\,\overline{\mu}}{\varrho_2^3} > 0.
\end{aligned}$$

The evaluation of the sign of Φ_{yy} requires a slightly more complicated calculation. By referring to case (b) of Sect. 4.3 (that is, to the point L_3), we have

$$\Phi_{yy} = 1 - \frac{1-\overline{\mu}}{\varrho_1^3} - \frac{\overline{\mu}}{\varrho_2^3} = 1 - \frac{1-\overline{\mu}}{\varrho_1^2}\frac{1}{\varrho_1} - \frac{\overline{\mu}}{\varrho_2^3}.$$

Since $\Phi_x = 0$,

$$x - \frac{(x+\overline{\mu})\,(1-\overline{\mu})}{\varrho_1^3} - \frac{\overline{\mu}\,(x+\overline{\mu}-1)}{\varrho_2^3} = 0,$$

$$\varrho_1 - \overline{\mu} - \frac{1-\overline{\mu}}{\varrho_1^2} + \frac{\overline{\mu}}{\varrho_2^2} = 0,$$

$$\varrho_1 - \overline{\mu} + \frac{\overline{\mu}}{\varrho_2^2} = \frac{1-\overline{\mu}}{\varrho_1^2}.$$

By substituting this in the expression for Φ_{yy}, we get

$$\begin{aligned}
\Phi_{yy} &= 1 - \left(\varrho_1 - \overline{\mu} + \frac{\overline{\mu}}{\varrho_2^2}\right)\frac{1}{\varrho_1} - \frac{\overline{\mu}}{\varrho_2^3} = \overline{\mu}\left(\frac{1}{\varrho_1} - \frac{1}{\varrho_1\,\varrho_2^2} - \frac{1}{\varrho_2^3}\right) \\
&= \overline{\mu}\,\frac{\varrho_2^3 - \varrho_2 - \varrho_1}{\varrho_1\,\varrho_2^3} = \overline{\mu}\,\frac{\varrho_2^3 - 1}{\varrho_1\,\varrho_2^3} = \frac{\overline{\mu}}{\varrho_1}\left(1 - \frac{1}{\varrho_2^3}\right) < 0, \qquad \varrho_2 < 1.
\end{aligned}$$

By means of analogous methods, one could verify that cases (a) and (c) give the same result. As regards the collinear solution we then have

$$\Phi_{xx} > 0, \qquad \Phi_{yy} < 0, \qquad \Phi_{xy} = 0.$$

If we put $\Lambda = \lambda^2$, the characteristic equation (4.68) in our case will be

$$\Lambda^2 + b\,\Lambda + c = 0,$$

with

$$b = 4 - \Phi_{xx} - \Phi_{yy}, \qquad c = \Phi_{xx}\,\Phi_{yy} - \Phi_{xy}^2 < 0.$$

Then $\lambda = \pm\sqrt{\Lambda}$ and the two real roots for Λ will have opposite signs:

$$\Lambda_1 = \frac{1}{2}\left(-b + \sqrt{b^2 - 4c}\right) > 0, \qquad \Lambda_2 = \frac{1}{2}\left(-b - \sqrt{b^2 - 4c}\right) < 0,$$

since $c < 0$, $\forall\, b/|b|$. Therefore $\lambda_{1,2}$ are real and $\lambda_{3,4}$ pure imaginary. Hence one of the four roots will have a positive real part; as a consequence, based upon what seen in Sect. 1.2, the collinear solution for system (4.50) is unstable for all the three points L_1, L_2, L_3.

In the case of the equilateral solution, by substituting the values $x = 1/2 - \overline{\mu}$, $y = \pm\sqrt{3}/2$ found in Sect. 4.3 into (4.69), we have

$$\Phi_{xx} = \frac{3}{4}, \qquad \Phi_{xy} = \pm\frac{3\sqrt{3}}{4}\left(1 - 2\overline{\mu}\right), \qquad \Phi_{yy} = \frac{9}{4}.$$

By substituting these into (4.68) one has the characteristic equation

$$\lambda^4 + \lambda^2 + \frac{27}{4}\overline{\mu}\left(1 - \overline{\mu}\right) = 0. \tag{4.70}$$

With the previous symbols,

$$\Lambda_{1,2} = \frac{1}{2}\left[-1 \pm \sqrt{1 - 27\,\overline{\mu}\left(1 - \overline{\mu}\right)}\right].$$

One can immediately verify that the only possibility for solutions that are *not unstable* is for

$$\overline{\mu}\left(1 - \overline{\mu}\right) < \frac{1}{27}. \tag{4.71}$$

In this case, one obtains

$$-\frac{1}{2} < \Lambda_1 < 0, \qquad -\frac{1}{2} > \Lambda_2 > -1,$$

and then all the four solutions λ_1, λ_2, λ_3, λ_4 will be purely imaginary. We have seen, however, that, in the case of purely imaginary solutions, the linear system is not sufficient to determine the behaviour of the solution of the non-linear system: we are facing a critical case. As we already anticipated at the beginning of this section, the question has remained unsolved for about two centuries. After a first result due to Littlewood,[15] who demonstrated the stability of the equilateral solution for a very long (depending on $\overline{\mu}$) but not infinite time, the stability for an infinite time has been demonstrated by applying a theorem, due to Arnold, concerning Hamiltonian systems with two degrees of freedom. Let us take a look at the contents of this theorem in our case. Go back at the Hamiltonian (4.66) whose form is

$$\mathcal{H} = \mathcal{H}_2 + \mathcal{H}_3 + \mathcal{H}_4 + \mathcal{H}_5 + \cdots$$

---

[15] J. E. Littlewood: On the equilateral configuration in the restricted problem of three bodies, *Proc. London Math. Soc.* (3) **9**, 343–372, (1959); The Lagrange configuration in celestial mechanics, *ibid.* **9**, 525–543, (1959).

For L_4, by substituting the values of the second derivative, one immediately gets

$$\mathcal{H}_2 = \frac{1}{2}\left(\eta_1^2 + \eta_2^2\right) + \xi_2\,\eta_1 - \xi_1\,\eta_2 + \frac{1}{8}\,\xi_1^2 - \frac{5}{8}\,\xi_2^2 - \frac{3\sqrt{3}}{4}\,(1 - 2\overline{\mu})\,\xi_1\,\xi_2, \quad (4.72)$$

which turns out to be an indefinite form. One can apply a linear canonical transformation[16] with constant coefficients to $\mathcal{H}_2$ such as to put it into the form

$$\mathcal{H}_2 = -i\,\omega_1\,Q_1\,P_1 + i\,\omega_2\,Q_2\,P_2,$$

with

$$\omega_1 = \sqrt{-\Lambda_1},\ \omega_2 = \sqrt{-\Lambda_2},$$

which is a *Birkhoff normal form*; since in our case the roots of the characteristic equation are simple and purely imaginary,

$$P_k = i\,\overline{Q}_k \quad (k = 1, 2),$$

and then, putting

$$Q_k = \alpha_k + i\,\beta_k,\qquad P_k = \beta_k + i\,\alpha_k,$$

we can write the second-degree term $\mathcal{H}_2$ as a sum of squares. Furthermore, still using a canonical transformation, one can find new variables, which we shall keep calling Q_k, P_k, such as to transform the subsequent terms to get

$$\mathcal{H} = -i\omega_1 U_1 + i\omega_2 U_2 + \frac{1}{2}\left(\mu_{11}U_1^2 + 2\mu_{12}U_1 U_2 + \mu_{22}U_2^2\right) + \mathcal{H}_5 + \cdots,$$

where $U_k = Q_k P_k$ $(k = 1, 2)$ and we know that $\omega_1/\omega_2 \neq m/n$ $(m, n = 1, 2, 3, 4)$ and that the coefficients μ_{ij} are real. Under the further condition that

$$\mathcal{H}_4 = \frac{1}{2}\left(\mu_{11}\,U_1^2 + 2\mu_{12}\,U_1 U_2 + \mu_{22}\,U_2^2\right)$$

is not divisible by $\mathcal{H}_2 = -i\,\omega_1\,U_1 + i\,\omega_2\,U_2$, it can be demonstrated that the origin is a stable solution for the corresponding Hamiltonian system. This is the conclusion of Arnold's theorem.[17] The application to the case of the equilateral solution is due to A. Leontovich (1962), A. Deprit and A. Deprit-Bartolomé (1967) and A. Markeev (1969).

---

[16]For a demonstration, see Siegel and Moser: op. cit., Sect. 15, or G. D. Birkhoff: *Dynamical Systems*, rev. edn. (Am. Math. Society, 1966), Chap. III. The actual determination of the transformation requires complicated calculations, see also K. R. Meyer, G. R. Hall: *Introduction to the Hamiltonian dynamical systems and the N-body problem*, (Springer, 1989), Chaps. II.G, VII.D.

[17]A demonstration of Arnold's theorem can be found in Siegel and Moser: op. cit., Sect. 35.

Our final conclusions[18] are as follows

The Lagrangian points of the equilateral solution are stable for all the ratios of masses which satisfy (4.71), with the exception of the two ratios

$$\overline{\mu} = \frac{15 - \sqrt{213}}{30} = 0.0135160\ldots \qquad \overline{\mu} = \frac{45 - \sqrt{1833}}{90} = 0.0242938\ldots$$

In the usual units

$$\overline{\mu} = \frac{m_2}{m_1 + m_2}, \qquad m_2 < m_1;$$

therefore, one can see that in the Moon–Earth case, where

$$\frac{m_{\mathrm{Moon}}}{m_{\mathrm{Moon}} + m_{\oplus}} = 0.01241\ldots,$$

we are a little below the smallest of the two forbidden values.

Inequality (4.71), resolved by taking into account that $\overline{\mu} < 1/2$, on the other hand yields

$$\overline{\mu} < \frac{1}{2} - \sqrt{\frac{23}{108}} = 0.0385\ldots$$

(this number has been named the *Routh value*), from which one can deduce that the upper limit of the ratio m_2/m_1 is about $1/25$: any planet (Jupiter included) of the solar system may then constitute together with the Sun a pair of primaries for an equilateral solution.

The discovery, in 1906, of an asteroid oscillating around the position which we have called L_5 in the Sun–Jupiter system was the first confirmation that the Lagrangian points were not simply a mathematical curiosity but, on the contrary, had applications even within the solar system. After this first asteroid (which was named Achilles), more than twelve other asteroids were discovered oscillating around L_4 and L_5 in the Sun–Jupiter system. The asteroids oscillating around L_5 were named after Greek heroes in the Trojan war, while those oscillating around L_4 were named after Trojan heroes; altogether the entire group is called *The Trojans*. Those oscillating around L_4 follow Jupiter in its motion, while those oscillating around L_5 precede it.

Another application of the equilateral solution was suggested in 1961 by the Polish astronomer Kordelewski for the Earth–Moon system: he discovered clusters of meteorites in the proximity of the triangular libration points. They can be observed, under the best conditions, as a faint luminosity. Eventually the *Voyager I* mission showed another example of the application of the equilateral solution to the Saturn–satellites of Saturn system: a satellite was discovered (called Dione B) which oscillated at the same distance from Saturn as the satellite Dione (now renamed Dione A). As far as the collinear solution is concerned, some time ago,[19] it was suggested that it could explain the

---

[18]See K. R. Meyer, G. R. Hall: op. cit. in Footnote 16.

[19]H. Gyldén: Sur un cas particulier du problème des trois corps, *Bull. Astron.* **1**, 361–369 (1884); F. R. Moulton: A meteoric theory of the "Gegenschein". *Astron. J.*, **21**, 17–22 (1900).

Gegenschein phenomenon, a faint luminosity which is observed after sunset in the plane of the ecliptic in a direction opposite to that of the Sun. This luminosity is ascribed to the illumination by the Sun of meteoric particles that have accumulated around point L_2. This is an interpretation which was subject to a number of criticisms, some of them well grounded,[20] and therefore not universally accepted.

Applications of the collinear solution can also occur in the domain of stellar dynamics.

4.5 The Delaunay Elements
for the Restricted Three-Body Problem

In Sect. 2.5, we introduced the Delaunay elements for the two-body problem; there, we were dealing with a separable periodic system and the Delaunay elements $(\ell, g, h; L, G, H)$ were the action–angle variables of the problem. However, with the restricted three-body problem we are no longer dealing with an integrable system; we can therefore regard the restricted problem as a problem generated by adding a second primary (of mass $\bar{\mu}$ in the dimensionless system) to the two-body problem, which consists of a primary of unit mass and a third body of negligible mass. In this way, the second primary turns out to be a "perturbation" of an integrable system. The elements, which before were constant, now vary with time and the differential equations which rule these variations can be written immediately, since they are just the canonical equations; for instance, the equations for the pair of conjugate variables ℓ, L will be

$$\frac{d\ell}{dt} = \frac{\partial \mathcal{H}}{\partial L}, \qquad \frac{dL}{dt} = -\frac{\partial \mathcal{H}}{\partial \ell},$$

with $\mathcal{H}$ given as a function of the constant elements.

We thus shall introduce the Delaunay elements, always bearing in mind the rule that in the limit for $\bar{\mu} \to 0$ we must recover what was already been established for the two-body problem. Furthermore, it must be emphasized that, since the restricted problem is by definition a planar problem, one is still confined to the elements (ℓ, L, g, G).

The Delaunay Elements in the Rotating System

Recall Hamiltonian (4.61), which can be rewritten as

$$\mathcal{H} = \frac{1}{2}\left(p_1^2 + p_2^2\right) - \left(q_1\,p_2 - q_2\,p_1\right) - U\left(q_1, q_2\right), \qquad (4.73)$$

---

[20]Among others, N. Moisseiev: Sur l'hypothèse de Gyldén–Moulton de l'origine de "Gegenschein". Partie 5 et Partie 6, *Astron. J. Sov. Union* **15**, 217–225, 226–231 (1938).

by omitting the constant term $-\overline{\mu}\left(1-\overline{\mu}\right)/2$, introduced for the sake of convenience in the definition of the function Φ and, in any case, irrelevant to subsequent developments. So far, we have worked with rectangular coordinates (such as q_1 and q_2); we shall now use polar coordinates (while still remaining in the synodic system). The relevant canonical transformation (see also Sect. 5.2) will give in our case ($\vartheta = \pi/2$)

$$q_1 = r\cos\varphi, \qquad q_2 = r\sin\varphi, \tag{4.74}$$

$$p_r = p_1\cos\varphi + p_2\sin\varphi, \qquad p_\varphi = -r\,p_1\sin\varphi + r\,p_2\cos\varphi. \tag{4.75}$$

From now onwards we will denote the canonical coordinates $Q_1 = r$, $Q_2 = \varphi$, $P_1 = p_r$, $P_2 = p_\varphi$. The new Hamiltonian will be therefore:

$$\mathcal{K} = \mathcal{H} = \mathcal{H}\left(r,\varphi,p_r,p_\varphi\right) = \frac{1}{2}\left(p_r^2 + \frac{1}{r^2}p_\varphi^2\right) - p_\varphi - U\left(r,\varphi\right), \tag{4.76}$$

with

$$U\left(r,\varphi\right) = \frac{1-\overline{\mu}}{\varrho_1} + \frac{\overline{\mu}}{\varrho_2},$$

$$\varrho_1 = \sqrt{\left(r\cos\varphi + \overline{\mu}\right)^2 + \left(r\sin\varphi\right)^2},$$

$$\varrho_2 = \sqrt{\left(r\cos\varphi + \overline{\mu} - 1\right)^2 + \left(r\sin\varphi\right)^2}.$$

The equations of motion are consequently

$$\begin{aligned}
\dot{Q}_1 &= \frac{\partial\mathcal{H}}{\partial P_1} &&\Leftrightarrow& \dot{r} &= p_r, \\[2mm]
\dot{Q}_2 &= \frac{\partial\mathcal{H}}{\partial P_2} &&\Leftrightarrow& \dot{\varphi} &= \frac{1}{r^2}p_\varphi - 1, \\[2mm]
\dot{P}_1 &= -\frac{\partial\mathcal{H}}{\partial Q_1} &&\Leftrightarrow& \dot{p}_r &= \frac{p_\varphi^2}{r^3} + \frac{\partial U}{\partial r}, \\[2mm]
\dot{P}_2 &= -\frac{\partial\mathcal{H}}{\partial Q_2} &&\Leftrightarrow& \dot{p}_\varphi &= \frac{\partial U}{\partial\varphi}.
\end{aligned} \tag{4.77}$$

From the second of equations (4.77),[21] one obtains

$$p_\varphi = r^2\left(\dot{\varphi} + 1\right) = r^2\frac{d}{dt}\left(\varphi + t\right).$$

The angle $\overline{\varphi} = \varphi + t$ is the angle with respect to the inertial system; hence $p_\varphi = r^2\dot{\overline{\varphi}}$. It now only remains to perform the last canonical transformation to obtain the Delaunay elements. Let us call the new canonical variables q_i', p_i' and let $W_2 = W_2(Q_i, p_i')$ be the generating function. Then

$$P_i = \frac{\partial W_2}{\partial Q_i}, \qquad q_i' = \frac{\partial W_2}{\partial p_i'}. \tag{4.78}$$

---

[21]If we were not in the dimensionless system, then we would have $r^2\frac{d}{dt}\left(\varphi + \omega t\right)$.

Keeping in mind our purpose, we impose the condition

$$p_\varphi = P_2 = p_2'. \tag{4.79}$$

From (4.78) and (4.79), one obtains $P_2 = p_2' = \partial W_2/\partial\varphi$, and integration thus yields $W_2 = p_2'\varphi + f(Q_1, p_1', p_2')$, with f an arbitrary function. Moreover, again from (4.78)

$$P_1 = p_r = \dot{r} = \frac{\partial W_2}{\partial r} = \frac{\partial f}{\partial Q_1}. \tag{4.80}$$

In the particular case $\bar{\mu} = 0$ (two-body problem), the energy integral yields $v^2 = 2/r - 1/a$. In the synodic system,

$$v^2 = \dot{r}^2 + \left(r\dot{\varphi}\right)^2, \qquad r^2\dot{\varphi} = p_\varphi = c = -\sqrt{a\left(1 - e^2\right)},$$

so that $v^2 = \dot{r}^2 + a\left(1 - e^2\right)/r^2$ and therefore

$$\dot{r} = \sqrt{-\frac{a\left(1 - e^2\right)}{r^2} + \frac{2}{r} - \frac{1}{a}}.$$

We shall rewrite this last relation, in general (with a and e variable), as

$$P_1 = \sqrt{-\frac{p_2'^2}{Q_1^2} + \frac{2}{Q_1} - \frac{1}{p_1'^2}}. \tag{4.81}$$

In (4.81) use has been made of (4.79); moreover we have put $p_1' = \sqrt{a}$. From (4.80) and (4.81),

$$\frac{\partial f}{\partial Q_1} = \sqrt{-\frac{p_2'^2}{Q_1^2} + \frac{2}{Q_1} - \frac{1}{p_1'^2}}.$$

By integrating this we get

$$f = \int \sqrt{-\frac{p_2'^2}{Q_1^2} + \frac{2}{Q_1} - \frac{1}{p_1'^2}}\, dQ_1 + g\left(p_1', p_2'\right),$$

where g is a new arbitrary function. Now

$$W_2(Q_1, Q_2, p_1', p_2') = p_2'Q_2 + \int \sqrt{-\frac{p_2'^2}{Q_1^2} + \frac{2}{Q_1} - \frac{1}{p_1'^2}}\, dQ_1 + g\left(p_1', p_2'\right). \tag{4.82}$$

So far we have made use of the first two equations of (4.78). In fact we have conjugate momenta $P_2 = p_2' = p_\varphi$ and P_1 given by (4.81). We must now utilize the other two equations, that is, determine q_1', q_2'. If we call the indefinite integral in (4.82) I, we can also write

$$I = \int_z^{Q_1} \left[-\frac{p_2'^2}{\xi^2} + \frac{2}{\xi} - \frac{1}{p_1'^2}\right]^{1/2} d\xi,$$

for arbitrary z. The variables still to be determined are

$$q_1' = \frac{\partial W_2}{\partial p_1'} = \frac{\partial I}{\partial p_1'} + \frac{\partial g}{\partial p_1'}, \qquad q_2' = \frac{\partial W_2}{\partial p_2'} = Q_2 + \frac{\partial I}{\partial p_2'} + \frac{\partial g}{\partial p_2'}.$$

We must then find the function $g\,(p_1', p_2')$; it is easier, and equivalent, to give the lower limit z of the integral I as a function of p_1' and p_2' and to put $g\,(p_1', p_2') \equiv 0$. What remains then is $W_2 = p_2' Q_2 + I$, with the lower limit of I to be determined. One takes

$$z = p_1' \left(p_1' - \sqrt{{p_1'}^2 - {p_2'}^2} \right),$$

which, since $p_1' = \sqrt{a}$ and ${p_2'}^2 = a(1-e^2)$, means that $z = a(1-e)$ (distance of the pericentre). Evaluating the integral and then differentiating gives

$$q_1' = \arccos\left\{ \left(1 - \frac{Q_1}{{p_1'}^2}\right)\left[1 - \left(\frac{p_2'}{p_1'}\right)^2\right]^{-1/2}\right\} - \frac{Q_1}{p_1'}\sqrt{\frac{2}{Q_1} - \frac{{p_2'}^2}{Q_1^2} - \frac{1}{{p_1'}^2}}$$

$$\tag{4.83}$$

$$q_2' = Q_2 - \arccos\left\{ \left(\frac{{p_2'}^2}{Q_1} - 1\right)\left[1 - \left(\frac{p_2'}{p_1'}\right)^2\right]^{-1/2}\right\}.$$

We now have to interpret the new variables. If in the first of equations (4.83) we substitute the explicit forms of p_1', p_2' and $\dot r$ we have

$$q_1' = \arccos\left(\frac{a - r}{ae}\right) - \frac{r\,\dot r}{\sqrt{a}}.$$

On the other hand, when a and e are constant (two-body problem),[22]

$$\frac{r\,\dot r}{\sqrt{a}} = e \sin u \tag{4.84}$$

is valid; therefore we can write the above expression for the coordinate q_1' in the form $r = a\,[1 - e\,\cos(q_1' + e\,\sin u)]$. By substituting (2.50), we finally obtain $u = q_1' + e \sin u$, that is $q_1' = u - e \sin u$, which is nothing other than Kepler's equation for the mean anomaly. Hence

$$q_1' = \ell. \tag{4.85}$$

From the second of equations (4.83), with suitable substitutions, we obtain

$$q_2' = \varphi - \arccos\left[\frac{a\,(1 - e^2)}{re} - \frac{1}{e}\right],$$

---

[22]Differentiating Kepler's equation yields $\dot \ell = n = \dot u\,(1 - e\,\cos u) = \dot u\,r/a$. As the mean motion is $n = a^{-3/2}$, we get $\dot u = 1/r\,\sqrt{a}$; differentiating then the relation $r = a\,(1 - e\,\cos u)$, so that $\dot r = a\,e\,\dot u\,\sin u$, gives (4.84).

from which

$$r = \frac{a\,(1 - e^2)}{1 + e\,\cos\,(\varphi - q_2')}.$$

By comparing this with the equation of the Keplerian orbit (2.21), one has

$$q_2' = \varphi - f. \tag{4.86}$$

The second of the new canonical variables is then the argument of the pericentre with respect to the synodic system measured from the q_1 axis (see Fig. 4.8).

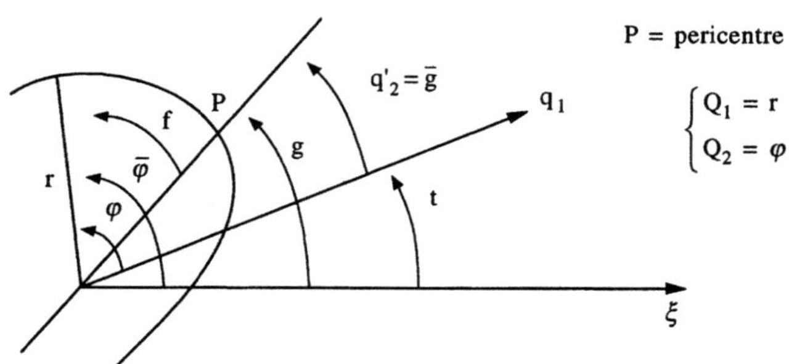

Fig. 4.8

To summarize, the generating function of the canonical transformation from polar coordinates to the Delaunay variables is given by

$$W_2 = Q_2 p_2' + \int_{p_1'\,[p_1' - (p_1'^2 - p_2'^2)^{1/2}]}^{Q_1} \left[-\frac{p_2'^2}{\xi^2} + \frac{2}{\xi} - \frac{1}{p_1'^2} \right]^{1/2} d\,\xi.$$

The transformation is from

$$\begin{cases} Q_1 = r, & P_1 = \dot{r}, \\ Q_2 = \varphi, & P_2 = r^2 \dfrac{d}{dt}\,(\varphi + t), \end{cases}$$

to

$$\begin{cases} q_1' = \ell, & p_1' = \sqrt{a}, \\ q_2' = \varphi - f, & p_2' = \sqrt{a\,(1 - e^2)}. \end{cases}$$

Conventionally, the following symbols are used:

$$\begin{aligned} q_1' &= \ell, & p_1' &= L, \\ q_2' &= \bar{g}, & p_2' &= G. \end{aligned} \tag{4.87}$$

We now have to calculate the new Hamiltonian and to write the canonical equations. By substituting (4.81) into (4.76), one has, upon calling the Hamiltonian in the new variables $\widetilde{\mathcal{H}}$,

$$\widetilde{\mathcal{H}} = \frac{1}{2}\left[-\frac{p_2'^2}{Q_1^2} + \frac{2}{Q_1} - \frac{1}{p_1'^2} + \frac{p_2'^2}{Q_1^2}\right] - p_2' - \widetilde{U} = \frac{1}{2}\left[\frac{2}{Q_1} - \frac{1}{p_1'^2}\right] - p_2' - \widetilde{U}.$$

Thus

$$\widetilde{U} = \frac{1-\overline{\mu}}{\varrho_1} + \frac{\overline{\mu}}{\varrho_2} = \frac{1}{\varrho_1} + \overline{\mu}\left(\frac{1}{\varrho_2} - \frac{1}{\varrho_1}\right)$$

when $\overline{\mu} = 0$, $\widetilde{U} = 1/\varrho_1 = 1/r = 1/Q_1$, so that

$$\widetilde{\mathcal{H}} = -\frac{1}{2}\frac{1}{p_1'^2} - p_2'.$$

For $\overline{\mu} \neq 0$, we write

$$\widetilde{U} = \frac{1}{Q_1} + R,$$

where R, the difference between $1/r$ and the actual potential $\widetilde{U}$, is the perturbing function of classical planetary theory. The Hamiltonian for the restricted three-body problem can then be written

$$\widetilde{\mathcal{H}} = -\frac{1}{2}\frac{1}{p_1'^2} - p_2' - R = -\frac{1}{2L^2} - G - R, \qquad (4.88)$$

and the canonical equations are

$$\dot{q}_1' = \frac{\partial\widetilde{\mathcal{H}}}{\partial p_1'} \Rightarrow \frac{d\ell}{dt} = \frac{1}{L^3} - \frac{\partial R}{\partial L}, \qquad \dot{p}_1' = -\frac{\partial\widetilde{\mathcal{H}}}{\partial q_1'} \Rightarrow \frac{dL}{dt} = \frac{\partial R}{\partial \ell},$$

$$\dot{q}_2' = \frac{\partial\widetilde{\mathcal{H}}}{\partial p_2'} \Rightarrow \frac{d\overline{g}}{dt} = -1 - \frac{\partial R}{\partial G}, \qquad \dot{p}_2' = -\frac{\partial\widetilde{\mathcal{H}}}{\partial q_2'} \Rightarrow \frac{dG}{dt} = \frac{\partial R}{\partial \overline{g}}. \qquad (4.89)$$

It is easy to check that for $R = 0$ (two-body problem) the first of equations (4.89) will again give us (2.84a) (that is, Kepler's third law), whilst the third and fourth show that a and e remain constant; from the second of these equations, we see that $\dot{\overline{g}} = -1$, that is, $\overline{g} = -t + \text{const.}$, and then $\varphi - f = -t + \text{const.}$ This means that the apse line rotates in the (rotating) synodic system with unit angular velocity with retrograde motion, and it is then fixed with respect to the inertial system. It is therefore verified that (4.89), in the limit for $R \to 0$, again give the equations of Keplerian motion.

The Delaunay Elements in the Inertial System

Let us now determine the Delaunay elements for the inertial system. We shall call the canonical variables in this system $\overline{Q}_i$, $\overline{P}_i$ and the new Hamiltonian $\overline{\mathcal{H}}$; because of what we said above regarding the limit of (4.89) for $R \to 0$, it must be the case that, while $q_2' = \overline{g}$ goes into $\overline{Q}_2 = g$ (with $g = \overline{g} + t$), the other elements are left invariant. Then the canonical transformation to be performed will be different from the identical transformation only as far as q_2' is concerned. In general,

$$\overline{W}_2 = \overline{W}_2\left(q_1', q_2'; \overline{P}_1, \overline{P}_2, t\right),$$

$$\overline{\mathcal{H}} = \widetilde{\mathcal{H}} + \frac{\partial \overline{W}_2}{\partial t}, \qquad p_i' = \frac{\partial \overline{W}_2}{\partial q_i'}, \qquad \overline{Q}_i = \frac{\partial \overline{W}_2}{\partial \overline{P}_i}. \qquad (4.90)$$

Taking into account that $\overline{Q}_2 = g = \overline{g} + t = q_2' + t$, by integrating the last of equations (4.90) for $i = 2$ one obtains

$$\overline{W}_2 = (q_2' + t)\,\overline{P}_2 + \psi\left(q_1', q_2', \overline{P}_1, t\right),$$

with ψ an arbitrary function. Similarly, from $\overline{Q}_1 = \ell = q_1'$, one obtains

$$\frac{\partial \overline{W}_2}{\partial \overline{P}_1} = q_1' = \frac{\partial \psi}{\partial \overline{P}_1}.$$

Integration then yields

$$\psi\left(q_1', q_2'; \overline{P}_1, t\right) = q_1'\,\overline{P}_1 + \chi\left(q_1', q_2', t\right),$$

with χ a new arbitrary function. In summary,

$$\overline{W}_2 = (q_2' + t)\,\overline{P}_2 + q_1'\,\overline{P}_1 + \chi\left(q_1', q_2', t\right).$$

Since also

$$p_1' = \overline{P}_1 = \frac{\partial \overline{W}_2}{\partial q_1'} = \overline{P}_1 + \frac{\partial \chi}{\partial q_1'},$$

$$p_2' = \overline{P}_2 = \frac{\partial \overline{W}_2}{\partial q_2'} = \overline{P}_2 + \frac{\partial \chi}{\partial q_2'},$$

one sees immediately that χ can only be a function of t; let us call this function $\gamma(t)$. Hence

$$\overline{W}_2 = (q_2' + t)\,\overline{P}_2 + q_1'\,\overline{P}_1 + \gamma(t). \qquad (4.91a)$$

Furthermore, the Hamiltonian $\overline{\mathcal{H}}$ must not be time dependent. From (4.90) and (4.91a), we obtain

$$\overline{\mathcal{H}} = \widetilde{\mathcal{H}} + \overline{P}_2 + \frac{d\gamma(t)}{dt};$$

the most obvious choice is then $\gamma = \text{const.}$, since the other possible choice $\gamma = \text{const.} \times t$ would give an additive and not essential constant in the Hamiltonian. Thus

$$\overline{W}_2 = (q_2' + t)\,\overline{P}_2 + q_1'\,\overline{P}_1 + a, \qquad (4.91b)$$

with a an arbitrary constant (an identical transformation would in this case have been $\overline{W}_2 = q_1'\,\overline{P}_1 + q_2'\,\overline{P}_2$). Using (4.88) the Hamiltonian is

$$\overline{\mathcal{H}} = \widetilde{\mathcal{H}} + \overline{P}_2 = -\frac{1}{2\,\overline{P}_1^2} - R = -\frac{1}{2\,L^2} - R. \qquad (4.92)$$

For $R \to 0$, (4 92) gives, as it should, the very same $\mathcal{H}$ as in Sect. 2.5 (without the factor μ^2 because of the particular system of units).

Let us add at this point the consideration that, besides the elements (ℓ, g, L, G), various combinations of the Delaunay variables are employed according to the case considered, including the exchange of the *coordinates* with the *conjugate momenta*. We list some of the combinations more frequently used:

1) $q_1 = L, \; q_2 = G\,;$ $p_1 = \ell, \; p_2 = g \; \mathcal{H} = \overline{\mathcal{H}}\,;$
2) $q_1 = L, \; q_2 = G - L\,;$ $p_1 = \ell + g, \; p_2 = g\,;$
3) $q_1 = L, \; q_2 = L - G\,;$ $p_1 = \ell + g, \; p_2 = -g\,;$
4) $q_1 = L - G, \; q_2 = G\,;$ $p_1 = \ell, \; p_2 = \ell + g.$

By applying to set (4) the transformation having as a generating function

$$W_4 = W_4\,(p_i,\, P_i) = \frac{P_1^2}{2}\,\cot p_1 + \frac{P_2^2}{2}\,\cot p_2$$

one obtains the Poincaré canonical variables:

$$Q_1 = \sqrt{2\,(L - G)}\,\cos\ell, \qquad P_1 = \sqrt{2\,(L - G)}\,\sin\ell,$$
$$Q_2 = \sqrt{2\,G}\,\cos(\ell + g), \qquad P_2 = \sqrt{2\,G}\,\sin(\ell + g).$$

By applying instead to set (3) the transformation having the generating function

$$W_3 = W_3\,(Q_i,\, p_i) = -Q_1 p_1 - \frac{1}{2}\,Q_2^2\,\tan p_2,$$

one obtains the following variables (half Delaunay and half Poincaré):

$$Q_1 = L, \qquad\qquad P_1 = \ell + g,$$
$$Q_2 = \sqrt{2\,(L - G)}\,\cos g, \qquad P_2 = -\sqrt{2\,(L - G)}\,\sin g.$$

4.6 The Regularization
of the Restricted Three-Body Problem

We now deal with the regularization of the system of order four relevant to the restricted problem. We shall formulate the problem in a general manner, by extending to non-natural systems the considerations of Sect. 2.6. In this way, even if at the end we shall carry out the regularization by means of the Levi-Cìvita transformation, the method can in any case be applied to any other regularizing transformation.

Without choosing dimensionless units, we consider, to maintain full generality, (4.48), which we rewrite as follows

$$\ddot{q}_1 - 2\omega\,\dot{q}_2 - \omega^2\,q_1 = \frac{\partial U}{\partial q_1},$$
$$\ddot{q}_2 + 2\omega\,\dot{q}_1 - \omega^2\,q_2 = \frac{\partial U}{\partial q_2}, \tag{4.93}$$

having put $x = q_1$, $y = q_2$. In Lagrangian form (4.93) become

$$\frac{d}{dt}\frac{\partial \mathcal{L}}{\partial \dot{q}_k} - \frac{\partial \mathcal{L}}{\partial q_k} = 0, \qquad k = 1, 2, \tag{4.94}$$

with

$$\mathcal{L} = \mathcal{T} + U = \frac{1}{2}\left(\dot{q}_1^2 + \dot{q}_2^2\right) + \frac{1}{2}\omega^2\left(q_1^2 + q_2^2\right) + \omega\left(q_1\dot{q}_2 - q_2\dot{q}_1\right) + U, \tag{4.95}$$

from which it appears that $\mathcal{L}$ contains the terms we called $\mathcal{T}_0(q)$ and $\mathcal{T}_1(q, \dot{q})$ in Sect. 1.5. From (4.95), one obviously obtains the momenta $p_1 = \dot{q}_1 - \omega q_2$ and $p_2 = \dot{q}_2 + \omega q_1$ and the Hamiltonian

$$\mathcal{H} = \frac{1}{2}\left(p_1^2 + p_2^2\right) - \omega\left(q_1 p_2 - q_2 p_1\right) - U, \tag{4.96}$$

which, if we put $\omega = 1$, coincides with (4.73). Let us now consider, as in Sect. 2.6, the complex variable $z = q_1 + i\,q_2$ and the transformation $z = f(w)$ with $w = Q_1 + i\,Q_2$. If we impose the validity of

$$\frac{\partial q_1}{\partial Q_1} = \frac{\partial q_2}{\partial Q_2}, \qquad \frac{\partial q_1}{\partial Q_2} = -\frac{\partial q_2}{\partial Q_1}, \tag{4.97}$$

then $z = f(w)$ will be a conformal transformation for every point where $f'(w) \neq 0$. The Jacobian of the transformation will be given by

$$J = \frac{\partial(q_1, q_2)}{\partial(Q_1, Q_2)} = \left(\frac{\partial q_1}{\partial Q_1}\right)^2 + \left(\frac{\partial q_1}{\partial Q_2}\right)^2 = \left(\frac{\partial q_2}{\partial Q_1}\right)^2 + \left(\frac{\partial q_2}{\partial Q_2}\right)^2 = \left|\frac{df}{dw}\right|^2. \tag{4.98}$$

By means of (4.97) and

$$\frac{d}{dt} = \dot{Q}_1\frac{\partial}{\partial Q_1} + \dot{Q}_2\frac{\partial}{\partial Q_2},$$

the expression for the kinetic energy becomes

$$\mathcal{T} = \frac{1}{2}\left(\dot{Q}_1^2 + \dot{Q}_2^2\right) J + \omega\,\dot{Q}_1\left(q_1\frac{\partial q_2}{\partial Q_1} - q_2\frac{\partial q_1}{\partial Q_1}\right)$$
$$+ \omega\,\dot{Q}_2\left(q_1\frac{\partial q_2}{\partial Q_2} - q_2\frac{\partial q_1}{\partial Q_2}\right) + \frac{1}{2}\omega^2\left(q_1^2 + q_2^2\right). \tag{4.99}$$

Furthermore,

$$\frac{\partial \mathcal{T}}{\partial \dot{Q}_1} = J\,\dot{Q}_1 + \omega\left(q_1\frac{\partial q_2}{\partial Q_1} - q_2\frac{\partial q_1}{\partial Q_1}\right),$$

$$\frac{d}{dt}\frac{\partial T}{\partial \dot{Q}_1} = \frac{d}{dt}(J\,\dot{Q}_1) - \omega\,J\,\dot{Q}_2$$

$$+ \omega\left[q_1\left(\frac{\partial^2 q_2}{\partial Q_1^2}\dot{Q}_1 + \frac{\partial^2 q_2}{\partial Q_1 \partial Q_2}\dot{Q}_2\right) - q_2\left(\frac{\partial^2 q_1}{\partial Q_1^2}\dot{Q}_1 + \frac{\partial^2 q_1}{\partial Q_1 \partial Q_2}\dot{Q}_2\right)\right],$$

$$\frac{\partial T}{\partial Q_1} = \frac{1}{2}\left(\dot{Q}_1^2 + \dot{Q}_2^2\right)\frac{\partial J}{\partial Q_1} + \omega\,\dot{Q}_1\left(q_1\frac{\partial^2 q_2}{\partial Q_1^2} - q_2\frac{\partial^2 q_1}{\partial Q_1^2}\right) + \omega\,J\,\dot{Q}_2$$

$$+ \omega\,\dot{Q}_2\left(q_1\frac{\partial^2 q_2}{\partial Q_1 \partial Q_2} - q_2\frac{\partial^2 q_1}{\partial Q_1 \partial Q_2}\right) + \frac{\partial}{\partial Q_1}\left[\frac{1}{2}\omega^2(q_1^2 + q_2^2)\right].$$

Analogously, one can obtain expressions for

$$\frac{\partial T}{\partial \dot{Q}_2}, \qquad \frac{d}{dt}\frac{\partial T}{\partial \dot{Q}_2}, \qquad \frac{\partial T}{\partial Q_2},$$

which we do not state for the sake of brevity.

At this point, we can rewrite Lagrange's equations in the new variables; they will be

$$\frac{d}{dt}\frac{\partial \mathcal{L}}{\partial \dot{Q}_k} - \frac{\partial \mathcal{L}}{\partial Q_k} = 0, \qquad k = 1, 2, \tag{4.100}$$

where $\mathcal{L} = T + U$ is written using for T expression (4.99) and for U the expression obtained by substituting for $q_1 = q_1\,(Q_1, Q_2)$ and $q_2 = q_2\,(Q_1, Q_2)$. By exploiting the derivatives evaluated as above, we get

$$\frac{d}{dt}\frac{\partial T}{\partial \dot{Q}_1} - \frac{\partial T}{\partial Q_1} = \frac{\partial U}{\partial Q_1},$$

from which

$$\frac{d}{dt}(J\,\dot{Q}_1) - 2\,\omega\,J\,\dot{Q}_2 - \frac{\partial}{\partial Q_1}\left[\frac{1}{2}\omega^2(q_1^2 + q_2^2)\right] - \frac{1}{2}\left(\dot{Q}_1^2 + \dot{Q}_2^2\right)\frac{\partial J}{\partial Q_1} = \frac{\partial U}{\partial Q_1}$$

and also

$$\frac{d}{dt}(J\,\dot{Q}_1) - 2\,\omega\,J\,\dot{Q}_2 = \frac{1}{2}\left(\dot{Q}_1^2 + \dot{Q}_2^2\right)\frac{\partial J}{\partial Q_1} + \frac{\partial}{\partial Q_1}\left[\frac{1}{2}\,\omega^2\,(q_1^2 + q_2^2) + U\right],$$

and lastly, h being constant,

$$\frac{d}{dt}(J\,\dot{Q}_1) - 2\,\omega\,J\,\dot{Q}_2 = \frac{1}{2}\left(\dot{Q}_1^2 + \dot{Q}_2^2\right)\frac{\partial J}{\partial Q_1} + \frac{\partial}{\partial Q_1}\left[\frac{1}{2}\,\omega^2\,(q_1^2 + q_2^2) + U + h\right].$$

Since, from

$$\mathcal{H} = \frac{1}{2}\left(\dot{q}_1^2 + \dot{q}_2^2\right) - \frac{1}{2}\omega^2\left(q_1^2 + q_2^2\right) - U = h,$$

one also has

$$\frac{1}{2}\omega^2\left(q_1^2 + q_2^2\right) + U + h = \frac{1}{2}\left(\dot{q}_1^2 + \dot{q}_2^2\right) = \frac{1}{2}\,J\left(\dot{Q}_1^2 + \dot{Q}_2^2\right).$$

we can finally write

$$\frac{d}{dt}(J\dot{Q}_1) - 2\,\omega\,J\dot{Q}_2 = \frac{1}{J}\frac{\partial}{\partial Q_1}\left\{J\left[\frac{1}{2}\omega^2\left(q_1^2 + q_2^2\right) + U + h\right]\right\}. \quad (4.101\text{a})$$

By means of analogous calculations, it can also be shown that

$$\frac{d}{dt}(J\dot{Q}_2) + 2\,\omega\,J\dot{Q}_1 = \frac{1}{J}\frac{\partial}{\partial Q_2}\left\{J\left[\frac{1}{2}\omega^2\left(q_1^2 + q_2^2\right) + U + h\right]\right\}. \quad (4.102\text{a})$$

If, as in (2.106), we put $dt = J\,d\tau$, (4.101a) and (4.102a) become

$$\frac{d^2Q_1}{d\tau^2} - 2\,\omega\,J\frac{dQ_2}{d\tau} = \frac{\partial}{\partial Q_1}\left\{J\left[\frac{1}{2}\omega^2\left(q_1^2 + q_2^2\right) + U + h\right]\right\}, \quad (4.101\text{b})$$

$$\frac{d^2Q_2}{d\tau^2} + 2\,\omega\,J\frac{dQ_1}{d\tau} = \frac{\partial}{\partial Q_2}\left\{J\left[\frac{1}{2}\omega^2\left(q_1^2 + q_2^2\right) + U + h\right]\right\}. \quad (4.102\text{b})$$

Equations (4.101b) and (4.102b) are the new equations of motion in place of (4.93). They are then the result of two transformations: the first is a conformal transformation and contains, so to speak, the geometric information about the orbit transformation; the second, which, as we have seen in the two-body problem, is the one that is essential for the regularization, controls the dynamics of the problem.

If we put

$$J\left[\frac{1}{2}\omega^2\left(q_1^2 + q_2^2\right) + U + h\right] = \Omega, \quad (4.103)$$

we can also rewrite the equations of motion as

$$\frac{d^2Q_1}{d\tau^2} - 2\,\omega\,J\frac{dQ_2}{d\tau} = \frac{\partial\Omega}{\partial Q_1},$$

$$\frac{d^2Q_2}{d\tau^2} + 2\,\omega\,J\frac{dQ_1}{d\tau} = \frac{\partial\Omega}{\partial Q_2}. \quad (4.104)$$

From these, one can immediately determine the first integral corresponding to the original Jacobi integral (note that the constant h for the Hamiltonian used in this section corresponds to $-C/2 + \overline{\mu}\,(1 - \overline{\mu})/2$ in the dimensionless system and with the definition of the potential given in Sect. 4.3). In fact, from

$$\frac{1}{2}\left(\dot{q}_1^2 + \dot{q}_2^2\right) = \frac{1}{2}J\left(\dot{Q}_1^2 + \dot{Q}_2^2\right) = \frac{1}{2}\frac{1}{J}\left[\left(\frac{dQ_1}{d\tau}\right)^2 + \left(\frac{dQ_2}{d\tau}\right)^2\right]$$

and from (4.103),

$$\left(\frac{dQ_1}{d\tau}\right)^2 + \left(\frac{dQ_2}{d\tau}\right)^2 = 2\,\Omega. \quad (4.105)$$

So that (4.104) turn out to be regularized, it will of course be necessary for Ω to be without singularities. We have already seen that the Levi-Civita transformation completely regularizes the two-body problem, making the product JU a constant in the case where the denominator of U vanishes in the origin, where the centre of attraction is located. In the restricted problem, things are different, because U consists of two fractions, and consequently there will be two possible singularities, corresponding to collisions with each of the two primaries separately. By means of the Levi-Civita transformation we can therefore regularize either singularity, choosing which primary we should deal with first. This will be then a *local regularization* (the term is due to Birkhoff). Furthermore, a translation should be performed as well, because the primaries are not located at the origin of the coordinate system. From these considerations it seems that if we want to regularize the equations in view of a collision with the primary of mass m_1 and coordinates $(x_1, 0)$, we must put:

$$q_1 - x_1 = Q_1^2 - Q_2^2, \qquad q_2 = 2 Q_1 Q_2. \tag{4.106}$$

The corresponding term in U will be

$$\frac{G m_1}{(Q_1^2 + Q_2^2)},$$

and multiplying this by $J = 4 \left(Q_1^2 + Q_2^2\right)$, will give a constant.

In the case where one of the primaries has a mass much greater than the other one (e.g. Sun–Jupiter–asteroid), by regularizing the term of the potential relevant to this primary à la Levi-Civita, one is able to schematize the problem as a (regularized) two-body problem perturbed by the presence of the smaller primary. One can therefore end up regarding the problem as a problem of a perturbated planar oscillator.[23]

It is clear that, to handle a real problem, that is, in the case of a numerical solution of a system which refers to a definite physical example, Levi-Civita regularization completely fulfils the requirements. To answer, instead, the question concerning the existence of the solution, *global regularization* is more appropriate, that is, the use of a transformation which cancels at the same time both singularities of the potential U. Regularizations of this sort have been suggested by various authors (Birkhoff, Thiele, Lemaître, Arenstorf, etc.). They are, as one could expect, more complicated transformations whose explanation can turn out to be very long. We therefore refer the reader, in addition to the celebrated paper by Birkhoff,[24] to the textbooks by Szebehely and Hagihara.[25]

---

[23]See, for instance, D. A. Pierce: A solution of the regularized equations of motion of the restricted problem, *Astron. J.* **71**, 545–561 (1966); Application of a solution of the regularized equations of motion of the restricted problem, *ibid* **71**, 562–565 (1966).

[24]G. D. Birkhoff: The restricted problem of three bodies, *Rend. Circ. Mat. Palermo* **39**, 265–334 (1915).

[25]V. Szebehely: op.cit., Chap. 3; Y. Hagihara: *Celestial Mechanics* (Japan Society for the Promotion of Science, 1975), Vol. IV, Chap. 17.

4.7 Extensions and Generalizations of the Restricted Problem

A great many of the bodies in the solar system have a very small inclination with respect to the plane of the ecliptic and in most cases the eccentricity of their elliptic orbits is also very small; the restricted problem can be regarded, therefore, in a first approximation, as a mathematical model admitting of real applications. Systems such as Sun–Earth–Moon or Sun–Jupiter–asteroid, can be handled successfully as physical realizations of the restricted three-body problem. Obviously, to make more accurate forecasts, one must take into account that the motion of the third body does not occur precisely in the plane where the primaries move and moreover that the orbits of the primaries are not exactly circular. For these reasons, although at first sight it may look inconsistent, it is more appropriate to consider certain problems as extensions and generalizations of the restricted problem, rather than as particular cases of the general three-body problem. From what we have said, it turns out that the extensions we can make are of two kinds (which may even be present at the same time):

a) to consider the primaries always moving in the same plane, but with orbits which are in general conics and not necessarily circles;
b) to consider the third body moving in (three-dimensional) space and not confined to the plane where the primaries move.

Proceeding step by step, let us deal first with the extension deriving only from point (b).

The Three-Dimensional Restricted Problem

In this case, (4.50) are replaced by

$$\ddot{x} - 2\,\dot{y} = \frac{\partial \Phi}{\partial x}\,, \qquad \ddot{y} + 2\,\dot{x} = \frac{\partial \Phi}{\partial y}\,, \qquad \ddot{z} = \frac{\partial \Phi}{\partial z}\,, \tag{4.107}$$

where the function $\Phi\,(x,\,y,\,z)$ is still given by (4.49), but with

$$U = \frac{1 - \bar{\mu}}{\varrho_1} + \frac{\bar{\mu}}{\varrho_2}\,, \tag{4.108}$$

where

$$\varrho_1 = \sqrt{\left(x + \bar{\mu}\right)^2 + y^2 + z^2}\,, \qquad \varrho_2 = \sqrt{\left(x + \bar{\mu} - 1\right)^2 + y^2 + z^2}\,,$$

now containing the coordinate z. For (4.107), the Jacobi integral still exists:

$$2\,\Phi\,(x,\,y,\,z) - \left(\dot{x}^2 + \dot{y}^2 + \dot{z}^2\right) = C. \tag{4.109}$$

The Hamiltonian form is obtained by extending (4.61), (4.62) and taking as the Hamiltonian

$$\mathcal{H} = \frac{1}{2}\left(p_1^2 + p_2^2 + p_3^2\right) - \left(q_1 p_2 - q_2 p_1\right) + \frac{1}{2}\left(q_1^2 + q_2^2\right) - \Phi\left(q_1, q_2, q_3\right) \quad (4.110)$$

and adding to (4.62)

$$\dot{q}_3 = p_3, \qquad \dot{p}_3 = \frac{\partial \Phi}{\partial q_3}, \qquad\qquad (4.111)$$

with $q_3 = z$.

For the Delaunay elements in three-dimensional space, we will still have Hamiltonian (4.92) in the inertial system, with $R = R(l, g, h; L, G, H)$, where

$$H = \sqrt{a\left(1 - e^2\right)}\,\cos i, \qquad h = \Omega. \qquad\qquad (4.112)$$

Also the combinations of the Delaunay elements listed at the end of Sect. 4.5 are generalized as follows

$$
\begin{aligned}
2) \quad & q_1 = L, & q_2 &= G - L, & q_3 &= H - G, \\
& p_1 = \ell + g + h, & p_2 &= g + h, & p_3 &= h; \\
3) \quad & q_1 = L, & q_2 &= L - G, & q_3 &= G - H, \\
& p_1 = \ell + g + h, & p_2 &= -(g + h), & p_3 &= -h; \\
4) \quad & q_1 = L - G, & q_2 &= G - H, & q_3 &= H, \\
& p_1 = \ell, & p_2 &= \ell + g, & p_3 &= \ell + g + h.
\end{aligned}
$$

Lastly the Poincaré canonical variables are extended to three-dimensional space through

$$
\begin{aligned}
Q_1 &= \sqrt{2\left(L - G\right)}\,\cos \ell, & P_1 &= \sqrt{2\left(L - G\right)}\,\sin \ell, \\
Q_2 &= \sqrt{2\left(G - H\right)}\,\cos(\ell + g), & P_2 &= \sqrt{2\left(G - H\right)}\,\sin(\ell + g), \\
Q_3 &= H, & P_3 &= \ell + g + h,
\end{aligned}
$$

and, analogously, the mixed ones are

$$
\begin{aligned}
Q_1 &= L, & P_1 &= \ell + g + h, \\
Q_2 &= \sqrt{2\left(L - G\right)}\,\cos\left(g + h\right), & P_2 &= -\sqrt{2\left(L - G\right)}\,\sin\left(g - h\right), \\
Q_3 &= \sqrt{2\left(G - H\right)}\,\cos h, & P_3 &= -\sqrt{2\left(G - H\right)}\,\sin h.
\end{aligned}
$$

If now, as in Sect. 4.3, we consider the manifold corresponding to the states of motion of the system, its singular points will be given by the equations $\dot{x} = \dot{y} = \dot{z} = 0$, $\partial\Phi/\partial x = \partial\Phi/\partial y = \partial\Phi/\partial z = 0$. The explicit expression of the z component of the gradient is

$$\frac{\partial \Phi}{\partial z} = -z\left(\frac{(1 - \overline{\mu})}{\varrho_1^3} + \frac{\overline{\mu}}{\varrho_2^3}\right) = 0,$$

from which, the quantity in square brackets certainly being positive when P_3 is at a finite distance, one deduces that $z = 0$. Therefore, the equilibrium points still lie in the xy plane: the Lagrangian points will have the same positions L_1, L_2, L_3, L_4, L_5 as in the planar case. However, an essential difference exists regarding the stability of the equilateral solution. In the planar problem, points L_4 and L_5 represent positions of stable equilibrium; by approximating a physical system (Sun–Jupiter–asteroid) to a planar system and taking into account the stability of the equilateral solution, we concluded that the physical system itself was stable. At this point, one might think that this conclusion is not valid, owing to the approximation. However, as Littlewood did in the case of the planar problem, one should instead ask whether the system is stable on a finite time scale, not for eternity. The answer is that the equilateral solution in the spatial problem is stable for a time which may be of the order of the age of the solar system.[26] One then speaks of *effective stability*.

Now, the locus of the zero velocity points is defined by

$$\Phi(x, y, z) = \frac{C}{2}; \tag{4.113}$$

it therefore consists of a family of surfaces embedded in three-dimensional space. Since both y and z appear only as squares in (4.113), there will be symmetry with respect to the xy and xz planes. The intersection with the xy plane will give rise to the zero velocity curves we have studied in Sect. 4.3. We saw there that, in the xy plane, the function $\Phi(x, y)$ attained its minimum value for $C = 3$. Now instead there are zero velocity surfaces for $C < 3$ which do not intersect the xy plane. For a detailed study, we refer the reader to the textbook[27] by Szebehely. Among the possible orbits in the restricted three-dimensional problem, there is a remarkable particular case:[28] the case of (one-dimensional) motions along the z axis. The importance of this case resides in the fact that, as we shall see, the system becomes equivalent to a real system with only one degree of freedom. It is the only exact solution, besides the equilibrium configurations, of the restricted three-body problem. The extension to the case where the primaries move on elliptic orbits then gives rise to the so-called *Sitnikov problem*, which we have already mentioned (Sect. 3.5, Footnote 37). First of all, let us see what are the conditions to be satisfied for the motion to occur along the z axis. From (4.107), by imposing the condition $x(t) = y(t) \equiv 0$, $\forall t$, we have

---

[26]See, for instance, A. Celletti, A. Giorgilli: On the stability of the Lagrangian points in the spatial restricted problem of three bodies, *Celestial Mech. and Dyn. Astron.* **50**, 31–58 (1991).

[27]V. Szebehely: op. cit., Chap. 10.

[28]G. Pavanini: Sopra una nuova categoria di soluzioni periodiche nel problema dei tre corpi, *Ann. Mat.* Serie III, **13**, 179–202 (1907); W. D. Mac Millan: An integrable case in the restricted problem of three bodies, *Astron. J.* **27**, 11–13 (1911).

$$\frac{\partial \Phi(0,0,z)}{\partial x} = 0, \qquad \frac{\partial \Phi(0,0,z)}{\partial y} = 0, \qquad \frac{\partial \Phi(0,0,z)}{\partial z} = \ddot{z}. \qquad (4.114)$$

From the first of these, we obtain $\varrho_1(0,0,z) = \varrho_2(0,0,z)$, and as a consequence, $\overline{\mu} = 1/2$. Hence, the masses of the two primaries must be equal and therefore also at the same distance from the origin where the barycentre is located. The circular orbits of the two primaries will then coincide (see Fig. 4.9).

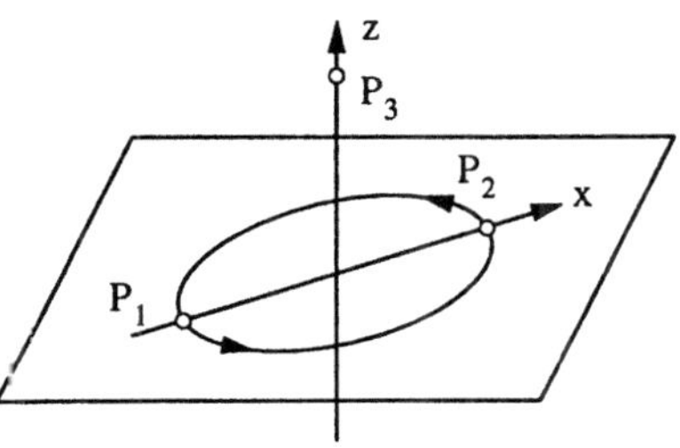

Fig. 4.9

In our dimensionless system, the distance of the masses from the origin will also be equal to $1/2$, so that the third of equations (4.114) is reduced to

$$\ddot{z} = -\frac{z}{\left(\frac{1}{4} + z^2\right)^{3/2}}, \qquad (4.115)$$

which is the equation of motion of a system that has only one degree of freedom. For the Jacobi integral, we have

$$\dot{z}^2 = 2\,\Phi - C = \frac{2}{\sqrt{\frac{1}{4} + z^2}} - \overline{C}, \qquad (4.116a)$$

with $\overline{C} = C - 1/4$. Equation (4.116a) can be rewritten in the form

$$\frac{1}{2}\dot{z}^2 - \frac{1}{\sqrt{\frac{1}{4} + z^2}} = E, \qquad (4.116b)$$

with $E = -\overline{C}/2$; it is evident from (4.116b) that E represents the total energy of the third body. Let us now see what information (4.116b) can provide. For $E > 0$, the right-hand side of (4.116b) never vanishes, and then P_3 will always move in the same direction, depending on the initial velocity. In any case, it will move indefinitely far away from the initial position at the time $t = 0$. Therefore, positive energy values cannot correspond to bounded motions (see Jacobi's stability criterion, Sect. 3.5). Since the potential energy term has values between 0 and 2, for $E < -2$, (4.116b) can never be satisfied by real values of z and $\dot{z}$.

If we consider $-2 < E < 0$, two values of z exist, which we shall call z_0 and $-z_0$, with

$$z_0 = \frac{\sqrt{4 - E^2}}{2\,|E|},$$

giving $\dot{z} = 0$. These two values for z correspond to the maximum and minimum values of the function $z(t)$ for $-2 < E < 0$, as can be seen from the sign of the second derivative, $\ddot{z}(z_0) = -z_0\,|E|^3$. Between $-z_0$ and z_0, the function $z(t)$ is always increasing or decreasing, and so the motion is inverted in the extreme positions only. Moreover, for $E = 0$, one has that z_0 and $-z_0$ equal $+\infty$ and $-\infty$ respectively (for $z \to \infty$, $\dot{z} \to 0$ and the motion is of parabolic type; see the table at the end of Sect. 3.5).

For $E = -2$, the extremes reduce to zero; then, in this case, the barycentre is a position of stable equilibrium for P_3. Owing to what we have said, the motion on the segment $[-z_0,\ z_0]$ turns out to be periodic, and the period will be given by

$$T = 2 \int_{-z_0}^{z_0} \frac{dz}{\sqrt{\frac{2}{r} + 2\,E}}.$$

To evaluate T, it is convenient to carry out a change of variable, by defining

$$u = \frac{1}{2r} = \frac{1}{\sqrt{1 + 4\,z^2}};$$

$u = 1$ corresponds to $z = 0$ and $u = |E|/2$ corresponds to $z = z_0$. For T we have then the expression

$$T = \int_{|E|/2}^{1} \frac{du}{u^2\,\sqrt{(1 - u)^2\,(u + E/2)}}. \tag{4.117}$$

A further change of variable, with

$$v^2 = \frac{1 - u}{1 + E/2}, \qquad k^2 = \frac{1 + E/2}{2},$$

then provides us with the possibility of transforming (4.117) into a complete elliptic integral in Legendre form of the third kind. The series expansion of the result is given by

$$T = \frac{\pi}{\sqrt{2}} \left(1 + \frac{9}{4}\,k^2 + \frac{345}{64}\,k^4 + \cdots \right). \tag{4.118}$$

The solution of the problem can be obtained only by inverting the elliptic integral corresponding to

$$t - t_0 = \int_{u}^{1} \frac{du}{4\,u^2\,\sqrt{(1 - u)^2\,(u + E/2)}}$$

and then expressing the variable z as a function of u obtained by means of elliptic functions. From (4.118), by making E tend to the limiting value -2 (and thus k to zero), one has $T = \pi/\sqrt{2}$. Therefore, when the motion degenerates into an equilibrium position, the period tends to the finite limit $\pi/\sqrt{2}$. The same result is obtained by linearizing the equation of motion (4.115); in fact, if we consider motions occurring near the origin by putting $z = \zeta$ (ζ is a small quantity), we obtain, neglecting second-order terms

$$\ddot{\zeta} + 8\zeta = 0,$$

that is the equation of an oscillator with period $\pi/\sqrt{2}$.

Furthermore, it has been shown by Pavanini[29] that, assuming for the primaries masses which differ by a small quantity, one still obtains, for energies not very different from $E = -2$, periodic solutions; they consist of closed orbits no longer confined to a segment of the z axis, although very close to it.

Regularization of the Three-Dimensional Restricted Problem

Let us now see what kind of difficulty arises when the study of the regularization of the planar case (Sect. 4.6) is extended to the three-dimensional case. We have already mentioned (both in Sect. 2.6 and in Sect. 4.2) that the Levi-Cìvita transformation cannot be extended to the three-dimensional case. It has been demonstrated[30] that a matrix endowed with the properties of the matrix $\boldsymbol{L}(\boldsymbol{u})$ of Sect. 2.6 can be only 2×2, 4×4 or 8×8. Kustaanheimo[31] worked out the idea, by analogy with spinor theory, of considering four dimensions by using column vectors with the fourth row equal to zero. If q_1, q_2, q_3 are the (Cartesian) coordinates of the third body ($m_3 \sim 0$), we introduce the four parameters u_1, u_2, u_3, u_4, three of which will be the new coordinates. We then define the K–S matrix:

$$\boldsymbol{K}(u) = \begin{pmatrix} u_1 & -u_2 & -u_3 & u_4 \\ u_2 & u_1 & -u_4 & -u_3 \\ u_3 & u_4 & u_1 & u_2 \\ u_4 & -u_3 & u_2 & -u_1 \end{pmatrix} \tag{4.119}$$

and the relevant transformation

$$\begin{pmatrix} q_1 \\ q_2 \\ q_3 \\ 0 \end{pmatrix} = \begin{pmatrix} u_1 & -u_2 & -u_3 & u_4 \\ u_2 & u_1 & -u_4 & -u_3 \\ u_3 & u_4 & u_1 & u_2 \\ u_4 & -u_3 & u_2 & -u_1 \end{pmatrix} \begin{pmatrix} u_1 \\ u_2 \\ u_3 \\ u_4 \end{pmatrix}. \tag{4.120a}$$

---

[29] G. Pavanini: loc. cit., 2nd part.

[30] The result, later on generalized by Jordan, Von Neumann and Wigner, is due to A. Hurwitz: Mathematische Werke II, (Birkhäuser, Basel, 1933), pp. 565–571.

[31] See P. Kustaanheimo. E. Stiefel: Perturbation theory of Kepler motion based on spinor regularization, *J. reine angew. Math.* **218**, 204–219 (1965).

In compact form this is

$$\boldsymbol{q} = \boldsymbol{K}(\boldsymbol{u})\,\boldsymbol{u}. \tag{4.120b}$$

In terms of components, we have

$$
\begin{aligned}
q_1 &= u_1^2 - u_2^2 - u_3^2 + u_4^2, \\
q_2 &= 2\left(u_1\,u_2 - u_3\,u_4\right), \\
q_3 &= 2\left(u_1\,u_3 + u_2\,u_4\right), \\
q_4 &= 0.
\end{aligned}
\tag{4.120c}
$$

As for matrix (4.119), the orthogonality property still applies

$$\boldsymbol{K}^T(\boldsymbol{u})\,\boldsymbol{K}(\boldsymbol{u}) = |\boldsymbol{u}|^2\,\mathbf{1}, \qquad \boldsymbol{K}^{-1}(\boldsymbol{u}) = \frac{1}{|\boldsymbol{u}|^2}\,\boldsymbol{K}^T(\boldsymbol{u}).$$

Moreover, properties (b) and (c) of Sect. 2.6 still apply, whereas the generalization of (2.97) requires a little more care. If we generalize it by writing its differential, we shall have:

$$d\boldsymbol{q} = 2\,\boldsymbol{K}(\boldsymbol{u})\,d\boldsymbol{u}. \tag{4.121}$$

The three equations obtained from the first three components of (4.121) give us total derivatives, which, upon integration, give (4.120c); the fourth, on the other hand, unlike (4.120a), does not give an identity but an equation:

$$u_4\,du_1 - u_3\,du_2 + u_2\,du_3 - u_1\,du_4 = 0, \tag{4.122}$$

which must be satisfied by the differentials of u. Consequently, whilst the transformation $\boldsymbol{u} \to \boldsymbol{q}$ is given by (4.120a), when the inverse transformation $\boldsymbol{q} \to \boldsymbol{u}$ is performed, (4.122) must also be satisfied.

From (4.120c), by squaring, adding and taking the square root, one gets

$$u_1^2 + u_2^2 + u_3^2 + u_4^2 = \sqrt{q_1^2 + q_2^2 + q_3^2} = r, \tag{4.123}$$

where r is the radial distance in the three-dimensional "physical" space. Also

$$d\,q_1^2 + d\,q_2^2 + d\,q_3^2 = 4\,r\left(d\,u_1^2 + d\,u_2^2 + d\,u_3^2 + d\,u_4^2\right),$$

if (4.122) applies. The independent variable is introduced afterwards in the usual way by writing $d\tau = dt/r$.

Let us now see how the transformation works. From (4.120c), one can immediately see that the points of the $u_1 u_2$ plane of the parametric space (which we shall call U^4) are transformed into points of the q_1, q_2 plane of the physical space $\mathbb{R}^3$ following the Levi-Cìvita relations (2.94a). Furthermore, if a point $\boldsymbol{u}$ of U^4 is transformed into $\boldsymbol{q}$ of $\mathbb{R}^3$, every point $\boldsymbol{v}$ which satisfies

$$
\begin{aligned}
v_1 &= u_1\cos\varphi - u_4\sin\varphi, & v_2 &= u_2\cos\varphi + u_3\sin\varphi, \\
v_4 &= u_1\sin\varphi + u_4\cos\varphi, & v_3 &= -u_2\sin\varphi + u_3\cos\varphi,
\end{aligned}
\tag{4.124}
$$

where φ is an arbitrary angle, is also transformed into the same point q. The image of a point in $\mathbb{R}^3$ is therefore a circle with radius $\sqrt{r}$ in the parametric space U^4. The tangent vector to this circle, which is obtained by differentiating (4.124) with respect to φ and evaluating the result at $\varphi = 0$, has as components $-u_4, u_3, -u_2, u_1$. Therefore, (4.122) commands the vector $d\,u$ originating from the point u to be orthogonal to the circle with radius $\sqrt{r}$ passing through that point. This circle is said to be a *fiber* of U^4, and (4.122) defines the *fibration* of the parametric space U^4. Two different fibres never intersect, because they correspond to two different points of $\mathbb{R}^3$.

If we now extend (4.122) to any two vectors u, v which define a plane $\mathbb{R}^2$ in U^4, that is, we require

$$u_4 v_1 - u_3 v_2 + u_2 v_3 - u_1 v_4 = 0, \qquad (4.125)$$

it can be demonstrated that the plane $\mathbb{R}^2$ is conformally transformed into a plane of $\mathbb{R}^3$, and moreover the transformation is of the Levi-Civita type,[32] meaning that it doubles the angles at the origin and squares the distances. Let us now look at the inverse transformation.

From the first of equations (4.120c) and from (4.123),

$$u_1^2 + u_4^2 = \frac{1}{2}\left(q_1 + r\right). \qquad (4.126)$$

Obtaining u_2 and u_3 from the last two equations of (4.120c) and by substituting (4.126) into them, one gets

$$u_2 = \frac{q_2\,u_1 + q_3\,u_4}{q_1 + r}, \qquad u_3 = \frac{q_3\,u_1 - q_2\,u_4}{q_1 + r}. \qquad (4.127a)$$

These relations are convenient for the case when $q_1 \geq 0$. For the case where q_1, on the other hand, is negative, the following are convenient:

$$u_2^2 + u_3^2 = \frac{1}{2}\left(r - q_1\right), \qquad u_1 = \frac{q_2\,u_2 + q_3\,u_3}{r - q_1}, \qquad u_4 = \frac{q_3\,u_2 - q_2\,u_3}{r - q_1}. \qquad (4.127b)$$

The Elliptic Restricted Problem in the Plane and in the Space

We shall confine ourselves here to a short outline. The planar elliptic problem, with respect to the circular one, shows a fundamental difference: the Jacobi integral no longer exists. The same occurs, obviously, for the spatial elliptic problem. The Lagrangian solutions, however, remain, as one would expect from the study of the general three-body problem. As the orbits of the primaries are elliptic, the rotating system which one has to refer to, in order to maintain the primaries in fixed positions, must be a system which rotates uniformly and with axes which expand and shrink. We shall not enter into the details of the calculations and we only remark that, in order to

---

[32]See E. L. Stiefel, G. Scheifele: op. cit., Chap. XI.

have dimensionless variables, the distance of the primaries, as in (2.21) with (2.22b),

$$r(t) = \frac{a\left(1 - e^2\right)}{1 + e \cos f},$$

(4.128)

will be used as unit lenght (a and e are the semimajor axis and the eccentricity of the ellipse that each primary describes around the barycentre and f is the true anomaly of that orbit). A particular case which has drawn considerable interest is Sitnikov's problem, (see Fig. 4.10), as has already been mentioned.

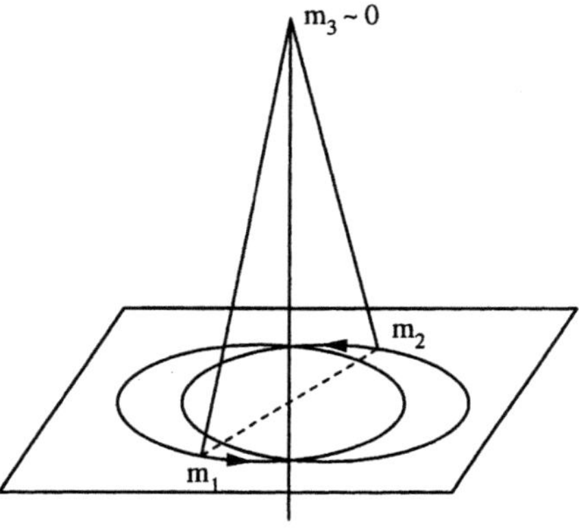

Fig. 4.10

The two masses $m_1 = m_2$ move along elliptic orbits around the barycentre located at the origin of the reference system. The third body, with mass $m_3 \sim 0$, moves along the straight line (z axis) orthogonal to the plane of motion of P_1 and P_2 and passing through the barycentre. The equation of motion along the z axis is

$$\frac{d^2 z}{d t^2} = -\frac{z}{\left[z^2 + r^2\left(t\right)\right]^{3/2}},$$

with $r(f(t))$ given by (4.128). The Hamiltonian

$$\mathcal{H} = \frac{1}{2}p^2 - \frac{1}{\sqrt{r^2\left(t\right) + z^2}},$$

with $p = \dot{z}$, is a function of the time and, as we have already said, no global integral exists. Sitnikov demonstrated the existence of solutions that are unbounded but pass an infinite number of times through the origin. The problem has also been studied by Moser,[33] who demonstrated the existence

[33] J. Moser: *Stable and Random Motions in Dynamical Systems*, (Princeton University Press, 1973), Chap. III, Sect. 5.

of chaotic orbits, and by Wodnar,[34] who confirmed Sitnikov's results through a more rigorous mathematical analysis.

Hill's Problem

We mentioned at the beginning of this section the Sun–Earth–Moon system as an example of the application of the restricted problem. We shall now try to explain, in broad outline, that particular version of the restricted problem known as Hill's problem and which is the basis of Hill's lunar theory,[35] later improved by Brown and others.

The starting point consists in the equations of the restricted problem, (4.50) and (4.51). This means that one assumes that the Moon's inclination is equal to zero and the orbits of the Sun and the Earth around their barycentre are circular. In addition, let us translate the y axis, by putting the origin of the rotating system at the position occupied by the body of mass $\overline{\mu}$ (in this case the Earth). If we call the new coordinates of the third body (the Moon) $\widetilde{x} = x - (1 - \overline{\mu})$ and $\widetilde{y} = y$, the function Φ will be given by

$$\Phi = (1 - \overline{\mu})\,\widetilde{x} + \frac{1}{2}\left(\widetilde{x}^2 + \widetilde{y}^2\right) + \frac{1 - \overline{\mu}}{\sqrt{(\widetilde{x} + 1)^2 + \widetilde{y}^2}} + \frac{\overline{\mu}}{\sqrt{\widetilde{x}^2 + \widetilde{y}^2}},$$

up to the constant terms which from now on we shall delete since they are irrelevant. Let us further simplify our model by considering the Earth–Moon distance small with respect to the Earth–Sun distance: we shall eliminate from the equations of motion powers of $\widetilde{x}$ and $\widetilde{y}$ higher than one. This is the same as suppressing in the expansion of Φ powers greater than two. What remains is

$$\Phi = \frac{\overline{\mu}}{2}\left(\widetilde{x}^2 + \widetilde{y}^2\right) + \frac{3}{2}(1 - \overline{\mu})\,\widetilde{x}^2 + \frac{\overline{\mu}}{\sqrt{\widetilde{x}^2 + \widetilde{y}^2}},$$

still up to constant terms. Furthermore, since the mass of the Earth is far smaller than the mass of the Sun, we can also neglect the first term and consider

$$\Phi = \frac{3}{2}(1 - \overline{\mu})\,\widetilde{x}^2 + \frac{\overline{\mu}}{\sqrt{\widetilde{x}^2 + \widetilde{y}^2}}. \tag{4.129}$$

Equations (4.50), in the coordinates $\widetilde{x}$ and $\widetilde{y}$, are now:

$$\ddot{\widetilde{x}} - 2\dot{\widetilde{y}} = \frac{\partial \Phi}{\partial \widetilde{x}}, \qquad \ddot{\widetilde{y}} + 2\dot{\widetilde{x}} = \frac{\partial \Phi}{\partial \widetilde{y}},$$

with Φ given by (4.129). If we operate a change of scale defined by

$$\widetilde{x} \to \alpha\,\widetilde{x}, \qquad \widetilde{y} \to \alpha\,\widetilde{y}, \qquad \overline{\mu} \to \beta\,\overline{\mu}, \qquad 1 - \overline{\mu} \to \beta\,(1 - \overline{\mu}). \tag{4.130}$$

---

[34]K. Wodnar: The original Sitnikov article – "New Insights", Master Thesis, University of Vienna, 1992.
[35]G. W. Hill: loc. cit., in Footnote 13, Sect. 4.3.

with

$$\alpha = \left(\frac{\bar{\mu}}{1-\bar{\mu}}\right)^{1/3} \qquad \beta = \frac{1}{1-\bar{\mu}}, \tag{4.131}$$

and again use the symbols x and y, we obtain just the system (4.50), with Φ reduced to the form:

$$\Phi = \frac{3}{2}x^2 + \frac{1}{\sqrt{x^2+y^2}}. \tag{4.132}$$

The system has the Jacobi integral (4.51).

If now we check what the relative equilibrium solutions for the problem schematized above are, we see that the equations

$$\frac{\partial \Phi}{\partial x} = 3x - \frac{x}{(x^2+y^2)^{\frac{3}{2}}} = 0, \qquad \frac{\partial \Phi}{\partial y} = -\frac{y}{(x^2+y^2)^{\frac{3}{2}}} = 0 \tag{4.133}$$

admit as unique solutions the two points

$$A = \left(\frac{1}{\sqrt[3]{3}}, 0\right), \qquad B = \left(-\frac{1}{\sqrt[3]{3}}, 0\right);$$

at these points

$$\Phi = \Phi_{\text{crit}} = \frac{3}{2}\sqrt[3]{3}. \tag{4.134}$$

The equilateral solution therefore no longer exists, and in addition one of the three points of the collinear solution (L_1) has disappeared. This is a consequence of our having considered x and y small and P_1 (the Sun) placed at an infinite distance (solar parallax equal to zero). A calculation similar to that performed in Sect. 4.4 then leads us to conclude that the two points A and B are positions of unstable equilibrium and the critical value (4.134) is neither a maximum nor a minimum. Hill's region, that is, the region where the motion is allowed, defined by

$$3x^2 + \frac{2}{\sqrt{x^2+y^2}} - C \geq 0,$$

is symmetric with respect to the x and y axes. For $C \leq 0$, it coincides with the whole xy plane. For $C > 0$, however Hill's region is included between two asymptotes parallel to the y axis and with abscissae $x = \pm\sqrt{C/3}$; depending on whether C greater, equal or smaller than $3^{4/3}$ $(= 2\,\Phi_{\text{crit}})$, one then has the situations shown in Fig. 4.11, 4.12 and 4.13 respectively.[36]

When $C > 3^{4/3}$, the motion can occur either inside the oval around the origin or inside the two infinite regions included between the curves and the asymptotes. For the motion of the Moon, the region concerned is the oval around the origin for $C > 3^{4/3}$. Hill's problem can be regularized by applying the Levi-Cìvita transformation, according to the method worked out in the

---

[36]For a detailed discussion, see V. Szebehely: op. cit., Sect. 10.4.

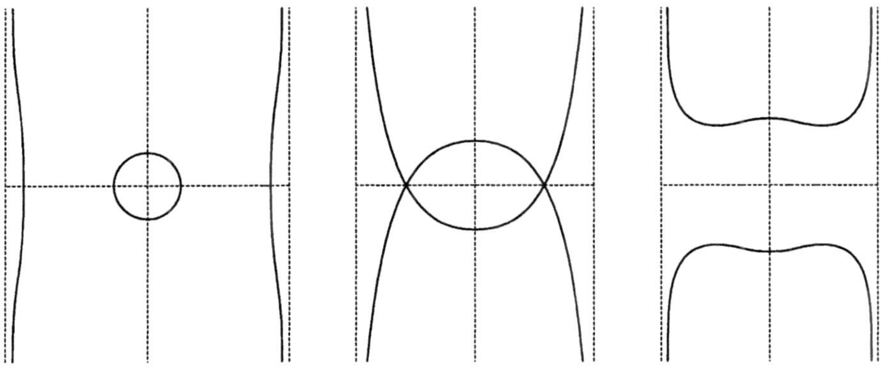

Fig. 4.11   Fig. 4.12   Fig. 4.13

previous section. One puts $z = x + iy$, $w = u + iv$, with $z = f(w)$ defined by $x = u^2 - v^2$ and $y = 2uv$. For the independent variable

$$d\tau = \frac{dt}{4\left(u^2 + v^2\right)}, \qquad \text{with} \qquad 4\left(u^2 + v^2\right) = J = |f'(u)|^2.$$

The following equations are obtained:

$$u'' - 8\left(u^2 + v^2\right) v' = \frac{\partial \Omega}{\partial u},$$
$$v'' + 8\left(u^2 + v^2\right) u' = \frac{\partial \Omega}{\partial v}, \tag{4.135}$$

with

$$\Omega = \left(\Phi - \frac{C}{2}\right) |f'|^2 = 6\left(u^2 - v^2\right)^2 \left(u^2 + v^2\right) + 4 - 2C\left(u^2 + v^2\right).$$

The integral which corresponds to the original Jacobi integral is given by

$$\frac{1}{2}\left(u'^2 + v'^2\right) - \Omega = 0. \tag{4.136}$$

From (4.135) and (4.136) it is easy to see how the singularity at the origin has been eliminated: $\Omega(0,0) = 4$, $\Omega_u(0,0) = \Omega_v(0,0) = 0$, and the velocity at the origin has modulus $\sqrt{8}$, as can be obtained from the Jacobi integral (4.136). The extremely simple form attained by the regularized problem allows us to study the behaviour of the solution of (4.135) near $(0,0)$, that is, near the Moon–Earth collision.

If $u(0) = v(0) = 0$ are the initial conditions for the coordinates at the time $t = \tau = 0$, since the modulus of the velocity at $(0,0)$ is $\sqrt{8}$, we have

$$u'(0) = \sqrt{8}\,\cos\alpha, \qquad v'(0) = \sqrt{8}\,\sin\alpha,$$

where α is the angle of the velocity with respect to the x axis. From (4.135), $u''(0) = 0$, $v''(0) = 0$ and, by taking and calculating their derivatives at $(0,0)$, we get

$$u'''(0) = -4\,C\,u'(0) = -4\,C\,\sqrt{8}\,\cos\alpha,$$
$$v'''(0) = -4\,C\,v'(0) = -4\,C\,\sqrt{8}\,\sin\alpha.$$

Hence

$$u = \left(\sqrt{8}\,\cos\alpha\right)\tau - \left(\frac{4}{3!}C\,\sqrt{8}\,\cos\alpha\right)\tau^3 + \cdots,$$
$$v = \left(\sqrt{8}\,\sin\alpha\right)\tau - \left(\frac{4}{3!}C\,\sqrt{8}\,\sin\alpha\right)\tau^3 + \cdots,$$

and also

$$t = 4\int_0^\tau \left(u^2 + v^2\right)d\tau = \frac{32}{3}\tau^3\left(1 - \frac{4}{5}C\,\tau^2 + \cdots\right).$$

The solution $u(\tau)$, $v(\tau)$ therefore admits an analytic continuation that is real and unique beyond $\tau = 0$.

Hill's lunar theory, radically changing the paradigm, which from Lagrange to Delaunay consisted in representing the orbit of the Moon as resulting from the perturbation of a Keplerian ellipse, assumes as its starting point a particularly simple periodic orbit. Going back to (4.131), (4.132), let us look for a periodic solution with initial conditions

$$x\,(0) > 0, \qquad y\,(0) = 0, \qquad \dot{x}\,(0) = 0, \qquad \dot{y}\,(0) > 0;$$

that is, at the time $t = 0$ the Moon is in conjunction with the Sun and has, in the rotating system, a velocity perpendicular to the line joining the Earth with the Sun. Furthermore, the solution must have the symmetry properties

$$x(t) = x(-t) = -x\left(t + \frac{T}{2}\right), \qquad y(t) = -x(-t) = -y\left(t + \frac{T}{2}\right),$$

where we have denoted the period by $T = 2\pi/\omega$. This particular solution is called *variational orbit*. In terms of Fourier series, we can write

$$x = \sum_{n=0}^\infty \left(A_n\,\cos n\omega t + B_n\,\sin n\omega t\right),$$
$$y = \sum_{n=0}^\infty \left(C_n\,\cos n\omega t + D_n\,\sin n\omega t\right),$$

and, because of the symmetries imposed, we also have $B_n = C_n = 0$, $\forall n$, and all the terms of even indices equal to zero, that is,

$$x = A_1\,\cos\omega t + A_3\,\cos 3\omega t + \cdots,$$
$$y = D_1\,\sin\omega t + D_3\,\sin 3\omega t + \cdots,$$

which, by putting $A_{2i+1} = a_i + a_{-i-1}$, $D_{2i+1} = a_i - a_{-i-1}$, we can rewrite as

$$x = \sum_{n=-\infty}^{\infty} a_n \cos\left(2\,n + 1\right) \omega t, \qquad y = \sum_{n=-\infty}^{\infty} a_n \sin\left(2\,n + 1\right) \omega t.$$

The solution depends on the sole parameter ω, having previously eliminated $\bar{\mu}$ by means of a scale transformation. Usually one puts $1/\omega = m$ (Hill's parameter) and then:

$$
\begin{aligned}
x &= \sum_{n=-\infty}^{\infty} a_n(m) \cos\left(\frac{2\,n + 1}{m} t\right), \\
y &= \sum_{n=-\infty}^{\infty} a_n(m) \sin\left(\frac{2\,n + 1}{m} t\right).
\end{aligned}
\tag{4.137}
$$

In the system that is not dimensionless,

$$\omega = \frac{n - n'}{n'}, \qquad m = \frac{n'}{n - n'}.$$

The observed value of n'/n (the ratio between the mean motion of the Sun around the barycentre of the Earth–Sun system and the mean motion of the Moon around the Earth) gives $m = 0.0808489338\dots$.

By substituting (4.137) in the equation of motion (4.50), (4.132), one obtains a system of infinite non-linear algebraic equations for the infinite unknowns given by the coefficients of series (4.137), depending on the parameter m. Hill demonstrated that this system admits a unique solution, at least for small values of m. The value $m = 0.0808489338\dots$ is included in the allowed values. For the coefficients a_n, one has the series expansions

$$a_0 = m^{2/3}\left(1 - \frac{2}{3}\,m + \frac{7}{18}\,m^2 - \cdots\right),$$

$$\frac{a_1}{a_0} = \frac{3}{16}\,m^2 + \frac{1}{2}\,m^3 + \cdots, \qquad \frac{a_{-1}}{a_0} = -\frac{19}{16}\,m^2 - \frac{5}{3}\,m^3 - \cdots,$$

$$\frac{a_2}{a_0} = \frac{25}{256}\,m^4 + \cdots, \qquad \frac{a_{-2}}{a_0} = m^4 + \cdots.$$

The solutions corresponding to different values of m, determined by Hill, are represented in Fig. 4.14. The first (inner) orbit corresponds to $m = 0.0808489338\dots$ and the last (outer) one to $m = 0.5609626\dots$. The first is the only one that is completely contained in the zero velocity oval. As can be inferred from the above expansions for the coefficients, it has the equations

$$x_0(t) = m^{2/3} \cos\frac{t}{m}, \qquad y_0(t) = m^{2/3} \sin\frac{t}{m},$$

which represent the motion of the Moon if the influence of the Sun is completely neglected. The coefficient $m^{2/3}$ comes from third Kepler's law. It is evident that, for small values of m, the circular orbit is only slightly deformed by the presence of the Sun.

Actually, as we know, the real motion of the Moon is very complicated[37] because of the presence of various perturbations (including the effects of the Earth's equatorial bulge). The lunar theory so far outlined concerns exclusively that which has been called by Brown the *main problem*, that is, a theory which takes into account only the presence of the three bodies Earth, Moon and Sun, and, in addition, considers them to be point masses. However, in this case, the presence of the Sun also gives rise to motions of the apsides and nodes.

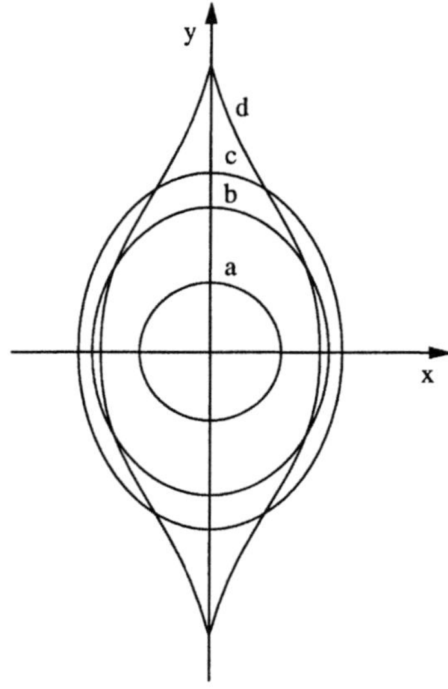

Fig. 4.14

We shall not deal with the last type of motion, but continue to consider the problem as a planar problem (otherwise one should consider a perturbation of an equation analogous to the third of equations (4.107), decoupled from the other ones). As to the motion of the perigee, we consider a displacement (a small first-order quantity) from the periodic orbit; calling the values corresponding to the periodic orbit itself x_0 and y_0, we have

$$x = x_0 + \delta x, \qquad y = y_0 + \delta y,$$

and we shall neglect the squares of δx and δy. By varying (4.131), to first order we obtain

---

[37]A curious description of the motion of the Moon was written by Brown for the *Encyclopedia Americana* (1949 edition, Vol. 19, p. 425). The entire passage can be found in S. Sternberg: *Celestial Mechanics*, Part I, (W. A. Benjamin, 1969), pp. XIV–XVI.

$$\frac{d^2}{dt^2}\,\delta x - 2\,\frac{d}{dt}\,\delta y = \left(\frac{\partial^2 \Phi}{\partial x^2}\right)\Bigg|_0 \delta x + \left(\frac{\partial^2 \Phi}{\partial x \,\partial y}\right)\Bigg|_0 \delta y,$$

$$\frac{d^2}{dt^2}\,\delta y + 2\,\frac{d}{dt}\,\delta x = \left(\frac{\partial^2 \Phi}{\partial x \,\partial y}\right)\Bigg|_0 \delta x + \left(\frac{\partial^2 \Phi}{\partial x^2}\right)\Bigg|_0 \delta y,$$

and, by varying the Jacobi integral (4.51), (4.132)

$$\left(\frac{dx}{dt}\right)\Bigg|_0 \frac{d}{dt}\,\delta x + \left(\frac{dy}{dt}\right)\Bigg|_0 \frac{d}{dt}\,\delta y - \left(\frac{\partial \Phi}{\partial x}\right)\Bigg|_0 \delta x - \left(\frac{\partial \Phi}{\partial y}\right)\Bigg|_0 \delta y = 0,$$

where $(\cdots)|_0$ mean the derivatives evaluated on the periodic orbit. If, instead of the variations δx and δy, we consider the variations δT and δN in the tangent and the normal to the orbit respectively, given by

$$\delta T = \delta x \, \cos\psi + \delta y \, \sin\psi,$$
$$\delta N = -\delta x \, \sin\psi + \delta y \, \cos\psi$$

(ψ is the angle between the tangent and the x axis), one succeeds, with some manipulation, in obtaining an equation like

$$\frac{d^2}{dt^2}\,\delta N + \Theta \, \delta N = 0. \tag{4.138}$$

Equation (4.138) is Hill's famous equation and one can show that the function Θ is an even periodic function with a period which is twice $2\pi/\omega$.[38] We confine ourselves to mentioning the fact that, if we represent the function Θ by means of a series of cosines, the solution of (4.138) turns out to be the solution of a system of an infinite number of homogeneous linear equations. Hill extended to this system the rules which are used for the solution of finite homogeneous linear systems and introduced (a bold idea in those days) an infinite determinant, that is, the determinant of a matrix with an infinite number of rows and columns.

---

[38]In addition to the already quoted papers, see G. W. Hill: On the part of the motion of the lunar perigee which is a function of the mean motions of the Sun and Moon, *Acta Math.* **8**, 1–36 (1886); E. W. Brown: *An Introductory Treatise on the Lunar Theory* (Cambridge University Press, 1896).

Chapter 5

Orbits in Given Potentials

The N-body problem presents unsurmountable difficulties even for N greater than three. It is clear, therefore, that, when N is big enough, the problem is definitely unmanageable, if considered from the point of view of the study of a dynamical system with $3N$ degrees of freedom. A possible way out of this impasse is to introduce the mean field potential generated by many bodies and to study the motion of a *single* particle, with the reasonable hypothesis that it does not appreciably perturb the external field. Further simplifying assumptions, like that of stationary equilibrium in the average distribution of the members of the system and its structural symmetry properties, circumscribe a dynamical problem with three degrees of freedom that presents many interesting aspects, especially with regard to its applications in galactic dynamics.

5.1 Introduction

In this chapter we shall study the orbits followed by a point particle subject to a force field arising from a given external potential. Unless otherwise specified, we shall assume that the motion of the particle occurs in Euclidean three-dimensional space, in which a potential function[1] $\Phi(r)$, $r \in \mathbb{R}^3$ is defined. The problem of the motion in a given potential (with two or three degrees of freedom), as we we have already remarked many times, is of fundamental importance in analytical mechanics since, with regard to symmetries or other properties exhibited by $\Phi(r)$, one tries to establish whether the associated dynamical system belongs to the class of integrable systems. In the treatment that follows, we also wish to lay emphasis on the astrophysical applications of this problem: if $\Phi(r)$ is the gravitational potential generated by a suitable mass distribution, the trajectories of the motions associated with it are a fairly good approximation of the orbits covered by the stars in a galaxy (and also, with almost the same accuracy, by a star in a globular cluster or by a galaxy in a large galactic cluster). The justification for this important property lies in the fact that the potential determining the orbit of a typical star in a galaxy is the mean field generated by the overall mass distribution, whereas the presence of the single stars themselves manifests itself as a slight perturbation, the effect of which becomes important only on a time-scale much longer than the average orbital period. The rigorous estimate of these time-scales is obviously impossible in general for systems composed of a large number of bodies ($N \sim 10^5 - 10^6$ in globular clusters and $N \sim 10^{10} - 10^{11}$ in ordinary galaxies). An approximate method of evaluation is based on the analogy between the self-gravitating N-body system and a particle gas:[2] the trajectory of a given test particle is changed by the deflections and the accelerations deriving from subsequent interactions with neighbouring particles. We can therefore introduce a time-scale after which the test particle has "forgotten" its initial motion, defining it as the time such that the quadratic variations in the velocity, due to the perturbations, sum up to reach the order of magnitude of the square of the initial velocity. In this very simplified picture, a parallel "braking" effect (the so-called *dynamical friction*) must be combined to the above random acceleration effect, so that the whole matter can be seen as the application of Langevin's theory of Brownian motion to the N-body gravitational problem.[3] In the case of binary interactions this time-scale is

---

[1] In this chapter we adopt the customary notation of works on stellar and galactic dynamics, so that we introduce the *potential* $\Phi = -U$ and use the symbols E for the total energy and L for the angular momentum instead of h and c.

[2] S. Chandrasekhar, J. Von Neumann: The statistics of the gravitational field arising from a random distribution of stars, I and II *Astrophys. J.* **95**, 489–531 (1941); **97**, 1–27 (1943).

[3] See, e.g., W. C. Saslaw: *Gravitational Physics of Stellar and Galactic Systems* (Cambridge University Press, Cambridge, 1985), pp. 3 ff.

$$\tau_{\mathrm{b}} \sim \frac{N}{\ln N}\tau_{\mathrm{d}}, \tag{5.1}$$

where τ_{d} is the *dynamical time* of the system, which means the *average crossing time* of the system for some of its typical members, or, more rigorously, the quantity (having the dimensions of a time) $\sqrt{D/a}$, where D characterizes the typical extension of the system (and can be identified, to orders of magnitude, with the $\bar{r}$ appearing in the virial relation of (3.73)) and a is the average acceleration felt by the typical member of the system. It has been suggested by many authors that the above estimate might be modified to take into account the strong inhomogeneity of a self-gravitating system and therefore of possible collective effects. In any case, the various models devised always give relations between the time-scale and the dynamical time of the kind

$$\tau \sim N^{\alpha}\tau_{\mathrm{d}}, \tag{5.2}$$

where $1/3 < \alpha < 1$. Since N is a big number, we see that the claim made that the time necessary to modify the phase-space distribution is much longer than the dynamical time is well grounded and we can treat large N-body systems in the framework of a *collisionless dynamics*. This means that the orbit covered by a typical star, for example, in an elliptical galaxy in a time of several billion years is, to a very good approximation, one of the trajectories admitted by the average potential of the galaxy.

In the light of this, the potentials that will be examined here are of the same kind as those generated by the simplest density distributions $\varrho(\boldsymbol{r})$ used to create galactic models: distributions sharing the property of a monotonic decrease outward and characterized by

a) spherical symmetry. $(\varrho = \varrho(r),\ r^2 = x^2 + y^2 + z^2)$;
b) spheroidal symmetry, $(\varrho = \varrho(\varpi, z),\ \varpi^2 = x^2 + y^2)$;
c) ellipsoidal symmetry.

We already know (Sect. 2.1) that the motion in a spherical potential can be reduced to a problem with only one degree of freedom and is therefore integrable. It is of great importance to note that even the ellipsoidal potentials exploited to build galactic models in many respects behave like integrable systems, as has been possible to verify in several reliable "numerical experiments". Therefore, the properties of the orbits in integrable spheroidal and ellipsoidal potentials are useful to illustrate some of the aspects of the structure of real stellar systems.

The possibility of showing analytically that a system is integrable is granted by the knowledge of a number of integrals of the motion, independent and in involution, equal to the number of the dimension of the configuration space of the system (Sect. 1.18). The explicit expression of the integrals of motion in terms of the coordinates of the phase space, as already remarked in Sect. 1.5, is possible only in a very limited number of cases and, in particular, when it is possible to associate the integral with a symmetry of the potential and with an invariance property of the coordinate system.

In the cases examined here an integral of the motion which is always present is the energy of the point particle: we are in fact interested in stationary configurations, so that the Hamiltonian does not depend explicitly on time, and then

$$I_1 = \mathcal{H}(\boldsymbol{p}, \boldsymbol{q}) = E$$

is the corresponding integral of the motion. The existence of $I_1 = E$ can be related to the symmetry of the homogeneity of time, which, expressed in the setting of Noether's theorem, corresponds to the conservation of the integral (1.B.60), owing to the invariance of the Lagrangian under the time translation. On the other hand, we are clearly not interested in the case of space translation symmetries, since realistic galactic potentials are never translationally invariant (unless in some cases of local approximation). But there will surely be some cases of rotational symmetry: in the case of spherical symmetry there will be the three integrals linked to the three possible spatial rotations which are the three components of the angular momentum of the particle along the orbit:

$$\Phi = \Phi(r) \quad \Longleftrightarrow \quad I_2 = L_x, \quad I_3 = L_y, \quad I_4 = L_z.$$

The three integrals I_2, I_3, I_4 are obviously not in involution. We shall see in Sect. 5.2 that a combination of them, the absolute value of the angular momentum,

$$|L| = \sqrt{L_x{}^2 + L_y{}^2 + L_z{}^2},$$

which is in involution with I_1, together with I_1 and one of the components, say L_z, is sufficient to describe the motion completely.

In the case of spheroidal symmetry, the invariance with respect to the rotation around the symmetry axis (the z axis, by definition) determines the conservation of the component of the angular momentum around this axis, which is therefore an integral of the motion:

$$\Phi = \Phi(\varpi, z) \quad \Longleftrightarrow \quad I_2 = L_z.$$

As we have already remarked, we know of the possible existence of an I_3 in case (b) above and of an I_2 and an I_3 in case (c) which, in general, are not associated with any symmetry of the kind envisaged above. Their explicit form, however, can be found only in some particular situations and by exploiting dedicated techniques that will be illustrated in what follows.

From the preliminary discussion of the Sect. 1.1 we know that the maximum number of independent conserved quantities in an autonomous system with n degrees of freedom is $2n - 1$. It is therefore necessary to establish a relation among the functions

$$F_k = F_k(\boldsymbol{p}, \boldsymbol{q}), \qquad k = 1, \ldots, 2n - 1 \tag{5.3}$$

and the integrals we spoke about in the case of integrability: the n independent integrals in involution I_a, $a = 1, \ldots, n$, for the nature of the motion in phase space, have the property of confining, or "isolating", the trajectory to the invariant tori. Following Wintner's denomination, in galactic dynamics they are named *isolating integrals*. It is important to note, however, that the number of isolating integrals can be even greater than n because they may not be in involution. For example, because of the commutation relations

$$[L_x, L_y] = L_z, \ [L_y, L_z] = L_x, \ [L_z, L_x] = L_y,$$

the three components of $\boldsymbol{L}$ are not in involution (the analysis of the motion in the spherical potential in Sect. 5.2 will shed further light on this aspect of the behaviour of an integrable system).

It is clear in addition that, in general, for isolating integrals no prescription exists that allows the determination of the form of their dependence on the phase-space coordinates. A simple but important case in which it is possible to find the explicit form of one or more isolating integrals is that in which the system is subject to resonances. Referring to the discussion of Sect. 1.3, we note that an integrable system is in resonance if two, or more, of its frequencies are commensurable. Given two integers $m_a, m_b \in N$, $a, b = 1, \ldots, n$ two frequencies are commensurable if

$$\omega_a m_a - \omega_b m_b = 0, \tag{5.4}$$

and we say that the resonance is of order m_b/m_a. The frequencies are functions of the actions, so that, if we vary the $\boldsymbol{J}$, the invariant torus T^n changes and, with it, the corresponding frequencies $\boldsymbol{\omega}(\boldsymbol{J})$. For suitable values of $\boldsymbol{J}$ condition (5.4) will be satisfied and the corresponding torus will be called "rational", or "resonant": in particular, if $n-1$ independent relations of type (5.4) are satisfied, the trajectory described by a solution of the form (1.C.149) after m_1 periods $T_1 = 2\pi/\omega_1$, $\ldots, m_n$ periods $T_n = 2\pi/\omega_n$, returns to the same values of $\boldsymbol{\theta}(0)$. It will therefore be closed, instead of covering densely the torus, and the trajectory will be "confined" to a manifold with a dimension less by one than in the general case, so that even in this case we can speak of other isolating integrals. The most interesting circumstance is that in which one, or more, relations of kind (5.4) are always satisfied (independently of the values of $\boldsymbol{J}$). For every resonance so established, the identically vanishing function

$$I_{n+1} = \boldsymbol{\theta}_a m_a - \boldsymbol{\theta}_b m_b \tag{5.5}$$

will be an integral in involution with the already known I_n. Standard examples of this situation are that of the Keplerian potential and the harmonic potential, as particular cases of spherical potentials (Sect. 5.2): in the first case the radial frequency $\omega_1 = \omega_r$ and the azimuthal one $\omega_2 = \omega_\psi$ coincide (order 1 resonance) and in the second case we have $\omega_1 = 2\omega_2$ (order 2 resonance).

From the previous discussion we see that, in general, the number of isolating integrals will be less than (and, only in some particular cases, equal to) the number of integrals of the motion, which, for (5.3), is always $k = 2n - 1$. The fundamental property that distinguishes, from the mathematical point of view, the isolating integrals $I_a(\boldsymbol{p}, \boldsymbol{q})$, $a = 1, \ldots, n_I$ from the other, non-isolating, integrals

$$F_A(\boldsymbol{p}, \boldsymbol{q}), \qquad A = n_I + 1, \ldots, k,$$

is that the $I_a(\boldsymbol{p}, \boldsymbol{q})$ are one-valued functions of the phase-space coordinates, whereas the F_A are many-valued functions. To see this, assume that we know n_I isolating integrals I_a. The level hypersurfaces of these n_I functions will define a submanifold of dimension $2n - n_I$ (see Sect. 1.2)

$$\mathcal{M}_I = \{(\boldsymbol{p}, \boldsymbol{q}) | \ I_a(\boldsymbol{p}, \boldsymbol{q}) = c_a\} . \tag{5.6}$$

Varying the parameters c_a we get an n_I–dimensional family of hypersurfaces $\mathcal{M}_I$, which makes up a *foliation* of the phase space, since these submanifolds cannot intersect themselves. In fact, if a trajectory of the dynamical system has a point on $\mathcal{M}_I = c_a$, it lies entirely on $\mathcal{M}_I$, because from the definition of integrals of motion, and the theorem of uniqueness of the solutions of differential equations, two of these hypersurfaces cannot have any point in common. Suppose we are given another integral $I'(\boldsymbol{p}, \boldsymbol{q})$; consider a point $(\boldsymbol{p}_0, \boldsymbol{q}_0)$ on $\mathcal{M}_I$ and a trajectory passing through it. $I'(\boldsymbol{p}, \boldsymbol{q})$ is isolating, with respect to this trajectory, if it is possible to find a subset of $\mathcal{M}_I$ of dimension $2n - n_I$ (the same as that of $\mathcal{M}_I$) in which no point is arbitrarily close to the hypersurface:

$$I'(\boldsymbol{p}, \boldsymbol{q}) = I'(\boldsymbol{p}_0, \boldsymbol{q}_0).$$

Otherwise, I' is non-isolating.[4] If therefore there are no more isolating integrals, the trajectory will densely cover the hypersurface $\mathcal{M}_I$: given a neighbourhood D of $(\boldsymbol{p}_0, \boldsymbol{q}_0)$, it will pass infinitely many times in every subset of D (see the recurrence theorem of Poincaré in Sect. 1.10). The F_A then establish some links among possible coordinates λ_A on $\mathcal{M}_I$, where $\lambda_A = \lambda_A(\boldsymbol{p}, \boldsymbol{q})$: making one of these (e.g. λ_1) explicit with respect to the others, we find that all the values of the other λ_A must correspond to one value of λ_1. Therefore $F_A(\boldsymbol{p}, \boldsymbol{q})$ are many-valued functions.

5.2 Orbits in Spherically Symmetric Potentials

In the previous section we noted that the potentials of interest in galactic dynamics are assumed to be integrable, and therefore the most powerful technique for studying these systems is that based on the use of the action–angle variables (see Sect. 1.16). Let us then study the simplest case, that of

---

[4] This definition of the isolating integral is due to J. Binney, S. Tremaine: *Galactic Dynamics* (Princeton University Press, 1987), p. 113.

the spherically symmetric potential, in the framework of the Hamilton–Jacobi theory. This approach offers the following advantages:

a) many results that are true in the spherical case are, in the action–angle formalism, qualitatively true also in other cases, and therefore they constitute a prototypical situation to which we will often refer;
b) in the particular case of Kepler's problem (Chap. 2), one obtains the Delaunay elements directly;
c) the action–angle settings are the best ones for constructing the perturbation theory (as will be seen in Volume 2).

We shall therefore uncertake the treatment of the motion in a central force field, closely following the exposition of the theory we gave in Sects. 1.16 and 1.17. This subject is of such great relevance for its physical applications that the number of treatments available in the literature is enormous. We think that one of the best expositions of all is still the beautiful book by Max Born,[5] whereas for applications in stellar dynamics the best up to date reference is the already cited book by Binney and Tremaine.[6]

We are given the Hamiltonian of a particle of unit mass in a gravitational potential that is invariant with respect to all rotations,

$$\mathcal{H} = \frac{1}{2}(p_x^2 + p_y^2 + p_z^2) + \Phi(\sqrt{x^2 + y^2 + z^2}),\tag{5.7}$$

where x, y, z are Cartesian coordinates, $p_x = \dot{x}$, $p_y = \dot{y}$, $p_z = \dot{z}$ their conjugate momenta and Φ a monotonically increasing function tending to zero if its argument goes to infinity. With a generating function of the third type (see (1.C.72)),

$$W_3(\boldsymbol{p}, \boldsymbol{Q}) = W_3(p_x, p_y, p_z, r, \vartheta, \varphi)$$
$$= -(p_x r \sin \vartheta \cos \varphi + p_y r \sin \vartheta \sin \varphi + p_z r \cos \vartheta),$$

we perform a canonical transformation to spherical coordinates

$$r = \sqrt{x^2 + y^2 + z^2},$$
$$\vartheta = \arccos \frac{z}{r},$$
$$\varphi = \arcsin \frac{y}{\sqrt{x^2 + y^2}},\tag{5.8}$$

whose conjugate momenta are

$$p_r = \dot{r},\tag{5.9a}$$
$$p_\vartheta = r^2 \dot{\vartheta},\tag{5.9b}$$
$$p_\varphi = r^2 \sin^2 \vartheta \dot{\varphi}.\tag{5.9c}$$

---

[5] M. Born: op. cit., in Sect. 1.17.
[6] J. Binney, S. Tremaine: op. cit., pp. 103 ff.

The Hamiltonian in the new coordinates is

$$\mathcal{H} = \frac{1}{2}\left(p_r^2 + \frac{1}{r^2}p_\vartheta^2 + \frac{1}{r^2\sin^2\vartheta}p_\varphi^2\right) + \Phi(r). \tag{5.10}$$

According to the theory of Sect. 1.16, to the Hamiltonian (5.10) there correspond the Hamilton–Jacobi equation

$$\left(\frac{\partial S}{\partial r}\right)^2 + \frac{1}{r^2}\left(\frac{\partial S}{\partial\vartheta}\right)^2 + \frac{1}{r^2\sin^2\vartheta}\left(\frac{\partial S}{\partial\varphi}\right)^2 + 2[\Phi(r) - I_1] = 0, \tag{5.11}$$

where the function $S(\boldsymbol{Q}, \boldsymbol{J})$ again generates a canonical transformation

$$(\boldsymbol{Q}, \boldsymbol{P}) = (r, \vartheta, \varphi, p_r, p_\vartheta, p_\varphi) \to (\boldsymbol{\theta}, \boldsymbol{J}), \tag{5.12}$$

by means of the partial derivatives

$$\boldsymbol{P} = \frac{\partial}{\partial\boldsymbol{Q}}S(\boldsymbol{Q}, \boldsymbol{J}), \qquad \boldsymbol{\theta} = \frac{\partial}{\partial\boldsymbol{J}}S(\boldsymbol{Q}, \boldsymbol{J}). \tag{5.13}$$

To frame the problem in the best setting we will solve partial differential equation (5.11), obtaining, at the small price of a some heavy mathematical work, the most enlightening expressions for the solution of the problem, since they will be expressed in terms of the variables that offer the most direct physical interpretation. We know, from the general discussion of Sect. 1.15 and the introductory treatment on isolating integrals in the previous section, that (5.11) is a *candidate* for a solution by separation of the variables. Let us therefore try to determine a *complete integral* of it, assuming a solution of the form

$$S(r, \vartheta, \varphi) = S_1(r) + S_2(\vartheta) + S_3(\varphi).$$

Putting this expression of S into (5.11), we can break it down into three ordinary differential equations:

$$\left(\frac{dS_1}{dr}\right)^2 + \frac{I_2^2}{r^2} + 2\left[\Phi(r) - I_1\right] = 0,$$

$$\left(\frac{dS_2}{d\vartheta}\right)^2 + \frac{I_3^2}{\sin^2\vartheta} = I_2^2,$$

$$\frac{dS_3}{d\varphi} = I_3,$$

which can be solved for the derivatives of S (according to (5.13)):

$$\frac{dS_1}{dr} = p_r = \sqrt{2\left[I_1 - \Phi(r)\right] - \frac{I_2^2}{r^2}}, \tag{5.14a}$$

$$\frac{dS_2}{d\vartheta} = p_\vartheta = \sqrt{I_2^2 - \frac{I_3^2}{\sin^2\vartheta}}, \tag{5.14b}$$

$$\frac{dS_3}{d\varphi} = p_\varphi = I_3. \tag{5.14c}$$

Of the three integration constants, we already know that I_1 is the integral of the motion corresponding to the energy: $I_1 = E$. To interpret the other two, we note that, in the spherical coordinate system, the angular momentum around the polar $(\vartheta = 0)$ axis is

$$L_z = r^2 \dot{\varphi} \sin^2 \vartheta \equiv p_\varphi, \tag{5.15}$$

and the absolute value of the total angular momentum is

$$|\boldsymbol{L}| = \sqrt{L_x^2 + L_y^2 + L_z^2} = r^2 \sqrt{\dot{\vartheta}^2 + \dot{\varphi}^2 \sin^2 \vartheta} \equiv \sqrt{p_\vartheta^2 + p_\varphi^2 \sin^{-2} \vartheta}. \tag{5.16}$$

Equations (5.14c) and (5.15) therefore say that $I_3 = L_z$ is the constant polar momentum. Equations (5.14b) and (5.16), however, say instead that $I_2 = |\boldsymbol{L}|$ is the constant absolute value of the total angular momentum (which, as is true in every central field, is constant also in direction). The orbit in space is always contained in a plane that has an inclination angle i with respect to the plane $\vartheta = \pi/2$ given by

$$i = \arccos \frac{I_3}{I_2}. \tag{5.17}$$

The line of intersection between the orbital plane and the plane $\vartheta = \pi/2$ is called the *line of the nodes*. From (5.14) it is also possible to deduce the qualitative aspect of the orbit in the orbital plane: the φ coordinate is cyclic and is therefore related to the rotation around the z axis. The ϑ coordinate is instead associated with a libration, symmetrical with respect to the plane $\vartheta = \pi/2$, whose limits are given by the zeroes of the argument in the square root in (5.14b), that is, by

$$\sin \vartheta = \frac{I_3}{I_2} = \cos i, \qquad \frac{\pi}{2} - i \leq \vartheta \leq \frac{\pi}{2} + i. \tag{5.18}$$

The radial character of the motion depends on the behaviour of the argument in the square root in (5.14a): if $\Phi(r)$ is a monotonically increasing function with r and such that

$$|\Phi(0)| = |\Phi_0| < \infty, \qquad \left.\frac{d\Phi}{dr}\right|_0 = 0, \qquad \lim_{r \to \infty} \Phi = 0,$$

the argument of the square root in (5.14a) will be positive for $r_{\min} \leq r \leq r_{\max}$ (where $r_{\min}$ and $r_{\max}$ are the only zeroes of the function inside the square root) and negative outside this ring, to which the orbit is therefore confined.

As can be easily verified, E, $|\boldsymbol{L}|$ and L_z are isolating integrals of the motion in involution, so that the motion of the system is quasi-periodic (Sect. 1.17). To find a complete charsterization of the motion, even to determine its progression in time, and to find the elements of the orbit, it is, however, better to pass to the action–angle variables $(\boldsymbol{J}, \boldsymbol{\theta})$. Starting from the canonical variables $(\boldsymbol{P}, \boldsymbol{Q})$, the action integrals, in virtue of (5.14), are

$$J_r = \frac{1}{2\pi} \oint p_r dr = \frac{1}{2\pi} \oint \sqrt{2\left[I_1 - \Phi(r)\right] - \frac{I_2^2}{r^2}}\, dr, \qquad (5.19a)$$

$$J_\vartheta = \frac{1}{2\pi} \oint p_\vartheta d\vartheta = \frac{1}{2\pi} \oint \sqrt{I_2^2 - \frac{I_3^2}{\sin^2 \vartheta}}\, d\vartheta, \qquad (5.19b)$$

$$J_\varphi = \frac{1}{2\pi} \oint p_\varphi d\varphi = \frac{1}{2\pi} \oint I_3 d\varphi, \qquad (5.19c)$$

where each integral is intended to be performed on a cycle (of libration in r and ϑ, of rotation in φ). The third integral immediately gives

$$J_\varphi = I_3 \equiv L_z. \qquad (5.20a)$$

By means of the substitution $\cos \vartheta = x \sin i$ the second integral can also be performed directly. In fact

$$J_\vartheta = -\frac{a I_2}{2\pi} \oint \frac{\sqrt{1 - x^2}}{1 - ax^2}\, dx,$$

where we have defined

$$a = \frac{I_2^2 - I_3^2}{I_2^2} < 1.$$

For the indefinite integral we have

$$\int \frac{\sqrt{1 - x^2}}{1 - ax^2}\, dx = \frac{1}{a} \arctan\left(\frac{x}{\sqrt{1 - x^2}}\right) - \frac{\sqrt{1 - a}}{a} \arctan\left[\sqrt{1 - a}\, \frac{x}{\sqrt{1 - x^2}}\right].$$

Being a libration, the integral must be multiplied by 2 and evaluated between $+1$ and -1, which are the integration limits corresponding to

$$\vartheta = \frac{\pi}{2} \pm i.$$

The result is simply

$$J_\vartheta = I_2 - I_3 \equiv |\boldsymbol{L}| - L_z. \qquad (5.20b)$$

If we know the explicit form of $\Phi(r)$, it is then possible to express the energy in terms of $\boldsymbol{J}$. The two results embodied in (5.20a) and (5.20b), inserted into the argument of the integral in (5.19a), enable us to reach the following important conclusion: the energy depends on J_r *and on the sum* $J_\vartheta + J_\varphi$. Therefore, the two frequencies

$$\omega_\vartheta = \frac{\partial \mathcal{H}}{\partial J_\vartheta}, \qquad \omega_\varphi = \frac{\partial \mathcal{H}}{\partial J_\varphi}$$

have the same value and the system is said to be *simply degenerate*, having a first order resonance (see Sects. 1.17 and 5.1). The degeneration can be removed by means of the simple canonical transformation

$$\theta_1 = \theta_r \qquad\qquad J_1 = J_r + J_\vartheta + J_\varphi \qquad \omega_1 = \omega_r \qquad\qquad \text{(5.21a)}$$
$$\theta_2 = \theta_\vartheta - \theta_r \qquad\quad J_2 = J_\vartheta + J_\varphi \qquad\qquad \omega_2 = \omega_\vartheta - \omega_r \qquad\quad \text{(5.21b)}$$
$$\theta_3 = \theta_\varphi - \theta_\vartheta \qquad\quad J_3 = J_\varphi \qquad\qquad\qquad \omega_3 = \omega_\varphi - \omega_\vartheta \equiv 0 \qquad \text{(5.21c)}$$

so that the energy is a function of the form

$$\mathcal{H} = \mathcal{H}(J_1, J_2).$$

From (5.20a) and (5.20b), we now see that

$$J_2 \equiv |\boldsymbol{L}|; \qquad J_3 \equiv L_z. \qquad\qquad\qquad \text{(5.20c)}$$

The angular variables ϑ_i, $i = 1, 2, 3$, rather than $\theta_r, \theta_\vartheta, \theta_\varphi$, are the variables that lead to the simplest representation of the motion on the orbit and that give physical meaning to some of the classical orbital elements. The (θ_i, J_i) are canonical coordinates, so that the transformation relations linking them to the (Q_i, P_i) coordinates are still given in the form of (5.13). The angular variables are then

$$\theta_i = \frac{\partial S}{\partial J_i} = \frac{\partial}{\partial J_i}\left[\int p_r(J_1, J_2)dr + \int p_\vartheta(J_2, J_3)d\vartheta + J_3\varphi\right], \qquad \text{(5.22)}$$

which, expressed in terms of the isolating integrals and of the coordinates (r, ϑ, φ), contain all the information on the orbits. Equation (5.22), for $i = 1$, gives

$$\theta_1 = \frac{\partial}{\partial J_1}\int p_r(J_1, J_2)dr = \int \frac{\partial p_r}{\partial \mathcal{H}}\frac{\partial \mathcal{H}}{\partial J_1}dr = \omega_1 \int \frac{\partial p_r}{\partial \mathcal{H}}dr = \omega_1 \int \frac{dr}{p_r}. \qquad \text{(5.23)}$$

where we have exploited (5.21a), by writing $\partial \mathcal{H}/\partial J_1 = \omega_1$, and where the last term of the equality is obtained by noting that p_r depends on the Hamiltonian through the argument of the square root appearing, for example, in (5.14a). On the other hand, since the expression of the motion in terms of the angular variables is trivially given by

$$\theta_1 = \omega_1(t - t_0), \qquad\qquad\qquad \text{(5.24)}$$

the explicit integration of (5.23), when possible, gives the link between t and r, that is the equation for the radial coordinate of the orbital motion:

$$\frac{dr}{p_r} = dt \longrightarrow \int_{r(t_0)}^{r} \frac{dr}{p_r} = t - t_0. \qquad\qquad \text{(5.25a)}$$

To a cycle in which the angle θ_1 increases by 2π there corresponds a whole libration cycle of the radial variable, so that

$$T_1 = T_r = \frac{2\pi}{\omega_1} \qquad\qquad\qquad \text{(5.25b)}$$

is the radial period of the motion. If we generalize the definition which belongs to the Keplerian case, θ_1 coincides with the *mean anomaly* (Sect. 2.4).

Equation (5.22), for $i = 2$, gives

$$\theta_2 = \frac{\partial}{\partial J_2}\left[\int p_r(J_1, J_2)dr + \int p_\vartheta(J_2, J_3)d\vartheta\right]$$
$$= \int\left(\frac{\partial p_r}{\partial \mathcal{H}}\frac{\partial \mathcal{H}}{\partial J_2} + \frac{\partial p_r}{\partial J_2}\right)dr + \int\frac{\partial p_\vartheta}{\partial J_2}d\vartheta,$$

which, using forms (5.14a) and (5.14b) of the momenta in terms of the actions (keeping in mind the links established by (5.20) and (5.21) between actions and integrals of the motion) and $\partial\mathcal{H}/\partial J_2 = \omega_2$, becomes

$$\theta_2 = \int\left(\omega_2 - \frac{J_2}{r^2}\right)\frac{dr}{p_r} + J_2\int\frac{d\vartheta}{p_\vartheta}. \tag{5.26}$$

Proceeding analogously, (5.22), for $i = 3$, gives

$$\theta_3 = \varphi - \int\frac{\partial p_\vartheta}{\partial J_3}d\vartheta = \varphi - \frac{J_3}{J_2}\int\frac{d\vartheta}{\sin^2\vartheta\sqrt{1 - \frac{J_3^2}{J_2^2\sin^2\vartheta}}}. \tag{5.27}$$

With a procedure similar to that for calculating (5.19b), we find that the two integrals in ϑ in (5.26) and (5.27) are respectively

$$J_2\int\frac{d\vartheta}{p_\vartheta} = \int\frac{d\vartheta}{\sqrt{1 - \frac{\cos^2 i}{\sin^2\vartheta}}} = \arcsin\left(\frac{\cos\vartheta}{\sin i}\right) + \text{const.}, \tag{5.28}$$

$$\int\frac{\partial p_\vartheta}{\partial J_3}d\vartheta = \int\frac{\cos i\, d\vartheta}{\sin^2\vartheta\sqrt{1 - \frac{\cos^2 i}{\sin^2\vartheta}}} = \arcsin\left(\cot i \cot\vartheta\right) + \text{const.} \tag{5.29}$$

Finally, to evaluate the integral in dr in (5.26), we can use equality (5.25a) written in the form

$$\omega_2\int_{r(t_0)}^r\frac{dr}{p_r} = \omega_2(t - t_0). \tag{5.30}$$

Let us now introduce the azimuthal angle ψ representing the angular coordinate of the point particle in the orbital plane, measured counterclockwise from the ascending node (see Fig. 5.1). From the definition of angular momentum and from (5.20c) and (5.25a) we have

$$|\boldsymbol{L}| = J_2 = r^2\frac{d\psi}{dt} = r^2 p_r\frac{d\psi}{dr},$$

or, in integrated form,

$$\psi - \psi(t_0) = |\boldsymbol{L}|\int_{r(t_0)}^r\frac{dr}{r^2 p_r}. \tag{5.31}$$

If even in this case we generalize the definition of the Keplerian case, the angle $\psi - \psi(t_0)$ turns out to be the *true anomaly*. With the symbols of Fig. 5.1, using

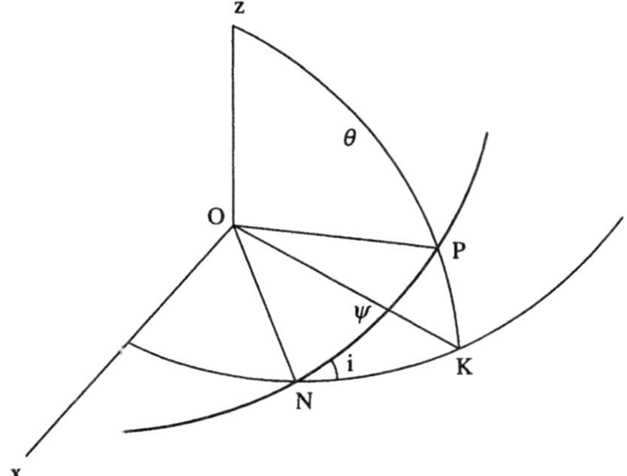

Fig. 5.1

relation (A.3) of spherical trigonometry (see the Mathematical Appendix) applied to the spherical triangle NPK, we can write the relation

$$\sin\psi\sin i = \sin\widehat{PKN}\sin\left(\frac{\pi}{2}-\vartheta\right).$$

Since $\widehat{PKN}$ is a right angle, this implies that the azimuthal angle can be written in the form

$$\psi = \arcsin\left(\frac{\cos\vartheta}{\sin i}\right). \tag{5.32}$$

If Ω is the longitude of the ascending node, observing that

$$\widehat{KOP} = \frac{\pi}{2}-\vartheta$$

and using (A.4) applied to the same triangle, we obtain the formula

$$\sin\widehat{NOK} = \sin(\varphi-\Omega) = \cot i\cot\vartheta. \tag{5.33}$$

We are now in a position to make a physical interpretation of the angle variables θ_i. Equations (5.33) and (5.29) put into (5.27) give

$$\theta_3 = \varphi - \arcsin\left[\sin(\varphi-\Omega)\right] \equiv \Omega, \tag{5.34}$$

so that θ_3 coincides with the longitude of the ascending node, *and is constant,* owing to the fact that ω_3 is zero, recalling (5.21c). Subsequently, substituting relations (5.28), (5.30), (5.31) and (5.32) into expression (5.26) simply gives

$$\theta_2 = \omega_2(t-t_0), \tag{5.35}$$

whose interpretation, in spite of its apparently trivial form, is given by the following line of reasoning: recalling (5.20c) and (5.31), we can also write (5.26) in the form

$$\theta_2 = \psi - \psi_0 + \int_{r_0}^{r} \left(\omega_2 - \frac{J_2}{r^2} \right) \frac{dr}{p_r} = \psi - \psi_0 + \frac{\partial}{\partial J_2} \int_{r_0}^{r} p_r dr,$$

where $\psi_0 = \psi(r(t_0))$ is the value of ψ at a fixed $r_0 = r(t_0)$. Now, the integral $\int p_r dr$ is a many-valued function of r that, in particular for a complete libration of the radial coordinate, increases exactly by J_1, and its derivative with respect to J_2 again takes its previous value. We see then that θ_2 is the angular distance (from the line of the nodes, for example) of the position of a point on the orbit with the radial coordinate fixed at the value r_0. If, in particular, we choose $r_0 = r_{\min}$ (that is, the point corresponding to the *pericentre*), we can give an interpretation of relation (5.35) by saying that θ_2 represents the angular advance of the pericentre and that ω_2 is the *precession* velocity of the pericentre. In a libration in r, which by (5.25b), happens in a radial period $T_r = T_1$, the pericentre advances by an azimuthal angle

$$\delta\theta_2 = \omega_2 T_1 = 2\pi \frac{\omega_2}{\omega_1}. \tag{5.36}$$

Recalling, from definition (5.21b), the relation among the azimuthal frequencies, we can write

$$\omega_2 = \omega_\vartheta - \omega_1 = \frac{2\pi}{T_\psi} - \omega_1 = \left(\frac{\Delta\psi}{2\pi} - 1 \right) \omega_1, \tag{5.37}$$

where we have introduced the azimuthal period

$$T_\psi = \frac{2\pi}{\Delta\psi} T_r = \frac{2\pi}{\omega_\vartheta} \equiv \frac{2\pi}{\omega_\varphi}, \tag{5.38}$$

defined as the time necessary for ψ to increase by 2π, and the change in the azimuthal angle corresponding to a radial period

$$\Delta\psi = |\boldsymbol{L}| \oint \frac{dr}{r^2 p_r} = 2|\boldsymbol{L}| \int_{r_{\min}}^{r_{\max}} \frac{dr}{r^2 p_r}. \tag{5.39}$$

In the case for which $\Delta\psi$ is a rational submultiple of 2π,

$$\Delta\psi = \frac{n}{k} 2\pi, \qquad n, k \in \boldsymbol{N},$$

the orbit is *closed* and therefore *periodic*. We then have a *resonance* that, expressed in the new frequencies, is given by the relation

$$m_1\omega_1 - m_2\omega_2 = 0, \qquad m_1 = k - n, \qquad m_2 = k.$$

After m_1 cycles in r and m_2 turns of the pericentre, the orbit closes in on itself.

In the light of the results obtained, the general properties of the orbits in a spherically symmetric potential can be summarized by saying that the generic orbit can be represented as an annulus (densely filled up after many periods) limited by $r = r_{\min}$ and $r = r_{\max}$. To specify the orbit completely we need to know four parameters (the *elements of the orbit*):

1) the inclination i;
2) the longitude of the node Ω;
3) the pericentre distance $r_{\min}$;
4) the apocentre distance $r_{\max}$.

The first two elements (i and Ω) are needed to fix the orientation of the plane of the orbit in space; the third and fourth ($r_{\min}$ and $r_{\max}$) its extension. To these four elements there correspond just as many isolating integrals: treatment by means of the action–angle variables gives them explicitly as an independent set formed by the three action integrals $\boldsymbol{J}$ and by the constant angle $\theta_3 = \Omega$. If from these one wants to come back to the standard integrals mentioned from the outset in the previous section (case (a), $I_a = (E, \boldsymbol{L})$), it is sufficient to use relations (5.20), the fact that $\tan \Omega = L_y/L_x$ and equation (5.19a) to find the function $E = E(\boldsymbol{J})$ explicitly.

If, as we have said above, there is a resonance of order m_1/m_2 between ω_1 and ω_2, then, in agreement with (5.5), we have a *fifth isolating integral*:

$$I_5 = m_1\theta_1 - m_2\theta_2,$$

which, for (5.23) and (5.26), can be written as

$$I_5 = m_1\omega_1 \int \frac{dr}{p_r} - m_2 \int \left(\omega_2 - \frac{J_2}{r^2}\right) \frac{dr}{p_r} - m_2 J_2 \int \frac{d\vartheta}{p_\vartheta},$$

and which, using the resonance condition $m_1\omega_1 - m_2\omega_2 = 0$, has the form

$$I_5(r, \vartheta) = m_2 \left[|\boldsymbol{L}| \int_{r_0}^{r} \frac{dr}{r^2 p_r} - \arcsin\left(\frac{\cos\vartheta}{\sin i}\right) \right]. \qquad (5.40)$$

I_5 is isolating because, if we explicitly write out the function $r = r(\vartheta)$, of which $I_5 = $ const. is the implicit form, for a given value of ϑ there corresponds an integer number m_2 of isolated values of r, and so I_5 is locally a one-valued function. If there is no resonance, then, as discussd in the previous section, a fifth integral of the motion exists, in this case given by

$$\psi_0(r, \vartheta) = \arcsin\left(\frac{\cos\vartheta}{\sin i}\right) - |\boldsymbol{L}| \int_{r_0}^{r} \frac{dr}{r^2 p_r},$$

but it is not isolating.

5.3 Orbits in Isochronal Potentials

Let us apply the theory of the orbits in spherically symmetric potentials from the previous section to the very interesting case of the family of isochronal potentials. This class of models exhibits many attractive features in stellar dynamics and has therefore been extensively studied since its introduction[7] and also applied as an *unperturbed* model to study the orbital structure of more complicated systems by means of perturbative[8] or other semi-analytical[9] techniques. The class of isochronal potentials is given by the function

$$\Phi_I(r) = -\frac{GM}{b + \sqrt{b^2 + r^2}}, \tag{5.41}$$

where the positive constant GM, the product of Newton's *gravitational constant* and a characteristic mass, must be identifyed with the constant μ used in Chap. 2 (see equation (2.16) and ff.). The function (5.41) is parametrized by the characteristic length b which gives the extent of the region where the potential resembles that of a homogeneous body, whereas, for $r \gg b$, it gives the typical behaviour of the potential far away from a finite body (of mass M), $\Phi \propto r^{-1}$. As a particular case of fundamental importance, function (5.41) comprises the Keplerian potential, when $b = 0$:

$$b = 0 \longrightarrow \Phi_I(r) \equiv \Phi_K(r) = -\frac{GM}{r}.$$

The aim of this section is to find the explicit form of the action–angle variables in a concrete case, along the lines proposed in the previous section. We shall also find, therefore, the classical *Delaunay elements* of the orbits of the two body problem, introduced in Sect. 2.5, since, in what follows, all the formulas of the Keplerian limit can be obtained by putting b equal to zero.

The isochronal potentials are so called because they are characterized by a very simple relation between the radial period and the energy:

$$T_r = \frac{2\pi GM}{(-2E)^{3/2}}, \tag{5.42a}$$

independent of b and the other integrals of the motion. Let us start then by giving the proof of this relation. The radial period, from (5.25), is given by the integral

$$T_r = 2 \int_{r_1}^{r_2} \frac{dr}{p_r} = 2 \int_{r_1}^{r_2} \frac{rdr}{\sqrt{2r^2\left[E - \Phi_I(r)\right] - L^2}}, \tag{5.43}$$

<hr>

[7] M. Hénon: L'amas isochrone I, *Annales d'Astrophys.* **22**, 126–139 (1959); L'amas isochrone II, ibid. 491-498.

[8] O. E. Gerhard, P. Saha: Recovering galactic orbits by perturbation theory, *Mon. Not. Roy. Astron. Soc.* **251**, 449–467, (1991).

[9] C. McGill, J. Binney: Torus construction in general gravitational potentials, *Mon. Not. Roy. Astron. Soc.* **244**, 634–645, (1990).

where r_1 and r_2 are the roots of the argument of the square root (from what we said above, there are two real roots). Let us introduce the auxiliary variable

$$s \equiv -\frac{GM}{b\Phi_I} = 1 + \sqrt{1 + \frac{r^2}{b^2}} \tag{5.44}$$

such that, solving for r, we have

$$\left(\frac{r}{b}\right)^2 = s^2\left(1 - \frac{2}{s}\right), \quad s \geq 2,$$

and

$$r\,dr = b^2(s-1)ds.$$

In terms of s (5.43) becomes

$$T_r = \frac{2b}{\sqrt{-2E}} \int_{s_1}^{s_2} \frac{(s-1)ds}{\sqrt{X}}, \tag{5.45}$$

where

$$X(s_1, s_2) = (s_2 - s)(s - s_1),$$

and we have introduced the new limits between which the argument of the square root is positive and that are defined by

$$s_2 + s_1 = 2 - \frac{GM}{Eb}, \tag{5.46a}$$

$$s_2 - s_1 = 2\sqrt{\left(1 + \frac{GM}{2Eb}\right)^2 + \frac{L^2}{2Eb^2}}, \tag{5.46b}$$

$$s_2 s_1 = -\frac{L^2 + 4GMb}{2Eb^2}. \tag{5.46c}$$

We also define the useful quantities

$$a = -\frac{GM}{2E} - b; \qquad e = \sqrt{1 + \frac{L^2}{2Ea^2}}, \tag{5.47}$$

for which, with hindsight, we will soon find an interpretation as orbital parameters and that allow us to write the relations in (5.46) as

$$s_2 = 2 + \frac{a}{b}(1 + e), \qquad s_1 = 2 + \frac{a}{b}(1 - e), \tag{5.48}$$

$$b(s_2 + s_1) = 2a + 4b, \qquad b(s_2 - s_1) = 2ae, \tag{5.49}$$

$$b^2 s_2 s_1 = 4b(a + b) + a^2(1 - e^2). \tag{5.50}$$

Exploiting the above relations (5.48–50) to write all expressions where s_2 and s_1 appear in terms of a, b and e, the indefinite integral appearing in (5.45) amounts to:

$$\tau(s) = \frac{a+b}{b} \arcsin\left(\frac{b(s-2)-a}{ae}\right) - \sqrt{X}, \tag{5.51}$$

so that the radial period becomes

$$T_r = \frac{2b}{\sqrt{-2E}}(\tau(s_2) - \tau(s_1)) = \frac{2\pi}{\sqrt{-2E}}(a+b) = \frac{2\pi GM}{(-2E)^{3/2}},$$

and so equality (5.42a) is proved. The function $\tau(s)$ defined by (5.51) allows us also to express the angular variable θ_1 in terms of the radial coordinate. In fact, recalling equation (5.23), we have

$$\theta_1 = \omega_1 \int \frac{dr}{p_r} = \frac{2\pi}{T_r} \frac{b}{\sqrt{-2E}} \tau(s).$$

It is convenient at this point to introduce the variable u (the *eccentric anomaly*), by means of

$$bs = 2b + a(1 - e\cos u), \tag{5.52}$$

so that the angle θ_1 can be written in the form

$$\theta_1 = \frac{\omega_1}{\sqrt{-2E}}[(a+b)u - ae\sin u] = u - \frac{ae}{a+b}\sin u. \tag{5.53}$$

The *true anomaly* $f(u) = \psi(u) - \psi_0$, defined such that ψ_0 corresponds to the pericentre, is obtained by integrating (5.31). Hence

$$\int \frac{dr}{r^2 p_r} = \frac{1}{\sqrt{-2Eb}} \int \frac{(s-1)ds}{(s-2)s\sqrt{X}}$$

$$= \frac{1}{\sqrt{-2Eb}}\frac{1}{2}\left[\int \frac{ds}{(s-2)\sqrt{X}} + \int \frac{ds}{s\sqrt{X}}\right]$$

$$= \frac{1}{2L}\arcsin\left(\frac{e - \cos u}{1 - e\cos u}\right)$$

$$+ \frac{1}{2\sqrt{L^2 + 4GMb}}\arcsin\left[\frac{ae - (2b+a)\cos u}{2b + a(1 - e\cos u)}\right].$$

By using the trigonometric relations

$$\arcsin\beta = \frac{\pi}{2} - 2\arctan\sqrt{\frac{1-\beta}{1+\beta}},$$

$$\tan\frac{u}{2} = \sqrt{\frac{1 - \cos u}{1 + \cos u}},$$

we can write the true anomaly as a function of the eccentric anomaly as follows

$$f(u) = \psi(u) - \psi_0 = |L| \int \frac{dr}{r^2 p_r}$$

$$= \arctan\left(\sqrt{\frac{1+e}{1-e}}\,\tan\frac{u}{2}\right)$$

$$+ \frac{|L|}{\sqrt{L^2 + 4GMb}}\,\arctan\left(\sqrt{\frac{a(1+e)+2b}{a(1-e)+2b}}\,\tan\frac{u}{2}\right).$$

$$(5.54)$$

Recalling equation (5.39), we can evaluate $\Delta\psi$ by means of

$$\Delta\psi = |L| \oint \frac{dr}{r^2 p_r} = f(u)\Big|_0^{2\pi},$$

so that, from the definition given in (5.37), the commensurability relation between the azimuthal and radial frequencies is

$$\frac{\omega_2}{\omega_1} = \frac{1}{2}\left(\frac{|L|}{\sqrt{L^2 + 4GMb}} - 1\right),$$

so that, taking into account that, from the isochronicity relation (5.42a) and the first of definitions (5.47), the radial frequency can be written as

$$\omega_1 = \frac{(-2E)^{3/2}}{GM} = \frac{\sqrt{GM}}{(a+b)^{3/2}},$$

$$(5.42b)$$

and the explicit form of the azimuthal frequency as

$$\omega_2 = \frac{\sqrt{GM}}{2(a+b)^{3/2}}\left(\frac{|L|}{\sqrt{L^2 + 4GMb}} - 1\right).$$

$$(5.55)$$

Exploiting the function $\tau(s)$ of (5.51), the time expressed in terms of the eccentric anomaly is

$$t - t_0 = \int_{r_{\min}}^{r} \frac{dr}{p_r} = \frac{b}{\sqrt{-2E}} \int_{s_{\min}}^{s} \frac{(s-1)ds}{\sqrt{X}}$$

$$= \frac{b}{\sqrt{-2E}}\left(\tau(s) - \tau(s_{\min})\right) = \sqrt{\frac{a+b}{GM}}\left[(a+b)u - ae\sin u\right],$$

$$(5.56)$$

whereas the equation of the orbit in the form $r = r(u)$ is, for (5.44) and (5.52),

$$r = a\sqrt{(1 - e\cos u)(1 - e\cos u + 2b/a)}.$$

$$(5.57)$$

It is useful to have the expression of the momenta in terms of the coordinates. By means of (5.56) and (5.57), we have

$$p_r = \frac{dr}{dt} = \frac{dr}{du}\frac{du}{dt} = \frac{dr}{du}\left(\frac{dt}{du}\right)^{-1} = \sqrt{\frac{GM}{a+b}}\,\frac{ae\sin u}{r},$$

$$(5.58)$$

whereas for p_ϑ the result is readily obtained by recalling the general expressions (5.14b) and (5.32):

$$p_\vartheta = \sqrt{L^2 - \frac{L_z^2}{\sin^2 \vartheta}} = |L| \sin i \, \frac{\cos \psi}{\sin \vartheta}. \tag{5.59}$$

As for the action integrals, for J_2 and J_3 the general results of the previous section give directly, from (5.20),

$$J_2 = J_\vartheta + J_\varphi = |L|, \qquad J_3 = J_\varphi = L_z = |L| \cos i = J_2 \cos i. \tag{5.60}$$

To find $J_1 = J_r + J_2$, we have to integrate the radial action integral (5.19a)

$$J_r = \frac{1}{\pi} \int_{r_{\min}}^{r_{\max}} \sqrt{2\left[E - \Phi_I(r)\right] - \frac{L^2}{r^2}} \, dr.$$

This can be done directly, exploiting, for example, the techniques of integration of complex analytic functions and the theorem of residues.[10] The result is

$$J_r = \frac{GM}{\sqrt{-2E}} - \frac{1}{2}\left(|L| + \sqrt{L^2 + 4GMb}\right). \tag{5.61}$$

It is interesting and instructive, however, to obtain this result following an alternative route.[11] From the fundamental property of the isochronal potential (5.42a), the radial frequency depends only on the energy, so that the partial derivative

$$\frac{\partial \mathcal{H}}{\partial J_r} = \frac{\partial E}{\partial J_r} = \omega_1(E) = \omega_r(E)$$

can be integrated in the form

$$J_r = \int \frac{dE}{\omega_r(E)} + \tilde{f}(L) = \frac{GM}{\sqrt{-2E}} + \tilde{f}(L), \tag{5.62}$$

where $\tilde{f}(L)$ is a function to be determined. To this end, we exploit the properties of circular orbits, that is, those orbits for which we have

$$v_r = \dot{r} = p_r \equiv 0 \longrightarrow J_r = J_r^{(c)} \equiv 0$$

and

$$v_\vartheta^2 + v_\varphi^2 = v_T^2 = r\frac{d\Phi_I}{dr} = \frac{GMr^2}{\sqrt{b^2 + r^2}\left(b + \sqrt{b^2 + r^2}\right)^2},$$

where $v_T^2 = r_c(d\Phi/dr)|_{r_c}$ represents the equilibrium between the centrifugal and the gravitational force (on the circular orbit $r = r_c$). The energy and the angular momentum of a circular orbit are respectively

---

[10]M. Born: op. cit. in Sect. 1.17, Appendix on Elementary and Complex Integration, pp. 303 ff.

[11]J. Binney, S. Tremaine, op. cit., p. 186

$$E^{(c)} = \frac{1}{2}v_T^2 + \Phi_I(r_c) = -\frac{GM}{2\sqrt{b^2 + r^2}}, \tag{5.63}$$

$$|L|^{(c)} = r_c v_T = r_c\left(E^{(c)}\right)\sqrt{2\left(E^{(c)} - \Phi_I\right)}$$
$$= \sqrt{GMb}\left(\sqrt{\frac{GM}{-2bE^{(c)}}} - \sqrt{\frac{-2bE^{(c)}}{GM}}\right). \tag{5.64}$$

So that J_r vanishes, the condition

$$\tilde{f}(L^{(c)}) = -\frac{GM}{\sqrt{-2E^{(c)}}} = -\sqrt{\frac{GMb}{\chi}}$$

must be satisfied for (5.62), where we have introduced the parameter

$$\chi = -\frac{2E^{(c)}}{GM}b, \quad 0 \le \chi \le 1.$$

Now, owing to (5.64), χ satisfies the equation

$$|L|^{(c)}/\sqrt{GMb} = (1 - \chi)/\sqrt{\chi},$$

admitting solutions

$$\sqrt{\chi} = \frac{-|L|^{(c)} \pm \sqrt{(L^{(c)})^2 + 4GMb}}{2\sqrt{GMb}}.$$

As for the values admitted by χ we must take the solution with the *plus* sign, so that the $\tilde{f}$ function that we seek, which *must be the same even if the orbit is not circular*, has the form

$$\tilde{f}(L) = \frac{2GMb}{|L| - \sqrt{L^2 + 4GMb}} = -\frac{1}{2}\left(|L| + \sqrt{L^2 + 4GMb}\right),$$

which, put into (5.62), provides (5.61), as was to be proved. For (5.60) and (5.61) the integral J_1 is then

$$J_1 = \frac{GM}{\sqrt{-2E}} + \frac{1}{2}\left(|L| - \sqrt{L^2 + 4GMb}\right). \tag{5.65}$$

The inversion of this last relation provides the explicit form of the Hamiltonian as a function of the action integrals

$$\mathcal{H}(J_1, J_2) = -\frac{G^2 M^2}{2\left[J_1 - \frac{1}{2}\left(J_2 - \sqrt{J_2^2 + 4GMb}\right)\right]^2}, \tag{5.66}$$

which, in one sense, is the main result of this treatment, since it gives a complete solution of the dynamical problem.

If we want to express the actions in terms of the model and orbital parameters, recalling definitions (5.47), we obtain

$$E = -\frac{GM}{2(a+b)}, \qquad |L| = \sqrt{\frac{GMa^2}{a+b}(1-e^2)}, \tag{5.67}$$

and inserting these into (5.65) yields

$$J_1 = \sqrt{GM(a+b)}\left[1 + \frac{a}{2(a+b)}\left(\sqrt{1-e^2} - \sqrt{1-e^2 + \frac{4b}{a}\left(1+\frac{b}{a}\right)}\right)\right], \tag{5.68}$$

whereas J_2 and J_3 can simply be obtained by inserting the second of equations (5.67) into (5.60). From (5.42b) and (5.55) the frequencies are given by

$$\omega_1 = \frac{\sqrt{GM}}{(a+b)^{3/2}}, \qquad \omega_2 = \frac{1}{2}\omega_1\left[\sqrt{\frac{1-e^2}{\left(1+\frac{2b}{a}\right)^2 - e^2}} - 1\right], \tag{5.69}$$

from which we see that, in general, the two frequencies are not commensurable. According to the discussion of the previous section, the orbit densely fills the annulus comprised between $r_{\min}$ and $r_{\max}$. Their explicit expression as orbit elements can be obtained from (5.48), coming back to the original radial coordinate by inverting the transformation (5.44). The angle–variables are the *mean anomaly* θ_1, given by (5.53), the angular advance of the pericentre

$$\theta_2 = \frac{\omega_2}{\omega_1}\theta_1 = \frac{1}{2}\left[\sqrt{\frac{1-e^2}{\left(1+\frac{2b}{a}\right)^2 - e^2}} - 1\right]\left(u - \frac{ae}{a+b}\sin u\right),$$

and the constant longitude of the ascending node θ_3 given by (5.34).

With these results we obtain a complete analytical description of the orbits admitted by the family of isochronal potentials. In the limit $b = 0$ (Keplerian potential) we again find the classical Delaunay elements of Sect. 2.5. The actions are

$$J_1 = \sqrt{GMa}, \quad J_2 = \sqrt{GMa(1-e^2)}, \quad J_3 = \sqrt{GMa(1-e^2)}\cos i. \tag{5.70}$$

The frequencies are

$$\omega_1 = \frac{\sqrt{GM}}{a^{3/2}}, \qquad \omega_2 = \omega_3 = 0. \tag{5.71}$$

The angles are

$$\theta_1 = \omega_1(t - t_0), \qquad \theta_2 = \psi_0, \qquad \theta_3 = \Omega. \tag{5.72}$$

Taking account of the slight change of notation, we see that equations (5.70) and (5.72) coincide with (2.82) defining the Delaunay elements for the Keplerian motion and the first of equations (5.71) has the same meaning of (2.84b).

Putting $b = 0$ into Hamiltonian (5.66) gives

$$\mathcal{H}(J_1) = -\frac{G^2 M^2}{2J_1^2} = -\frac{G^2 M^2}{2(J_r + J_\vartheta + J_\varphi)^2} = -\frac{GM}{2a} = E, \qquad (5.73)$$

which accounts for the complete degeneracy of Kepler's problem, whereas $b = 0$ into (5.56) and (5.57) gives *Kepler's equation* (2.53),

$$\theta_- = \frac{\sqrt{GM}}{a^{3/2}}(t - t_0) = u - e\sin u,$$

and the equation of the orbit (2.50)

$$r = a(1 - e\cos u).$$

5.4 Elliptical Coordinates and Stäckel's Theorem

In this section we introduce some of the curvilinear coordinate systems most useful for their application to analytical dynamics: in essence they are families of *elliptical* (in two dimensions) and *ellipsoidal* (in three dimensions) coordinates and, owing to Stäckel's theorem, they play a very important role in the application of the Hamilton–Jacobi theory to the study of the motion in potentials endowed with spheroidal and ellipsoidal symmetries. In the case of motion in two dimensions we will also introduce conformal transformations in the complex plane, discussing the link between elliptical coordinates and a conformal transformation that is particularly useful for applications in mechanics. For an exhaustive treatment we also give the relation between parabolic coordinates (as another class of curvilinear coordinates in the plane) and the Levi-Civita transformation, previously used for the regularization of the two-body problem (Sect. 2.6). A complete account of the application of ellipsoidal coordinates can be found in the classic work by Jacobi,[12] whereas the general properties of curvilinear coordinate systems are reviewed in the textbooks by Whittaker and Watson,[13] Morse and Feshbach,[14] and Jeffreys and Swirles.[15] As ellipsoidal coordinates have come back in vogue for their applications in galactic dynamics, there are very good expositions of their relation to the motion in galactic potentials by Lynden-Bell[16] and de Zeeuw.[17]

[12]K. G. J. Jacobi: op. cit., Lecture 26.

[13]E. T. Whittaker, G. N. Watson: *A Course of Modern Analysis* (Cambridge University Press, 1927), pp. 547 ff.

[14]P. M. Morse, H. Feshbach: *Methods of Theoretical Physics* (McGraw Hill, New York, 1953), Chap. 5.

[15]H. Jeffreys, B. Swirles: *Methods of Mathematical Physics* (Cambridge University Press, 1956), pp. 157 ff., 532 ff.

[16]D. Lynden-Bell: Stellar Dynamics – Potentials with isolating integrals, *Mon. Not. Roy. Astron. Soc.* **124**, 95–123 (1962).

[17]P. T. de Zeeuw: Elliptical galaxies with separable potentials, *Mon. Not. Roy. Astron. Soc.* **216**, 273–334 (1985).

Let x, y and z be the Cartesian coordinates in space. Consider the equation in the variable τ

$$f(\tau) = \frac{x^2}{\tau + \alpha} + \frac{y^2}{\tau + \beta} + \frac{z^2}{\tau + \gamma} - 1 = 0, \tag{5.74}$$

where α, β and γ are constants such that

$$\alpha < \beta < \gamma < 0.$$

For $\tau = 0$ (5.74) becomes the equation of an ellipsoid with semiaxes $a = \sqrt{-\alpha}$, $b = \sqrt{-\beta}$, $c = \sqrt{-\gamma}$ referred to its own axes, whereas, if x, y and z are the coordinates of a point which is external to the ellipsoid, we have

$$\frac{x^2}{a^2} + \frac{y^2}{b^2} + \frac{z^2}{c^2} - 1 > 0. \tag{5.75}$$

If in (5.74), having fixed the coordinates x, y and z of a point P, the variable τ is considered to be the unknown, then we arrive at an equation of third degree in τ:

$$x^2(\tau + \beta)(\tau + \gamma) + y^2(\tau + \alpha)(\tau + \gamma) + z^2(\tau + \alpha)(\tau + \beta)$$
$$-(\tau + \alpha)(\tau + \beta)(\tau + \gamma) = 0. \tag{5.76}$$

All the roots of this equation are real. To find them we observe, first of all, that

$$\lim_{\tau \to \infty} f(\tau) = -1.$$

Then, letting $\tau = -\alpha + \varepsilon$ in (5.74), we find that the quantity

$$f(-\alpha + \varepsilon) = \frac{x^2}{\varepsilon} + \frac{y^2}{\beta - \alpha + \varepsilon} + \frac{z^2}{\gamma - \alpha + \varepsilon} - 1$$

is positive if ε is sufficiently small. Therefore, equation (5.74) possesses a positive root lying between $-\alpha$ and $+\infty$, which we denote by λ.

If then, in the equation $f(\tau) = 0$, we alternatively put first $\tau = -\beta + \varepsilon$ and then $\tau = -\alpha - \varepsilon$, for ε sufficiently small we have the inequalities

$$f(-\beta + \varepsilon) > 0 \quad \text{and} \quad f(-\alpha - \varepsilon) < 0,$$

so that we can state that a second root, μ, of $f(\tau) = 0$ lies between $-\beta$ and $-\alpha$. If, finally, in (5.74), we put first $\tau = -\gamma + \varepsilon$ and then $\tau = -\beta - \varepsilon$, so that for ε sufficiently small we have the inequalities

$$f(-\gamma + \varepsilon) > 0 \quad \text{and} \quad f(-\beta - \varepsilon) < 0,$$

we can conclude that a third root, ν, of $f(\tau) = 0$ lies between $-\gamma$ and $-\beta$. To summarize, the three roots of (5.74) satisfy the chain of inequalities

$$-\gamma \leq \nu \leq -\beta \leq \mu \leq -\alpha \leq \lambda. \tag{5.77}$$

Given the Cartesian coordinates x, y and z of a point, the three roots λ, μ and ν turn out to be uniquely determined: they can be interpreted as the *ellipsoidal coordinates* of the point. In turn, given the ellipsoidal coordinates, the inverse transformation exists, which allows us to find from the latter coordinates the corresponding Cartesian coordinates, but it turns out that this inverse transformation is not unique. Hereafter we establish the form of both these transformations and discuss briefly the principal properties of the ellipsoidal coordinates.

To obtain x, y and z, as functions of λ, μ and ν we observe that (5.76) can certainly be written as the product of binomials of the variable and the roots, so that

$$-(\tau - \lambda)(\tau - \mu)(\tau - \nu) = 0.$$

Equating the left-hand sides of this equation and of (5.76) and dividing by $(\tau + \alpha)(\tau + \beta)(\tau + \gamma)$, we find that

$$f(\tau) = \frac{x^2}{\tau + \alpha} + \frac{y^2}{\tau + \beta} + \frac{z^2}{\tau + \gamma} - 1 = -\frac{(\tau - \lambda)(\tau - \mu)(\tau - \nu)}{(\tau + \alpha)(\tau + \beta)(\tau + \gamma)}. \tag{5.78}$$

To find the coordinate, say, x, in terms of the ellipsoidal coordinates we multiply both sides of (5.78) by $\tau + \alpha$ and, after that, we impose the condition $\tau = -\alpha$; proceeding in an analogous manner for the other coordinates we obtain

$$x^2 = \frac{(\lambda + \alpha)(\mu + \alpha)(\nu + \alpha)}{(\alpha - \beta)(\alpha - \gamma)},$$

$$y^2 = \frac{(\lambda + \beta)(\mu + \beta)(\nu + \beta)}{(\beta - \alpha)(\beta - \gamma)}, \tag{5.79}$$

$$z^2 = \frac{(\lambda + \gamma)(\mu + \gamma)(\nu + \gamma)}{(\gamma - \alpha)(\gamma - \beta)}.$$

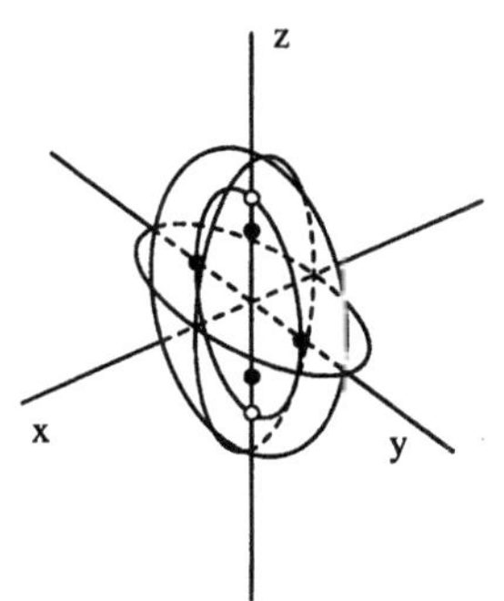

Fig. 5.2a

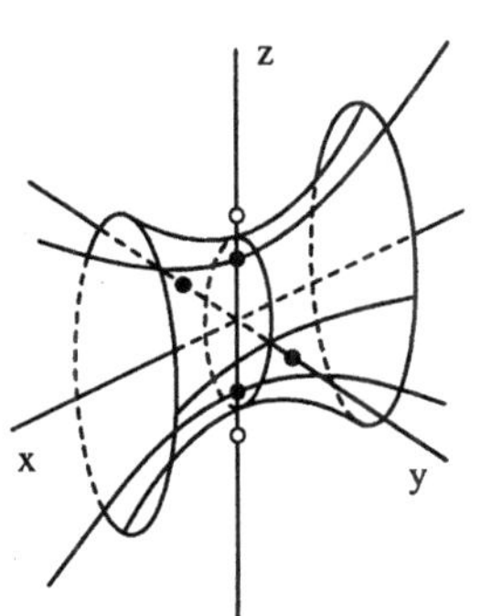

Fig. 5.2b

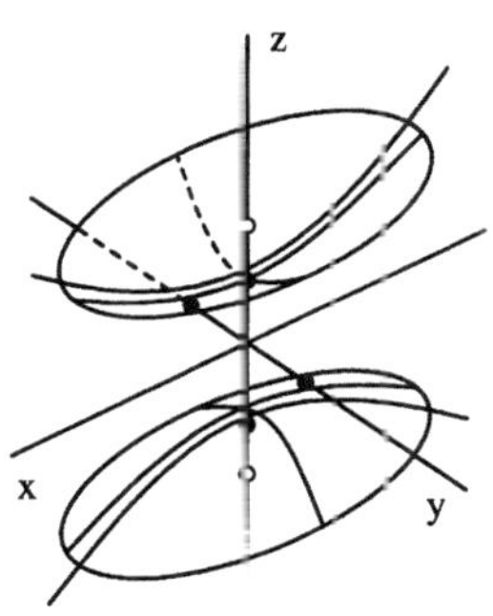

Fig. 5.2c

To each set (λ, μ, ν), lying in the intervals given by (5.77), there therefore correspond eight points, one for each octant. The surfaces $\lambda(x, y, z) = \text{const.}$ are *ellipsoids* (see Fig. 5.2a), since for inequalities (5.77) we have that

$$\lambda + \alpha > 0, \qquad \lambda + \beta > 0, \qquad \lambda + \gamma > 0,$$

and so, for $\tau = \lambda$, the denominators of (5.74) are positive and the equation is therefore that of an ellipsoid: for large values of λ this ellipsoid tends to the sphere of radius $\sqrt{\lambda + \alpha}$, whereas as λ tends to the limit $-\alpha$ the ellipsoid shrinks more rapidly in the direction of the x axis (its major axis always remains in the direction of the z axis) until at $\lambda = -\alpha$ its minor axis becomes zero and the ellipsoid degenerates into the area delimited by the *focal ellipse* with the equation

$$\frac{y^2}{(\beta - \alpha)} + \frac{z^2}{(\gamma - \alpha)} = 1, \qquad (5.80a)$$

whose foci are at the points $z = \pm\sqrt{\gamma - \beta}$ and which intersects the z axis in the points $z = \pm\sqrt{\gamma - \alpha}$ and the y axis at the points $y = \pm\sqrt{\beta - \alpha}$.

The surfaces $\mu(x, y, z) = \text{const.}$ are *hyperboloids constituted of one sheet* (see Fig. 5.2b) around the x axis, since inequalities (5.77) now give

$$\mu + \alpha < 0, \qquad \mu + \beta > 0, \qquad \mu + \gamma > 0.$$

For $\mu = -\alpha$ they degenerate into the surface given by the part of the plane $x = 0$ external to the focal ellipse defined above. With μ tending to $-\beta$ the ellipse by which the hyperboloid cuts the $x = 0$ plane shrinks and, in the limit $\mu = -\beta$, it coincides with the segment on the z axis limited by the foci $z = \pm\sqrt{\gamma - \beta}$. At the same time the hyperboloid degenerates into the part of the $y = 0$ plane lying between the two branches of the *focal hyperbola* given by

$$\frac{x^2}{(\alpha - \beta)} + \frac{z^2}{(\gamma - \beta)} = 1. \qquad (5.80b)$$

This hyperbola has its foci at the points $z = \pm\sqrt{\gamma - \alpha}$ and cuts the z axis at the foci of the focal ellipse.

The surface $\nu = -\beta$ coincides with the portion of the $y = 0$ plane external to the two branches of the focal hyperbola. The surfaces $\nu(x, y, z) = \text{const.}$, diminishing the value of ν, are *hyperboloids consisting of two sheets* (see Fig. 5.2c), since inequalities (5.77) give in this case

$$\nu + \alpha < 0, \qquad \nu + \beta < 0, \qquad \nu + \gamma > 0.$$

In the limit $\nu = -\gamma$ these hyperboloids coincide with the $z = 0$ plane.

Through each point x, y and z pass *one* ellipsoid λ, *one* hyperboloid of one sheet μ and *one* hyperboloid of two sheets ν. Given the three values of x, y and z, the three corresponding values of λ, μ and ν are found by means of the relations

$$\lambda + \mu + \nu = -\alpha - \beta - \gamma + x^2 + y^2 + z^2,$$
$$\lambda\mu + \mu\nu + \nu\lambda = \alpha\beta + \beta\gamma + \gamma\alpha - (\beta + \gamma)x^2 - (\gamma + \alpha)y^2 - (\alpha + \beta)z^2,$$
$$\lambda\mu\nu = -\alpha\beta\gamma + \beta\gamma x^2 + \gamma\alpha y^2 + \alpha\beta z^2,$$

which are obtained by equating, in equality (5.78), the powers one, two and zero respectively of τ.

To determine the line element in ellipsoidal coordinates we can use the fact that the metric tensors of the two coordinate frames are linked by the relation

$$g_{AB} = \frac{\partial x^i}{\partial \tau^A} \frac{\partial x^j}{\partial \tau^B} \delta_{ij}; \qquad A, B = 1, 2, 3; \qquad i, j = 1, 2, 3;$$

where $\tau^A = \{\lambda, \mu, \nu\}$ and the matrix elements of the coordinate transformation are defined as in (A.7). We find then that, for e.g. $A = 1$,

$$g_{11} = \sum_i \left(\frac{\partial x^i}{\partial \lambda}\right)^2 = \frac{1}{4}\left(\frac{x^2}{(\lambda + \alpha)^2} + \frac{y^2}{(\lambda + \beta)^2} + \frac{z^2}{(\lambda + \gamma)^2}\right) = G_1^2,$$

$$g_{12} = \sum_{ij} \frac{\partial x^i}{\partial \lambda} \frac{\partial x^j}{\partial \mu} = 0,$$

$$g_{13} = \sum_{ij} \frac{\partial x^i}{\partial \lambda} \frac{\partial x^j}{\partial \nu} = 0,$$

as can be verified by explicitly performing the derivatives of equalities (5.79); this allows us also to check the orthogonality of the coordinate system. To obtain expressions written only in terms of the τ^A, we can differentiate both sides of equation (5.78) with respect to τ so that we have

$$\frac{x^2}{(\tau + \alpha)^2} + \frac{y^2}{(\tau + \beta)^2} + \frac{z^2}{(\tau + \gamma)^2} = \frac{(\tau - \mu)(\tau - \nu)}{(\tau + \alpha)(\tau + \beta)(\tau + \gamma)}$$
$$+ (\tau - \lambda)\frac{\partial}{\partial \tau}\frac{(\tau - \mu)(\tau - \nu)}{(\tau + \alpha)(\tau + \beta)(\tau + \gamma)}.$$

Putting $\tau = \lambda$, this equality becomes

$$\frac{x^2}{(\lambda + \alpha)^2} + \frac{y^2}{(\lambda + \beta)^2} + \frac{z^2}{(\lambda + \gamma)^2} = \frac{(\lambda - \mu)(\lambda - \nu)}{(\lambda + \alpha)(\lambda + \beta)(\lambda + \gamma)},$$

so that in the end we find that

$$G_1^2 = P^2 = \frac{1}{4}\frac{(\lambda - \mu)(\lambda - \nu)}{(\lambda + \alpha)(\lambda + \beta)(\lambda + \gamma)}. \tag{5.81a}$$

Following the same route for the cases $A = 2$ and $A = 3$, we obtain respectively

$$G_2^2 = Q^2 = \frac{1}{4}\frac{(\mu - \nu)(\mu - \lambda)}{(\mu + \alpha)(\mu + \beta)(\mu + \gamma)},\tag{5.81b}$$

$$G_3^2 = R^2 = \frac{1}{4}\frac{(\nu - \lambda)(\nu - \mu)}{(\nu + \alpha)(\nu + \beta)(\nu + \gamma)},\tag{5.81c}$$

so that the line element, with general form (A.6), in ellipsoidal coordinates becomes

$$ds^2 = P^2 d\lambda^2 + Q^2 d\mu^2 + R^2 d\nu^2.$$

It is clear that a three-dimensional coordinate system induces on given surfaces, in particular on planes, two-dimensional coordinate systems. The simplest and most important example we can describe is that of *elliptical coordinates in the plane*: putting $z = 0$ in (5.74) (and assuming, without loss of generality, that γ also is zero) we can define the elliptical coordinates (λ, μ) on the xy plane as the roots for τ of the equation

$$\frac{x^2}{\tau + \alpha} + \frac{y^2}{\tau + \beta} - 1 = 0.\tag{5.82a}$$

Without going through all the previous treatment of the three-dimensional case, to find all the relevant properties of the elliptical coordinates it suffices to put $z = 0$ (and $\gamma = 0$) in all the expressions from (5.75) to (5.81). We easily find then that the generalization of inequalities (5.77) in this case is

$$-\beta \leq \mu \leq -\alpha \leq \lambda,\tag{5.82b}$$

and that the relations between the Cartesian and the elliptical coordinates in the plane are

$$x^2 = \frac{(\lambda + \alpha)(\mu + \alpha)}{\alpha - \beta}, \qquad y^2 = \frac{(\lambda + \beta)(\mu + \beta)}{\beta - \alpha},\tag{5.83a}$$

from which we can find

$$\lambda + \mu = -\alpha - \beta + x^2 + y^2, \qquad \lambda\mu = \alpha\beta - \beta x^2 - \alpha y^2\tag{5.83b}$$

and, finally, that the line element is given by the expression

$$ds^2 = P^2 d\lambda^2 + Q^2 d\mu^2,$$

where the components of the metric tensor are

$$P^2 = \frac{1}{4}\frac{\lambda - \mu}{(\lambda + \alpha)(\lambda + \beta)}, \qquad Q^2 = \frac{1}{4}\frac{\mu - \lambda}{(\mu + \alpha)(\mu + \beta)}.\tag{5.84}$$

The curves $\lambda = $ constant are ellipses with the major axis in the direction of the y axis and foci at the points $y = \pm\sqrt{\beta - \alpha}$. The degenerate ellipse $\lambda = -\alpha$ is the segment of the y axis lying between the foci. The curves $\mu = $ constant are hyperbolas with the same foci. For $\mu = -\beta$ they coincide with the x axis; the value $\mu = -\alpha$ corresponds to the part of the y axis outside the foci. Near the origin the elliptical coordinates are approximatively similar to the Cartesian ones, whereas at large distances from the origin ($\lambda \to \infty$) the ellipses become progressively almost circular and the hyperbolas tend to coincide with their asymptotes, which are straight lines passing through the origin. Therefore, at large distances from the origin, the elliptical coordinates tend to the polar ones. However, it is easy to verify that, in the particular case for which $\alpha = \beta$, the elliptical coordinates in the plane coincide exactly with the polar ones (r, φ).

By using curvilinear coordinates in the plane we could very much enlarge the setting of the treatment, since their range of applications is very large and there are many links with other fields of analytical mechanics. The opportunity offered by the introduction of elliptical coordinates just made allows us, without too much digression, to briefly treat the coordinate systems generated by conformal transformations in the complex plane and to shed light on their usefulness in applications. Further details can be found in the textbooks by Jeffreys and Swirles[18] and Markushevich.[19]

Let us denote by $z = x + iy$ the points on the complex plane and let $w = F(z)$ be an analytic function. $F(z)$ is therefore a single-valued function and its derivative $F'(z)$ is continuous all over the plane: if we indicate with $u(x, y)$ and $v(x, y)$, respectively, the real and the imaginary parts of w, they must satisfy the Cauchy–Riemann relations (2.95), where the change of notation $x_1 \to x$, $x_2 \to y$, $u_1 \to u$, $u_2 \to v$ is assumed. The transformation of the complex plane generated by the function $F(z)$ can be interpreted as a transformation of coordinates of the kind described in Appendix 2. Applying relations (2.95) we soon see that the Jacobian of the transformation is not zero and that, using the metric tensor transformation (A.7), the line element in the two systems can be written

$$dx^2 + dy^2 = \left[\left(\frac{\partial x}{\partial u} \right)^2 + \left(\frac{\partial y}{\partial u} \right)^2 \right] (du^2 + dv^2)$$

$$= \left| \frac{\partial x}{\partial u} + i \frac{\partial y}{\partial u} \right|^2 (du^2 + dv^2)$$

$$= |f'(w)|^2 (du^2 + dv^2) ,$$

where we have introduced the inverse function $z = f(w)$, so that, clearly, we have

$$f'(w) = \frac{1}{F'(z)}. \tag{5.85}$$

Two line elements that are related to each other by a multiplicative factor are said to be conformal; the function $F(z)$ therefore generates a *conformal transformation*. Without entering into the general exposition of the theory

<hr>

[18]H. Jeffreys, B. Swirles: op. cit., pp. 419 ff.

[19]A. I. Markushevich: *The Theory of Analytic Functions: a brief course* (MIR, Moscow, 1983), pp. 60 ff.

of conformal transformations, in the following we describe two particularly useful examples.

The first one represents a different way of introducing elliptical coordinates. Consider the analytical function

$$z = f(w) = \Delta \sinh w, \tag{5.86}$$

with Δ some real constant, which, exploiting the explicit expression for the hyperbolic function in terms of the complex variable

$$\sinh(w) = \sinh(u + iv) = \sinh u \cos v + i \cosh u \sin v,$$

provides the coordinate transformation

$$\begin{aligned} x &= \Delta \sinh u \cos v, \\ y &= \Delta \cosh u \sin v, \end{aligned} \tag{5.87}$$

with conformal factor

$$|f'(w)|^2 = \Delta^2 |\cosh(w)|^2 = \Delta^2 \left(\sinh^2 u + \cos^2 v \right).$$

It is then easy to verify that the further transformation

$$\sinh^2 u = \frac{\lambda + \alpha}{\beta - \alpha}, \qquad \sin^2 v = \frac{\mu + \beta}{\beta - \alpha}, \qquad \Delta^2 = \beta - \alpha$$

reduces expressions (5.87) to the standard form of equation (5.83a) of the relation between plane Cartesian and elliptical coordinates. It can also be easily verified that the properties of the conformal elliptical coordinates u and v are consistent with those examined above, noting in particular that the curves $u = $ const. are the ellipses and the ones at $v = $ const. are the hyperbolas.

The second example of great relevance is that obtained by using the function $z = f(w) = w^2$. In this case we have the coordinate transformation

$$\begin{aligned} x &= u^2 - v^2, \\ y &= 2uv, \end{aligned}$$

with conformal factor

$$|f'(w)|^2 = 4|w|^2 = 4(u^2 + v^2).$$

This coordinate transformation was introduced in Sect. 2.6, where it was used to discuss the problem of the regularization of the two-body problem, and is known as the Levi-Civita transformation (2.94). It is easy to see that the curves with constant u and v are parabolas with the axis given by the x axis, focus at the origin and concavity respectively towards the left and the right. If afterwards we introduce the further transformation

$$\lambda = (u + v)^2, \qquad \mu = (u - v)^2,$$

we obtain the *parabolic coordinates* (λ, μ), already introduced in Sect. 1.15 for the study of the resonant anisotropic oscillator, as can be verified by observing that the relation between these coordinates and the Cartesian ones is

$$x = \sqrt{\lambda\mu},$$
$$y = \frac{1}{2}(\lambda - \mu), \qquad \lambda, \mu > 0,$$

in agreement with (1.C.113).

If two or three of the constants α, β or γ in (5.74) are equal, the ellipsoidal coordinates assume particular forms in which these degenerations reflect some further symmetry.

If $\beta = \gamma$ the coordinates (λ, μ, ν) become *oblate spheroidal coordinates* (λ, μ, γ). The surfaces for which $\lambda = $ const. are oblate spheroids with the minor axis in the direction of the x axis. The surfaces at $\mu = $ const. are hyperboloids of revolution around the x axis. The interval in which ν can range vanishes, so that it can no longer be used as a coordinate. The surfaces at $\nu = $ const. in the general case (hyperboloids with two sheets around the z axis), are now the planes that contain the x axis and that are now labelled by a constant value of the azimuthal angle γ. In every plane at $\gamma = $ const. (*meridional plane*) the coordinates (λ, μ) are elliptical coordinates, defined now as the roots for τ of the equation

$$\frac{x^2}{\tau + \alpha} + \frac{\tilde{z}^2}{\tau + \beta} - 1 = 0, \qquad \tilde{z}^2 = y^2 + z^2, \tag{5.88}$$

where $-\beta \leq \mu \leq -\alpha \leq \lambda$. The foci are at $\tilde{z} = \pm\sqrt{\beta - \alpha}$ and the relations between (λ, μ) and $(x, \tilde{z})$ can be deduced from (5.82) and (5.83). In the equatorial plane $x = 0$ the coordinates are polar $(\tilde{z}, \chi)$ with $\tilde{z}^2 = \tau + \beta$ $(\tau = \lambda$ or $\mu)$. The focal ellipse $\lambda = \mu = -\alpha$ in the yz plane of the general case, now degenerates into the *focal circle* of radius $\tilde{z} = \sqrt{\beta - \alpha}$. The line element is now

$$ds^2 = P^2 d\lambda^2 + Q^2 d\mu^2 + \tilde{z}^2 d\chi^2,$$

where the components of the metric tensor are the same as in (5.84)

If $\beta = \alpha$, the coordinates (λ, μ, ν) become *prolate spheroidal coordinates* (λ, ϕ, ν). The surfaces with $\lambda = $ const. are now prolate spheroids elongated in the direction of the z axis. The surfaces at $\nu = $ const. are hyperboloids of revolution with two sheets around the z axis. The interval in which μ can range vanishes $(\mu = -\beta = \alpha)$, so that it can no longer be used as a coordinate. The surfaces with $\mu = $ const. of the general case (hyperboloids with one sheet) are now planes that contain the z axis and that are therefore labelled by a constant value of the azimuthal angle ϕ. In every meridional plane at $\phi = $ const. the variables (λ, ν) are elliptical coordinates, now defined as the roots for τ of the equation

$$\frac{\varpi^2}{\tau + \alpha} + \frac{z^2}{\tau + \gamma} - 1 = 0, \qquad \varpi^2 = x^2 + y^2, \tag{5.89}$$

where $-\gamma \leq \nu \leq -\alpha \leq \lambda$. The foci are at $\varpi = 0$, $z = \pm\sqrt{\gamma - \alpha}$ and the relations between (λ, ν) and (ϖ, z) can be deduced from (5.82) and (5.83). In the equatorial plane the coordinates (ϖ, ϕ), with $\varpi^2 = \lambda + \alpha$, are polar coordinates. The line element is now

$$ds^2 = P^2 d\lambda^2 + \varpi^2 d\phi^2 + R^2 d\mu^2,$$

where the components of the metric tensor P and R are given by expressions analogous to (5.84).

If, finally, $\gamma = \beta = \alpha$, we are back with the spherical coordinates $(\lambda, \vartheta, \varphi)$ of (A.8), with $r^2 = \lambda + \alpha$.

We now come to the question of the separability of the Hamilton–Jacobi equation. As already mentioned in Sect. 1.15 this problem is related both to the properties of the potential and to the coordinate system used. If we limit ourselves to orthogonal coordinate systems, that is, those given by a line element of type (A.8), Stäckel's theorem (see Sect. 1.15) provides the necessary and sufficient conditions which must be satisfied at the same time by the potential and the coordinate system in order that the dynamical system under study be separable.[20] From considerations which will be made in Sect. 5.8 it turns out that some potentials of interest in astrophysics can satisfy with good approximation the conditions of Stäckel's theorem if ellipsoidal coordinates are used.

Stäckel's Theorem. Let a Hamiltonian dynamical system be given with a Hamilton function[21]

$$\mathcal{H}(p_i, q^i) = \mathcal{T} + \Phi(q^i), \qquad i = 1, 2, 3,$$

where the kinetic energy is given by the last of equations (A.9). The Hamiltonian system is separable if and only if there exists a regular 3×3 matrix $\mathcal{U}^i{}_j$, such that the elements on the i-th row are functions only of the coordinate q^i, and a vector $\mathcal{W}^i$, whose i-th component is a function of only the coordinate q^i, such that

$$\sum_i G_i^{-2} \mathcal{U}^i{}_j = \delta^1{}_j, \qquad \Phi = \sum_i G_i^{-2} \mathcal{W}^i. \tag{5.90}$$

It can be observed that, if we denote by $\bar{\mathcal{U}}^i{}_j$ the inverse matrix of $\mathcal{U}^i{}_j$, so that

$$\sum_k \mathcal{U}^i{}_k \bar{\mathcal{U}}^k{}_j = \delta^i{}_j,$$

then relations (5.90) imply that

$$G_i^{-2} = \bar{\mathcal{U}}^1{}_i; \qquad \Phi = \sum_i \bar{\mathcal{U}}^1{}_i \mathcal{W}^i. \tag{5.91}$$

---

[20] H. Goldstein: *Classical Mechanics* (Addison Wesley, 1980), Appendix D.

[21] The theorem can be generalized to N dimensions, but we will limit ourselves to the $N = 3$ dimensions of ordinary space.

Proof. Let us first show that conditions (5.90) are *necessary* for the separability. Assume, therefore, that the system is separable, that is that the partial differential equation (which is the specialization to the present situation of general equation (1.C.90))

$$\frac{1}{2}\sum_i G_i^{-2}\left(\frac{\partial S^\star}{\partial q^i}\right)^2 + \Phi = \alpha_1 \tag{5.92}$$

admits a complete integral of the form

$$S^\star = \sum_i S_i^\star, \qquad S_i^\star = S_i^\star(q^i, \alpha_1, \alpha_2, \alpha_3). \tag{5.93}$$

By then substituting the complete integral (5.93) into (5.92), this equation is satisfied identically for every value of q^k and α_i in the domain of interest. Differentiating with respect to any α_i we obtain

$$\sum_i G_i^{-2}\frac{\partial S^\star}{\partial q^i}\frac{\partial^2 S^\star}{\partial\alpha_1\partial q^i} = 1, \quad \sum_i G_i^{-2}\frac{\partial S^\star}{\partial q^i}\frac{\partial^2 S^\star}{\partial\alpha_a\partial q^i} = 0, \quad a = 2, 3. \tag{5.94}$$

The coefficient of G_i^{-2} in any one of equations (5.94) is a function of q^i only, since $S^\star$ is of the form (5.93) and the determinant of these coefficients is

$$\mathcal{D} = \frac{\partial S^\star}{\partial q_1}\frac{\partial S^\star}{\partial q_2}\frac{\partial S^\star}{\partial q_3}\left|\frac{\partial^2 S^\star}{\partial\alpha\partial q}\right|.$$

By choosing a set $\{\alpha_1, \alpha_2, \alpha_3\}$ such that $\mathcal{D}$ does not vanish, the algebraic system (5.94) takes just the required form of the first of (5.90) with

$$\mathcal{U}^i{}_j = \frac{\partial S^\star}{\partial q^i}\frac{\partial^2 S^\star}{\partial\alpha_j\partial q^i}, \qquad |\mathcal{U}^i{}_j| \neq 0,$$

whereas (5.92) can be inverted to give

$$\Phi = \alpha_1 - \frac{1}{2}\sum_i G_i^{-2}\left(\frac{\partial S^\star}{\partial q^i}\right)^2 = \sum_i G_i^{-2}\left[\alpha_1\mathcal{U}^i{}_1 - \frac{1}{2}\left(\frac{\partial S^\star}{\partial q^i}\right)^2\right]$$

which is just the form required by the second of (5.90). This completes the proof that conditions (5.90) are necessary for the separability.

Let us now prove that (5.90) are sufficient: in fact, they allow us to write the Hamilton–Jacobi equation (5.92) in the form

$$\sum_i G_i^{-2}\left[\left(\frac{\partial S^\star}{\partial q^i}\right)^2 - 2\left(\sum_{k=1}^3 \alpha_k\mathcal{U}^i{}_k - \mathcal{W}^i\right)\right] = 0,$$

where α_2 and α_3 are arbitrary constants. It is then clear that a principal function of the form given by (5.93) is a complete integral if $S_i^\star$ satisfy the equalities

$$S_i^\star(q^i) = \int^{q^i} \sqrt{\mathcal{F}_i(x)}\, dx, \qquad \mathcal{F}_i(q^i) = 2\left(\sum_k \alpha_k \mathcal{U}^i{}_k - \mathcal{W}^i\right),$$

where the lower bound of the integration interval is chosen either as a constant or as a simple root of $\mathcal{F}_i$. The solution of the dynamical problem is therefore reduced to *quadratures* by integrating

$$t - t_0 = \frac{\partial S^\star}{\partial \alpha_1} = \sum_i \int \frac{\mathcal{U}^i{}_1 dq^i}{\sqrt{\mathcal{F}_i(q^i)}}, \tag{5.95}$$

$$-\beta_a = \frac{\partial S^\star}{\partial \alpha_a} = \sum_i \int \frac{\mathcal{U}^i{}_a dq^i}{\sqrt{\mathcal{F}_i(q^i)}}, \quad a = 2, 3, \tag{5.96}$$

$$p_i = \frac{\partial S^\star}{\partial q^i} = \sqrt{\mathcal{F}_i(q^i)}. \tag{5.97}$$

Equations (5.96) determine the trajectory in the q^i space (without taking account of the time dependence). Together with (5.95), they provide the solution of Lagrange's problem. Equations (5.97) give the components of the momenta as functions of the coordinates. This completes the proof of Stäckel's theorem.

5.5 Planar Potentials

We have emphasized in various circumstances above that the Hamilton–Jacobi method (see Sect. 1.15) is the more powerful technique for solving problems in mechanics. For this reason and also with the aim of giving a unified view of the subject, the study of orbits in fixed potentials will be made by exploiting, as far as possible, the separability of the Hamilton–Jacobi equation. It is therefore obvious that Stäckel's theorem is an important result in this framework, since (5.95–97) give, at least formally, the complete solution to the problem (provided that the potential and the coordinate system obey suitable conditions). In Sect. 5.1 we also focussed our attention on the important role played by the isolating integrals, whose direct determination (when possible) allows a complete analysis of the motion. The aim of the present section is therefore the search for the second integral in a conservative system with two degrees of freedom, without the assumption that the conditions of Stäckel's theorem are satisfied. The result of the analysis is then that the existence of a second integral (quadratic in the momenta) implies that the potential and the coordinate system *do satisfy* Stäckel conditions, which are then recovered in a completely independent way.

In this section, at variance with the rest of the chapter, we study the motion of a point mass in the plane, where a two-dimensional potential

$$\Phi(x, y), \qquad \{x, y\} \in \mathbb{R}^2$$

is defined. The general problem of a Hamiltonian systems with two degrees of freedom is of fundamental importance by itself, but the following considerations demonstrate the usefulness of studying motion in two-dimensional potentials also for the help given in cases of interest in the three-dimensional world. First of all there are cases in which a symmetry of the potential allows us, in a suitable coordinate system, to reduce the number of degrees of freedom by means of the *effective potential* (an important example is that of rotational symmetry, which is examined in Sect. 5.8). A second point is the existence of planar orbits in every triaxial potential provided with reflection symmetry planes: their analysis, even if seldom possible analytically, sheds light on the general properties of spatial orbital motion. Finally, mass density distributions exist (for example the central regions of barred spiral galaxies) that give rise to gravitational potentials fairly well approximated by planar potentials.

We do not have a general method for solving the equations of motion of a system with two degrees of freedom. This problem is related to the impossibility of verifying the existence of (and finding explicitly) a second isolating integral. Therefore, in what follows we consider *prototypical* cases in which it is possible to integrate the system in the sense of Liouville's theorem (Sect. 1.19), postponing to Volume 2 the study of those cases in which the only viable route is that of perturbation theory.

The prototype of integrable dynamical systems with two (or more) degrees of freedom is the *anisotropic harmonic oscillator* (see Sects. 1.3, 1.7, 1.15). In strict analogy, every potential of the form

$$\Phi(x,y) = \Phi_1(x) + \Phi_2(y), \tag{5.98}$$

supports the two isolating integrals

$$I_x = p_x^2 + 2\Phi_1(x), \qquad I_y = p_y^2 + 2\Phi_2(y). \tag{5.99}$$

The motion of the system is given by the superposition of the librations along the two axes obtained as the solutions of the decoupled system of equations

$$\ddot{x} + \frac{d\Phi_1}{dx} = 0, \qquad \ddot{y} + \frac{d\Phi_2}{dy} = 0,$$

which is the generalization of (1.A.38), the equation of the two-dimensional harmonic oscillator. In agreement with Liouville's theorem on integrable systems, it is possible to *decouple* the two degrees of freedom in a general system if we are able to find a second isolating integral, namely a function $I_2(p_x, p_y, x, y)$, which is constant along the trajectories of motion, that is $\dot{I}_2 = 0$ (obviously, the *first* isolating integral, I_1, is the energy). A direct method for fulfilling the quest for the second integral is that of assuming a given functional form for I_2 and enforcing its conservation along the trajectories by means of the condition expressed by (1.C.40) that the Poisson bracket with the Hamiltonian

$$\mathcal{H} = \frac{1}{2}\left(p_x^2 + p_y^2\right) + \Phi(x,y) \equiv I_1,$$

be zero,

$$(I_2, \mathcal{H}) = 0.$$

In other words, the partial differential equation

$$\frac{\partial I_2}{\partial x}p_x + \frac{\partial I_2}{\partial y}p_y = \frac{\partial I_2}{\partial p_x}\frac{\partial \mathcal{H}}{\partial x} + \frac{\partial I_2}{\partial p_y}\frac{\partial \mathcal{H}}{\partial y} \tag{5.100}$$

must be satisfied. A general review of the results obtained by this method is that by Hietarinta:[22] let us examine here some simple cases with applications in galactic dynamics.

We will limit ourselves to the case in which the second integral is a polynomial in the momenta. Let us first examine the very simple case of an integral linear in the momenta

$$I_2 = A(x,y)p_x + B(x,y)p_y + C(x,y). \tag{5.101}$$

We have to find what kind of potentials admit an integral of this form, explicitly determining the functions A, B, C and Φ. The condition expressed by (5.100) must hold for any value of the phase space coordinates. Equating to zero the coefficients of the polynomial expression obtained by inserting (5.101) in (5.100), we find the following set of equations:

$$\frac{\partial A}{\partial x} = 0, \quad \frac{\partial B}{\partial y} = 0, \quad \frac{\partial C}{\partial x} = \frac{\partial C}{\partial y} = 0,$$

$$\frac{\partial A}{\partial y} + \frac{\partial B}{\partial x} = 0, \qquad A\frac{\partial \Phi}{\partial x} + B\frac{\partial \Phi}{\partial y} = 0. \tag{5.102}$$

The set of equations (5.102) has the following solutions: the function C is simply a constant, which can be put equal to zero without loss of generality. The functions A and B are given by

$$A = ay + b, \qquad B = -ax + c,$$

where a, b and c are arbitrary constants (perhaps complex). With these expressions for A and B, the form of the second invariant can be written explicitly and the last of equations (5.102) can be solved for Φ. There are two distinct cases: $a = 0$ (the uniform force field), for which we find that

$$\Phi(x,y) = f(cx - by), \qquad I_2 = bp_x + cp_y,$$

and $a \neq 0$ (the central force field), with

$$\Phi(x,y) = f(x^2 + y^2), \qquad I_2 = yp_x - xp_y,$$

giving the conservation of angular momentum (f is an arbitrary function of its argument).

---

[22] J. Hietarinta: Direct methods for the search of the second invariant, *Phys. Rep.* **147**, 87–154, (1987).

Let us move on to the case of the invariant quadratic in the momenta, namely

$$I_2 = Ap_x^2 + Bp_xp_y + Cp_y^2 + D, \tag{5.103}$$

where A, B, C and D are functions of x and y. Proceeding exactly as in the linear case, equating to zero the coefficients of the polynomial expression obtained by inserting (5.103) in (5.100), we find the following set of equations:

$$\frac{\partial A}{\partial x} = 0, \qquad \frac{\partial C}{\partial y} = 0, \qquad \frac{\partial A}{\partial y} + \frac{\partial B}{\partial x} = 0, \qquad \frac{\partial B}{\partial y} + \frac{\partial C}{\partial x} = 0, \tag{5.104}$$

$$\frac{\partial D}{\partial x} = 2A\frac{\partial \Phi}{\partial x} + B\frac{\partial \Phi}{\partial y}, \qquad \frac{\partial D}{\partial y} = B\frac{\partial \Phi}{\partial x} + 2C\frac{\partial \Phi}{\partial y}. \tag{5.105}$$

The set of equations (5.104) has the following solutions:

$$A = ay^2 + by + c, \qquad B = -2axy - bx - dy + e, \qquad C = ax^2 + dx - f, \tag{5.106}$$

where a, b, c, d, e and f are arbitrary constants (perhaps complex). Inserting these functions in set (5.105), we obtain the following integrability condition for D:

$$(2axy + bx + dy + e) \left(\frac{\partial^2 \Phi}{\partial x^2} - \frac{\partial^2 \Phi}{\partial y^2} \right) + 3(2ay + b)\frac{\partial \Phi}{\partial x}$$
$$- 2 \left(a(x^2 - y^2) + dx - by + f - c \right) \frac{\partial^2 \Phi}{\partial x \partial y} - 3(2ax + d)\frac{\partial \Phi}{\partial y} = 0. \tag{5.107}$$

This equation can be solved for all possible choices of the constants. Let us therefore examine the most interesting cases.

1) $a \neq 0$, $f \neq c$, $b = d = e = 0$. This case was solved by Darboux[23] by the method of characteristic curves (see, e.g., Courant and Hilbert[24]) in the case for which $a = 1$ (simply equivalent to the general case $a \neq 0$, since I_2 contains a as a constant factor). To accord with the treatment of elliptical coordinates we define the new constants $\alpha = -f$ and $\beta = -c$. For a partial differential equation with the structure of (5.107), the corresponding differential equation for the characteristics is

$$xy \left(dy^2 - dx^2 \right) + \left(x^2 - y^2 + \beta - \alpha \right) dx dy = 0.$$

This equation, with the change of variables $t = x^2, s = y^2$, is equivalent to Clairaut's equation

$$t\frac{ds}{dt} - s = (\alpha - \beta)\frac{ds}{dt}\left(\frac{ds}{dt} + 1 \right)^{-1},$$

<hr>

[23]G. Darboux: Sur un problème de mécanique, *Archives Néerlaindaises* (ii), **6**, 371–377 (1901).

[24]R. Courant, D. Hilbert: *Methods of Mathematical Physics* (Interscience, New York, 1953).

whose solution is given by the family of curves

$$s(t; m) = mt - (\alpha - \beta)\,\frac{m}{m+1},$$

where the arbitrary parameter m labels a single curve in the family. Returning to the Cartesian coordinates, we can represent the integral curves of the characteristic equation in implicit form by the expression

$$(m+1)(mx^2 - y^2) = (\alpha - \beta)\,m.$$

By again changing the parameter of the family from m to τ according to the relation

$$\tau = -\frac{m\alpha + \beta}{m+1}$$

we are finally able to write the family of integral curves in the elegant and simple form

$$\frac{x^2}{\tau + \alpha} + \frac{y^2}{\tau + \beta} = 1,$$

in which we recognize (5.82a), and which leads us to the introduction of elliptical coordinates λ and μ in the plane as the roots for τ of (5.82a). We have then arrived at the interesting fact that the characteristic curves of (5.105) are in practice *two families of confocal conics*. The general theory of partial differential equations thus shows that, on passing to the characteristic variables λ and μ, (5.107) can be cast in the canonical form

$$\frac{\partial^2 \Phi}{\partial\lambda\partial\mu} + \Lambda\frac{\partial\Phi}{\partial\lambda} + \mathrm{M}\frac{\partial\Phi}{\partial\mu} = 0, \tag{5.108}$$

where Λ and M are functions of λ and μ, which can be determined by explicitly changing variables from Cartesian coordinates to elliptical ones by means of transformation rules (5.83a). However, this boring calculation can be avoided if we observe, following Darboux, that (5.108) possesses, as particular solutions, the cases of the spherically symmetric potential and of the potentials which are already separated in Cartesian coordinates, that is, those given by (5.98). Let us consider, for example, the two simple cases

$$\Phi_a = x^2 + y^2 = \lambda + \mu + \alpha + \beta,$$

$$\Phi_b = x^{-2} = \frac{\alpha - \beta}{(\lambda + \alpha)(\mu + \alpha)},$$

chosen for the straigthforward exploitation of the transformation of coordinates (5.83a). By substitution of Φ_a and Φ_b in equation (5.108) we obtain the following equations for Λ and M:

$$\Lambda + \mathrm{M} = 0, \qquad (\mu + \alpha)\Lambda + (\lambda + \alpha)\mathrm{M} = 1,$$

so that (5.108) becomes

$$(\lambda - \mu)\frac{\partial^2 \Phi}{\partial\lambda\partial\mu} - \left(\frac{\partial\Phi}{\partial\lambda} - \frac{\partial\Phi}{\partial\mu}\right) = 0. \tag{5.109}$$

By observing that (5.109) implies the condition

$$\frac{\partial^2}{\partial\lambda\partial\mu}\left[(\lambda - \mu)\,\Phi\right] = 0,$$

we can quickly find its solution in the form

$$\Phi(\lambda,\mu) = \frac{F(\lambda) - G(\mu)}{\lambda - \mu}, \tag{5.110}$$

where $F(\lambda)$ and $G(\mu)$ are arbitrary functions of their arguments. The previous expression, displaying the form required to the potential, agrees with the expression of the potential in the statement of Stäckel's theorem (see (5.91)), as can be easily verified remembering that the elements of the metric for elliptical coordinates are given by (5.84). To find the explicit form of the function D we must integrate (5.105), with Φ given by solution (5.110). Exploiting coordinate transformation rules (5.79), the transformation formulas for the gradient can be established:

$$\frac{\partial}{\partial x} = \frac{\partial\lambda}{\partial x}\frac{\partial}{\partial\lambda} + \frac{\partial\mu}{\partial x}\frac{\partial}{\partial\mu} = \frac{2x}{\lambda - \mu}\left[(\lambda + \beta)\frac{\partial}{\partial\lambda} - (\mu + \beta)\frac{\partial}{\partial\mu}\right],$$

$$\frac{\partial}{\partial y} = \frac{\partial\lambda}{\partial y}\frac{\partial}{\partial\lambda} + \frac{\partial\mu}{\partial y}\frac{\partial}{\partial\mu} = \frac{2y}{\lambda - \mu}\left[(\lambda + \alpha)\frac{\partial}{\partial\lambda} - (\mu + \alpha)\frac{\partial}{\partial\mu}\right].$$

Using these differential operators and expressions (5.106) for A and B, we find that one or another of (5.105) gives

$$2\left[(\lambda + \beta)\mu\frac{\partial}{\partial\lambda} - (\mu + \beta)\lambda\frac{\partial}{\partial\mu}\right]\Phi = \left[(\lambda + \beta)\frac{\partial}{\partial\lambda} - (\mu + \beta)\frac{\partial}{\partial\mu}\right]D,$$

and it is easy to verify that the function D, to satisfy the integrability condition, must assume the following form:

$$D(\lambda,\mu) = 2\frac{\mu F(\lambda) - \lambda G(\mu)}{\lambda - \mu}.$$

The integral of motion (5.103) is therefore

$$I_2 = (y^2 - \beta)p_x^2 + (x^2 - \alpha)p_y^2 - 2xyp_xp_y + 2\frac{\mu F - \lambda G}{\lambda - \mu}.$$

Better still, by exploiting the transformation rules (A.11) for momenta, to find the variables conjugate to the elliptical coordinates,

$$p_\lambda = P^2\frac{d\lambda}{dt} = P^2\left(\frac{\partial\lambda}{\partial x}p_x + \frac{\partial\lambda}{\partial y}p_y\right),$$

$$p_\mu = Q^2\frac{d\mu}{dt} = Q^2\left(\frac{\partial\mu}{\partial x}p_x + \frac{\partial\mu}{\partial y}p_y\right),$$

we can express I_2 as a function of the canonical set of variables $(\lambda, \mu, p_\lambda, p_\mu)$:

$$I_2 = P^{-2}\mu p_\lambda^2 + Q^{-2}\lambda p_\mu^2 + 2\frac{\mu F - \lambda G}{\lambda - \mu}. \qquad (5.111)$$

2) $a = 1$, $f = c$, $b = d = e = 0$. This is not really a different case, since it is included in the previous general discussion of $a \neq 0$, but it can be cast into a particular simple form by observing that it corresponds to the condition $\alpha = \beta$, which, on the basis of the treatment of elliptical coordinates in the preceding section (see the discussion following (5.84)), implies the choice, as coordinate system, of polar coordinates (r, φ). It is now possible to show, following a line completely analogous to the above, that the potential must be of the form

$$\Phi(r, \varphi) = F(r) + \frac{G(\varphi)}{r^2}, \qquad (5.112a)$$

which is known, in galactic dynamics, as the *Eddington potential*. The corresponding integral of motion is

$$I_2 = p_\varphi^2 + 2G(\varphi), \qquad (5.112b)$$

where $p_\varphi = r^2\dot\varphi$, which can be obtained from (5.9c) if we put, as a natural two-dimensional reduction of spherical coordinates, $\vartheta \equiv \pi/2$. In particular, in the case of axial symmetry $(G(\varphi) \equiv 0)$, the fact that the integral (5.112b) is constant expresses the conservation of angular momentum, $p_\varphi = $ const.

3) $a = 0$, $d(\text{or}b) \neq 0$, $f \neq c$, $e = 0$. Looking at (5.107), we see that this case the integrability condition is

$$2x\frac{\partial^2\Phi}{\partial x\partial y} + y\left(\frac{\partial^2\Phi}{\partial y^2} - \frac{\partial^2\Phi}{\partial x^2}\right) + 3\frac{\partial\Phi}{\partial y} = 0,$$

whose solution is

$$\Phi(x, y) = \frac{1}{r}\left[F(r + x) + G(r - x)\right], \qquad (5.113a)$$

where F and G are arbitrary functions of their arguments. The second integral is of the form

$$I_2 = (yp_x - xp_y)p_y + \frac{1}{r}\left[(r + x)G - (r - x)F\right]. \qquad (5.113b)$$

An example in this class was the resonant anisotropic oscillator (1.C.114).

4) $a = 0$, $f \neq c$, $b = d = e = 0$. Looking at (5.107), we see that this case implies the condition

$$\frac{\partial^2\Phi}{\partial x\partial y} = 0,$$

whose solution is of the form given in (5.98). Therefore the integral is

$$I_2 = c\left(p_x^2 + 2\Phi_1\right) + f\left(p_y^2 + 2\Phi_2\right).$$

To write the integrals in the form given by (5.99), it is sufficient to observe that

$$I_x = \frac{2fE - I_2}{f - c}, \qquad I_y = \frac{I_2 - 2cE}{f - c},$$

where the total energy E is the value assigned to the first integral I_1.

5.6 The Problem of Two Fixed Centres in the Plane

The problem of two fixed centres of attraction was already studied and recognized as being separable by Euler in 1760 and was completely solved by Jacobi. It is the most renowned of the plane integrable problem (after, of course, the Kepler's and the harmonic oscillator problems) and is an example well suited to illustrating the integration of the equations of motion by means of the method of the separation of the Hamilton–Jacobi equation in elliptical coordinates, in agreement with the technique presented in Sect. 5.4.

We state the problem in the following manner:[25] we consider the motion of a body (m_0) attracted, under the laws of Newtonian gravitation, by two *fixed* bodies of masses m_1 and m_2, restricting the study to the case in which the motion is confined to a plane containing the two centres of attraction.[26] Assuming that their Cartesian coordinates are $(0, c)$ and $(0, -c)$, and that those of m_0 are in general (x, y), the potential acting on m_0 will be

$$\Phi(x, y) = -G\left(\frac{m_1}{r_1(x, y)} + \frac{m_2}{r_2(x, y)}\right), \tag{5.114}$$

where the distances of the moving body from the centres are

$$r_1 = \sqrt{x^2 + (y - c)^2}, \quad r_2 = \sqrt{x^2 + (y + c)^2}.$$

We attack the problem from the outset by passing to elliptical coordinates: the aim is to express the potential in the form of (5.91), as required in the statement of Stäckel's theorem, so that the solution of the dynamical problem is reduced to the integration of equations of the form of (5.95–97).

The sum and the difference of the squares of the distances of the moving body from the centres are respectively

$$r_2^2 + r_1^2 = 2(x^2 + y^2 + c^2), \quad r_2^2 - r_1^2 = 4cy,$$

so that, recalling formulas (5.83) for the transformation between Cartesian and elliptical coordinates, the last equalities can be rewritten as

---

[25] L. A. Pars: *A Treatise on Analytical Dynamics* (Heinemann, London, 1965).

[26] This is clearly not a severe restriction since the problem is invariant with respect to rotation around the axis connecting the two centres: in the following section we will exploit this fact to study the full three-dimensional case.

$$r_2^2 + r_1^2 = 2(\lambda + \mu + 2\beta),$$
$$r_2^2 - r_1^2 = 4\sqrt{(\lambda + \beta)(\mu + \beta)},$$

$$(5.115)$$

with the definition

$$c = \sqrt{\beta - \alpha}. \tag{5.116}$$

According to the discussion on the properties of elliptical coordinates following (5.84), it is easy to see that the position expressed by (5.116) implies putting the two attracting bodies at the *foci* of the coordinate system. Decomposing the difference of the squares of the distances, we find that it is straightforward to obtain the simple relation

$$r_2 = \sqrt{\lambda + \beta} + \sqrt{\mu + \beta},$$
$$r_1 = \sqrt{\lambda + \beta} - \sqrt{\mu + \beta}.$$

$$(5.117)$$

Putting these into expression (5.114), we see that the potential indeed has the form required by Stäckel's theorem:

$$\Phi(\lambda, \mu) = \frac{m_{\mathrm{d}}\sqrt{\mu + \beta} - m_{\mathrm{s}}\sqrt{\lambda + \beta}}{\lambda - \mu},$$

where the two new "masses"

$$m_{\mathrm{d}} = G(m_2 - m_1), \qquad m_{\mathrm{s}} = G(m_2 + m_1)$$

have been introduced. We assume, by definition, that $m_2 > m_1$, so that we have the inequalities

$$m_{\mathrm{s}} > m_{\mathrm{d}} > 0. \tag{5.118}$$

Exploiting form (5.84) of the metric components in elliptical coordinates, the Hamilton–Jacobi equation

$$\mathcal{H}\left(q, \frac{\partial S^\star}{\partial q}\right) = \frac{1}{2}\left[P^{-2}\left(\frac{\partial S^\star}{\partial \lambda}\right)^2 + Q^{-2}\left(\frac{\partial S^\star}{\partial \mu}\right)^2\right] + \Phi(\lambda, \mu) = \alpha_1$$

(which is equivalent, in two dimensions and in this coordinate system, to (5.92)) becomes

$$2(\lambda + \alpha)(\lambda + \beta)\left(\frac{\partial S^\star}{\partial \lambda}\right)^2 - 2(\mu + \alpha)(\mu + \beta)\left(\frac{\partial S^\star}{\partial \mu}\right)^2$$
$$+ m_{\mathrm{d}}\sqrt{\mu + \beta} - m_{\mathrm{s}}\sqrt{\lambda + \beta} = (\lambda - \mu)E,$$

$$(5.119)$$

where $\alpha_1 = E$, the total energy, is the first integration constant. If the complete integral (5.93) is written in the form

$$S^\star = S_\lambda(\lambda) + S_\mu(\mu),$$

(5.119) separates into the set of ordinary equations

$$2(\lambda + \alpha)(\lambda + \beta)\left(\frac{dS_\lambda}{d\lambda}\right)^2 = E(\lambda + \beta) + m_s\sqrt{\lambda + \beta} - I_2,$$
$$2(\mu + \alpha)(\mu + \beta)\left(\frac{dS_\mu}{d\mu}\right)^2 = E(\mu + \beta) + m_d\sqrt{\mu + \beta} - I_2 \tag{5.120}$$

where I_2 is the second integration constant.[27] The equations of motion, corresponding to (5.95–97), general solutions according to the hypothesis of Stäckel's theorem, are now

$$t - t_0 = \frac{1}{2}\int \frac{\sqrt{\lambda + \beta}d\lambda}{\sqrt{2(\lambda + \alpha)L(\lambda)}} + \frac{1}{2}\int \frac{\sqrt{\mu + \beta}d\mu}{\sqrt{2(\mu + \alpha)M(\mu)}}, \tag{5.121}$$

$$\beta_2 = \frac{1}{2}\int \frac{d\lambda}{\sqrt{2(\lambda + \alpha)(\lambda + \beta)L(\lambda)}} + \frac{1}{2}\int \frac{d\mu}{\sqrt{2(\mu + \alpha)(\mu + \beta)M(u)}}, \tag{5.122}$$

$$p_\lambda = \sqrt{\frac{L(\lambda)}{2(\lambda + \alpha)(\lambda + \beta)}}, \qquad p_\mu = \sqrt{\frac{M(\mu)}{2(\mu + \alpha)(\mu + \beta)}}, \tag{5.123}$$

where have been introduced the two new functions

$$L = E(\lambda + \beta) + m_s\sqrt{\lambda + \beta} - I_2,$$
$$M = E(\mu + \beta) + m_d\sqrt{\mu + \beta} - I_2, \tag{5.124}$$

which, if we look at the right-hand sides of set (5.120) and remember that the coordinates must satisfy inequalities (5.82b), are such that, during the motion,

$$L \geq 0, \qquad M \leq 0. \tag{5.125}$$

It is instructive, at this point, to perform a qualitative analysis of the orbits in this problem. With this aim in mind it is sufficient to find the zeroes of the functions appearing in the denominators of (5.121–123) and to determine the intervals that the coordinates fall in. To facilitate this study it is convenient to make the change of variables

$$\ell = \sqrt{\lambda + \beta}, \qquad \iota = \sqrt{\mu + \beta}, \tag{5.126}$$

such that the inequalities mentioned turn out to be

$$\ell \geq c, \qquad c \geq \iota \geq -c. \tag{5.127}$$

Definitions (5.124) simply become[28]

---

[27]The choice of the negative sign in front of I_2 is made with the intent that it coincide with integral (5.111) of the previous section.

[28]Recalling the relations written after (5.87), we see that the coordinates (ℓ, ι) are related to the coordinates (u, v), generated by the conformal transformation of (5.86), by $\ell = \Delta\cosh u$, $\iota = \Delta\sin v$.

$$L = E\ell^2 + m_{\mathrm{s}}\ell - I_2,$$
$$M = E\iota^2 + m_{\mathrm{d}}\iota - I_2.$$

Let us denote by ℓ_1, ℓ_2 the zeroes of L, and by ι_1, ι_2 the zeroes of M; ℓ_1, ℓ_2 are certainly real numbers. If, in fact, they were complex, the coefficients of the trinomial $L(\ell)$ should be such that

$$m_{\mathrm{s}}^2 < -4EI_2,$$

and accordingly, as a result of (5.118), the coefficients of $M(\iota)$ should be such that

$$m_{\mathrm{d}}^2 < -4EI_2,$$

so that ι_1, ι_2 too would be complex. But this is impossible, since if both L and M had complex zero, they would both have the same sign as E, and so one of inequalities (5.125) would be violated. The condition that ℓ_1, ℓ_2 be real can be used as the basis for the orbit classification, instead of using, for example, the values of the constants of the motion E, I_2.

We limit ourselves to the case $E < 0$ (*bound* orbits). Inequalities (5.125) then imply that, since the first coefficients of the trinomial are negative, ℓ must be inside the interval (ℓ_1, ℓ_2), whereas ι must be outside the interval (ι_1, ι_2) (in the case for which ι_1 and ι_2 are real). Let us therefore look for the critical curves in the $\ell_1 \ell_2$ plane, assuming, by definition, that $\ell_1 \geq \ell_2$ always. This assumption and the fact that the sum of the zeroes

$$\ell_1 + \ell_2 = -\frac{m_{\mathrm{s}}}{E} \tag{5.128}$$

must be positive for the bound orbits, eliminate, in the $\ell_1 \ell_2$ plane, the whole region lying on the left of the two straight lines $\ell_2 \pm \ell_1 = 0$ (see Fig. 5.3). Putting (5.128) and the equation giving the product of the zeroes

$$\ell_1 \ell_2 = \frac{I_2}{E} \tag{5.129}$$

into the equation $M(\iota) = 0$, we find that ι_1 and ι_2 are the roots of the equation

$$x^2 - \frac{m_{\mathrm{d}}}{m_{\mathrm{s}}}(\ell_1 + \ell_2)x + \ell_1\ell_2 = 0.$$

The condition $\iota_1 = \iota_2$ is satisfied if

$$(\ell_1 + \ell_2)^2 = 4\ell_1\ell_2(m_{\mathrm{s}}/m_{\mathrm{d}})^2.$$

This, if we introduce the variable θ, such that

$$\frac{m_{\mathrm{s}}}{m_{\mathrm{d}}} = \cosh\theta, \qquad \frac{m_1}{m_2} = \tanh^2\frac{1}{2}\theta,$$

becomes

$$\frac{\ell_1}{\ell_2} = e^{\pm 2\theta}, \tag{5.130}$$

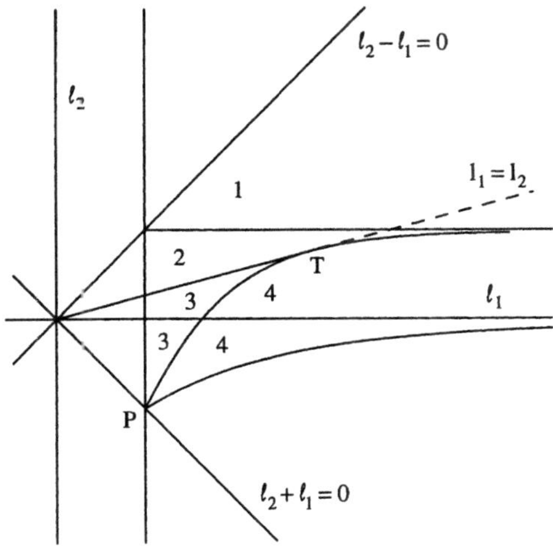

Fig. 5.3

representing a pair of straight lines with the same slope with respect to the axes of the $\ell_1 \ell_2$ plane (denoted by $\iota_1 = \iota_2$ in Fig. 5.3).

Other critical curves, useful for orbit classification, are given by the conditions $\iota_1 = c$ or $\iota_2 = c$, which, proceeding as before, correspond to the equation

$$c^2 - \frac{m_{\mathrm d}}{m_{\mathrm s}}(\ell_1 + \ell_2)c + \ell_1\ell_2 = 0,$$

that is, to the hyperbola of equation

$$(\ell_1 - cm_{\mathrm d}/m_{\mathrm s})(\ell_2 - cm_{\mathrm d}/m_{\mathrm s}) = -c^2\tanh^2\theta. \tag{5.131}$$

The tangents to this hyperbola are the straight lines (5.130); in particular, the one with the positive exponent touches the lower branch of hyperbola (5.131) at the point

$$T \equiv (ce^\theta,\ ce^{-\theta}).$$

To determine which part of (5.158) corresponds to $\iota_1 = c$ and which to $\iota_2 = c$, we assume first of all that $\iota_1 \geq \iota_2$. Now, if $\iota_2 = c$, then $\iota_1 + \iota_2 \geq 2c$ implies the inequality

$$\ell_1 + \ell_2 \geq 2cm_{\mathrm s}/m_{\mathrm d} = 2c\cosh\theta,$$

so that ι_2 is equal to c on the right of T. An analogous argument proves that $\iota_1 = -c$ or $\iota_2 = -c$ correspond to parts of the lower branch of the hyperbola

$$(\ell_1 + cm_{\mathrm d}/m_{\mathrm s})(\ell_2 + cm_{\mathrm d}/m_{\mathrm s}) = -c^2\tanh^2\theta. \tag{5.132}$$

Both hyperbolas (5.131) and (5.132) pass through the point $P \equiv (c, -c)$. This completes the determination of the critical curves, although it remains to eliminate some other regions in the $\ell_1 \ell_2$ plane. From the first of equations (5.127) we see that the semi-plane $\ell_1 < c$ must be removed, since, by (5.125), ℓ must lie inside the interval (ℓ_1, ℓ_2). Moreover, from the second of equations (5.127), and since, from what we said above, in the case of real zeroes for the trinomial $M(\iota)$, ι must lie outside the interval (ι_1, ι_2), we must exclude the region for which $\iota_1 \geq c \geq -c \geq \iota_2$. Once the regions not consistent with all the above conditions have been removed, there remain, for the bound orbits, the regions numbered from 1 to 4 in Fig. 5.3, representing four different types of orbits. In region 1 we have $\ell_1 > \ell_2 > c$, so that the ℓ motion is given by a libration between $\ell = \ell_1$ and $\ell = \ell_2$. In regions 2, 3 and 4 we have $\ell_2 < c$, so that the ℓ motion is given by a libration between $\ell = \ell_1$ and $\ell = c$. In regions 1 and 2 the zeroes of M are either complex or real but with $\iota_1 > \iota_2 > c$, so that ι can take any value in the range between $-c$ and c. In region 3 we have

$$c > \iota_1 > \iota_2 > -c,$$

so that there are two possibilities for the ι motion, either a libration between c and ι_1 or a libration between ι_2 and $-c$. In region 4 we have

$$\iota_1 > c > \iota_2 > -c,$$

so that the only possibility for the ι motion is a libration between ι_2 and $-c$. To summarize, in region 1 the orbit is then a curve (in general not closed) which, after sufficient time, densely fills up the *oval annular ring* contained between the ellipses $\ell = \ell_1$ and $\ell = \ell_2$ (see Fig. 5.4). In region 2 the orbit is *figure-eight shaped* around both centres and, after enough time, densely fills up the ellipse $\ell = \ell_1$ (see Fig. 5.5). In region 3, the moving body becomes the *satellite* of one or the other of the fixed masses (see Fig. 5.6), since, shifting from parameters of the region 2 to those of region 3, the *figure-eight* has broken into two distinct curves. Shifting then from region 3 to 4, we find that one of them disappears and the moving body can only be a satellite of m_1.

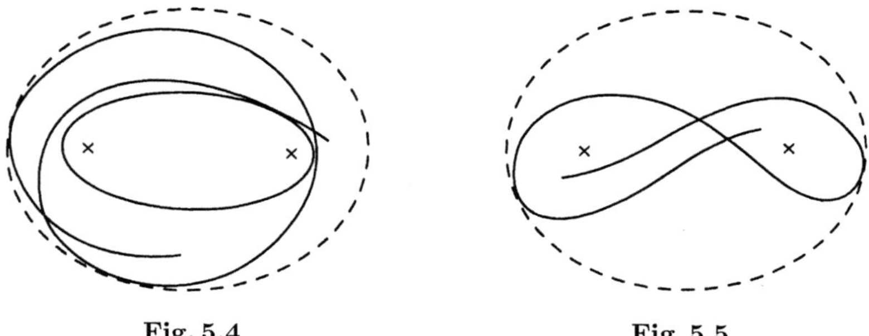

Fig. 5.4   Fig. 5.5

Having worked out the qualitative analysis, we can move on to the explicit integration of the orbits. We limit ourselves to a specific example: the case

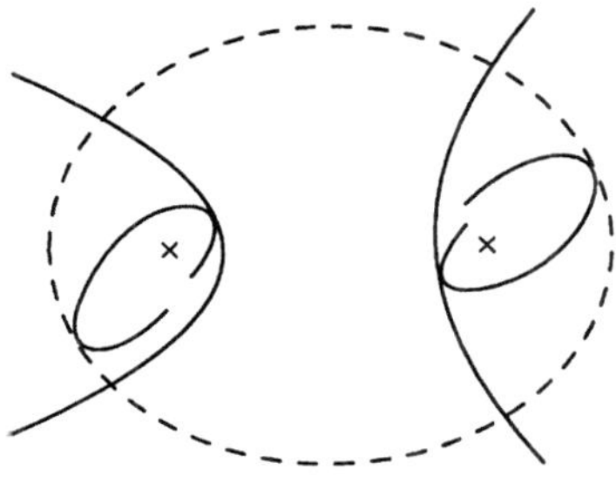

Fig. 5.6

of the elliptical annulus of Fig. 5.4. From (5.122) we see that the trajectory is determined by the equation

$$\frac{d\lambda}{\sqrt{(\lambda+\alpha)(\lambda+\beta)L(\lambda)}} = \frac{d\mu}{\sqrt{(\mu+\alpha)(\mu+\beta)M(\mu)}}. \tag{5.133}$$

The relation between momenta and velocities (A.11) and the form of the metric coefficients (5.84) lead to a parametrization of the trajectory in the form $\lambda = \lambda(\tau)$, $\mu = \mu(\tau)$, where the new parameter τ is introduced by means of

$$d\tau = \frac{dt}{\sqrt{2}(\lambda-\mu)}.$$

However, the coordinates ℓ, ι are more useful also for the integration of the equations of motion. In fact, the functions $L(\lambda)$ and $M(\mu)$ are irrational, whereas, expressing (5.133) in terms of the coordinates ℓ, ι, we find that

$$\frac{d\ell}{\sqrt{(\ell_1-\ell)(\ell-\ell_2)(\ell-c)(\ell+c)}} = \frac{d\iota}{\sqrt{(c-\iota)(\iota+c)(\iota_1-\iota)(\iota_2-\iota)}}, \tag{5.134}$$

where the intervals for the variables are those found before. Equating both sides of (5.134) to $2\,d\tau$, where now, from (5.126),

$$d\tau = \frac{dt}{\sqrt{2}(\ell^2-\iota^2)}$$

and using Jacobi elliptical functions, we find that we can parametrize the orbits in the form $\ell = \ell(\tau)$, $\iota = \iota(\tau)$. One method is the following: by means of the change of variable

$$\frac{\ell-\ell_2}{\ell_1-\ell} = \frac{\ell_2-c}{\ell_1-c}\frac{x^2}{1-x^2},$$

the left-hand side of (5.134) takes the form

$$\frac{2}{\sqrt{(\ell_1-c)(\ell_2+c)}}\frac{dx}{\sqrt{(1-x^2)(1-k^2x^2)}},$$

where

$$k^2 = \frac{2c(\ell_1 - \ell_2)}{(\ell_1 - c)(\ell_2 + c)}.$$ (5.135)

Introducing the *Legendre elliptic function of the first kind*[29]

$$u = \int_0^x \frac{d\xi}{\sqrt{(1 - \xi^2)(1 - k^2\xi^2)}},$$

where u is related to the parameter τ by the relation

$$u = \sqrt{(\ell_1 - c)(\ell_2 + c)}\,\tau,$$

we know that the functions $\operatorname{sn} u$ and $\operatorname{cn} u$, called, respectively, the *Jacobi sine–amplitude* and *cosine–amplitude* functions, result implicitly defined by

$$x = \operatorname{sn} u, \qquad \sqrt{1 - x^2} = \sqrt{1 - \operatorname{sn}^2 u} = \operatorname{cn} u.$$

Therefore, the solution for ℓ can be expressed in the form

$$\ell(u(\tau)) = \frac{\ell_1(\ell_2 - c)\operatorname{sn}^2 u + \ell_2(\ell_1 - c)\operatorname{cn}^2 u}{(\ell_2 - c)\operatorname{sn}^2 u + (\ell_1 - c)\operatorname{cn}^2 u}.$$ (5.136)

Analogously, if we use the change of variable

$$\frac{\iota + c}{c - \iota} = \frac{\iota_1 + c}{\iota_1 - c}\frac{y^2}{1 - y^2},$$

the right hand side of (5.134) takes the form

$$\frac{2}{\sqrt{(\iota_1 - c)(\iota_2 + c)}}\frac{dy}{\sqrt{(1 - y^2)(1 - \kappa^2 y^2)}},$$

where

$$\kappa^2 = \frac{2c(\iota_1 - \iota_2)}{(\iota_1 - c)(\iota_2 + c)}.$$ (5.137)

Therefore, following the same procedure as above,

$$y = \operatorname{sn} v, \qquad v = \sqrt{(\iota_1 - c)(\iota_2 + c)}\,(\tau - \tau_0)$$

and

$$\iota(v(\tau)) = c\frac{(\iota_1 + c)\operatorname{sn}^2 v - (\iota_1 - c)\operatorname{cn}^2 v}{(\iota_1 + c)\operatorname{sn}^2 v + (\iota_1 - c)\operatorname{cn}^2 v}.$$ (5.138)

The orbit is now parametrized in the form $\ell = \ell(\tau)$, $\iota = \iota(\tau)$. The parameter k^2 for the Jacobi functions appearing in (5.136) is defined by (5.135), whereas the parameter κ^2 of the functions appearing in (5.138) is defined in (5.137).

---

[29]See, for instance, E. T. Whittaker, G. N. Watson: *A Course of Modern Analysis* (Cambridge University Press, 1927), pp. 512 ff.

5.7 Axially Symmetric Potentials – Motion in the Potential of the Earth

We now return to the study of fully three-dimensional orbits, considering potentials of the kind $\Phi = \Phi(\varpi, z)$, where, having introduced the cylindrical coordinates defined in (A.8a), we assume that the potential does not depend on the azimuthal angle φ and is endowed with rotational symmetry around the z axis. This case is of paramount importance for astrophysical applications, since potentials of this form are generated by mass distributions provided with rotational motions around the symmetry axis, typical of solid celestial bodies, stars, disk galaxies and also common, if not dominating, in elliptical galaxies.

A glance at the equations of motion (A.17), suggests from the outset some of their peculiarities. The angular momentum around the z axis

$$L_z = p_\varphi = \varpi v_\varphi \tag{5.139}$$

is constant, since the corresponding conjugate variable is ignorable. Equation (A.17) then reduces to

$$\ddot{\varpi} = \frac{L_z^2}{\varpi^3} - \frac{\partial \Phi}{\partial \varpi},$$
$$\ddot{z} = -\frac{\partial \Phi}{\partial z}, \tag{5.140}$$

which are the equations of the motion in a symmetry plane rotating with angular velocity given by $\dot{\varphi} = L_z/\varpi^2$ (*meridional plane*) and can also be interpreted as the equations of motion associated with a new *effective potential*

$$\Phi_{\text{eff}} = \Phi + \frac{L_z^2}{2\varpi^2}. \tag{5.141}$$

Since the problem is in this way reduced to two degrees of freedom, the motion in the meridional plane is a problem to which the general methods illustrated in the previous sections apply. Let us assume then that the reduced problem is separable and that, in addition to the energy E, there exist a second integral I_2. Choosing a suitable coordinate system, for example the prolate spheroidal coordinates (λ, φ, ν) defined by equations (5.89), given the values of E, I_2 and L_z, we can find the area in the meridional plane allowed to the orbit. This area, say A, can be found, as usual, cutting the level curves of the effective potential (5.141) with the line E constant. The axially symmetric volume obtained by making A rotate around the symmetry axis defines the orbit in the space, because the orbit densely fills all the allowed volume the frequencies of the motions in λ, ν and φ being in general incommensurable. Once the motion in λ and ν has been determined, that in φ follows from (5.139) written in the form

$$\varpi^2 \left(\frac{d\varphi}{dt} \right)^2 = L_z^2.$$

A particularly interesting example of motion in a spheroidal potential is that of motion in the potential of the Earth, which finds a useful application in the determination of the orbits of artificial satellites.

In general, if we have a finite mass distribution endowed with rotational symmetry, the external potential can be written as the series expansion

$$\Phi = -\frac{GM}{r}\left[1 + \sum_{k=2}^{\infty} A_k \left(\frac{R}{r}\right)^k P_k(\sin b)\right], \qquad (5.142)$$

where M is the total mass, R is the equatorial radius (defined by $b = \pi/2$, where b is the latitude), r is the distance from the centre of mass, where the origin of the coordinate system is set, P_k are the Legendre polynomials and A_k are the coefficients of the expansion. In the case of the Earth, from studies of the motion of satellites, it has been found that the first three coefficients are

$$A_2 = -1.0826 \times 10^{-3}, \quad A_3 = 2.5 \times 10^{-6}, \quad A_4 = 1.6 \times 10^{-6}. \qquad (5.143)$$

Since in the case of spheroidal symmetry the odd terms are missing in expansion (5.142), we see that the Earth has a non-negligible *pear-shaped* distortion. It is clear that we cannot hope to solve the problem of the motion in the potential (5.142) exactly if the order of the expansion is arbitrarily high, but, if it is truncated at a sufficiently low order, it can be solved by an ingenious trick. Let us again take the two fixed centres problem of the previous section and suppose that the two masses m_1 and m_2 are put at the points $z = c_1$ and $z = -c_2$. We now want to study the *three-dimensional* problem, so that we write the potential in the form

$$\Phi(x, y, z) = -G\left(\frac{m_1}{r_1(x, y, z)} + \frac{m_2}{r_2(x, y, z)}\right)$$

$$= -G\left(\frac{m_1}{\sqrt{x^2 + y^2 + (z - c_1)^2}} + \frac{m_2}{\sqrt{x^2 + y^2 + (z + c_2)^2}}\right).$$

Exploiting the series expansions

$$\frac{r}{r_2} = \sum_{n=0}^{\infty}\left(-\frac{c_2}{r}\right)^n P_n(\sin b), \qquad \frac{r}{r_1} = \sum_{n=0}^{\infty}\left(\frac{c_1}{r}\right)^n P_n(\sin b),$$

the potential can be written as

$$\Phi = -\frac{G(m_1 + m_2)}{r}\left[1 + \sum_{k=1}^{\infty} \frac{\gamma_k}{r^k} P_k(\sin b)\right], \qquad (5.144)$$

where the coefficients γ_k are defined by

$$\gamma_k = \frac{m_1 c_1^k + m_2(-c_2)^k}{m_1 + m_2}. \qquad (5.145)$$

In view of the resemblance of (5.142) and (5.144), the procedure is apparently straightforward: if we succeed in identifying the coefficients γ_k with $A_k R^k$, we have a complete coincidence of the two cases and therefore the problem of the motion of the satellite is solved, since it has been reduced to an integrable problem. Unfortunately, in expression (5.145), there are only four really free parameters which we can use to make γ_k, and not infinitely many as A_k. We can therefore realize the coincidence only up to four terms in the expansion: if, as is obvious, we choose the first four terms, we have the relations

$$m_1 + m_2 = M,$$
$$m_1 c_1 - m_2 c_2 = 0,$$
$$m_1 c_1^2 + m_2 c_2^2 = M A_2 R^2, \tag{5.146}$$
$$m_1 c_1^3 - m_2 c_2^3 = M A_3 R^3,$$

which can be solved for c,

$$c_1 = \frac{1}{2} R \left[\frac{A_3}{A_2} + \sqrt{\left(\frac{A_3}{A_2}\right)^2 + 4A_2} \right], \quad c_2 = -\frac{1}{2} R \left[\frac{A_3}{A_2} - \sqrt{\left(\frac{A_3}{A_2}\right)^2 + 4A_2} \right] \tag{5.147}$$

and for m,

$$m_1 = M \frac{c_2}{c_1 + c_2}, \qquad m_2 = M \frac{c_1}{c_1 + c_2}. \tag{5.148}$$

Recalling the numerical values of A_2 and A_3, we see that in the argument of the square roots of (5.147), the negative term $4A_2$ dominates by three orders of magnitude: the distances c_1 and c_2 (and therefore also the masses m_1 and m_2) turn out to be *complex numbers*! This result, surprising at first glance, need not worry us: what matters is that the potential is real and thus the related motion. We thus arrive at the conclusion that the motion of a satellite in the field of the Earth is reduced to Euler's classic problem, with complex masses put at complex distances (real masses at real distances would give a *prolate* spheroid instead of an *oblate* one). If, moreover, in (5.147) and (5.148), A_3 is neglected, that is, not taking into account the asymmetry of the two terrestrial hemispheres with respect to the equatorial plane, we obtain

$$m_1 = m_2 = \frac{1}{2} M, \qquad c_1 = ic = iR\sqrt{|A_2|}, \qquad c_2 = -ic = -iR\sqrt{|A_2|},$$

so that the problem becomes simply

$$\Phi(\varpi, z) = -\frac{GM}{2} \left[\frac{1}{\sqrt{\varpi^2 + (z - ic)^2}} + \frac{1}{\sqrt{\varpi^2 + (z + ic)^2}} \right].$$

Without entering into the details, we can summarize the results obtained as follows. There are many autonomous dynamical systems (of very real interest in celestial mechanics or in other fields of analytical mechanics) in which the existence of non-classical isolating integrals, such as

$$I_2 = P^{-2}\mu p_\lambda^2 + Q^{-2}\lambda p_\mu^2 - 2\frac{m_s\mu\sqrt{\lambda+\beta} - m_d\lambda\sqrt{\mu+\beta}}{\lambda - \mu},$$

of the two fixed centres problem, is suggested by several peculiarities displayed by the orbits of the system, but of which *it is impossible to find the explicit form.*

Still in the class of the potentials we are examining in this section, the most interesting case is certainly that of the potential of our Galaxy, which is imagined to be a rotationally symmetrical mass distribution, very roughly represented as a thin disc embedded in a huge spheroidal halo. Without going into all the details, we simply wish to recall that, from the classical works by Kapteyn, Jeans and Oort,[30] the study of the kinematics of stellar motion in the neighbourhood of the Sun and the first attempts to construct equilibrium models for the Galaxy provided compelling evidence for the existence of a non-classical integral of the motion (the celebrated *problem of the third integral in galactic dynamics*): in particular, the triaxial anisotropy of the velocity dispersion of the stars in the disc cannot be described by galactic models constructed on the basis of distribution functions that depend only on the energy and the angular momentum. The potentials we have examined (the Stäckel cases, essentially) do not allow great flexibility for careful modelling of the galactic disc. Nevertheless, they can be considered to be the paradigm of the dynamical structure of the galaxies. To deal specifically with some real cases, we need to shift to perturbative techniques, which we will study in Volume 2.

5.8 Orbits in Triaxial Potentials

When studying motion in potentials endowed with axial symmetry, we mentioned some of the peculiarities displayed by the orbits supported by them. It is clear that, if we remove the axial symmetry, but maintain the symmetry of reflection with respect to every coordinate plane and remain within the conditions in which the Hamilton–Jacobi equation is separable, the characteristics of the orbits are even more varied and interesting: we dedicate this section to the analysis of motion in a generic triaxial Stäckel potential, that is, a potential of the form introduced in the statement of Stäckel's theorem in the second of equations (5.90). Other aspects of the orbits in the axially symmetric case will emerge as limiting cases, without the necessity of boring recapitulations. The general expression for the potential is

---

[30] J. C. Kapteyn: First attempt at a theory of the arrangement and the motion of the sidereal system, *Astrophys. J.* **55**, 302–327 (1922); J. H. Jeans: *Astronomy and Cosmogony* (Cambridge University Press, 1928); J. H. Oort: Stellar dynamics, in *Galactic Structure*, ed. by A. Blaauw, M. Schmidt (University of Chicago Press, 1965), pp. 455–530.

$$\Phi(\lambda,\mu,\nu) = -\frac{F(\lambda)}{(\lambda-\mu)(\lambda-\nu)} - \frac{F(\mu)}{(\mu-\nu)(\mu-\lambda)} - \frac{F(\nu)}{(\nu-\lambda)(\nu-\mu)}, \quad (5.149)$$

where, for the functions in the numerators, we have used, for simplicity, the same symbol since, from (5.77), the three variables λ, μ, ν each ranges over its own interval, and these intervals do not overlap and cover all the real axis from $-\gamma$ to $+\infty$. We can also use the notation $F = F(\tau)$, with τ representing, in turn, the three coordinates λ, μ, ν. For the solution of the Hamilton–Jacobi equation, the definition of the isolating integrals and the qualitative analysis of the orbits, we will follow closely the treatment by de Zeeuw,[31] whose elegant and exhaustive work can be considered, within reason, the standard reference for this problem.

The Hamiltonian is

$$\frac{1}{2}\left(P^{-2}p_\lambda^2 + Q^{-2}p_\mu^2 + R^{-2}p_\nu^2\right) + \Phi(\lambda,\mu,\nu) = E, \quad (5.150)$$

where P, Q and R are given by (5.81). Putting $p_\tau = \partial S^\star/\partial\tau$, $\tau = \lambda$, μ, ν into (5.150), we obtain the Hamilton–Jacobi equation. With the potential of (5.149), after multiplying it by $(\lambda-\mu)(\mu-\nu)(\nu-\lambda)$, we find

$$(\mu-\nu)\left[2(\lambda+\alpha)(\lambda+\beta)(\lambda+\gamma)\left(\frac{\partial S^\star}{\partial\lambda}\right)^2 - F(\lambda) - \lambda^2 E\right]$$

$$+ (\nu-\lambda)\left[2(\mu+\alpha)(\mu+\beta)(\mu+\gamma)\left(\frac{\partial S^\star}{\partial\mu}\right)^2 - F(\mu) - \mu^2 E\right] \quad (5.151)$$

$$+ (\lambda-\mu)\left[2(\nu+\alpha)(\nu+\beta)(\nu+\gamma)\left(\frac{\partial S^\star}{\partial\nu}\right)^2 - F(\nu) - \nu^2 E\right] = 0.$$

Assuming, as usual, that the solution is of the form

$$S^\star(\lambda,\mu,\nu) = S_\lambda(\lambda) + S_\mu(\mu) + S_\nu(\nu),$$

and defining the function

$$U(\tau) = 2(\tau+\alpha)(\tau+\beta)(\tau+\gamma)\left(\frac{dS_\tau}{d\tau}\right)^2 - F(\tau) - \tau^2 E, \quad \tau = \lambda,\ \mu,\ \nu, \quad (5.152)$$

we see that the Hamilton–Jacobi equation (5.151) then simply becomes

$$(\mu-\nu)U(\lambda) + (\nu-\lambda)U(\mu) + (\lambda-\mu)U(\nu) = 0.$$

This equation must be satisfied for every value of λ, μ and ν. Its partial derivatives with respect to each coordinate imply that $U'(\tau)$ is a constant, so that

$$U(\tau) = i_2\tau - i_3,$$

---

[31]de Zeeuw: loc. cit., Footnote 17, Sect. 5.4.

where i_2 and i_3 are two constants. Putting this into the Hamilton–Jacobi equation, we arrive at the equations of motion:

$$\frac{dS_\tau}{d\tau} = p_\tau = \sqrt{\frac{\tau^2 E - \tau i_2 + i_3 + F(\tau)}{2(\tau + \alpha)(\tau + \beta)(\tau + \gamma)}}, \tag{5.153}$$

which is the realization of (5.97) in the present case.

The constants i_2 and i_3 appearing in the equations of motion are the values of the two isolating integrals, I_2 and I_3, admitted by the problem in addition to the energy $H = E$. Their explicit expressions in terms of phase space coordinates can be found in the same way that we found I_2 in the axially symmetric case before: we solve (5.153) for i_2 and i_3 and substitute in place of E the explicit form of Hamiltonian (5.150). The easiest way to write the ensuing expressions is, however, by means of the quantities X, Y and Z defined by

$$X = \frac{p_\lambda^2}{2P^2} - \frac{F(\lambda)}{(\lambda - \mu)(\lambda - \nu)}, \tag{5.154a}$$

$$Y = \frac{p_\mu^2}{2Q^2} - \frac{F(\mu)}{(\mu - \nu)(\mu - \lambda)}, \tag{5.154b}$$

$$Z = \frac{p_\nu^2}{2R^2} - \frac{F(\nu)}{(\nu - \lambda)(\nu - \mu)}. \tag{5.154c}$$

With a little algebra, we then find the following relations giving the isolating integrals:

$$I_1 = H = X + Y + Z, \tag{5.155a}$$

$$I_2 = (\mu + \nu)X + (\nu + \lambda)Y + (\lambda + \mu)Z, \tag{5.155b}$$

$$I_3 = \mu\nu X + \nu\lambda Y + \lambda\mu Z. \tag{5.155c}$$

We know that every function of H, I_2, I_3 is again an integral of the motion. The freedom of choosing three arbitrary independent functions of H, I_2 and I_3 can be exploited to simplify the qualitative analysis of the motion. To this end it is convenient to keep the energy $H = E$ as an integral, but, instead of I_2 and I_3, it is more useful to introduce the quantities J and K defined by

$$J = \frac{\alpha^2 H + \alpha I_2 + I_3}{\alpha - \gamma}, \qquad K = \frac{\gamma^2 H + \gamma I_2 + I_3}{\gamma - \alpha}. \tag{5.156}$$

Among other things, with these functions it is easier to establish the links between the triaxial case and its axial and spherical limits. In terms of the values of j and k of the integrals J and K, the equations of motion (5.153) become

$$\frac{dS_\tau}{d\tau} = p_\tau = \sqrt{\frac{(\tau + \alpha)(\tau + \gamma)E - (\tau + \gamma)j - (\tau + \alpha)k + F(\tau)}{2(\tau + \alpha)(\tau + \beta)(\tau + \gamma)}}. \tag{5.157}$$

For the qualitative analysis of the bound ($E < 0$) orbits, by analogy with what was said in Sect. 5.7, we have to find the conditions such that, for given values E, j and k of the integrals, the three quantities p_λ^2, p_μ^2 and p_ν^2 are not all negative. As a consequence, the allowed intervals for the coordinates λ, μ and ν will be determined and the motion will result as the combination of *librations* between the *inversion points* $p_\tau^2 = 0$ and, if $p_\tau^2 > 0$ for every τ, *rotations*. According to the type of combination of librations and rotations, a specific *family* of orbits will ensue. In particular, the intervals delimit a volume in physical space, with different shapes for the various families. Since in general the frequencies of the motion in each of the three coordinates are not commensurable, the orbits densely fill the allowed volume. A complete classification can be performed determining, by means of (5.157), for every combination of E, j and k, whether an orbit exists, and if so to what family it belongs. To this end it is convenient to rewrite (5.157) in the form

$$p_\tau^2 = \frac{E - j/(\tau + \alpha) - k/(\tau + \gamma) + G(\tau)}{2(\tau + \beta)}, \qquad (5.158)$$

where the function $G(\tau)$ is defined as

$$G(\tau) = \frac{F(\tau)}{(\tau + \alpha)(\tau + \gamma)}.$$

Rearranging its form, we can cast (5.158) in the enlightening form

$$E = 2(\tau + \beta)p_\tau^2 + V_{\text{eff}}(\tau), \qquad (5.159)$$

where we have introduced the *effective potential*

$$V_{\text{eff}}(\tau) = \frac{j}{\tau + \alpha} + \frac{k}{\tau + \gamma} - G(\tau). \qquad (5.160)$$

Equation (5.159) has the familiar form of the energy equation for one-dimensional motion in the effective potential (5.160). It differs from the standard form in one important respect: the usual condition, for which the motion is permitted in the regions in which the energy is greater than the effective potential, is true for λ and μ, *but not for* ν, as can easily be deduced by considering that the factor multiplying p_τ^2 would be negative in such a case. Summarizing, we can say that p_τ^2 is positive (and therefore the motion is possible) if ν, μ and λ satisfy the inequalities

$$E \le V_{\text{eff}}(\nu), \qquad E \ge V_{\text{eff}}(\mu), \qquad E \ge V_{\text{eff}}(\lambda). \qquad (5.161)$$

The procedure is therefore to plot, given the values of J and K, the curve of the effective potential. With this in mind, it is necessary to assume a form for $G(\tau)$: since our main interest is in self-gravitating systems, it is sensible to assume that $G(\tau)$ is a finite positive function, monotonically decreasing over the whole range $-\gamma \le \tau \le +\infty$. Comparison with the line $E = $ const.

taking into account conditions (5.161), gives directly the intervals available to the motion. A glance at the plots in Fig. 5.7 shows that the orbits can exist only in the case $k > 0$, but since we have to distinguish between the cases $E > V_{\text{eff}}(-\beta)$ and $E < V_{\text{eff}}(-\beta)$, there are four possible combinations giving rise to four different types of orbits.

a) Boxes. In the cases

$$j < 0, \qquad k > 0, \qquad V_{\text{eff}}(-\beta) < E < 0,$$

from Fig. 5.7a, it is easy to see that the allowed intervals for the coordinates are

$$-\gamma \leq \nu \leq \nu_{\max}, \qquad -\beta \leq \mu \leq \mu_{\max}, \qquad -\alpha \leq \lambda \leq \lambda_{\max}.$$

Recalling the properties of ellipsoidal coordinates (Sect. 5.4), we see that the projection of the orbit on the xz plane must always be inside the area bounded by the focal hyperbola (5.80b), whereas the projection on the yz plane must always be inside the area bounded by the focal ellipse of (5.80a) (see Fig. 5.8). The motion is given by the superposition of three librations in each coordinate so that, following Schwarzschild,[32] we call this type of orbit a *box-orbit*.

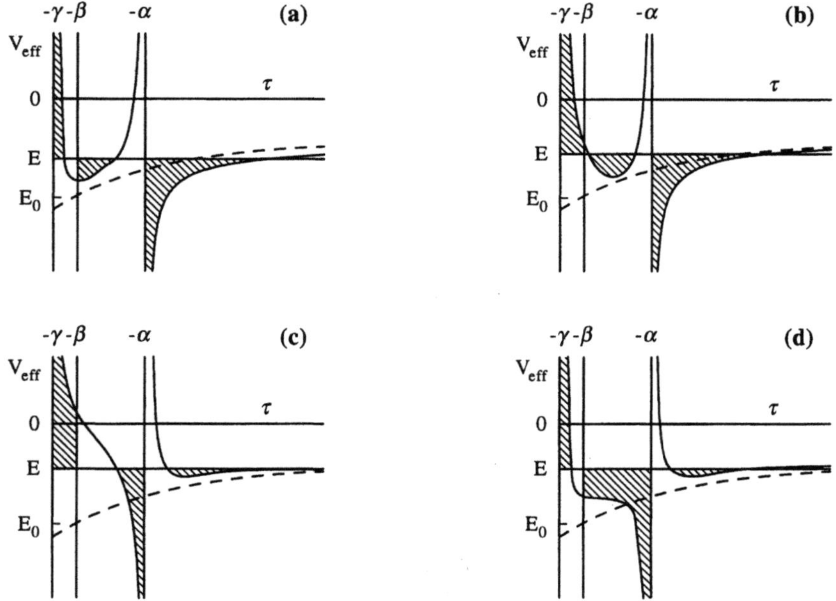

Fig. 5.7

---

[32]M. Schwarzschild: A numerical model for a triaxial stellar system in dynamical equilibrium, *Astrophys. J.* **232**, 236–247 (1979)

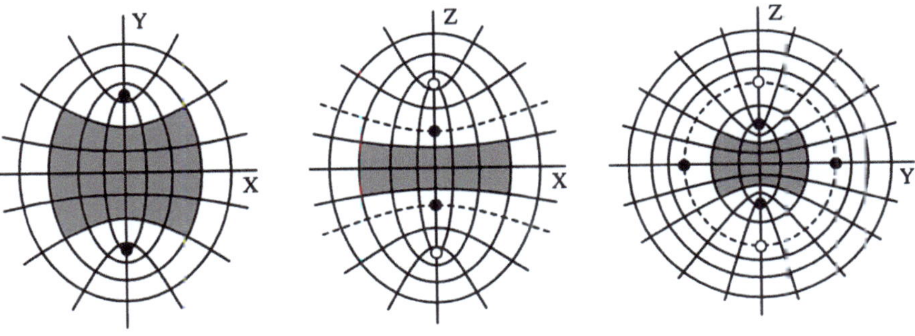

Fig. 5.8

b) Inner Long Axis Tubes. In the cases

$$j < 0, \qquad k > 0, \qquad V_{\text{eff}}(\mu_0) < E < V_{\text{eff}}(-\beta),$$

where μ_0 is the value of μ for which V_{eff} has a minimum, from Fig. 5.7b we see that the allowed intervals for the coordinates are

$$-\gamma \le \nu \le -\beta, \qquad \mu_{\min} \le \mu \le \mu_{\max}, \qquad -\alpha \le \lambda \le \lambda_{\max}.$$

The orbit is always around the x axis. Its projection on the xz plane must be outside the area bounded by the focal hyperbola (5.80b), whereas the projection on the yz plane must always be inside the area bounded by the focal ellipse (5.80a) (see Fig. 5.9). The motion in ν is a rotation and we call it a *tube orbit* around the x axis. Since it crosses the z axis between the *inner foci* (those of the focal hyperbola) at $z = \pm\sqrt{\gamma - \beta}$ and the *outer foci* (those of the focal ellipse) at $z = \pm\sqrt{\gamma - \alpha}$, we call it an *inner* long axis tube.

c) Outer Long Axis Tubes. In the cases

$$j > 0, \qquad k > 0, \qquad V_{\text{eff}}(\lambda_0) < E < V_{\text{eff}}(-\beta),$$

where λ_0 is the value of λ for which V_{eff} has a minimum, from Fig. 5.7c we see that the allowed intervals for the coordinates are

$$-\gamma \le \nu \le -\beta, \qquad \mu_{\min} \le \mu \le -\alpha, \qquad -\lambda_{\min} \le \lambda \le \lambda_{\max}.$$

The orbit is always around the x axis (see Fig. 5.10). Its projection on the xz plane must again be outside the area bounded by the focal hyperbola (5.80b) and the projection on the yz plane must always be outside the area bounded by the focal ellipse (5.80a). The motion in ν is again a rotation around the x axis, and since it crosses the z axis outside the *outer foci*, we call this kind of orbit an *outer* long axis tube.

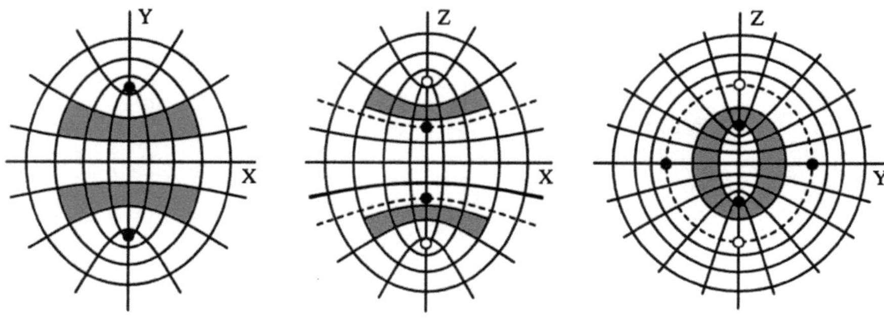

Fig. 5.9

d) Short Axis Tubes. In the cases

$$j > 0, \qquad k > 0, \qquad V_{\text{eff}}(\beta) \ \ (\text{or } V_{\text{eff}}(\lambda_0)) < E < 0,$$

where λ_0 is the value of λ for which V_{eff} has a minimum, from Fig. 5.7d, we see that the allowed intervals for the coordinates are

$$-\gamma \le \nu \le -\nu_{\max}, \qquad -\beta \le \mu \le -\alpha, \qquad -\lambda_{\min} \le \lambda \le \lambda_{\max}.$$

The orbit is always outside and around the z axis (see Fig. 5.11). Its projection on the xz plane must be inside an area bounded by the focal hyperbola (5.80b) and the projection on the yz plane must be always outside the area bounded by the focal ellipse (5.80a). The motion in μ is now a rotation around the z axis and we call this kind of orbit a *short* axis tube.

In summary, the point to emphasize about the above results is that the "zoology" of orbits we have illustrated above is highly representative of the orbits we observe in triaxial potentials considered as realistic approximations of elliptical galaxies.

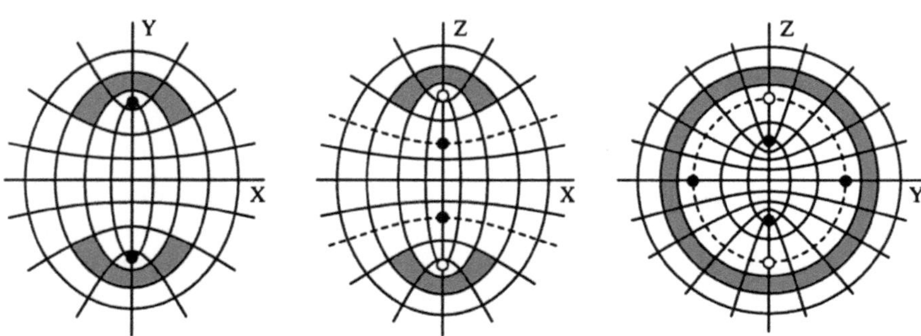

Fig. 5.10

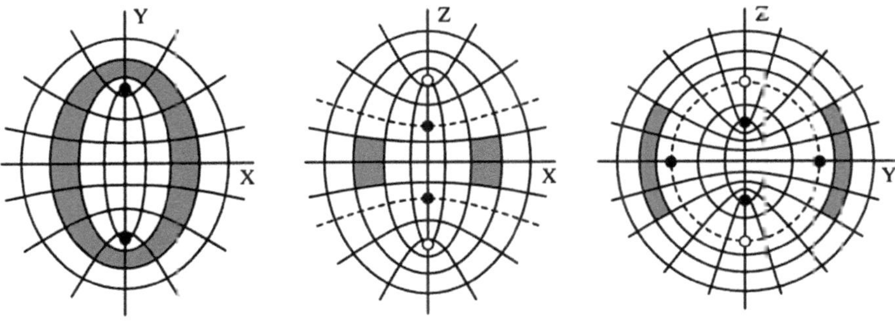

Fig. 5.11

5.9 Configurational Invariants

Up to now all that we have done in *the quest for integrability*[33] is intended to be valid for all the values of the energy of the system. This point is implicit in the general condition for the existence of first integrals $I(\boldsymbol{p}, \boldsymbol{q})$, namely that their Poisson bracket with the Hamiltonian vanishes,

$$(I, \mathcal{H}) = 0.$$

A function on phase space is a first integral if and only if this condition is satisfied and this is true for every value of the energy. We can therefore speak of *integrability at an arbitrary energy.*

Let us see how things change, with respect to what was discussed in Sect. 5.5 regarding integrability at arbitrary energies, if we search for integrability at a fixed energy in the case of the classic two-dimensional system. For the sake of simplicity we consider only the case of the second invariant linear in the momenta, but we soon enforce the condition

$$\frac{1}{2}\left(p_x^2 + p_y^2\right) + \Phi(x, y) = E_0, \tag{5.162}$$

and use it consistently from now on. This is accomplished by eliminating one of the momenta; we choose p_y, by means of relation (5.162) written in the form

$$p_y^2 = 2(E_0 - \Phi) - p_x^2. \tag{5.163}$$

The second invariant of the form (5.101), with the expression of the conservation of energy explicitly included, is now written

$$I_2 = A(x, y)p_x + B(x, y)\sqrt{2(E_0 - \Phi) - p_x^2}. \tag{5.164}$$

---

[33]See M. Tabor: *Chaos and Integrability in Nonlinear Dynamics*, (Wiley, New York, 1989), p. 322.

Computing its Poisson parenthesis with $\mathcal{H}$, namely the condition that it be identically zero for every value of x, y and p_x, we arrive at the following relations:

$$\frac{\partial A}{\partial x} - \frac{\partial B}{\partial y} = 0, \qquad \frac{\partial A}{\partial y} + \frac{\partial B}{\partial x} = 0, \qquad (5.165)$$

$$A\frac{\partial \Phi}{\partial x} + B\frac{\partial \Phi}{\partial y} - 2(E_0 - \Phi)\frac{\partial B}{\partial y} = 0. \qquad (5.166)$$

It is clear that this set of equations allows for a larger class of potentials if compared with (5.102). It is important to observe that if (5.166) were required to hold for every value of the energy, this would imply that $\partial B/\partial y = 0$, so that the system of (5.165) and (5.166) would reduce to system (5.102). This simple but important point further clarifies the issue of integrability at fixed energy. In other words, this means that (5.165–166) are the necessary and sufficient conditions for the existence of an integral at fixed energy, but they are only necessary (and not sufficient) for the existence of an integral at arbitrary energies.

A simple example which is general enough to shed light on the whole matter is the following. Let us choose as a simple solution of (5.165) that with $A = x$ and $B = y$, so that the second integral is

$$I_2 = xp_x + yp_y. \qquad (5.167)$$

For definiteness let us fix the energy at the value $E_0 = 0$, so that, for (5.166), the potential must satisfy the equation

$$x\frac{\partial \Phi}{\partial x} + y\frac{\partial \Phi}{\partial y} + 2\Phi = 0.$$

The solution of this equation is

$$\Phi(x, y) = x^{-2}f(y/x),$$

with f an arbitrary function, so that it is straightforward to verify that

$$\dot{I}_2 = (I_2, \mathcal{H}) = 2E_0, \qquad (5.168)$$

where $\mathcal{H}$ is taken to assume arbitrary values. This is direct verification that I_2 is a first integral only on the fixed energy surface $E_0 = 0$. An expression like (5.168) can be seen as the condition for a *weak involution property*,[34] which is reminiscent of the more general situation of Liouville's theorem for systems with a number of integrals in *strong* involution equal to the number of degrees of freedom (see Sect. 1.18). I_2 is said to be a *configurational invariant*.

---

[34]This definition was introduced by W. Sarlet, P. G. L. Leach, F. Cantrijn: First integrals versus configurational invariants and a weak form of complete integrability, *Physica* **17D**, 87–98 (1985).

It is very interesting to develop the treatment of configurational invariants in a geometric framework, instead of the analytical approach followed before. The appeal of the geometric picture is twofold: first, in the simplest cases the conditions for the existence of the invariants are related to the symmetries of the dynamical problem through Noether's theorem (see Sect. 1.6); second, if we represent the trajectories of the system at a fixed energy E_C as the geodesics on the manifold endowed with the Jacobi metric, the Killing equations (1.B.41) and (1.B.42) are directly applicable and show all their power. If we exploit the Jacobi form of mechanics (see Sect. 1.8), we can obtain the equations of motion for a conservative dynamical system from the Lagrangian

$$\mathcal{L} = \frac{1}{2} g_{ik} \dot{q}_i \dot{q}_k, \qquad g_{ik} = \eta_{ik}(E_0 - \Phi). \tag{5.169}$$

Requiring, for example, the invariance of the action integral with respect to a transformation of the form

$$\bar{q}_k = q_k + \xi_k(\boldsymbol{q})\, \varepsilon,$$

following the general analysis developed in Sect. 1.6, we find that the invariance is guaranteed by the condition

$$\frac{\partial \mathcal{L}}{\partial q_i} \xi_i + \frac{\partial \mathcal{L}}{\partial \dot{q}_i} \frac{\partial \xi_i}{\partial q_j} \dot{q}_j = 0. \tag{5.170}$$

With a Lagrangian of the "kinetic" form of (5.169) it is easy to verify that (5.170) can be written in the standard form of the Killing equation:

$$\frac{\partial g_{ik}}{\partial q_j} \xi_j + g_{ij} \frac{\partial \xi_j}{\partial q_k} + g_{kj} \frac{\partial \xi_j}{\partial q_i} = 0, \tag{5.171}$$

which is well-known in Riemannian geometry, and that the invariant can be written in the form

$$I = g_{ij} \xi_j \dot{q}_k = p_k \xi_k, \tag{5.172}$$

where we have used the duality relation between the velocity and momentum. We therefore see that, in the case of a symmetry related to the existence of the Killing vector $\boldsymbol{\xi}$, the invariant *linear* in the momentum (and so of the form of (5.101) or (5.164)) emerges in a straightforward way. It is then natural to think that the conditions imposed by the Killing equations (5.171) are equivalent to system (5.102) (or (5.165–166)). In fact, with the Jacobi metric of (5.169), Killing equation (5.171) gives the set of conditions

$$2(E_0 - \Phi)\xi_{(i,k)} - \eta_{ik}\Phi_{,j}\xi_j = 0, \tag{5.173}$$

where we have used the simplified notation of commas to denote partial derivatives and

$$2\xi_{(i,k)} = \xi_{i,k} + \xi_{k,i}.$$

Consider again the case of two degrees of freedom. If we want conditions (5.173) to be satisfied for every value of the energy, we find that

$$\xi_{(i,k)} = 0, \qquad \Phi_{,j}\xi_j = 0, \tag{5.174}$$

which, by the way, are valid for any number of degrees of freedom. But if we require that conditions (5.173) are satisfied on the fixed E_0 surface, we find that

$$\xi_{1,2} + \xi_{2,1} = 0, \qquad \xi_{1,1} = \xi_{2,2}, \qquad (E_0 - \Phi)\xi_{2,2} - \Phi_{,j}\xi_j = 0. \tag{5.175}$$

Defining $\xi_1 = A$ and $\xi_2 = B$, so that (5.172) agrees explicitly with (5.101), we see that (5.174) coincides with the general case (5.102), whereas (5.175) are the same as (5.165–166). The geometric interpretation of this situation is that the "family" of surfaces endowed with the Jacobi metric, at some particular value of E_0, "takes a shape" that allows for the existence of a Killing vector (which is the generator of a transformation symmetry of the surface) to which a configurational invariant linear in the momentum is associated.

It is also possible to generalize this geometrical approach to higher-order invariants. If the second-order invariant defined by (5.103) is written in the form

$$I = K_{ij}p_i p_j + D,$$

where K_{ij} is a symmetrical second-rank matrix, it is easy to see that the commutativity of I with $\mathcal{H}$, expressed by the set of equations (5.104) and (5.105), can be written in the compact form

$$K_{(ij,k)} = 0, \qquad 2\Phi_j K_{ji} = D_{,i}. \tag{5.176}$$

These equations can be interpreted as the conditions for the existence of a second-order *Killing tensor*[35] for the metric g_{ij}. The geometric interpretation of Killing tensor is not so simple as before (in particular they cannot generate simple point transformations,[36] but must be treated as generalized transformations, like those introduced in Sect. 1.6), but, as before, (5.176) can also be solved at a specified value of the energy, giving rise to a configurational quadratic invariant.

We shall not dwell on this question. However, it gives us the opportunity to stress that research activity in differential and Riemannian geometry can still be very fruitfully applied in classical mechanics.

---

[35]See, e.g., R. Wald: *General Relativity* (University of Chicago Press, 1984), p. 440.
[36]K. Rosquist: Killing tensor conservation laws and their generators, *J. Math. Phys.* **30**, 2319–2321 (1989).

Mathematical Appendix

Notation

Throughout this book the summation convention on repeated indices has been adopted; the summation symbol $\sum$ is used only when confusion is possible. At the same time, very often we do not distinguish between covariant and contravariant components of vectors and tensors, the question being irrelevant in our treatment. The Bachmann–Landau convention (o, O) is used to express the notion of "order of" (see, for instance, E. T. Whittaker, G. N. Watson: *A Course of Modern Analysis* (Cambridge University Press, 1927), p. 11.).

A.1 Spherical Trigonometry

Some simple relations of spherical trigonometry are useful: given a spherical triangle with vertex angles $\hat{A}, \hat{B}$ and $\hat{C}$ and sides given by segments of arc a, b and c respectively (see Fig. A.1), the relations

$$\cos a = \cos b \, \cos c + \sin b \, \sin c \, \cos \hat{A}, \tag{A.1}$$

$$\cos \hat{B} \, \sin a = \cos b \, \sin c - \sin b \, \cos c \, \cos \hat{A}, \tag{A.2}$$

$$\frac{\sin \hat{A}}{\sin a} = \frac{\sin \hat{B}}{\sin b} = \frac{\sin \hat{C}}{\sin c}. \tag{A.3}$$

hold. From (A.1), exchanging the roles of a and b, we have

$$\cos b = \cos a \, \cos c + \sin a \, \sin c \, \cos \hat{B},$$

and, further, exchanging a by c and making some calculations, we have

$$\sin c \cos \hat{B} = \cos b \sin a.$$

One of the equalities contained in relation (A.3) can be written in the form

$$\sin c \sin \hat{B} = \sin b \sin \hat{C}.$$

Dividing by this relation the last obtained we can write

$$\cot \hat{B} \sin \hat{C} = \cot b \sin a, \tag{A.4}$$

a formula of use in the application of the theory of orbits in spherical potentials.

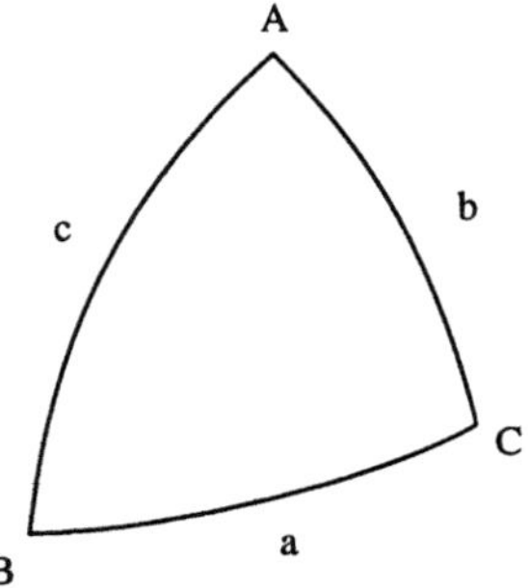

Fig. A.1

A.2 Curvilinear Coordinate Systems

For the study of orbits in potentials in two and three dimensions it is very useful to be able to express the equations of motion in arbitrary coordinate systems, so as to explicit at their best the methods for solving dynamical problems presented in Chaps. 1 and 5. There are several curvilinear coordinate systems in which the Hamilton–Jacobi equation is separable with potentials of real interest: we present in what follows the essential properties of curvilinear orthogonal coordinates.

Three scalar functions

$$q^i = f^i(x, y, z), \qquad i = 1, 2, 3 \tag{A.5}$$

of the Cartesian coordinates $\{x, y, z\}$ define, if the parameters q^i are varied, a curvilinear coordinate system. The functions f^i are assumed to be invertible, so that it is possible to express the Cartesian coordinates in terms of the curvilinear coordinates: $x = x(q^i)$, $y = y(q^i)$ and $z = z(q^i)$. The fact that the correspondence between the coordinates of a point in the two systems is one-to-one is warranted by the condition that the Jacobian determinant[1]

$$\left| \frac{\partial(x, y, z)}{\partial(q_1, q_2, q_3)} \right| = \begin{vmatrix} \frac{\partial x}{\partial q_1} & \frac{\partial x}{\partial q_2} & \frac{\partial x}{\partial q_3} \\ \frac{\partial y}{\partial q_1} & \frac{\partial y}{\partial q_2} & \frac{\partial y}{\partial q_3} \\ \frac{\partial z}{\partial q_1} & \frac{\partial z}{\partial q_2} & \frac{\partial z}{\partial q_3} \end{vmatrix}$$

does not vanish. The squared line element in a general coordinate system is

$$ds^2 = \sum_{i,j=1}^{3} g_{ij} dq^i dq^j, \tag{A.6}$$

where the elements of the *metric tensor* g_{ij} are given by

$$g_{ij} = \frac{\partial x}{\partial q^i} \frac{\partial x}{\partial q^j} + \frac{\partial y}{\partial q^i} \frac{\partial y}{\partial q^j} + \frac{\partial z}{\partial q^i} \frac{\partial z}{\partial q^j}. \tag{A.7}$$

In the following we present the properties of those coordinate systems where the metric tensor is diagonal: $g_{ij} = G_i^2 \delta_{ij}$. In this case the coordinate frame is called *orthogonal*. Orthogonal coordinates (besides Cartesian coordinates) include cylindrical coordinates:

---

[1] In formulae where curvilinear coordinates appear there is a slight notational inconsistency, since the coordinates are indicated by q^i (with upper indices) for generic i, whereas they appear with lower indices when this is one of the values 1,2,3. The notation with the upper index is preferable for various reasons (it is of ease in sums such as that in (A.6) and for consistency with the usual notation of differential geometry) so that we try to use it consistently. To avoid confusion with exponential powers, however, we have kept the lower indices when they take specific values. Finally, the summation symbol $\sum$ is used, since, in formulae like (A.10), no sum is intended.

$$q_1 = \varpi = \sqrt{x^2 + y^2}, \qquad q_2 = \varphi = \arctan\frac{y}{x}, \qquad q_3 = z;$$
$$ds^2 = d\varpi^2 + \varpi^2 d\varphi^2 + dz^2; \qquad G_1 = 1, \ G_2 = \varpi, \ G_3 = 1, \tag{A.8a}$$

and the spherical ones

$$q_1 = r = \sqrt{x^2 + y^2 + z^2}, \quad q_2 = \vartheta = \arccos\left(z/r\right), \quad q_3 = \varphi = \arctan(y/x);$$
$$ds^2 = dr^2 + r^2 d\vartheta^2 + r^2 \sin^2\vartheta d\varphi^2; \quad G_1 = 1, \ G_2 = r, \ G_3 = r\sin\vartheta. \tag{A.8b}$$

One more important aspect of the use of curvilinear coordinates is that of the relations among the components of the velocities and the momenta and the various expressions of the equations of the motion. The metric introduced in (A.6) allows us to define in a natural way (see Sect. 1.4) the quadratic form of the kinetic energy

$$\mathcal{T} = \frac{1}{2}\sum_i G_i^2(\dot{q}^i)^2 = \frac{1}{2}\sum_i (v^i)^2 = \frac{1}{2}\sum_i G_i^{-2}(p_i)^2, \tag{A.9}$$

where we have put the mass equal to unity. Associated with the coordinate components $\dot{q}^i$ of the velocity there are components $v^i = G_i \dot{q}^i$ in the orthonormal basis consisting of the vectors

$$\hat{e}_i = G_i^{-1}\frac{\partial}{\partial q^i}, \tag{A.10}$$

and the components of the momentum[2]

$$p_i = G_i^2 \dot{q}^i, \tag{A.11}$$

introduced by means of a Legendre transformation with $\mathcal{T}$ in the role (see (1.C.1) of Lagrangian. In cylindrical and spherical coordinates, using the metric elements appearing in (A.8), we have respectively

$$p_\varpi = \dot{\varpi}, \ p_\varphi = \varpi\dot{\varphi}, \ p_z = \dot{z} \tag{A.12a}$$

and

$$p_r = \dot{r}, \ p_\vartheta = r^2\dot{\vartheta}, \ p_\varphi = r^2\sin^2\vartheta\dot{\varphi}. \tag{A.12b}$$

The equations of motion of a unit mass particle in a potential field Φ can then be written in the general form

$$\ddot{q}^i + \sum_{j,k}\Gamma^i{}_{jk}\dot{q}^j\dot{q}^k + \sum_j g^{ij}\frac{\partial\Phi}{\partial q^j} = 0, \tag{A.13}$$

where

$$g^{ij} = G_i^{-2}\delta^{ij}$$

---

[2] Even in this case the lower index is consistent with the standard notation of differential geometry: in the old language of tensorial algebra p_i are the components of a *covariant* vector, whereas X^i are the components of a *controvariant* vector.

is the metric inverse of the one in (A.6) and the quantities $\Gamma^i{}_{jk}$, defined by[3]

$$\Gamma^i{}_{jk} = \frac{1}{2} \sum_m g^{im} \left(\frac{\partial g_{mj}}{\partial q^k} + \frac{\partial g_{mk}}{\partial q^j} - \frac{\partial g_{jk}}{\partial q^m} \right), \qquad (A.14)$$

are the *coefficients of the connection* given by the natural basis associated with curvilinear coordinates. For an orthogonal metric it is easy to verify that

$$\Gamma^i{}_{jk} = 0, \qquad \Gamma^i{}_{jj} = -\frac{1}{2} G_i^{-2} \frac{\partial G_j^2}{\partial q^j},$$
$$\Gamma^i{}_{ij} = G_i^{-1} \frac{\partial G_i}{\partial q^j}, \qquad \Gamma^i{}_{ii} = G_i^{-1} \frac{\partial G_i}{\partial q^i}, \qquad (A.15)$$

where $i \neq j \neq k$ and, as always in this section, there is no sum over repeated indices. For example, in cylindrical coordinates the (A.15) become

$$\Gamma^1{}_{22} = -\varpi, \qquad \Gamma^2{}_{12} = \Gamma^2{}_{21} = \frac{1}{\varpi}, \qquad \text{all the others zero,} \qquad (A.16)$$

and (A.13) assume the two equivalent forms

$$\begin{cases} \ddot{\varpi} - \varpi \dot{\varphi}^2 + \dfrac{\partial \Phi}{\partial \varpi} = 0, \\[2mm] \ddot{\varphi} + 2 \dfrac{\dot{\varpi}}{\varpi} \dot{\varphi} + \dfrac{1}{\varpi^2} \dfrac{\partial \Phi}{\partial \varphi} = 0, \\[2mm] \ddot{z} + \dfrac{\partial \Phi}{\partial z} = 0, \end{cases} \qquad (A.17a)$$

and

$$\begin{cases} \dot{v}_\varpi - \dfrac{v_\varphi^2}{\varpi} + \dfrac{\partial \Phi}{\partial \varpi} = 0, \\[2mm] \dot{v}_\varphi + \dfrac{1}{\varpi} v_\varpi v_\varphi + \dfrac{1}{\varpi} \dfrac{\partial \Phi}{\partial \varphi} = 0, \\[2mm] \dot{v}_z + \dfrac{\partial \Phi}{\partial z} = 0, \end{cases} \qquad (A.17b)$$

where the transformation of the components of the velocities $v^i = G_i \dot{q}^i$ has been used, giving

$$v_\varpi = \dot{\varpi}, \qquad v_\varphi = \varpi \dot{\varphi}, \qquad v_z = \dot{z}. \qquad (A.18)$$

Analogously in the case of spherical coordinates, the coefficients of the connection are

---

[3] For a detailed discussion of the foundations of geometry in curved spaces we refer to D. Laugwitz, *Differential and Riemannian Geometry* (Academic Press, 1965).

$$\Gamma^1{}_{22} = -r, \qquad \Gamma^1{}_{33} = -r\sin^2\vartheta,$$
$$\Gamma^2{}_{12} = \Gamma^2{}_{21} = \Gamma^3{}_{13} = \Gamma^3{}_{31} = \frac{1}{r},$$
$$\Gamma^2{}_{33} = -\sin\vartheta\cos\vartheta,$$
$$\Gamma^3{}_{23} = \Gamma^3{}_{32} = \frac{\cos\vartheta}{\sin\vartheta}, \qquad \text{all the others zero,} \tag{A.19}$$

and the equations of motion are

$$\begin{cases} \ddot{r} - r\dot{\vartheta}^2 - r\sin^2\vartheta\dot{\varphi}^2 + \dfrac{\partial\Phi}{\partial r} = 0, \\[2mm] \ddot{\vartheta} + 2\dfrac{\dot{r}}{r}\dot{\vartheta} - \sin\vartheta\cos\vartheta\dot{\varphi}^2 + \dfrac{1}{r^2}\dfrac{\partial\Phi}{\partial\vartheta} = 0, \\[2mm] \ddot{\varphi} + 2\dfrac{\dot{r}}{r}\dot{\varphi} + 2\cot\vartheta\dot{\vartheta}\dot{\varphi} + \dfrac{1}{r^2\sin^2\vartheta}\dfrac{\partial\Phi}{\partial\varphi} = 0, \end{cases} \tag{A.20a}$$

or, equivalently,

$$\begin{cases} \dot{v}_r - \dfrac{v_\vartheta^2 + v_\varphi^2}{r} + \dfrac{\partial\Phi}{\partial r} = 0, \\[2mm] \dot{v}_\vartheta + \dfrac{1}{r}(v_r v_\vartheta - \cot\vartheta v_\varphi^2) + \dfrac{1}{r}\dfrac{\partial\Phi}{\partial\vartheta} = 0, \\[2mm] \dot{v}_\varphi + \dfrac{1}{r}(v_r v_\varphi + \cot\vartheta v_\vartheta v_\varphi) + \dfrac{1}{r\sin\vartheta}\dfrac{\partial\Phi}{\partial\varphi} = 0, \end{cases} \tag{A.20b}$$

where the transformation of the components of the velocities is given by

$$v_r = \dot{r}, \qquad v_\vartheta = r\dot{\vartheta}, \qquad v_\varphi = r\sin\vartheta\dot{\varphi}. \tag{A.21}$$

In the general case, (A.13) and (A.15) give

$$\ddot{q}^i + 2G_i^{-1}\sum_{j=1}^{3}\frac{\partial G_i}{\partial q^j}\dot{q}^i\dot{q}^j - G_i^{-2}\sum_{j=1}^{3}G_j\frac{\partial G_j}{\partial q^i}\left(\dot{q}^j\right)^2 + G_i^{-2}\frac{\partial\Phi}{\partial q^i} = 0. \tag{A.22}$$

Wanting to express this general equation of motion in terms of the velocities v^i or the momenta, we need the components of the connection in the orthonormal basis (A.10), that is, the quantities

$$\hat{\Gamma}^i{}_{jk} = \left\langle \hat{\omega}^i, \frac{\partial\hat{e}_j}{\partial q^k}\right\rangle, \tag{A.23}$$

where the basis forms $\hat{\omega}^i$, duals of the orthonormal basis (A.10), are given by the relations

$$\hat{\omega}^i = G_i dq^i, \qquad \left\langle\hat{\omega}^i, \hat{e}_j\right\rangle = \delta^i{}_j. \tag{A.24}$$

Substituting (A.10) and (A.24) into (A.23) and making use of the definition

$$\Gamma^i{}_{jk} = \left\langle dq^i, \frac{\partial^2}{\partial q^j \partial q^k}\right\rangle,$$

which is the analogue of (A.23) in the coordinate basis and generates (A.14), we can explicitly determine $\hat{\Gamma}$. With the agreement, in the following, that the indices are such that $i \neq j$ and that $F_{,i}$ represents the partial derivative of the generic function F with respect to the coordinate q^i, we have

$$\hat{\Gamma}^i{}_{jj} = \left\langle G_i dq^i, G_j^{-1} \frac{\partial}{\partial q^j}\left(G_j^{-1}\frac{\partial}{\partial q^j}\right)\right\rangle$$

$$= G_i\left\langle dq^i, -\frac{G_{j,j}}{G_j^3}\frac{\partial}{\partial q^j} + G_j^{-2}\frac{\partial^2}{\partial q^j \partial q^j}\right\rangle = G_i G_j^{-2}\Gamma^i{}_{jj}.$$

Through similar calculations we find all the components of the connection in the orthonormal basis:

$$\hat{\Gamma}^i{}_{ij} = G_j^{-1}\left(\Gamma^i{}_{ij} - G_i^{-1}G_{i,j}\right),$$

$$\hat{\Gamma}^i{}_{ji} = G_j^{-1}\Gamma^i{}_{ji},$$

$$\hat{\Gamma}^i{}_{ii} = G_i^{-1}\left(\Gamma^i{}_{ii} - G_i^{-1}G_{i,i}\right).$$

Exploiting coordinate expressions (A.15), these can be written as

$$\hat{\Gamma}^i{}_{jj} = -G_i^{-1}G_j^{-1}G_{j,i}, \quad \hat{\Gamma}^i{}_{ji} = G_i^{-1}G_j^{-1}G_{i,j}, \quad \text{all others zero.} \qquad (A.25)$$

The general form of the equations of motion in the orthonormal frame is therefore

$$\dot{v}^i + \sum_{i\neq j} G_i^{-1}G_j^{-1}\left(G_{i,j}v^i v^j - G_{j,i}(v^j)^2\right) + G_i^{-1}\frac{\partial \Phi}{\partial q^i} = 0. \qquad (A.26)$$

Using the relation $p_i = G_i^2 \dot{q}^i = G_i v^i$ and exploiting the equations

$$\dot{v}^i = \frac{d}{dt}(G_i^{-1}p_i) = G_i^{-1}\dot{p}_i + \sum_j \frac{\partial}{\partial q^j}(G_i^{-1}p_i)\frac{dq^j}{dt},$$

the equations of motion in terms of the momenta are

$$\dot{p}_i + \frac{1}{2}\frac{\partial}{\partial q^i}\left[\sum_{j=1}^3 (G_j^{-1}p_j)^2\right] + \frac{\partial \Phi}{\partial q^i} = 0. \qquad (A.27)$$

It is worth remarking that these equations, once one has introduced a Hamilton function $\mathcal{H} = \mathcal{T} + \Phi$ and remembering (A.9), are the expression of the canonical equations (1.C.3) for the momenta in a general curvilinear coordinate system.

A.3 Riemannian Geometry

One more reason for introducing the kinematics along the lines sketched in the previous section is that, without any change in the notation, they allow us to treat the general case of motion in a non-Euclidean space. We shall restrict ourselves here to the case of Riemannian space, that is, space in which the metric element is determined by g_{ab}, which is a positive definite tensor. Our interest in Riemannian geometry is justified by the fact that we have encountered several "geometric" ways of formulating the laws of mechanics (such as Jacobi's approach, described in Sect. 1.8), so that in many circumstances a complete geometric treatment of orbit theory can be useful.

The principal aspect to emphasize is that the metric element given by (A.6) describes, in general, the properties of non-Euclidean spaces in which the distance between infinitesimally close points is a quadratic form in the increments of the coordinates that reflects *intrinsic* geometric properties, rather then a simple change of coordinates in flat space. It is therefore obvious that it is essential to be able to characterize in an *invariant* way (that is, in a way independent of the chosen coordinates) the intrinsic *curvature* properties of the space: we will devote the final part of this section to this aim. For the moment, we want to consider kinematics in curved space.

The equations of motion in a generic curved space are essentially the same as (A.13), which we write here again, but without the potential term to better examine only the geometric aspects of the motion:[4]

$$\frac{d^2 q^a}{ds^2} + \Gamma^a{}_{bc} \frac{dq^b}{ds} \frac{dq^c}{ds} = 0. \tag{A.28}$$

The only differences with respect to the formulae of the previous section are that now the independent variable is no longer the time but an arbitrary affine parameter along the curve and that the convention is adopted of the sum over repeated indices. Equations (A.28) are usually called the *equations of the geodesics*, since their solutions give the curves of extremal length in the geometry given by the metric element (A.6). If the space is flat (and the potential is zero) these solutions are obviously straight lines. If one defines the velocity vector $u^a = dq^a/ds$, then (A.28) can be written in the form

$$\frac{du^a}{ds} + \Gamma^a{}_{bc} u^b u^c = 0. \tag{A.29}$$

Writing the differentiation with respect to s as the total derivative

$$\frac{d}{ds} = \frac{dq^a}{ds} \frac{\partial}{\partial q^a} = u^a \frac{\partial}{\partial q^a},$$

---

[4] Actually, the Jacobi geometry just referred to has the effect of *incorporating* the potential into the metric tensor, so that the Γ's convey dynamics into the geometry of a suitable space.

we can interpret the left-hand side of (A.29) as the action of a differential operator on a generic vector field X^a,

$$\frac{DX^a}{ds} = \frac{dq^b}{ds}\left(\frac{\partial X^a}{\partial q^b} + \Gamma^a{}_{bc}X^c\right). \tag{A.30}$$

If the result of this action vanishes (that is, if $DX^a/ds = 0$), the vector X^a is said to be *parallely transported along the integral curves of u^a*. The operation in the brackets in (A.30) can be performed in general without reference to a curve: it is called the *covariant derivative* and is denoted for the sake of brevity as follows

$$X^a{}_{;b} = X^a{}_{,b} + \Gamma^a{}_{bc}X^c. \tag{A.31}$$

It can be applied to tensors of arbitrary rank and to covectors defined by the general duality relation

$$X_b = g_{ab}X^a, \tag{A.32}$$

the generalization of (A.11) in the previous section. As an example which enables us to present a quite generic situation, consider the covariant derivative of a mixed second-rank tensor $A^a{}_b$; it is written

$$A^a{}_{b;c} = A^a{}_{b,c} - \Gamma^d{}_{bc}A^a{}_d + \Gamma^a{}_{dc}A^d{}_b. \tag{A.33}$$

In the metric tensor g_{ab} all the information about the intrinsic geometrical structure of the space considered is contained, so that, for example, the explicit expression of the connection coefficients is given by (A.14). Since all the formulae involving geometric objects (vectors, tensors, curves, etc.) and algebraic and differential operations with them, are the same as are found working with generic coordinates in Euclidean space, one needs an invariant way to discriminate between the two situations. In other words, we need some tool that enables us to know if the metric (and its connection) we are using, represents a genuine non-Euclidean (or, as it is usually said, "curved") space or whether it is simply related to some, maybe very complicated, change of coordinates in flat space. This tool is the *Riemann curvature tensor*, defined as

$$R^a{}_{bcd} = \Gamma^a{}_{bd,c} - \Gamma^a{}_{bc,d} + \Gamma^a{}_{kc}\Gamma^k{}_{bd} - \Gamma^a{}_{kd}\Gamma^k{}_{bc}. \tag{A.34}$$

This tensor enjoys the fundamental property of being identically zero if and only if the space is flat: since this property does not depend on the coordinates used, the Riemann tensor is the invariant intrinsic representative of the geometric properties of the space in which motion happens. Several other geometric objects can be constructed starting from the Riemann tensor; an important quantity is the *Gaussian curvature* in the plane identified by two unit vectors n^a and m^a:

$$C(n,m) = R_{abcd}n^a m^b n^c m^d. \tag{A.35}$$

There are then two tensors (actually a tensor and a scalar) that play an important role in Riemannian geometry:

$$R_{ab} = R^c{}_{acb} = g^{cd} R_{cadb} \tag{A.36}$$

and

$$R = g^{ab} R_{ab} = g^{ab} g^{cd} R_{cadb}. \tag{A.37}$$

These are respectively the *Ricci tensor* and the *Ricci curvature scalar*. There are simple but important relations between these tensors and the Gaussian curvatures: taking as unit vectors the orthonormal basis vectors $\hat{e}_i^a$ (where the lower index labels which vector it is) to compute the quantities in (A.35), we have the *Gaussian sectional curvatures*

$$C_{ij} = R_{abcd} \hat{e}_i^a \hat{e}_j^b \hat{e}_i^c \hat{e}_j^d. \tag{A.38}$$

It is quite easy to verify that the sum of the sectional curvatures in the direction of $\hat{e}_j^b$ is given by

$$\sum_j C_{ij} = R_{ab} \hat{e}_i^a \hat{e}_i^b,$$

so that it is related to the Ricci tensor, whereas the sum of all the curvatures is related to the Ricci scalar by

$$\sum_{i,j} C_{ij} = R.$$

An application of the Riemann tensor which is of particular interest in the study of dynamical systems is to the analysis of the relative behaviour of geodesics, namely their stability properties and their long time behaviour. Let us assume that we are given a solution of equation (A.29), that is, a geodesic taken as a "reference" and let n^a be a vector representing a small displacement with respect to the reference geodesic. If u^a is the tangent vector of the reference, the evolution of the perturbation is governed by the equation

$$\frac{D^2 n^a}{ds^2} + R^a{}_{bcd} u^b n^c u^d = 0. \tag{A.39}$$

This fundamental equation is called the *equation of the geodesic deviation* or *Jacobi–Levi-Civita equation*[5] and provides all the information necessary to determine the influence of slight changes in the initial conditions on the geodesic flow.

One final argument that is worth examining for its applicability in analytical dynamics is the question of the symmetries of the metric. Let us consider the transformation given by a one-parameter group of motions generated by a vector field ξ^a and ask for the conditions such that the metric tensor is invariant under such a transformation. It turns out that these conditions are

$$g_{ab,c} \xi^c + g_{ac} \xi^c{}_{,b} + g_{cb} \xi^c{}_{,a} = 0,$$

---

[5] See T. Levi-Civita: Sur l'écart géodésique, *Math. Ann.* **97**, 291–320 (1927).

which, looking at the relations between the connection coefficients and the derivatives of the metric tensor (A.14), can be written in the equivalent forms

$$\xi_{a,b} + \xi_{b,a} - \Gamma^c{}_{ab}\xi_c = 0,$$

or

$$2\xi_{(a;b)} = \xi_{a;b} + \xi_{b;a} = 0. \tag{A.40}$$

These last two equations are known as *Killing equations* in Riemannian geometry and the vector ξ^a as a *Killing vector*. Its fundamental importance rests on the fact that its existence guarantees the existence of conserved quantities along the geodesics. Let us consider the scalar quantity $J = \xi_a u^a$ and compute its derivative along the geodesic whose tangent vector is u^a itself:

$$\frac{DJ}{ds} = \xi_{a;b} u^a u^b + \xi_a u^a{}_{;b} u^b = 0. \tag{A.41}$$

We find that the derivative vanishes, since the first term is zero by Killing equation (A.40) and the second is also zero from the equation of the geodesics. J is therefore conserved along the motion. A straightforward generalization of this circumstance[6] is to conserved quantities of higher order in the velocities: suppose that we want to find the condition such that the scalar

$$I = A_{ab} u^a u^b$$

is conserved along the geodesics. A calculation in the same spirit as that in (A.41) shows that I is conserved if and only if the equations

$$A_{(ab;c)} = 0 \tag{A.42}$$

are satisfied. Generalizing the above definition, the symmetric tensor A_{ab}, which satisfies the condition that its symmetrized covariant derivatives vanish, can be called a *Killing tensor*.

---

[6] See, for instance, L. P. Eisenhart: *Riemannian Geometry* (Princeton University Press, 1950), Sect. 39.

Bibliographical Notes

The purpose of these notes is twofold. On the one hand, they are intended to provide the reader with a list of the sources relevant to each section and of the books or papers which are deemed fundamental for deepening one's knowledge of the topic under consideration: on the other hand, occasionally the discussion is broadened by trying to locate the topic in a wider and clearer context, while having in mind the aim of stimulating a non-passive understanding.

In these notes, any works already mentioned in the text are not repeated, except in the case of further specific discussion.

Introduction

As far as we know, there exists no study on the evolution of mental schemes or models (we do not know how better to describe them) and on their reappearance from time to time in the history of scientific thought, borrowing from the contingent cultural context the forms by which they can be expressed. One case we have tried to emphasize, and which is well known to the community of experts in celestial mechanics, is that of epicycles and Fourier series, but certainly others exist–less macroscopic cases–which, from Ptolemaic theory to modern celestial mechanics, have continued equally unchanged in their inner nature of mental schemes.

To deepen the "techniques" used in ancient times for the study of the kinematics of celestial bodies, a textbook that is still useful is that monumental one by P. Duhem: *Le Système du Monde – Histoire des Doctrines Cosmologiques de Platon à Copernic* (Hermann, Paris, 1913). Another interesting textbook is G. V. Schiaparelli: *Scritti sulla Storia dell'Astronomia Antica* (Zanichelli, Bologna, 1927; available only in Italian). Also important, in order to follow the evolution of scientific thought in the pre-Copernican era and the history of ancient astronomy, is J. L. E. Dreyer: *History of the Planetary Systems from Thales to Kepler* (Cambridge University Press, 1906; reprinted by Dover).

The studies on ancient astronomy by Neugebauer deserve a separate mention. Except for O. Neugebauer: *The Exact Sciences in Antiquity* (Princeton University Press, 1952), a work suitable for a reader of average scientific education, they are texts addressed essentially to specialists. We confine ourselves to mentioning O. Neugebauer: *A History of Ancient Mathematical Astronomy*, Parts 1–3 (Springer, 1975).

For the Copernican revolution, a well-known text is T. S. Kuhn: *The Copernican Revolution. Planetary Astronomy in the Development of Western Thought* (Harvard University Press, Cambridge, 1957). The period from ancient times up to Kepler is dealt with by the above-mentioned history of astronomy. For readers who can understand Italian, a small book which is a mine of information as well as being very useful is R. Marcolongo: *Il Problema dei Tre Corpi* (Hoepli, Milan, 1914).

As far as Newton's works are concerned, besides Cajori's edition, we note *Isaac Newton's Philosophiae Naturalis Principia Mathematica. The third edition (1726) with variant readings*, ed. by A. Koyré and I. B. Cohen (Cambridge University Press, 1972). Moreover, since it would be foolish to pretend to list a bibliography, however summary, of the studies on Newtonian thought, we only mention R. S. Westfall: *Never at Rest. A Biography of I. Newton* (Cambridge University Press, 1980) and A. Koyrè: *Newtonian Studies* (Chapman & Hall, London, 1965), where the reader can find the desired references.

Obviously, our interest is centred on the evolution of theories related to the motion of the celestial bodies; therefore our indications ought always to be aimed at this goal, rather than to acquiring a broad knowledge of the history of astronomy or mechanics. Hence it is perhaps more appropriate with regard to modern authors to refer directly to their works. For instance, in the case of Poincaré it is convenient to approach his masterpiece itself. Here we mention the English translation, H. Poincaré: *New Methods in Celestial Mechanics* (Hilger, 1990). Furthermore, the following book is helpful, R. Dugas: *Histoire de la Mécanique* (Éditions du Griffon, Neuchâtel, 1955; reprinted by Dover as *A History of Mechanics*, 1988).

In order not to repeat ourselves, we refer to the relevant chapters in our Volume 2 the bibliography on perturbation theory and related problems, and on chaotic motions. On this last topic, we suggest the agile booklet of D. Ruelle, which affords a pleasant and intelligent reading: *Chance & Chaos* (Princeton University Press, 1991); this represents one of the rare examples of educated imparting of knowledge. We also suggest the paper by V. Szebehely: Is Celestial Mechanics deterministic?, in *Applications of Modern Dynamics to Celestial Mechanics and Astrodynamics*, ed. by V. Szebehely (Reidel, 1982).

Chapter 1: Dynamics and Dynamical Systems – Quod Satis

A. Newtonian Dynamics

Sects. 1.1, 1.2: An ever-growing number of books about dynamical systems is being produced. Since we are interested in a very narrow area, we confine ourselves to the following.

J. Guckenheimer, P. Holmes: *Nonlinear Oscillations, Dynamical Systems and Bifurcations of Vector Fields* (Springer, 1983).

A. J. Lichtenberg, M. A. Lieberman: *Regular and Stochastic Motion* (Springer, 1983); 2nd edn: *Regular and Chaotic Dynamics* (Springer, 1992).

F. Verhulst: *Nonlinear Differential Equations and Dynamical Systems* (Springer, 1990).

P. Hagedorn: *Nonlinear Oscillations*, 2nd edn (Clarendon Press, Oxford, 1988).

M. Tabor: *Chaos and Integrability in Nonlinear Dynamics* (Wiley, 1989).

Also, the "classic" textbooks on differential equations, such as

E. A. Coddington, N. Levinson: *Theory of Ordinary Differential Equations* (McGraw-Hill, New York, 1955).

J. K. Hale: *Ordinary differential equations* (Wiley-Interscience, New York, 1969).

L. Cesari: *Asymptotic Behaviour and Stability Problems in Ordinary Differential Equations* (Springer, 1971).

M. W. Hirsch, S. Smale: *Differential Equations, Dynamical Systems and Linear Algebra* (Academic Press, 1974).

Still interesting today is Liapunov's book: *Problème Général de la Stabilité du Mouvement* (Princeton University Press, 1947; a reproduction of the French translation, 1907, of the Russian original, 1892).

Sect. 1.3: The problem of the linear oscillator is dealt with more or less broadly in all the treatises on mechanics. Our treatment is taken from V. I. Arnold: *Mathematical Methods of Classical Mechanics*, 2nd edn (Springer, 1989). Also interesting is the paper by K. R. Meyer: The geometry of harmonic oscillators, *Amer. Math. Monthly* **97**, 457–465 (1990), reported afterwards in the book K. R. Meyer, G. R. Hall: *Introduction to Hamiltonian Dynamical Systems and the N-Body Problem* (Springer, 1992).

B. Lagrangian Dynamics

Sects. 1.4, 1.5: In order to remain in the frame of the "classics", besides the books of Whittaker and Synge already mentioned, a book we like more than any other which emphasizes the "geometric" approach to mechanics is C. Lanczos: *The Variational Principles of Mechanics*, 4th edn (Toronto University Press, 1970; reprinted by Dover, 1986). We add L. A. Pars: *A Treatise on Analytical Dynamics* (Heinemann, London, 1964).

Sects. 1.6, 1.7: The theorem stated in 1918 by Emmy Noether is doubtless one of the most important achievements of mathematical physics in this century. Even today, perhaps, it is difficult to fully evaluate its implications in the various fields of applied mathematics. We have chosen the version which we consider most suitable for application to discrete systems, which are implied in celestial mechanics and stellar dynamics. The point of view at the heart of this formulation was proposed for the first time by E. Rund and D. Logan (see below). Lack of space prevents us from mentioning the enormous number of papers which have been devoted to Noether's theorem from 1918 onward. We shall confine ourselves solely to those which have been brought into consideration for our exposition:

E. Bessel-Hagen: Über die erhaltungssätze der Elektrodynamik, *Math. Ann.* **84**, 258–276 (1921).

E. L. Hill: Hamilton's principle and the conservation theorems of mathematical physics, *Rev. of Mod. Phys.* **23**, 253–260 (1951).

A. Trautman: Noether equations and conservation laws, *Commun. Math. Phys.* **6**, 248–261 (1967).

H. Rund: A direct approach to Noether's theorem in the calculus of variations, *Utilitas Mathematica* **2**, 205–214 (1972).

J. D. Logan: On some invariance identities of H. Rund, *Utilitas Mathematica* **7**, 281–286 (1975).

Dj. S. Djukic: A procedure for finding first integrals of mechanical systems with gauge-variant Lagrangians, *Int. J. Non-linear Mech.* **8**, 479–488 (1973).

Dj. S. Djukic, B. D. Vujanovic: Noether's theory in classical nonconservative mechanics, *Acta Mechanica* **23**, 17–27 (1975).

J. D. Logan: *Invariant Variational Principles* (Academic Press, New York, 1977).

J. A. Kobussen: On a systematic search for integrals of the motion, *Helvetica Physica Acta* **53**, 183–200 (1980).

Sect. 1.8: The principle of least action in Jacobi form is dealt with in various textbooks on mechanics; among those already mentioned, we recall the books of Lanczos and Pars and, in addition, L. Brillouin: *Les Tenseurs en Mécanique et en Élasticité* (Masson et Cie, Paris, 1945).

Today, the range of applications of "Jacobi geometry" has been considerably enlarged. We mention a few papers

V. G. Gurzadyan, G. K. Savvidy: Collective relaxation of stellar systems, *Astron. Astrophys.* **160**, 203–213 (1986).

D. Boccaletti, G. Pucacco, R. Ruffini: Multiple relaxation time-scales in stellar dynamics, *Astron. Astrophys.* **244**, 48–51 (1991).

P. Cipriani, G. Pucacco: Jacobi geometry and chaos in N-body systems, in *Proceedings of the Workshop "From Newton to Chaos"*, ed. by A. Roy (NATO ASI, Cortina, 1993).

M. Pettini: Geometrical hints for a nonperturbative approach to Hamiltonian dynamics, *Phys. Rev. E* **47**, 828–850 (1993).

M. Szydlowski: Curvature of gravitationally bound mechanical Systems, *J. Math. Phys.* **35**, 1850–1880 (1994).

S. Bazanski, P. Jaranowski: The inverse Jacobi problem, *J. Phys. A: Math. Gen.* **27**, 3221–3234 (1994).

C. Hamiltonian Dynamics and Hamilton-Jacobi Theory

Sects. 1.9–1.15: The relevant textbooks have been mentioned in the footnotes. In addition, we mention a stimulating and, at the same time, simple approach to a modern view of classical mechanics, I. Percival, D. Richards: *Introduction to Dynamics* (Cambridge University Press, 1982), and an interesting opportunity for casting a glance at topics tightly bound to classical dynamics, D. Park: *Classical dynamics and its quantum analogues*, 2nd edn (Springer, 1990).

Sects. 1.16, 1.17: The metod of action–angle variables can be found in almost all the textbooks mentioned so far; nevertheless, we confirm our conviction that Born's is the clearest exposition from the point of view of applications. Historically, the first systematic treatment is due to K. Schwarzschild in his paper Zur Quantumhypothese, *Sitzber. Berl. Akad. Wiss.* 548–568 (1916), where terms action variables and angle variables (in German *Wirkungsvariable* and *Winkelvariable* respectively) appeared for the first time

Sect. 1.18: A proof of Liouville's theorem in terms of modern differential geometry can be found, for instance, in B. A. Dubrovin, A. T. Fomenko, S. P. Novikov: *Modern Geometry–Methods and Applications*, Vol. II (Springer, 1991).

Sect. 1.19: On the Painlevé property, see M. D. Kruskal, P. A. Clarkson: The Painlevé-Kowalevski and Poly-Painlevé tests for integrability, *Studies in Applied Mathematics* **86**, 87–165 (1992).

Chapter 2: The Two-Body Problem

Sect. 2.1: The main source of this section is a small but "classic" book

H. Pollard: *Mathematical Introduction to Celestial Mechanics* (Prentice-Hall, 1966; reprinted in *Carus Mathematical Monographs*, 1976).

Obviously, with regard to this problem, the reader has only an embarassment of choice. Besides the book of Danby, already mentioned in the Preface, and a classic like that of Moulton, we mention

D. Brouwer, G. Clemence: *Methods of Celestial Mechanics* (Academic Press, 1961).

H. C. Plummer: *An introductory Treatise on Dynamical Astronomy* (Cambridge University Press, 1918).

A. E. Roy: *Orbital Motion*, 3rd edn (Adam Hilger, 1988).

W. M. Smart: *Celestial Mechanics* (Longmans, Green and Co., 1953).

L. G. Taff: *Celestial Mechanics. A Computational Guide for the Practitioner* (Wiley, 1985).

Sect. 2.2: For the history of the L–R–L vector, see the papers by H. Goldstein: Prehistory of the Runge–Lenz vector, *Am. J. Phys.* **43**, 735–738 (1975), and: More on the prehistory of the Laplace–Runge–Lenz vector, *Am. J. Phys.* **44**, 1123–1124 (1976) and D. Park, op. cit., pp. 73–75.

The literature on the L–R–L vector and its generalizations is at present so plentiful that we cannot even try to give a representative list. The main stream of research has concerned the Lie-group approach. In the early times, we have

V. Fock: Zur Theorie des Wasserstoffatoms, *Z. Physik* **98**, 145–154 (1935).

V. Bargmann: Zur Theorie des Wasserstoffatom, *Z. Physik* **99**, 576–582 (1936).

V. Bargmann: Irreducible representations of the Lorentz group, *Annals of Math.* **48**, 568–640 (1947).

And then

H. Bacry, H. Ruegg: Dynamical groups and spherical potentials in classical mechanics, *Commun. Math. Phys.* **3**, 323–333 (1966).

V. A. Dulock, H. V. McIntosh: On the degeneracy of the Kepler problem, *Pac. J. Math.* **19**, 39–55 (1966).

D. M. Fradkin: Existence of the dynamic symmetries 0_4 and SU_3 for all classical central potential problems, *Prog. Theor. Phys.* **37**, 798–812 (1967).

G. Györgyi: Kepler's equation, Fock variables, Bacry's generators and Dirac Brackets, *Il Nuovo Cimento A* **53**, 717–736 (1968).

G. E. Prince, C. J. Eliezer: On the Lie symmetries of the classical Kepler problem, *J. Phys. A, Math. Gen.* **14**, 587–596 (1981).

V. M. Gorringe, P. G. Leach: The first integrals and their Lie algebra of the most general autonomous Hamiltonian of the form $H = T + V$ possessing a Laplace-Runge-Lenz vector, *J. Austral. Math. Soc. Ser. B* **34**, 511–522 (1993).

V. M. Gorringe, P. G. Leach: Kepler's third law and the oscillator's isochronism, *Am. J. Phys.* **61**, 991–995 (1993).

The last two authors have devoted many interesting papers to the L–R–L vector and its generalizations in Kepler-like problems; the reader can find the references in the two papers quoted. For the approach using Noether's theorem, we mention the above cited paper by Kobussen.

Sect. 2.3: Our treatment follows the suggestions contained in Arnold's book. See also

J. Roels, C. Aerts: Central forces depending on the distance only–case where all the bounded orbits are periodic, *Celestial Mech.* **44**, 77–85 (1988).

E. Onofri, M. Pauri: Search for periodic Hamiltonian flows: A generalized Bertrand's theorem, *J. Math. Phys.* **19**, 1850–1858 (1978).

Sect. 2.4: For recent research on Kepler's equation, see

T. M. Burkardt, J. M. A. Danby: The solution of the Kepler's Equation–I, *Celestial Mech.* **31**, 95–107 (1983); The solution of the Kepler's Equation–II, *ibid.*, 317–328 (1983).

P. Colwell: Kepler's equation and Newton's method, *Celestial Mech.* **52**, 203–204 (1991).

R. H. Gooding, A. W. Odell: Procedures for solving Kepler's Equation, *Celestial Mech.* **38**, 307–334 (1986); The hyperbolic Kepler's Equation (and the elliptic equation revisited), *ibid.* **44**, 267–282 (1988).

S. Mikkola: A cubic approximation for Kepler's Equation, *Celestial Mech.* **40**, 329–334 (1987).

E. W. Ng: A general algorithm for the solution of Kepler's Equation for elliptic orbits, *Celestial Mech.* **20**, 243–249 (1979).

A. Nijenhuis: Solving Kepler's Equation with high efficiency and accuracy, *Celestial Mech.* **51**, 319–330 (1991).

R. A. Serafin: Bounds on the solution to Kepler's Equation, *Celestial Mech.* **38**, 111–121 (1986).

G. R. Smith: A simple, efficient starting value for the iterative solution of Kepler's Equation, *Celestial Mech.* **19**, 163–166 (1979).

L. G. Taff, T. A. Brennan: Solving Kepler's Equation, *Celestial Mech.* **46**, 163–176 (1989).

Sect. 2.5: The books mentioned for Sect. 2.1 are good references for this section also.

Sect. 2.6: Additional bibliography on the problem of regularization can be found in the footnotes to Sects. 3.4, 4.2, 4.6. Interesting readings for a mathematically oriented reader are represented by:

R. Easton: Regularization of vector fields by surgery, *J. Diff. Eq.* **10**, 92–99 (1971).

J. Milnor: On the geometry of the Kepler problem, *Amer. Math. Monthly* **90**, 353–365 (1983).

J. Moser: Regularization of the Kepler's problem and the averaging method on a manifold, *Commun. Pure Appl. Math.* **23**, 609–636 (1970).

E. A. Belbruno: Two-body motion under the inverse square central force and equivalent geodesic flows, *Celestial Mech.* **15**, 465–476 (1977)

Yu. Osipov: The Kepler problem and geodesic flows in spaces of constant curvature, *Celestial Mech.* **16**, 191–208 (1977).

M. D. Vivarelli: On the connection among three classical mechanical problems via the hypercomplex KS transformation, *Celestial Mech.* **50**, 109–124 (1991).

J. F. Cariñena, C. Lopez, M. A. Del Olmo, M. Santander: Conformal geometry of the Kepler orbit space, *Celestial Mech.* **52**, 307–343 (1991).

Many papers on the subject of this section can be considered relevant for Sect. 2.2 as well. Furthermore, note that many things have been discovered and subequently rediscovered; see, for instance,

V. I. Arnold, V. A. Vasilév: Newton's Principia read 300 years later, *Not. Amer. Math. Soc.* **36**, 1148–1154 (1989).
L. Mittag, M. J. Stephen: Conformal transformations and the application of complex variables in mechanics and quantum mechanics, *Am. J. Phys.* **60**, 207–211 (1992).
T. Needhan: Newton and the transmutation of force, *Amer. Math. Monthly* **100**, 119–137 (1993).

Sect. 2.7: We have summarized the paper by W. Kaplan: Topology of the two-body problem, *Amer. Math. Monthly* **49**, 316–323 (1942). See also S. Smale: Topology of mechanics–I, *Inventiones Math.* **10**, 305–331 (1970); Topology of mechanics–II, *ibid.* **11**, 45–64 (1971).

Chapter 3: The N-Body Problem

Sects. 3.1–3.5: As additional reading, we suggest a paper by G. Benettin, L. Galgani, A. Giorgilli: On the Poincaré's non-existence theorem, in *Advances in Nonlinear Dynamics and Stochastic Processes*, ed. by R. Livi, A. Politi (World Scientific, 1985), and a review on Painlevé's conjecture by F. N. Diacu: *The Mathematical Intelligencer* **15**, 6–12 (1993).

Sect. 3.7: For this section, the reader can also consult Y. Hagihara: *Celestial Mechanics*, Vol. I (MIT Press, 1970), Chap. 3. This book is particularly useful for its rich bibliography.

Sect. 3.8: The central configurations of the N-body problem have always provoked great interest among mathematicians. We confine ourselves to mentioning only three papers, in which the reader can find further references

D. G. Saari: On the role and the properties of N-body central configurations, *Celestial Mech.* **21**, 9–20 (1980).
F. Pacella: Central configurations of the N-body problem via equivalent Morse theory, *Archive for Rat. Mech. and Anal.* **97**, 59–74 (1987).
J. Llibre: On the number of central configurations in the N-body problem, *Celestial Mech.* **50**, 89–96 (1981).

As a final reading, we suggest the book by F. N. Diacu: *Singularities of the N-body problem. An Introduction to Celestial Mechanics* (Université de Montréal, Centre de Récherches Mathématiques, Montréal, PQ, 1992). Unfortunately, this beautiful book was not available to us while we were writing this chapter.

Chapter 4: The Three-Body Problem

Sects. 4.1, 4.2: As a general reference, the reader can refer to

C. Marchal: *The three-body problem* (Elsevier, 1990).
Y. Hagihara: *Celestial Mechanics*, Vol. IV (Japan Society for the Promotion of Science, 1975).

Sect. 4.3: For the restricted problem, the main source is still the magnificent book by Szebehely, which we have already made abundant use of.

Sect. 4.4: To complete the bibliography given in the footnotes, see

K. G. J. Jacobi: Sur le mouvement d'un point et sur un cas particulier du problème des trois corps, *Compte Rendu de l'Académie des Sciences, Paris*, pp. 59–61 (1836).
A. M. Leontovich: On the stability of Lagrange's periodic solutions of the restricted three-body problem, *Soviet Math. Dokl.* **3**, 425–429, (1962).
A. Deprit, A. Deprit Bartholomé: Stability of the triangular Lagrangian points, *Astron. J.* **72**, 173–179 (1967).

Sect. 4.5: The source is Szebehely's book.

Sect. 4.6: More complete references are in Y. Hagihara: *Celestial Mechanics*, Vol. IV (see above).

Sect. 4.7: For the motion of the Moon, see also

A. Cook: *The Motion of the Moon* (Adam Hilger, 1988).

Chapter 5: Orbits in Given Potentials

Sects. 5.1–5.3: The subject of these sections is so general that the literature devoted to it is enormous. Just to complete the references on isolating integrals we cite, besides Wintner's book,

K. F. Ogorodnikov: *Dynamics of Stellar Systems* (Pergamon, Oxford, 1965).
E. Onofri, M. Pauri: Constants of motion and degeneration in Hamiltonian Systems, *J. Math. Phys.* **14**, 1106–1115 (1973).

Sect. 5.4: As to the problem of the separability of the Hamilton–Jacobi equation, we also refer to

P. Stäckel: Eine charakteristiche Eigenschaft der Flächen, deren Linienelement ds durch $ds^2 = (\chi(q_1) + \lambda(q_2))\,(dq_1^2 + dq_2^2)$ gegeben wird. *Math. Ann.* **35**, 91–101 (1890).
T. Levi-Civita: Sulla integrazione della equazione di Hamilton–Jacobi per separazione di variabili, *Math. Ann.* **66**, 383–397 (1904).
P. Burgatti: Determinazione delle equazioni di Hamilton-Jacobi integrabili mediante la separazione delle variabili, *Rendiconti Acc. Naz. Lincei* **20**, 108–111 (1911).

F. A. Dall'Acqua: Le equazioni di Hamilton–Jacobi che si integrano per separazione di variabili, *Rendiconti Circ. Mat. Palermo* **33**, 341–351 (1912).

L. P. Eisenhart: Separable Systems of Stäckel, *Ann. Math.* **35**, 284–305 (1934).

Sect. 5.5: The complete generalization of Darboux's results has been given by B. Dorizzi, B. Grammaticos and A. Ramani: A new class of integrables systems, *J. Math. Phys.* **24**, 2282–2288 (1983).

Sect. 5.7: Interesting reading is provided by V. Béletski: *Essais sur le Mouvement des Corps Cosmiques*, 2nd edn (Éditions Mir, Moscou, 1986).

Sect. 5.9: The study of configurational invariants starts with Birkhoff's book, G. Birkhoff: *Dynamical Systems* (1927; reprinted by the American Mathematical Society, Providence, Rhode Island 1966).

Invariants at fixed energy are very common in field theory since the pioneering work of Dirac:

P. A. M. Dirac: Homogeneous variables in classical dynamics, *Proc. Cambridge Phil. Soc.* **29**, 389–400 (1933).

P. A. M. Dirac: Generalized Hamiltonian dynamics, *Canad. J. Math.* **2**, 129–148 (1950).

P. A. M. Dirac: *Lectures on Quantum Mechanics* (Yeshiva University, New York, 1964).

More on the use of Killing tensors can be found in

P. Sommers: On Killing tensors and constants of motion, *J. Math. Phys.* **14**, 787–790 (1973).

E. G. Kalnins, W. Miller, Jr.: Killing tensors and variable separation for Hamilton–Jacobi and Helmholtz equations, *SIAM J. Math. Anal.* **11**, 1011–1026 (1980).

K. Rosquist, C. Uggla: Killing tensors in two-dimensional space-times with applications to cosmology, *J. Math. Phys.* **32**, 3412–3422 (1991).

K. Rosquist, G. Pucacco: Invariants at fixed and arbitrary energy – A unified geometric approach, *J. Phys. A: Math. Gen.* **28**, 3235–3252 (1995).

S. Benenti: Intrinsic characterization of the variable separation in the Hamilton–Jacobi equation, *J. Math. Phys.* **38**, 6578–6602 (1997).

M. Karlovini, K. Rosquist: A unified treatment of cubic invariants at fixed and arbitrary energy, *J. Math. Phys.* **41**, 370–384 (2000).

Name Index

Page numbers in italics indicate that the name is referred to in a footnote or in the Bibliographic Notes.

Aerts, C. *380*
Apollonius 2
Appel, P. *146*
Arenstorf, R. F. 283
Arnold, V. I. *8*, 10, 144, 269, *270*, *377*, *380*, *382*

Bacry, H. *380*
Bargmann, V. *380*
Bauer, F. *23*
Bazanski, S. *378*
Belbruno, E. A. *381*
Béletski, V. *384*
Benenti, S. *384*
Benettin, G. *382*
Bertrand, J. 131, 141, *141*, 146
Bessel, F. W. 152, 154
Bessel-Hagen, E. *188*, *378*
Binet A. 134
Binney, J. J. *306*, 307, *307*, *316*, *320*
Birkhoff, G. D. 11, 270, *270*, 283, *283*, *384*
Boccaletti, D. *378*
Born, M. 106, *106*, *109*, *113*, 307, *307*, *320*, *379*
Brennan, T. A. *381*
Brillouin, L. *178*, *378*
Brouwer, D. *379*
Brown, E. W. *298*, *299*
Bruns, E. H. 184, 187, *187*
Burkardt, T. M. *381*
Burgatti, P. 92, *383*

Cajori, F. *5*, *376*
Cantrijn, F. *43*, *360*
Cariñena, J. F. *381*
Cartan, E. 65, *65*, 69, *69*, 70, *70*, 75

Casati, G. *10*
Cauchy, A. L. 155, 165, 163, 180, 182, 184, 192, 196, 244, 247
Cavendish, H. 8
Celletti, A. *286*
Cesari, L. *377*
Chandrasekhar, S. *V*, *302*
Chang, Y. F. *123*
Chazy, J. 201, *201*, 205, *205*, *208*
Chebyshev, P. L. 55
Cipriani, P. *378*
Clairaut, A. C. 6, 338
Clarkson, P. A. *379*
Clausius, R. 209, *209*, 211
Clemence, G. *379*
Coddington, E. A. *377*
Cohen, I. B. *376*
Colwell, P. *381*
Cook, A. *383*
Copernicus, N. 3, 4
Courant, R. 337, *337*

D'Alembert, J. le R. 6
Dall'Acqua, F. A. 92, *384*
Damour, T. *178*
Danby, J. M. A. *VII*, *379*, *381*
Dante 9
Darboux, G. 92, 146, 337, *337*, 338, *384*
Darwin, G. H. *265*
Del Olmo, M. A. *381*
Delaunay, Ch.-E. 161, 162 *171*, 272, 273, 285, 296, 316, 322
Deprit, A. 270, *383*
Deprit Bartholomé, A. 270, *383*
de Vries, G. 117
de Zeeuw, P. T. 323, *323*, 353, *353*

Subject Index

ASTRONOMY AND
ASTROPHYSICS LIBRARY

Series Editors: I. Appenzeller · G. Börner · A. Burkert · M.A. Dopita
T. Encrenaz · M. Harwit · R. Kippenhahn · J. Lequeux
A. Maeder · V. Trimble

The Stars By E. L. Schatzman and F. Praderie

Modern Astrometry 2nd Edition
By J. Kovalevsky

**The Physics and Dynamics of Planetary
Nebulae** By G. A. Gurzadyan

Galaxies and Cosmology By F. Combes, P.
Boissé, A. Mazure and A. Blanchard

Observational Astrophysics 2nd Edition
By P. Léna, F. Lebrun and F. Mignard

Stellar Interiors. Physical Principles,
Structure, and Evolution
By C. J. Hansen and S. D. Kawaler

Physics of Planetary Rings Celestial
Mechanics of Continuous Media
By A. M. Fridman and N. N. Gorkavyi

Tools of Radio Astronomy 4th Edition
By K. Rohlfs and T. L. Wilson

Astrophysical Formulae 3rd Edition
(2 volumes)
Volume I: Radiation, Gas Processes
and High Energy Astrophysics
Volume II: Space, Time, Matter
and Cosmology
By K. R. Lang

Tools of Radio Astronomy Problems and
Solutions By T. L. Wilson and S. Hüttemeister

Galaxy Formation By M. S. Longair

Astrophysical Concepts 2nd Edition
By M. Harwit

Astrometry of Fundamental Catalogues
The Evolution from Optical to Radio
Reference Frames
By H. G. Walter and O. J. Sovers

Compact Stars. Nuclear Physics, Particle
Physics and General Relativity
By N. K. Glendenning

The Sun from Space By K. R. Lang

Stellar Physics (2 volumes)
Volume 2: Stellar Evolution and Stability
By G. S. Bisnovatyi-Kogan

Theory of Orbits (2 volumes)
Volume 1: Integrable Systems
and Non-perturbative Methods
Volume 2: Perturbative
and Geometrical Methods
By D. Boccaletti and G. Pucacco

Black Hole Gravitohydromagnetics
By B. Punsly

Stellar Structure and Evolution
By R. Kippenhahn and A. Weigert

Gravitational Lenses By P. Schneider,
J. Ehlers and E. E. Falco

Reflecting Telescope Optics (2 volumes)
Volume I: Basic Design Theory and its
Historical Development. Second Edition
Volume II: Manufacture, Testing, Alignment,
Modern Techniques
By R. N. Wilson

Interplanetary Dust
By E. Grün, B. Å. S. Gustafson, S. Dermott
and H. Fechtig (Eds.)

The Universe in Gamma Rays
By V. Schönfelder

Astrophysics. A Primer By W. Kundt

Cosmic Ray Astrophysics
By R. Schlickeiser

Astrophysics of the Diffuse Universe
By M. A. Dopita and R. S. Sutherland

The Sun An Introduction. Second Edition.
By M. Stix

Order and Chaos in Dynamical Astronomy
By G. J. Contopoulos

Astronomical Image and Data Analysis
By J.-L. Starck and F. Murtagh

The Early Universe Facts and Fiction
4th Edition By G. Börner

The Solar System 4th Edition
By T. Encrenaz, J.-P. Bibring, M. Blanc,
M. A. Barucci, F. Roques, Ph. Zarka

If you have any concerns about our products,
you can contact us on
ProductSafety@springernature.com

In case Publisher is established outside the EU,
the EU authorized representative is:
Springer Nature Customer Service Center GmbH
Europaplatz 3, 69115 Heidelberg, Germany

Printed by Libri Plureos GmbH
in Hamburg, Germany